**Quantile Regression**

# Quantile Regression

Applications on Experimental and Cross Section Data
Using EViews

*I Gusti Ngurah Agung*
The Ary Suta Center
Jakarta, Indonesia

*Registered Offices*
John Wiley & Sons, Inc., 111 River Street, Hoboken, NJ 07030, USA
John Wiley & Sons Ltd, The Atrium, Southern Gate, Chichester, West Sussex, PO19 8SQ, UK

*Editorial Office*
9600 Garsington Road, Oxford, OX4 2DQ, UK

For details of our global editorial offices, customer services, and more information about Wiley products visit us at www.wiley.com.

Wiley also publishes its books in a variety of electronic formats and by print-on-demand. Some content that appears in standard print versions of this book may not be available in other formats.

*Library of Congress Cataloging-in-Publication Data*

Names: Agung, I Gusti Ngurah, author.
Title: Quantile regression : applications on experimental and cross section
    data using EViews / I Gusti Ngurah Agung.
Description: First edition. | Hoboken : Wiley, [2021] | Includes
    bibliographical references.
Identifiers: LCCN 2020025365 (print) | LCCN 2020025366 (ebook) | ISBN
    9781119715177 (cloth) | ISBN 9781119715160 (adobe pdf) | ISBN
    9781119715184 (epub)
Subjects: LCSH: Quantile regression. | Mathematical statistics. | EViews
    (Computer file)
Classification: LCC QA278.2 .A32 2021 (print) | LCC QA278.2 (ebook) | DDC
    519.5/36—dc23
LC record available at https://lccn.loc.gov/2020025365
LC ebook record available at https://lccn.loc.gov/2020025366

Cover Design: Wiley
Cover Images: © Kertlis/iStock/Getty Images

Set in 9.5/12.5pt STIXTwoText by Straive, Chennai, India
Printed and bound by CPI Group (UK) Ltd, Croydon, CR0 4YY

C9781119715177_020621

*Dedicated to my wife Anak Agung Alit Mas, our children Martiningsih, Ratnaningsing, and Darma Putra, as well as all our generation*

# Contents

**Preface**   *xiii*
**About the Author**   *xvii*

**1**       **Test for the Equality of Medians by Series/Group of Variables**   *1*
1.1       Introduction   *1*
1.2       Test for Equality of Medians of *Y1* by Categorical Variables   *1*
1.3       Test for Equality of Medians of *Y1* by Categorical Variables   *3*
1.4       Testing the Medians of *Y1* Categorized by *X1*   *6*
1.5       Testing the Medians of *Y1* Categorized by *RX1 = @Ranks(X1,a)*   *10*
1.6       Unexpected Statistical Results   *10*
1.7       Testing the Medians of *Y1* by *X1* and Categorical Factors   *12*
1.8       Testing the Medians of *Y* by Numerical Variables   *16*
1.8.1     Findings Based on Data_Faad.wf1   *16*
1.8.2     Findings Based on Mlogit.wf1   *16*
1.9       Application of the Function *@Mediansby(Y,IV)*   *20*

**2**       **One- and Two-way ANOVA Quantile Regressions**   *25*
2.1       Introduction   *25*
2.2       One-way ANOVA Quantile Regression   *25*
2.3       Alternative Two-way ANOVA Quantile Regressions   *27*
2.3.1     Applications of the Simplest Equation Specification   *27*
2.3.2     Application of the Quantile Process   *29*
2.3.3     Applications of the Models with Intercepts   *31*
2.4       Forecasting   *37*
2.5       Additive Two-way ANOVA Quantile Regressions   *40*
2.6       Testing the Quantiles of *Y1* Categorized by *X1*   *40*
2.7       Applications of QR on Population Data   *43*
2.7.1     One-way-ANOVA-QRs   *43*
2.7.2     Application of the Forecasting   *46*
2.7.3     Two-way ANOVA-QRs   *49*
2.8       Special Notes and Comments on Alternative Options   *58*

**3     *N*-Way ANOVA Quantile Regressions**   65
3.1     Introduction   65
3.2     The Models Without an Intercept   65
3.3     Models with Intercepts   72
3.4     $I \times J \times K$ Factorial QRs Based on susenas.wf1   81
3.4.1   Alternative ESs of *CWWH* on *F1, F2,* and *F3*   82
3.4.1.1   Applications of the Simplest ES in (3.5a)   82
3.4.1.2   Applications of the ES in (3.5b)   84
3.4.1.3   Applications of the ES in (3.5c)   89
3.5     Applications of the *N*-Way ANOVA-QRs   96
3.5.1   Four-Way ANOVA-QRs   96

**4     Quantile Regressions Based on (X*1*,Y*1*)**   107
4.1     Introduction   107
4.2     The Simplest Quantile Regression   107
4.3     Polynomial Quantile Regressions   109
4.3.1   Quadratic Quantile Regression   109
4.3.2   Third Degree Polynomial Quantile Regression   112
4.3.3   Forth Degree Polynomial Quantile Regression   114
4.3.4   Fifth Degree Polynomial Quantile Regression   115
4.4     Logarithmic Quantile Regressions   115
4.4.1   The Simplest Semi-Logarithmic QR   116
4.4.2   The Semi-Logarithmic Polynomial QR   118
4.4.2.1   The Basic Semi-Logarithmic Third Degree Polynomial QR   118
4.4.2.2   The Bounded Semi-Logarithmic Third Degree Polynomial QR   118
4.5     QRs Based on MCYCLE.WF1   119
4.5.1   Scatter Graphs of *(MILL,ACCEL)* with Fitted Curves   119
4.5.2   Applications of Piecewise Linear QRs   121
4.5.3   Applications of the Quantile Process   123
4.5.4   Alterative Piecewise Linear QRs   127
4.5.5   Applications of Piecewise Quadratic QRs   129
4.5.6   Alternative Piecewise Polynomial QRs   135
4.5.7   Applications of Continuous Polynomial QRs   140
4.5.8   Special Notes and Comments   144
4.6     Quantile Regressions Based on SUSENAS-2013.wf1   147
4.6.1   Application of *CWWH* on *AGE*   147
4.6.1.1   Quantile Regressions of CWWH on AGE   148
4.6.1.2   Application of Logarithmic QRs   154
4.6.2   An Application of Life-Birth on *AGE* for Ever Married Women   160
4.6.2.1   QR(Median) of LBIRTH on AGE as a Numerical Predictor   161

**5     Quantile Regressions with Two Numerical Predictors**   165
5.1     Introduction   165
5.2     Alternative QRs Based on Data_Faad.wf1   165
5.2.1   Alternative QRs Based on *(X1,X2,Y1)*   165

5.2.1.1   Additive QR   *165*
5.2.1.2   Semi-Logarithmic QR of *log(Y1)* on *X1* and *X2*   *166*
5.2.1.3   Translog QR of *log(Y1)* on *log(X1)* and *log(X2)*   *168*
5.2.2     Two-Way Interaction QRs   *168*
5.2.2.1   Interaction QR of *Y1* on *X1* and *X2*   *168*
5.2.2.2   Semi-Logarithmic Interaction QR Based on *(X1,X2,Y1)*   *171*
5.2.2.3   Translogarithmic Interaction QR Based on *(X1,X2,Y1)*   *172*
5.3       An Analysis Based on Mlogit.wf1   *174*
5.3.1     Alternative QRs of *LW*   *176*
5.3.2     Alternative QRs of *INC*   *183*
5.3.2.1   Using *Z*-Scores Variables as Predictors   *184*
5.3.2.2   Alternative QRs of *INC* on Other Sets of Numerical Predictors   *185*
5.3.2.3   Alternative QRs Based on Other Sets of Numerical Variables   *188*
5.4       Polynomial Two-Way Interaction QRs   *191*
5.5       Double Polynomial QRs   *197*
5.5.1     Additive Double Polynomial QRs   *197*
5.5.2     Interaction Double Polynomial QRs   *200*

**6**         **Quantile Regressions with Multiple Numerical Predictors**   *207*
6.1       Introduction   *207*
6.2       Alternative Path Diagrams Based on *(X1,X2,X3,Y1)*   *207*
6.2.1     A QR Based on the Path Diagram in Figure 6.1a   *207*
6.2.2     A QR Based on the Path Diagram in Figure 6.1b   *208*
6.2.3     QR Based on the Path Diagram in Figure 6.1c   *208*
6.2.3.1   A Full Two-Way Interaction QR   *209*
6.2.3.2   A Full Three-Way Interaction QR   *209*
6.2.4     QR Based on the Path Diagram in Figure 6.1d   *210*
6.3       Applications of QRs Based on Data_Faad.wf1   *210*
6.4       Applications of QRs Based on Data in Mlogit.wf1   *217*
6.5       QRs of *PR1* on *(DIST1,X1,X2)*   *223*
6.6       Advanced Statistical Analysis   *227*
6.6.1     Applications of the Quantiles Process   *227*
6.6.1.1   An Application of the Process Coefficients   *227*
6.6.1.2   An Application of the Quantile Slope Equality Test   *228*
6.6.1.3   An Application of the Symmetric Quantiles Test   *229*
6.6.2     An Application of the Ramsey RESET Test   *229*
6.6.3     Residual Diagnostics   *232*
6.7       Forecasting   *232*
6.7.1     Basic Forecasting   *233*
6.7.2     Advanced Forecasting   *237*
6.8       Developing a Complete Data_LW.wf1   *240*
6.9       QRs with Four Numerical Predictors   *242*
6.9.1     An Additive QR   *242*
6.9.2     Alternative Two-Way Interaction QRs   *245*
6.9.2.1   A Two-Way Interaction QR Based on Figure 6.35a   *245*

6.9.2.2    A Two-Way Interaction QR Based on Figure 6.35b    *245*
6.9.2.3    A Two-Way Interaction QR Based on Figure 6.35c    *245*
6.9.2.4    A Two-Way Interaction QR Based on Figure 6.35d    *245*
6.9.3      Alternative Three-Way Interaction QRs    *246*
6.9.3.1    Alternative Models Based on Figure 6.36a    *246*
6.9.3.2    Alternative Models Based on Figure 6.36b    *246*
6.9.3.3    Alternative Models Based on Figure 6.36c    *247*
6.9.3.4    Alternative Models Based on Figure 6.36d    *247*
6.10       QRs with Multiple Numerical Predictors    *254*
6.10.1     Developing an Additive QR    *254*
6.10.2     Developing a Simple Two-Way Interaction QR    *258*
6.10.3     Developing a Simple Three-Way Interaction QR    *259*

**7          Quantile Regressions with the Ranks of Numerical Predictors**    *267*
7.1        Introduction    *267*
7.2        NPQRs Based on a Single Rank Predictor    *267*
7.2.1      Alternative Piecewise NPQRs of *ACCEL* on *R_Milli*    *268*
7.2.2      Polynomial NPQRs of *ACCEL* on *R_Milli*    *274*
7.2.3      Special Notes and Comments    *276*
7.3        NPQRs on Group of *R_Milli*    *278*
7.3.1      An Application of the *G_Milli* as a Categorical Variable    *278*
7.3.2      The *k*th-Degree Polynomial NPQRs of *ACCEL* on *G_Milli*    *280*
7.4        Multiple NPQRs Based on Data-Faad.wf1    *283*
7.4.1      An NPQR Based on a Triple Numerical Variable (*X1,X2,Y*)    *283*
7.4.2      NPQRs with Multi-Rank Predictors    *283*
7.5        Multiple NPQRs Based on MLogit.wf1    *286*

**8          Heterogeneous Quantile Regressions Based on Experimental Data**    *291*
8.1        Introduction    *291*
8.2        HQRs of *Y1* on *X1* by a Cell-Factor    *291*
8.2.1      The Simplest HQR    *291*
8.2.2      A Piecewise Quadratic QR    *304*
8.2.3      A Piecewise Polynomial Quantile Regression    *311*
8.3        HLQR of *Y1* on (*X1,X2*) by the Cell-Factor    *312*
8.3.1      Additive HLQR of *Y1* on (*X1,X2*) by CF    *313*
8.3.2      A Two-Way Interaction Heterogeneous-QR of *Y1* on (*X1,X2*) by CF    *317*
8.3.3      An Application of Translog-Linear QR of *Y1* on (*X1,X2*) by *CF*    *319*
8.4        The HLQR of *Y1* on (*X1,X2,X3*)by a Cell-Factor    *321*
8.4.1      An Additive HLQR of *Y1* on (*X1,X2,X3*) by *CF*    *321*
8.4.2      A Full Two-Way Interaction HQR of *Y1 on* (*X1,X2,X3*) by *CF*    *321*
8.4.3      A Full Three-Way Interaction HQR of *Y1* on (*X1,X2,X3*) by *CF*    *322*

**9          Quantile Regressions Based on CPS88.wf1**    *333*
9.1        Introduction    *333*
9.2        Applications of an ANOVA Quantile Regression    *334*
9.2.1      One-Way ANOVA-QR    *334*
9.2.2      Two-Way ANOVA Quantile Regression    *338*

9.2.2.1    The Simplest Equation of Two-Way ANOVA-QR    *338*
9.2.2.2    A Special Equation of the Two-Way ANOVA-QR    *339*
9.2.2.3    An Additive Two-Way ANOVA-QR    *343*
9.2.3      Three-Way ANOVA-QRs    *345*
9.3        Quantile Regressions with Numerical Predictors    *349*
9.3.1      QR of *LWAGE* on *GRADE*    *349*
9.3.1.1    A Polynomial QR of *LWAGE* on *GRADE*    *349*
9.3.1.2    The Simplest Linear QR of *Y1* on a Numerical *X1*    *353*
9.3.2      Quantile Regressions of *Y1* on *(X1,X2)*    *356*
9.3.2.1    Hierarchical and Nonhierarchical Two-Way Interaction QRs    *356*
9.3.2.2    A Special Polynomial Interaction QR    *358*
9.3.2.3    A Double Polynomial Interaction QR of *Y1* on *(X1,X2)*    *361*
9.3.3      QRs of *Y1* on Numerical Variables *(X1,X2,X3)*    *362*
9.3.3.1    A Full Two-Way Interaction QR    *362*
9.3.3.2    A Full-Three-Way-Interaction QR    *363*
9.4        Heterogeneous Quantile-Regressions    *366*
9.4.1      Heterogeneous Quantile Regressions by a Factor    *366*
9.4.1.1    A Heterogeneous Linear QR of *LWAGE* on *POTEXP* by *IND1*    *366*
9.4.1.2    A Heterogeneous Third-Degree Polynomial QR of *LWAGE* on *GRADE*    *369*
9.4.1.3    An Application of QR for a Large Number of Groups    *373*
9.4.1.4    Comparison Between Selected Heterogeneous QR(Median)    *378*

**10**         **Quantile Regressions of a Latent Variable**    *381*
10.1       Introduction    *381*
10.2       Spearman-rank Correlation    *381*
10.3       Applications of ANOVA-QR($\tau$)    *384*
10.3.1     One-way ANOVA-QR of *BLV*    *384*
10.3.2     A Two-Way ANOVA-QR of *BLV*    *391*
10.3.2.1   The Simplest Equation of a Two-Way ANOVA-QR of *BLV*    *392*
10.3.2.2   A Two-way ANOVA-QR of *BLV* with an Intercept    *392*
10.3.2.3   A Special Equation of Two-Way ANOVA-QR of BLV    *393*
10.4       Three-way ANOVA-QR of *BLV*    *394*
10.5       QRs of *BLV* on Numerical Predictors    *396*
10.5.1     QRs of *BLV* on *MW*    *396*
10.5.1.1   The Simplest Linear Regression of *BLV* on *MW*    *396*
10.5.1.2   Polynomial Regression of *BLV* on *MW*    *399*
10.5.2     QRs of *BLV* on Two Numerical Predictors    *402*
10.5.2.1   An Additive QR of *BLV*    *402*
10.5.2.2   A Two-Way Interaction QR of BLV on MW and AGE    *402*
10.5.2.3   A Two-way Interaction Polynomial QR of BLV on *MW* and *AGE*    *405*
10.5.3     QRs of BLV on Three Numerical Variables    *407*
10.5.3.1   Additive QR of BLV on MW, AGE, and HB    *407*
10.5.3.2   A Full Two-Way Interaction QR of *BLV* on *MW, GE,* and *HB*    *409*
10.5.3.3   A Full Three-Way Interaction QR of *BLV* on *AGE*, *HB*, and *BW*    *410*
10.6       Complete Latent Variables QRs    *411*
10.6.1     Additive Latent Variable QRs of *BLV*    *412*
10.6.2     Advanced Latent Variables QRs    *416*
10.6.2.1   The Two-Way-Interaction QR of *PBLV* on *(PLV1,PLV2)*    *416*

10.6.2.2    Two-Way-Interaction QRs of *PBLV* on (*PLV1,PLV2,PLV3*)   *418*
10.6.2.3    A Special Full-Two-Way-Interaction QR of *PBLV*   *420*
10.6.2.4    An Application of a Nonlinear QR of PBLV   *422*
10.6.2.5    An Application of Semi-Log Polynomial QR of log(PBLV)   *424*
10.7       An Application of Heterogeneous Quantile-regressions   *428*
10.7.1     A Heterogeneous QR of *BLV* by a Categorical Factor   *430*
10.7.1.1   A Two-level Heterogeneous QR of *BLV*   *430*
10.7.1.2   A Three-Level Heterogeneous QR of PBLV   *432*
10.7.2     Heterogeneous QR of *BLV* by Two Categorical Factors   *433*
10.8       Piecewise QRs   *441*
10.8.1     The Simplest PW-QR of *BLV* on *MLV*   *441*
10.8.2     An Extension of the SPW-QR of *BLV* on *MLV* in (10.29)   *444*
10.8.3     Reduced Models of the Previous PW-QRs of *BLV*   *446*
10.8.3.1   A Reduced Model of the SPW-QR of *BLV* in (10.29)   *446*
10.8.3.2   A Reduced Model of the PW-2WI-QR of *BLV* in (10.31)   *448*

**Appendix A  Mean and Quantile Regressions**   *453*
A.1        The Single Parameter Mean and Quantile Regressions   *453*
A.1.1      The Single-Parameter Mean Regression   *453*
A.1.2      The Single-Parameter Quantile Regression   *454*
A.2        The Simplest Conditional Mean and Quantile Regressions   *454*
A.2.1      The Simplest Conditional Mean Regression   *454*
A.2.2      The Simplest Conditional Quantile Regressions   *455*
A.3        The Estimation Process of the Quantile Regression   *455*
A.3.1      Applications of the Quantile Process   *456*
A.3.1.1    An Application of Quantile Process/Process Coefficients   *457*
A.3.1.2    An Application of Quantile Process/Slope Equality Test   *458*
A.3.1.3    An Application of the Quantile Process/Symmetric Quantile Test   *458*
A.3.1.4    The Impacts of the Combinations of Options   *459*
A.4        An Application of the Forecast Button   *460*

**Appendix B  Applications of the t-Test Statistic for Testing Alternative
             Hypotheses**   *461*
B.1        Testing a Two-Sided Hypothesis   *461*
B.2        Testing a Right-Sided Hypothesis   *461*
B.3        Testing a Left-Sided Hypothesis   *463*

**Appendix C  Applications of Factor Analysis**   *465*
C.1        Generating the BLV   *465*
C.2        Generating the MLV   *465*

**References**   *469*
**Index**   *471*

# Preface

This book presents various Quantile Regressions (QR), based on the cross-section and experimental data. It has been found that the equation specification or the estimation equation of the LS-Regression or Mean-Regression (MR) can be applied directly for the Quantile-Regression. Hence, this book can be considered as an extension or a modification of all Mean-Regressions presented in all books, and papers, such as Agung (2008, 2009b, 2011a,b), Gujarati (2003), Wooldridge (2002), Huitema (1980), Kementa (1980), and Neter and Wasserman (1974).

In addition, compare to the Mean-Regression (MR), the Quantile-Regression is a robust regression having critical advantages over the MR, mainly for the robustness to outliers, no normal distribution assumption, and it can present more complete distribution of the objective, criterion or dependent random variable, using the linear programing estimation method (Davino et al. 2014; Koenker 2005; Koenker and Hallock 2001), As a more detail comparison between the Conditional QR and the Condional MR, see Appendix A.

The models presented in this book in fact are the extension or modification of all mean-regression presented in my second book: "Cross Section and Experimental Data Analysis Using EViews" (Agung 2011a). For this reason, it is recommended the readers to use also the models in the book to conduct the quantile-regression analysis, using their own data sets.

This book contains ten chapters.

Chapter 1 presents the applications of the test for medians of any response or criterion variable $Y_i$, by series/group of categorical variables, numerical categorical or the ranks of a numerical variable.

Chapter 2 presents the applications of One-Way and Two-Way ANOVA Quantile-Regressions. In addition, the application of the object *Quantile Process* having three alternative options are introduced. As the modification of the parametric DID (Difference-In-Differences) of the means of any variable $Y_i$ by two factors or categorical variables, this chapter presents the DID of the Quantile($\tau$) of any variable $Y_i$, by two categorical variables, starting with two dichotomous variables or $2 \times 2$ factorial QR. It has been well known that the value of a DID is representing the two-way interaction effect of the corresponding two factors, indicating the effect of a factor on the criterion variable $Y_i$ depends on the other factor. Then a special $2 \times 3$ factorial QR to demonstrate how to compute its DID.

Chapter 3 presents *N*-Way ANOVA Quantile-Regressions. Specific for N = 3, alternative equation specifications are presented, starting with a $2 \times 2 \times 2$ factorial ANOVA QR without an intercept, to show how compute conditional two-way interactions factors, and a three-way interaction factor, which should be tested using the Wald Test. As an extension the $2 \times 2 \times 2$ factorial ANOVA QR without an intercept, a $2 \times 2 \times 3$ factorial ANOVA QR without an intercepts to show how to compute the 3-way interaction factor. Then selected $I \times J \times K$ factorial ANOVA QRs with an intercepts are presented using alternative equation specifications, to show the advantages in using each equation specification.

Chapter 4 presents quantile regressions based on bivariate numerical variable $(X_i, Y_i)$, starting with the simplest linear quantile-regression. Then it is extended quadratic QR and alternative polynomial QRs, and alternative logarithmic QRs with lower and upper bounds.

Chapter 5 presents various QRs based on triple variables $(X1_t, X2_t, Y1_t)$ as the extension or modification of each model presented in chapter 4. Three alternative path diagrams based on the triple variables are presented, as the guide to defined equation specifications of alternative QRs. An additive QR is presented as the simplest QR, which is extended to semi-logarithmic and translog QRs, with the examples presented based on an experimental data in Data_Faad.wf1. Then they are extended to interaction QRs. As additional illustrative examples, the statistical results of alternative QRs are presented based on MLogit.wf1, which I consider as a special data in EViews work file. In addition special Quantile Slope Equality Test also is presented as an illustration.

Chapter 6 presents various QRs with multi numerical predictors. Four alternative path diagrams based on the variables *(X1,X2,X3,Y1)* are presented as a guide to define specific equations of alternative QRs., such as additive, two-way interaction and three-way interaction QRs. The Data_Faad.wf1 having four numerical variables, namely *X1, X2, Y1,* and *Y2,* can be used to replace the causal or up-and-down relationships presented in Figure 6.1. Hence, this chapter presents statistical results of additive, two-way interaction and three-way interaction QRs of *Y2* on *Y1, X1* and *X2* with its possible reduced QRs, which are develop using the *manual multistage selection method* (MMSM) and the trial-and error method, since the STEPWISE method is not applicable for the QR. As additional empirical statistical results, selected sets of four variables in Mlogit.wf1 are used.

Furthermore, more advanced statistical analyses, such as the application of Quantile Process, Residual Analysis, Stability Test, and the "Static Forecast Method", are presented, with special notes and comments.

Chapter 7 presents various QRs with te ranks of numerical variables as the predictors or independent variables, which is considered as *nonparametric-quantile- regressions* (NPQR). They are the modification of the QRs having numerical independent variables presented in previous chapters, which are defined as the *semiparametric-quantile-regressions*. Statistical results of alternative QRs are presented based on the MCycle.wf1, Data_Faad.wf1, and Mlogit.wf1.

Chapter 8 presents illustrative examples on *Heterogeneous Quantile Regressions* (HQRs) based on experimental data. In this chapter it is introduced a symbol *CF* = Cell-Factor to represent one or more categorical variable, which also can be generated using one or more numerical variables Then various HQRs based on *(X1,Y1)* are presented, starting with the simplest HQR of *Y1* on a numerical variable *X1* by a dichotomous factor or a dummy variable using three alternative equation specification. As the extension of HLQRs,

Heterogeneous Quadratic QR and Heterogeneous Polynomial QRs are presented. Then they are extended to Heterogeneous Linear QR (HLQR) and Heterogeneous Polynomial QR (HPQR) using one or more factors. Furthermore, they are extended to Additive and Two-Way-Interaction HQRs based on numerical variables *(X1, X2, Y1)* and *(X1, X2, X3, Y1)*. Whenever the *CF* is generated, based on the independent variable *X1*, then we have a *Peace-Wise-QR* of *Y1* on *X1* or *X2* by *CF*.

Chapter 9 presents various QRs, as the applications of selected QRs presented in previous chapters, based on selected sets of variables in CPS88.wf1, which is a special work-file in EViews 10.

Chapter 10 presents various QRs based on a health data set, specifically QRs of a *baby latent variable (BLV)* on selected sets of eight selected set of mother indicators in *BBW*.wf1. The *BLV* is generated using three baby indicators, namely *BBW* = baby birth weight, *FUNDUS*, and *MUAC* = mid upper arm circumference. Two special One-Way ANOVA-QRs of *BLV* on *@Expand(@Round(MW)* or *@Expand(AGE)*, and selected types of QRs, which have been presented in previous chapters. As additional QRs, a *mother latent variable (MLV)* is generated based on four mother indicators, namely based on the two ordinal variables *ED* (Education level) and *SE* (Social Economic levels), and two dummy variables *SMOKING*, which is defined *SMOKING = 1* for the smoking mothers, and *NO_ANC*, which is defined as *NO_ANC = 1* for the mothers don't have antenatal care. Then outputs of selected QRs of *BLV* on *MLV* and other mother indicators are presented as illustrative empirical the full latent variable QRs.

I wish to express my gratitude to the graduate School of Management, Faculty of Economics and Business, University of Indonesia for providing rich intellectual environment while I was at the University of Indonesia from 1987 to 2018, that were indispensable for supporting in writing this text. I also would like to thank The Ary Suta Center, Jakarta for providing intellectual environment and facilities, while I am an advisor from 2008 till now.

Finally, I would like to thank the reviewers, editors, and all staffs at John Wiley for their hard works in getting this book to a publication.

# About the Author

With regards to the request from Wiley: *"Please provide us with a brief biography including details that explain why you are the ideal person to write this book,"* I would present my background, experiences and findings in doing statistical data analysis, as follows:

I have a Ph.D. degree in Biostatistics (1981) and a master degree in Mathematical Statistics (1977) from the North Carolina University at Chapel Hill, NC. USA, a master degree in Mathematics from New Mexico State University, Las Cruces, NM. USA, a degree in Mathematical Education (1962) from Hasanuddin University, Makassar, Indonesia, and a certificate from *"Kursus B-I/B-II Ilmu Pasti"* (B-I/B-II Courses in Mathematics), Yogyakarta, which is a five year non-degree program in advanced mathematics. So that I would say that I have a good background knowledge in mathematical statistics as well as applied statistics. In my dissertation in Biostatistics, I present new findings, namely the Generalized Kendall's tau, the Generalized Pair Charts, and the Generalized Simon's Statistics, based on the data censored to the right.

Supported by my knowledge in mathematics, mathematical functions in particular, I can evaluate the limitation, hidden assumptions or the unrealistic assumption(s), of all regression functions, such as the fixed effects models based on panel data, which in fact are ANCOVA models. As a comparison, Agung (2011a, 2011b) presents several alternative acceptable ANCOVA models, in the statistical sense, and the worst ANCOVA models, in both theoretical and statistical senses.

Furthermore, based on my exercises and experiments in doing data analyses of various field of studies; such as finance, marketing, education and population studies since 1981 when I worked at the Population Research Center, Gadjah Mada University, 1985–1987; and while I have been at the University of Indonesia, 1987–2018, I have found unexpected or unpredictable statistical results based on various time series, cross-section and panel data models, which have been presented with special notes and comments in Agung (2019, 2014, 2011, and 2009), compare to the models which are commonly applied. Since 2008, I have been an advisor at the Ary Suta Center, Jakarta, and give free consultation on the statistical analysis for the students from any university. So far, there were six students of the Graduate School of Business, Universiti Kebangsaan Malaysia, where four are Malaysian, had visited the Ary Suta Center for consultation, several times. Even though, a student can have consultation by email igustinagung32@gmail.com.

For this book, I have been doing many exercises and experiments to apply the QREG – Quantile Regression (including LAD) estimation method in EViews, so I can present various alternative statistical results based on the equation specification of any quantile-regression, with special notes and comment. Because, unexpectedly the statistical results may present the statement *"Estimation successful but the solution may not be unique"* which should be accepted. On the other hand, a problem of *"incomplete outputs"* having the NA's for their Std. error and t-statistic, can be obtained, but with a complete or acceptable quantile-regression function, and the most serious problem is the *"error messages"*. appeared on-screen, which should be overcome by using the trial-and-method, since there is no standardized method to solve such problems.

# 1

# Test for the Equality of Medians by Series/Group of Variables

## 1.1 Introduction

I consider the test for equality of medians of a numerical or ordinal endogenous or objective variable by series/group of categorical variables as the earliest *nonparametric test* statistic. It also is called the *distribution free test* because it does not rely on the population distribution.

EViews provides a specific test for equality of the medians of any numerical or ordinal categorical variable, say *Y1,* categorized or classified by series/group of variables, either numerical or categorical. However, the numerical predictors will be directly transformed to ordinal categorical factors. The steps of the analysis are as follows:

1) Present the values of the variable *Y1* on-screen.
2) Select *View/Descriptive Statistics & Tests/Equality Test by Classification*…. The dialog shown in Figure 1.1 appears on-screen.
3) Insert the *Series/Group for classify,* select the *Median* option, and then click *OK* to obtain the output of several nonparametric test statistics.

This chapter presents various examples of alternative tests, either unconditional or conditional tests, with categorical and/or numerical predictors, which will be directly transformed to ordinal variables.

## 1.2 Test for Equality of Medians of *Y1* by Categorical Variables

This section presents examples of alternative test statistics for *Y1* by categorical variables using the Data_Faad.wf1 file, from one of my student's thesis, as presented in Figure 1.2. The data are experimental data using four classes of high school students. They are classified as experimental groups by the two factors, *A* and *B,* with two covariates, namely $X1$ = learning motivation in mathematics and $X2$ = test scores of mathematics before the experiment, and two final tests as the response variables, *Y1* and *Y2.*

Figure 1.2 presents Data_Faad.wf1, with four additional ordinal variables, generated using the following equations. The four variables include *G2* and *G4,* which are generated based on the exogenous variable $X1$, and *H2* and *H3*, which are generated based on the exogenous variable $X2$. Therefore, we could present alternative *N*-way quantile

*Quantile Regression: Applications on Experimental and Cross Section Data Using EViews,* First Edition.
I Gusti Ngurah Agung.
© 2021 John Wiley & Sons Ltd. Published 2021 by John Wiley & Sons Ltd.

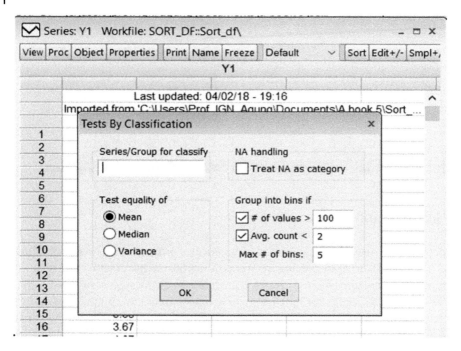

**Figure 1.1** The options for doing the test for equality of means, medians, or variances of *Y1* categorized by values of series/group.

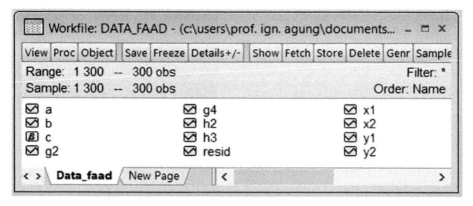

**Figure 1.2** Data_Faad.wf1, with additional ordinal variables *G2, G4, H2*, and *H4*.

regressions, for $N = 2$. See the following examples.

$$G2 = 1 + 1 * (X1 >= @\text{Quantile}(X1, 0.50)) \tag{1.1}$$

$$G4 = 1 + 1 * (X1 >= @\text{Quantile}(X1, 0.25)) + 1 * (X1 >= @\text{Quantile}(X1, 0.50))$$
$$+ 1 * (X1 >= @\text{Quantile}(X1, 0.75)) \tag{1.2}$$

$$H2 = 1 + 1 * (X2 >= @\text{Quantile}(X2, 0.50)) \tag{1.3}$$

$$H3 = 1 + 1 * (X2 >= @\text{Quantile}(X2, 0.30)) + 1 * (X2 >= @\text{quantile}(X2, 0.70))$$
$$\tag{1.4}$$

## 1.3   Test for Equality of Medians of *Y1* by Categorical Variables

This section presents examples of alternative test statistics for those applied based on the *N*-way QRs.

**Example 1.1   *Test for Medians of Y1 by the Dichotomous Factors A and B***
Figure 1.3 presents the statistical results of the medians test of *Y1* by the factors *A* and *B*. Based on these results, the following findings and notes are presented:

1) This figure presents four nonparametric test statistics, with $df = (4 − 1) = 3$, for testing the differences between the medians of the four cells generated by the factors *A* and *B*.
2) Based on each of the six test statistics, the joint effects of *A* and *B* on *Y1* or *Med(Y1)* are significant, at the 1% level of significance. In other words, the medians of *Y1* have significant differences between the cells/groups generated by the factors *A* and *B*.
3) Note the different ordering of the factors *A* and *B* in the Category Statistics and the ordering to insert the factors *B* and *A*, as presented in the statement "Categorized by values of *B* and *A*".

Test for Equality of Medians of Y1

Categorized by values of B and A

Date: 04/15/18   Time: 17:35

*Sample: 1 300*

Included observations: 300

| Method | df | Value | Probability |
|---|---|---|---|
| Med. Chi-square | 3 | 19.27602 | 0.0002 |
| Adj. Med. Chi-square | 3 | 17.64706 | 0.0005 |
| Kruskal-Wallis | 3 | 36.05044 | 0.0000 |
| Kruskal-Wallis (tie-adj.) | 3 | 36.36917 | 0.0000 |
| van der Waerden | 3 | 42.53425 | 0.0000 |

**Category Statistics**

| A | B | Count | Median | > Overall Median | Mean Rank | Mean Score |
|---|---|---|---|---|---|---|
| 1 | 1 | 75 | 6.500000 | 48 | 200.5267 | 0.612669 |
| 1 | 2 | 75 | 5.670000 | 32 | 147.2667 | −0.047770 |
| 2 | 1 | 75 | 5.670000 | 24 | 129.2400 | −0.248392 |
| 2 | 2 | 75 | 5.500000 | 26 | 124.9667 | −0.311851 |
|  | All | 300 | 5.670000 | 130 | 150.5000 | 0.001164 |

**Figure 1.3**   The statistical results of the medians test by the factors *A* and *B*.

4) The disadvantage of this test is that the Wald test can't be applied for testing hypotheses. However, we can test the hypothesis on medians differences of *Y1* between levels of selected factor or factors, conditional for one or two other factors, by selecting an appropriate sample. See the following example.

**Example 1.2** *Joint Effects of A and B on* **Med(Y1),** *Conditional for G4 = 2*
Based on the output in Figure 1.4, at the 1% level, we can conclude that the medians of *Y1* have significant differences between the cells generated by the factors *A* and *B*, conditional for the level of *G4* = 2.

**Example 1.3** *Joint Effects of A and B on* **Med(Y1),** *Conditional for G4 = 2 and H3 = 1*
Figure 1.5 presents the statistical results for testing the medians differences of *Y1*, between the four cells (*A* = *i*, *B* = *j*), or the joint effects of *A* and *B* on *Med(Y1)*, based on a sub-sample *G4* = 2 and *H3* = 1.
  Based on these results, the findings and notes are as follows:

1) Note that the cell (*A* = 1, *B* = 1) has no observation. So, the two-way tabulation is an incomplete table.

---

Test for equality of medians of Y1

Categorized by values of B and A

Date: 04/15/18    Time: 19:10

**Sample: 1 300 IF G4 = 2**

Included observations: 66

| Method | df | Value | Probability |
|---|---|---|---|
| Med. Chi-square | 3 | 15.64050 | 0.0013 |
| Adj. Med. Chi-square | 3 | 12.18812 | 0.0068 |
| Kruskal-Wallis | 3 | 16.15117 | 0.0011 |
| Kruskal-Wallis (tie-adj.) | 3 | 16.34152 | 0.0010 |
| van der Waerden | 3 | 16.69569 | 0.0008 |

**Category statistics**

| A | B | Count | Median | > Overall Median | Mean rank | Mean score |
|---|---|---|---|---|---|---|
| 1 | 1 | 12 | 6.500000 | 10 | 50.33333 | 0.892667 |
| 1 | 2 | 17 | 6.000000 | 11 | 38.26471 | 0.160849 |
| 2 | 1 | 10 | 5.585000 | 3 | 25.60000 | −0.369159 |
| 2 | 2 | 27 | 5.500000 | 8 | 25.94444 | −0.360331 |
| | All | 66 | 5.670000 | 32 | 31.50000 | 0.000393 |

**Figure 1.4** The statistical results for testing the joint effects of *A* and *B* on *Y1*, based on a sub-sample *G4* = 2.

Test for equality of medians of Y1

Categorized by values of B and A

Date: 04/15/18   Time: 19:40

**Sample: 1 300 IF G4 = 2 AND H3 = 1**

Included observations: 17

| Method | df | Value | Probability |
|---|---|---|---|
| Med. Chi-square | 2 | 2.198176 | 0.3332 |
| Adj. Med. Chi-square | 2 | 0.648194 | 0.7232 |
| Kruskal-Wallis | 2 | 0.416993 | 0.8118 |
| Kruskal-Wallis (tie-adj.) | 2 | 0.428008 | 0.8073 |
| van der Waerden | 2 | 0.479449 | 0.7868 |

**Category statistics**

| A | B | Count | Median | > Overall Median | Mean rank | Mean score |
|---|---|---|---|---|---|---|
| 1 | 1 | 0 | NA | NA | NA | NA |
| 1 | 2 | 3 | 4.670000 | 1 | 7.333333 | −0.280521 |
| 2 | 1 | 5 | 5.330000 | 2 | 9.100000 | 0.011235 |
| 2 | 2 | 9 | 5.500000 | 5 | 9.500000 | 0.104619 |
|  | All | 17 | 5.330000 | 8 | 9.000000 | 0.009187 |

**Figure 1.5**   The statistical results for testing the joint effects of *A* and *B* on *Y1*, based on a sub-sample *G4* = 2 and *H3* = 1.

2) At the 10% level, the five nonparametric test statistics show that the medians of *Y1* between the three cells—$(A = 1, B = 2)$, $(A = 2, B = 1)$, and $(A = 2, B = 2)$—have significant differences, conditional for cell $(G4 = 2, H3 = 1)$, which are indicated by $df = (3 - 1) = 2$.

**Example 1.4   *The Medians Difference of Y1 Between Two Levels G4 = 3 and G4 = 4, Conditional for $(A = 1, B = 2)$***

Figure 1.6 presents the statistical results for testing the medians difference of *Y1*, between the levels $G4 = 3$ and $G4 = 4$, conditional for $(A = 1, B = 2)$, based on the sub-sample $A = 1$ and $B = 2$ and $G4 > 2$. Based on these results, the following findings and notes are presented:

1) These results in fact show the test of the medians difference between the two cells $(A = 1, B = 2, G4 = 3)$ and $(A = 1, B = 2, G4 = 4)$ in a $2 \times 2 \times 4$ factorial QR of *Y1*.
2) Compared with the previous examples, specific for testing the medians difference between a pair of cells, the results present seven nonparametric test statistics, where the first two are the Wilcoxon/Mann–Whitney and the Wilcoxcon/Mann–Whitney (tie adj.), without the *df* (degree of freedom). And the other five test statistics are the

Test for equality of medians of Y1

Categorized by values of G4

Date: 04/16/18   Time: 05:36

**Sample: 1 300 IF A = 1 AND B = 2 AND G4 > 2**

Included observations: 32

| Method | df | Value | Probability |
|---|---|---|---|
| Wilcoxon/Mann-Whitney | | 2.260218 | 0.0238 |
| Wilcoxon/Mann-Whitney (tie-adj.) | | 2.274436 | 0.0229 |
| Med. Chi-square | 1 | 5.039053 | 0.0248 |
| Adj. Med. Chi-square | 1 | 1.555556 | 0.0593 |
| Kruskal-Wallis | 1 | 5.194805 | 0.0227 |
| Kruskal-Wallis (tie-adj.) | 1 | 5.260367 | 0.0218 |
| van der Waerden | 1 | 5.474734 | 0.0193 |

**Category statistics**

| G4 | Count | Median | > Overall Median | Mean rank | Mean score |
|---|---|---|---|---|---|
| 3 | 14 | 5.670000 | 3 | 12.21429 | −0.422409 |
| 4 | 18 | 6.415000 | 11 | 19.83333 | 0.331767 |
| All | 32 | 6.300000 | 14 | 16.50000 | 0.001815 |

**Figure 1.6**   The statistical results for testing the means difference between two levels, $G4 = 3$ and $G4 = 4$, based on a sub-sample $A = 1$ and $B = 2$ and $G4 > 2$.

same as the statistics in previous examples with $df = (2 - 1) = 1$. See Hardle (1999) and Conover (1980) for the characteristics of the non-parametric statistics.

3) Note that at the 5% level of significance, the two medians have insignificant difference, based on one of the test statistics: namely the *Adj. Med Chi-square* statistic of 1.5556 with $df = 1$ and *Prob.* $= 0.0593$. But based on the other statistics, they have significant difference.

## 1.4   Testing the Medians of *Y1* Categorized by *X1*

EViews provides a program that can transform a numerical predictor directly to an ordinal categorical variable having an unexpected number of levels, where the values of the numerical variables are classified into closed-open intervals having the same length. See the following selected examples.

**Example 1.5    *Testing Medians of Y1 by X1, Using the Default Option***
Figure 1.7 presents the statistical results acquired by selecting the option *Median*, inserting the numerical variable *X1* as the series for classify (refer to Figure 1.1), using 5 as the *Max # of Bins*, and then clicking *OK*.

Based on these results, the findings and notes are as follows:

1) The results present only four categories for the values of *X1,* by using the default options and 5 as the *Max # of Bins*. The four categories are generated directly by the software, with the same length of closed-open intervals. See the following alternative examples.
2) The results present five different nonparametric test statistics with $df = (4 - 1) = 3$, for the medians differences of *Y1*, with the same conclusion that they have significant differences.

**Example 1.6    *Testing Medians of Y1 by X1, Using 15 as the* Max # of Bins**
Figure 1.8 presents the statistical results for testing the medians of *Y1* by *X1*, using 15 as the *Max # of Bins*. In this case, the output presents exactly 15 levels of the values of *X1,* indicated by $df = (15 - 1) = 14$ of the test statistics.

Test for equality of medians of Y1
Categorized by values of X1
Date: 04/15/18    Time: 13:53
*Sample: 1 300*
Included observations: 300

| Method | df | Value | Probability |
|---|---|---|---|
| Med. Chi-square | 3 | 36.54586 | 0.0000 |
| Adj. Med. Chi-square | 3 | 32.70783 | 0.0000 |
| Kruskal-Wallis | 3 | 71.68461 | 0.0000 |
| Kruskal-Wallis (tie-adj.) | 3 | 74.33609 | 0.0000 |
| van der Waerden | 3 | 82.61869 | 0.0000 |

Category statistics

| X1 | Count | Median | > Overall Median | Mean rank | Mean score |
|---|---|---|---|---|---|
| [0.5, 1) | 32 | 4.330000 | 2 | 54.87500 | −1.175696 |
| [1, 1.5) | 146 | 5.670000 | 54 | 136.4932 | −0.134818 |
| [1.5, 2) | 116 | 6.330000 | 69 | 189.4612 | 0.431201 |
| [2, 2.5) | 6 | 7.000000 | 5 | 248.0833 | 1.272583 |
| All | 300 | 5.670000 | 130 | 150.5000 | 0.001164 |

**Figure 1.7**    Statistical results of the test for medians of *Y1*, categorized by the values of *X1*, using 5 as the *Max # of Bins*.

Test for equality of medians of Y1
Categorized by values of X1
Date: 04/15/18    Time: 11:48
**Sample: 1 300**
Included observations: 300

| Method | df | Value | Probability |
|---|---|---|---|
| Med. Chi-square | 14 | 47.31285 | 0.0000 |
| Adj. Med. Chi-square | 14 | 34.17616 | 0.0019 |
| Kruskal-Wallis | 14 | 80.91499 | 0.0000 |
| Kruskal-Wallis (tie-adj.) | 14 | 81.63039 | 0.0000 |
| van der Waerden | 14 | 90.22555 | 0.0000 |

Category statistics

| X1 | Count | Median | > Overall Median | Mean rank | Mean score |
|---|---|---|---|---|---|
| [0.7, 0.8) | 5 | 4.000000 | 0 | 29.90000 | −1.642366 |
| [0.8, 0.9) | 7 | 4.330000 | 0 | 57.21429 | −1.205984 |
| [0.9, 1) | 20 | 4.500000 | 2 | 60.30000 | −1.048428 |
| ... | | | | | |
| [1.9, 2) | 4 | 6.750000 | 3 | 220.5000 | 0.853188 |
| [2, 2.1) | 1 | 7.500000 | 1 | 291.0000 | 1.933650 |
| [2.1, 2.2) | 5 | 7.000000 | 4 | 239.1000 | 1.140370 |
| All | 300 | 5.670000 | 130 | 150.5000 | 0.001164 |

**Figure 1.8** A part of the results for testing the medians of *Y1*, categorized by the values of *X1*, using 15 as the *Max # of Bins*.

### Example 1.7   *Testing Medians of Y1 by X1, Using 30 as the Max # of Bins*

Figure 1.9 presents the statistical results for testing the medians of *Y1* by *X1*, using 30 as the *Max # of Bins*. Based on these results, the findings and notes are as follows:

1) Unexpectedly, the output presents 28 levels or intervals of *X1,* which are indicated by $df = (28 - 1) = 27$ of the test statistics.
2) The graphs of *Count* and *Median* of *Y1* are shown in Figure 1.10, with a minimum value of 4.000 and a maximum value of 8.000. Note that 3 of the 28 intervals—[11.95, 2), [2.05, 2.1), and [2.15, 2.2)—or the *X1*-levels of 13, 14 and 16, have only one observation each.
3) The graph can be developed using Microsoft Excel. The steps are as follows:
   3.1 Copy the 28 complete scores of the two variables, *Count* and *Median*, in Figure 1.9 to an Excel file.

Test for equality of medians of Y1
Categorized by values of X1
Date: 04/16/18   Time: 07:58
**Sample: 1 300**
Included observations: 300

| Method | df | Value | Probability |
|---|---|---|---|
| Med. Chi-square | 27 | 58.25747 | 0.0004 |
| Adj. Med. Chi-square | 27 | 34.42433 | 0.1541 |
| Kruskal-Wallis | 27 | 91.84382 | 0.0000 |
| Kruskal-Wallis (tie-adj.) | 27 | 92.65584 | 0.0000 |
| van der Waerden | 27 | 102.9577 | 0.0000 |

Category statistics

| X1 | Count | Median | > Overall Median | Mean rank | Mean score |
|---|---|---|---|---|---|
| [0.75, 0.8) | 5 | 4.000000 | 0 | 29.90000 | −1.642366 |
| [0.8, 0.85) | 3 | 4.670000 | 0 | 74.16667 | −0.792185 |
| [0.85, 0.9) | 4 | 4.000000 | 0 | 44.50000 | −1.516333 |
| [0.9, 0.95) | 12 | 4.500000 | 1 | 55.75000 | −1.113574 |
| [0.95, 1) | 8 | 4.500000 | 1 | 67.12500 | −0.950710 |
| ... | | ... | | | |
| [1.9, 1.95) | 3 | 6.500000 | 2 | 201.1667 | 0.700101 |
| [1.95, 2) | 1 | 7.000000 | 1 | 272.5000 | 1.312449 |
| [2.05, 2.1) | 1 | 7.500000 | 1 | 291.0000 | 1.933650 |
| [2.1, 2.15) | 4 | 7.000000 | 3 | 224.2500 | 0.826663 |
| [2.15, 2.2) | 1 | 8.000000 | 1 | 298.5000 | 2.395200 |
| All | 300 | 5.670000 | 130 | 150.5000 | 0.001164 |

**Figure 1.9**   A part of the results for testing the medians of *Y1*, categorized by values of *X1*, using 30 as the *Max # of Bins*.

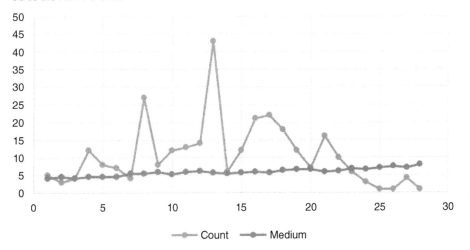

**Figure 1.10**   The graphs of *Count* and *Median* of *Y1*, in Figure 1.9.

3.2 Select the two variables *Count* and *Median* and their scores. Then select *Insert* and click one of the chart types in the *Charts* group. I used the *Line graph with markers* option to produce the graphs of *Count* and *Median* of *Y1* in Figure 1.10.

## 1.5 Testing the Medians of *Y1* Categorized by *RX1* = @*Ranks(X1,a)*

In this section, I want to explore testing for the medians of *Y1* categorized by *RX1* = @Ranks(*X1,a*). At first, I was expecting the variable would be treated as a categorical variable. However, the following example shows that *RX1* is treated as a numerical variable.

**Example 1.8** *Testing Medians of Y1 by* **RX1**, *Using 300 as the Max # of Bins*
Figure 1.11 presents the statistical results for testing the medians of *Y1* by *X1*, using 300 as the *Max # of Bins*, to make *RX1* be a categorical variable. Based on these results, the findings and notes are as follows:

1) The output presents 104 intervals with a length of 1 (one). So, we can say that *RX1* has 104 different scores, indicated by $df = (104 - 1) = 103$ of the test statistics. In addition, the output presents the values of Count and Median for 104 intervals of *RX1*, with their graphs presented in Figure 1.12. In fact *X1* also 104 different scores. For instance, *RX1* has four ranks of 2.5 in the first close-open interval [2,3), as presented in Figure 1.13.
2) However, if other values for *Max # of Bins* are used, such as 15 and 30, the variable *RX1* is treated as a numerical variable. It is unexpected that when using 15 for the *Max # of Bins*, the output presents only 16 intervals of *RX1*. Do these as exercises.

## 1.6 Unexpected Statistical Results

Compare the previous testing of the medians of *Y1* categorized by *X1* with the following example, which presents unexpected statistical results from testing the medians of *Y1* categorized by the values of *X2*.

**Example 1.9** *Testing the Medians of Y1 by X2*
Figure 1.14 presents the unexpected statistical results for testing the medians of *Y1* categorized by the values of *X2,* using several different values for *Max # of Bins*. I have used the *Max # of Bins* values of 2, 10, 15, and 300. Based on these results, the following findings and notes are presented:

1) The numerical variable *X2* has 27 different values, indicated by $df = (27 - 1) = 26$ of the test statistics.
2) I can't find any explanation for these differential results, compared with the testing for the medians of *Y1* categorized by the values of *X1*, presented in several previous examples.

Test for equality of medians of Y1
Categorized by values of RX1
Date: 04/17/18   Time: 06:46
**Sample: 1 300**
Included observations: 300

| Method | df | Value | Probability |
|---|---|---|---|
| Med. Chi-square | 103 | 120.7935 | 0.1111 |
| Adj. Med. Chi-square | 103 | 38.87847 | 1.0000 |
| Kruskal-Wallis | 103 | 144.5314 | 0.0044 |
| Kruskal-Wallis (tie-adj.) | 103 | 145.8092 | 0.0035 |
| van der Waerden | 103 | 151.7712 | 0.0009 |

Category statistics

| RX1 | Count | Median | > Overall Median | Mean rank | Mean score |
|---|---|---|---|---|---|
| [2, 3) | 4 | 1.835000 | 0 | 10.62500 | −1.960317 |
| [5, 6) | 1 | 5.500000 | 0 | 107.0000 | −0.370562 |
| [6, 7) | 1 | 4.670000 | 0 | 52.00000 | −0.943324 |
| ... | | | | | |
| ... | | | | | |
| [295, 296) | 1 | 7.500000 | 1 | 291.0000 | 1.933650 |
| [296, 297) | 2 | 6.165000 | 1 | 176.0000 | 0.340876 |
| [298, 299) | 1 | 7.000000 | 1 | 272.5000 | 1.312449 |
| [299, 300) | 1 | 7.000000 | 1 | 272.5000 | 1.312449 |
| [300, 301) | 1 | 8.000000 | 1 | 298.5000 | 2.395200 |
| All | 300 | 5.670000 | 130 | 150.5000 | 0.001164 |

**Figure 1.11**   A part of the results of the test for equality of medians of *Y1*, categorized by *RX1*, using 300 as the *Max # of Bins*.

3) Special notes:
   If we want to have a fixed categories or levels for *X2*, then we can generate an ordinal variable based on *X2*, such as the factors *H2* and *H3*, or *G4* based on *X1*. Another method is to generate three closed-open intervals, namely *H3a* or *O3X2*, based on values of *X2*, from 1.3 to 3.9, using the following equation:

$$O3X2 = 1 + 1 * (X2 > = 2) + 1 * (X2 > = 3)$$

4) I also have tried doing additional analyses to explore similar problems. I found only one other unexpected result, shown in Figure 1.15, which had the same test statistics as presented in Figure 1.14. This was the case when using selected pairs of the seven

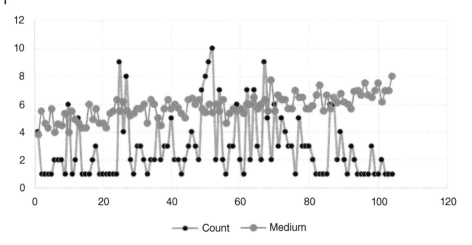

**Figure 1.12** The graphs of *Count* and *Median* of *Y1*, by the levels of *X1*, in Figure 1.11.

| Obs # | X1 | RX1 | Y1 | | Obs # | X1 | RX1 | Y1 |
|-------|------|-----|------|---|-------|------|-----|------|
| 151 | 0.78 | 2.5 | 1.67 | | 226 | 0.78 | 2.5 | 1.5 |
| 152 | 0.78 | 2.5 | 4.33 | | 151 | 0.78 | 2.5 | 1.67 |
| 226 | 0.78 | 2.5 | 1.5 | | 227 | 0.78 | 2.5 | 4 |
| 227 | 0.78 | 2.5 | 4 | | 152 | 0.78 | 2.5 | 4.33 |
| RX1 = (1 + 2 + 3 + 4)/4 = 2.5 | | | | | Med(Y1) = (1.67 + 4)/2 = 1.835 | | | |

**Figure 1.13** The four observations in the first interval of *RX1*: [2,3).

numerical variables in the Mlogit.wf1 EViews data file, with the dependent variable *INC*, categorized by values of each of the numerical variables *LW, X1, X2, Z1, Z2,* and *Z1*. See examples 1.13, 1.14, and 1.17 based on the data in Mlogit.wf1.

## 1.7  Testing the Medians of *Y1* by *X1* and Categorical Factors

The following examples present the testing for the medians of *Y1* by *X1*, with either one or two categorical variables.

### Example 1.10  *Testing the Medians of Y1 by X1 and A*
Figure 1.16 shows the results for testing the medians of *Y1* by *X1* and *A*, in Data_Faad.wf1. The results were obtained using 5 as the *Max # of Bins*.
Based on these results, the findings and notes are as follows:

1) The results present four levels of *X1*, and one of them has no observation, conditional for the level $A = 2$. Hence, the test statistics have $df = (7 - 1) = 6$.

Test for equality of medians of Y1
Categorized by values of X2
Date: 10/15/20 Time: 17:46
**Sample: 1 300**
Included observations: 300

| Method | df | Value | Probability |
|---|---|---|---|
| Med. Chi-square | 26 | 56.76601 | 0.0004 |
| Adj. Med. Chi-square | 26 | 36.33805 | 0.0856 |
| Kruskal-Wallis | 26 | 72.25183 | 0.0000 |
| Kruskal-Wallis (tie-adj.) | 26 | 72.89063 | 0.0000 |
| van der Waerden | 26 | 75.25417 | 0.0000 |

Category statistics

| X2 | Count | Median | > Overall Median | Mean rank | Mean score |
|---|---|---|---|---|---|
| 1.3 | 5 | 4.330000 | 0 | 32.70000 | −1.558832 |
| 1.4 | 5 | 3.670000 | 0 | 34.50000 | −1.600223 |
| 1.5 | 3 | 5.330000 | 0 | 70.33333 | −0.799002 |
| 1.6 | 7 | 5.670000 | 2 | 142.9286 | −0.140681 |
| 1.7 | 7 | 4.330000 | 0 | 57.57143 | −0.990623 |
| 1.8 | 10 | 4.670000 | 0 | 62.85000 | −0.873197 |
| 1.9 | 10 | 5.500000 | 2 | 113.5000 | −0.366951 |
| 2.0 | 11 | 5.500000 | 2 | 111.0909 | −0.393099 |
| 2.1 | 14 | 5.670000 | 5 | 148.4286 | −0.007211 |
| 2.2 | 14 | 5.670000 | 6 | 142.9286 | −0.040499 |
| 2.3 | 15 | 5.670000 | 6 | 148.6000 | −0.017457 |
| 2.4 | 13 | 5.670000 | 5 | 140.8846 | −0.143019 |
| 2.5 | 32 | 5.500000 | 13 | 138.4844 | −0.134828 |
| 2.6 | 15 | 5.670000 | 5 | 149.5000 | −0.010591 |
| 2.7 | 21 | 5.670000 | 8 | 148.9524 | 0.061708 |
| 2.8 | 21 | 6.300000 | 13 | 192.6667 | 0.466994 |
| 2.9 | 18 | 6.300000 | 12 | 183.2222 | 0.385050 |
| 3.0 | 13 | 6.330000 | 9 | 184.4615 | 0.315660 |
| 3.1 | 18 | 6.330000 | 14 | 202.7222 | 0.509875 |
| 3.2 | 15 | 6.000000 | 8 | 183.6000 | 0.399907 |
| 3.3 | 8 | 5.835000 | 4 | 172.6875 | 0.321973 |
| 3.4 | 9 | 6.300000 | 5 | 176.7778 | 0.229228 |
| 3.5 | 5 | 6.000000 | 3 | 154.0000 | 0.017109 |
| 3.6 | 4 | 6.000000 | 2 | 169.0000 | 0.170129 |
| 3.7 | 4 | 7.000000 | 3 | 228.3750 | 0.993454 |
| 3.8 | 2 | 6.915000 | 2 | 261.5000 | 1.221907 |
| 3.9 | 1 | 7.000000 | 1 | 272.5000 | 1.312449 |
| All | 300 | 5.670000 | 130 | 150.5000 | 0.001164 |

**Figure 1.14**  Unexpected results for testing the medians of *Y1* categorized by *X2*.

Test for equality of medians of Y1
Categorized by values of Y2
Date: 04/18/18   Time: 13:18
**Sample: 1 300**
Included observations: 300

| Method | df | Value | Probability |
|---|---|---|---|
| Med. Chi-square | 26 | 131.7459 | 0.0000 |
| Adj. Med. Chi-square | 26 | 99.72207 | 0.0000 |
| Kruskal-Wallis | 26 | 175.9275 | 0.0000 |
| Kruskal-Wallis (tie-adj.) | 26 | 177.4830 | 0.0000 |
| van der Waerden | 26 | 181.6164 | 0.0000 |

Category statistics

| Y2 | Count | Median | > Overall Median | Mean rank | Mean score |
|---|---|---|---|---|---|
| 1.50 | 8 | 1.670000 | 0 | 11.18750 | −2.016399 |
| 1.67 | 2 | 4.500000 | 0 | 38.25000 | −1.169536 |
| 4.00 | 18 | 4.330000 | 0 | 29.05556 | −1.379970 |
| *** | | | | | |
| *** | | | | | |
| 7.50 | 19 | 6.500000 | 17 | 231.7368 | 0.839504 |
| 7.75 | 1 | 6.670000 | 1 | 255.0000 | 1.024397 |
| 8.00 | 3 | 6.670000 | 3 | 254.8333 | 1.044731 |
| All | 300 | 5.670000 | 130 | 150.5000 | 0.001164 |

**Figure 1.15**   Unexpected results for testing the medians of Y1 categorized by Y2.

2) As a comparison using the dependent variables X2 and A, the results would present $2 \times 27 = 54$ cells, but only 47 have observations, even when using 2 as the *Max # of Bins*. The test statistics are presented in Figure 1.17, with $df = (47 − 1) = 46$.

### Example 1.11   *Testing the Medians of Y1 by X1, B, and A*

Figure 1.18 shows the test statistics for testing the medians *Y1* by *X1, B,* and *A*, in Data_Faad.wf1, using 5 as the *Max # of Bins*. However, the results present only four levels of *X1*.

On the other hand, the test statistics have $df = (12 − 1) = 11$. This indicates there are only 12 non-empty cells of the 16 generated by the three categorical variables *A, B,* and *X1*, as shown by the complete output of the test.

Test for equality of medians of Y1
Categorized by values of X1 and A
Date: 04/16/18    Time: 16:58
**Sample: 1 300**
Included observations: 300

| Method | df | Value | Probability |
|---|---|---|---|
| Med. Chi-square | 6 | 71.80200 | 0.0000 |
| Adj. Med. Chi-square | 6 | 62.21680 | 0.0000 |
| Kruskal-Wallis | 6 | 85.42606 | 0.0000 |
| Kruskal-Wallis (tie-adj.) | 6 | 86.18134 | 0.0000 |
| van der Waerden | 6 | 95.58094 | 0.0000 |

Category statistics

| A | X1 | Count | Median | > Overall Median | Mean rank | Mean score |
|---|---|---|---|---|---|---|
| 1 | [0.5, 1) | 9 | 4.330000 | 0 | 71.11111 | −0.857987 |
| 1 | [1, 1.5) | 69 | 5.670000 | 27 | 144.5870 | −0.039232 |
| 1 | [1.5, 2) | 66 | 6.500000 | 48 | 211.8106 | 0.684254 |
| 1 | [2, 2.5) | 6 | 7.000000 | 5 | 248.0833 | 1.272583 |
| 2 | [0.5, 1) | 23 | 4.330000 | 2 | 48.52174 | −1.300018 |
| 2 | [1, 1.5) | 77 | 5.500000 | 27 | 129.2403 | −0.220472 |
| 2 | [1.5, 2) | 50 | 5.670000 | 21 | 159.9600 | 0.097171 |
| 2 | [2, 2.5) | 0 | NA | NA | NA | NA |
| | All | 300 | 5.670000 | 130 | 150.5000 | 0.001164 |

**Figure 1.16**  Results for testing the medians of *Y1* categorized by *X1* and *A*.

Categorized by values of X2 and A
**Sample: 1 300**
Included observations: 300

| Method | df | Value | Probability |
|---|---|---|---|
| Med. Chi-square | 46 | 161.7451 | 0.0000 |
| Adj. Med. Chi-square | 46 | 101.1943 | 0.0000 |
| Kruskal-Wallis | 46 | 117.3116 | 0.0000 |
| Kruskal-Wallis (tie-adj.) | 46 | 118.3488 | 0.0000 |
| van der Waerden | 46 | 119.2714 | 0.0000 |

**Figure 1.17**  Results for testing the medians of *Y1* categorized by *X2* and *A*.

Test for equality of medians of Y1
Categorized by values of X1 and B and A
Date: 04/18/18   Time: 17:12
*Sample: 1 300*
Included observations: 300

| Method | df | Value | Probability |
|---|---|---|---|
| Med. Chi-square | 11 | 118.6199 | 0.0000 |
| Adj. Med. Chi-square | 11 | 95.80868 | 0.0000 |
| Kruskal-Wallis | 11 | 90.01479 | 0.0000 |
| Kruskal-Wallis (tie-adj.) | 11 | 90.81065 | 0.0000 |
| van der Waerden | 11 | 101.8929 | 0.0000 |

**Figure 1.18**   The test statistics for testing the medians of *Y1* by *X1*, *B*, and *A*.

## 1.8   Testing the Medians of *Y* by Numerical Variables

As a comparative study, the following two subsections present examples based on Data_Faad.wf1 and Mlogit.wf1.

### 1.8.1   Findings Based on Data_Faad.wf1

**Example 1.12**   *Testing the Medians of Y1 categorized by X1 and X2*
Figure 1.19 presents the test statistics for testing the medians *Y1* categorized by *X1* and *X2*, in Data_Faad.wf1, using 5 as the *Max # of Bins*. However, it is unexpected that the results present only four levels of *X1*, but many levels of *X2*.

Referring to the problem with the categorized of *X2* presented in Figure 1.19, the results present $27 \times 4 = 108$ cells generated by *X2* and *X1*, but the test statistics only have $df = (67 - 1) = 66$, which indicates that there are 42 empty cells.

### 1.8.2   Findings Based on Mlogit.wf1

**Example 1.13**   *Testing the Medians of INC Categorized by X1 and X2*
Figure 1.20 presents the test statistics for testing the medians *INC* categorized by *X1* and *X2*, in Mlogit.wf1, using 2 as the *Max # of Bins* .

It is unexpected that the output presents exactly the same statistical results as using 4 as the *Max # of Bins*. I have tried several times to ensure I didn't mistype, but I can't find the problem. However, by using 5 as the *Max # of Bins*, the test statistics shown in Figure 1.21 are obtained, with $df = (25 - 1) = 24$.

Test for equality of medians of Y1
Categorized by values of X1 and X2
Date: 04/18/18   Time: 11:48
**Sample: 1 300**
Included observations: 300

| Method | df | Value | Probability |
|---|---|---|---|
| Med. Chi-square | 66 | 322.7169 | 0.0000 |
| Adj. Med. Chi-square | 66 | 335.9800 | 0.0000 |
| Kruskal-Wallis | 66 | 131.9326 | 0.0000 |
| Kruskal-Wallis (tie-adj.) | 66 | 131.0991 | 0.0000 |
| van der Waerden | 66 | 142.9591 | 0.0000 |

Category statistics

| X2 | X1 | Count | Median | > Overall Median | Mean rank | Mean score |
|---|---|---|---|---|---|---|
| 1.300000 | [0.5, 1) | 3 | 1.670000 | 0 | 10.66667 | −2.009283 |
| 1.300000 | [1, 1.5) | 2 | 4.915000 | 0 | 65.75000 | −0.883155 |
| 1.300000 | [1.5, 2) | 0 | NA | NA | NA | NA |
| 1.300000 | [2, 2.5) | 0 | NA | NA | NA | NA |
| *** | | | | | | |
| *** | | | | | | |
| 1.900000 | [0.5, 1) | 0 | NA | NA | NA | NA |
| 1.900000 | [1, 1.5) | 0 | NA | NA | NA | NA |
| 1.900000 | [1.5, 2) | 1 | 7.000000 | 1 | 272.5000 | 1.312449 |
| 1.900000 | [2, 2.5) | 0 | NA | NA | NA | NA |
| | All | 300 | 5.670000 | 130 | 150.5000 | 0.001164 |

**Figure 1.19** A part of the results for testing the medians of *Y1* by *X2* and *X1*.

**Example 1.14**  *Testing the Medians of INC Categorized by LW, X2, and X1*
Figure 1.22 presents the test statistics for testing the medians *INC* categorized by *LW, X2,* and *X1,* in Mlogit.wf1, using 2 as the *Max # of Bins*. Based on the output, the following findings and notes are presented:

1) As expected, the results present a complete $2 \times 2 \times 2$ tabulation of the medians.
2) Based on the test statistics, we can conclude that the dichotomous variables *X1, X2,* and *LW* have insignificant joint effects on *INC,* or *Med(INC)*. In other words, the eight medians of *INC* categorized by the values *X1, X2,* and *LW* have insignificant differences.
3) As an additional illustration, Figure 1.23 shows the test for the equality of the medians of *LW* categorized by *X1, X2,* and *Z1,* using 2 as the *Max # of Bins* . It shows the eight

Test for equality of medians of INC
Categorized by values of X2 and X1
Date: 04/18/18   Time: 18:06
**Sample: 1 1000**
Included observations: 1000

| Method | df | Value | Probability |
|---|---|---|---|
| Med. Chi-square | 3 | 1.056003 | 0.3831 |
| Adj. Med. Chi-square | 3 | 2.694832 | 4411 |
| Kruskal-Wallis | 3 | 0.554064 | 0.9069 |
| Kruskal-Wallis (tie-adj.) | 3 | 0.554064 | 0.9069 |
| van der Waerden | 3 | 0.345048 | 0.9513 |

Category statistics

| X1 | X2 | Count | Median | > Overall Median | Mean rank | Mean score |
|---|---|---|---|---|---|---|
| [0, 0.5) | [0, 0.5) | 247 | 5.230930 | 134 | 507.0729 | −0.002831 |
| [0, 0.5) | [0.5, 1) | 244 | 4.999364 | 123 | 499.1967 | −0.026009 |
| [0.5, 1) | [0, 0.5) | 250 | 4.292503 | 118 | 489.8920 | 0.001347 |
| [0.5, 1) | [0.5, 1) | 259 | 4.738205 | 125 | 505.6988 | 0.025902 |
| | All | 1000 | 4.839388 | 500 | 500.5000 | −1.51E−17 |

**Figure 1.20** The statistical results for testing the medians of Y1 by X1 and X2, using 2 as the *Max # of Bins*.

Test for equality of medians of INC
Categorized by values of X2 and X1
Date: 04/18/18   Time: 18:39
**Sample: 1 1000**
Included observations: 1000

| Method | df | Value | Probability |
|---|---|---|---|
| Med. Chi-square | 24 | 28.37237 | 0.2446 |
| Adj. Med. Chi-square | 24 | 22.75337 | 0.5344 |
| Kruskal-Wallis | 24 | 14.90306 | 0.9236 |
| Kruskal-Wallis (tie-adj.) | 24 | 14.90306 | 0.9236 |
| van der Waerden | 24 | 15.79986 | 0.8952 |

**Figure 1.21** The test statistics for testing the medians of Y1 by X2 and X1, using 5 as the *Max # of Bins*.

Test for equality of medians of INC
Categorized by values of LW and X2 and X1
Date: 04/18/18   Time: 16:32
*Sample (adjusted): 1 998*
Included observations: 790 after adjustments

| Method | df | Value | Probability |
|---|---|---|---|
| Med. Chi-square | 7 | 4.483507 | 0.7227 |
| Adj. Med. Chi-square | 7 | 2.907499 | 0.8934 |
| Kruskal-Wallis | 7 | 4.125189 | 0.7653 |
| Kruskal-Wallis (tie-adj.) | 7 | 4.125189 | 0.7653 |
| van der Waerden | 7 | 4.501008 | 0.7206 |

Category statistics

| X1 | X2 | LW | Count | Median | > Overall Median | Mean rank |
|---|---|---|---|---|---|---|
| [0, 0.5) | [0, 0.5) | [0, 10) | 188 | 5.177043 | 101 | 390.1064 |
| [0, 0.5) | [0, 0.5) | [10, 20) | 6 | 1.890972 | 3 | 388.6667 |
| [0, 0.5) | [0.5, 1) | [0, 10) | 190 | 5.199538 | 98 | 404.5842 |
| [0, 0.5) | [0.5, 1) | [10, 20) | 5 | 2.814888 | 1 | 228.0000 |
| [0.5, 1) | [0, 0.5) | [0, 10) | 189 | 4.703461 | 89 | 381.7831 |
| [0.5, 1) | [0, 0.5) | [10, 20) | 5 | 4.356085 | 2 | 398.4000 |
| [0.5, 1) | [0.5, 1) | [0, 10) | 200 | 4.842086 | 97 | 406.8300 |
| [0.5, 1) | [0.5, 1) | [10, 20) | 7 | 5.037541 | 4 | 409.8571 |
| | | All | 790 | 4.947504 | 395 | 395.5000 |

**Figure 1.22**  The statistical results for testing the medians of *Y1* by *X1*, *X2*, and *LW,*using 2 as the *Max # of Bins*.

Test for equality of medians of LW
Categorized by values of Z1 and X2 and X1
Date: 04/18/18   Time: 20:17
*Sample (adjusted): 1 998*
Included observations: 790 after adjustments

| Method | df | Value | Probability |
|---|---|---|---|
| Med. Chi-square | 7 | 48.90831 | 0.0000 |
| Adj. Med. Chi-square | 7 | 45.20125 | 0.0000 |
| Kruskal-Wallis | 7 | 64.31390 | 0.0000 |
| Kruskal-Wallis (tie-adj.) | 7 | 64.31390 | 0.0000 |
| van der Waerden | 7 | 65.18911 | 0.0000 |

**Figure 1.23**  Test statistics for testing medians of *LW* categorized by *Z1* and *X2* and *X1*, using 2 as the *Max # of Bins*.

medians of *LW* have a significant difference, based on each non-parametric test with $df = (8 - 1) = 7$ and *p*-value = 0.0000.

## 1.9  Application of the Function @*Mediansby(Y,IV)*

As an alternative method, EViews provides a function, @Mediansby(*Y1,IV*), which we can use to generate a new variable for the medians of *Y1* by the values of an independent variable *IV*, either as a categorical or a numerical variable.

**Example 1.15**  *Application of @Medianby(Y1,G4) in Data_Faad.wf1*
By clicking *Quick/Generate series* and inserting the following equation, *MedY1_G4* is inserted directly as a new variable in the file:

$$MedY1\_G4 = @Mediansby(Y1, G4) \tag{1.5}$$

As an illustration, Figure 1.24 presents the descriptive statistics of *MedY1_G4* by *G4* and its one-way tabulation. Based on these results, the following findings and notes are presented:

1) The means and medians of *MedY1_G4* by *G4* have exactly the same values, because *MedY1_G4* has the same scores and values at each level of *G4*.
2) The one-way tabulation of *MedY1_G4* presents only three values of the medians of *Y1* because the medians of *Y1* by *G4* have only three different values. In general, for a categorical *Gk* with many levels, the one-way tabulation of *MedY1_Gk* would always result in a smaller number of distinct medians than the *MedY1_Gk* by *Gk*. See the following example.

**Example 1.16**  *Application of @Mediansby(Y1,X1), in Data_Faad.wf1*
This example generates the variable *MedY1_X1* = @Mediansby(*Y1,X1*) for the numerical variable *X1*.

| Descriptive Statistics for MEDY1_G4 Categorized by values of G4 Date: 04/26/18  Time: 17:10 Sample: 1 300 Included observations: 300 | | | | | Tabulation of MEDY1_G4 Date: 04/26/18  Time: 17:02 Sample: 1 300 Included observations: 300 Number of categories: 3 | | |
|---|---|---|---|---|---|---|---|
| G4 | Mean | Median | Obs. | | Value | Count | Percent |
| 1 | 5.330000 | 5.330000 | 75 | | 5.330000 | 75 | 25.00 |
| 2 | 5.670000 | 5.670000 | 66 | | 5.670000 | 149 | 49.67 |
| 3 | 5.670000 | 5.670000 | 83 | | 6.500000 | 76 | 25.33 |
| 4 | 6.500000 | 6.500000 | 76 | | Total | 300 | 100.00 |
| All | 5.795267 | 5.670000 | 300 | | | | |

**Figure 1.24**  Descriptive statistics of *MedY1_G4* by *G4* and its one-way tabulation.

| X1 | Mean | Median | Obs. | No. |
|---|---|---|---|---|
| [0.78, 0.785) | 3.835 | 3.835 | 4 | 1 |
| [0.79, 0.795) | 5.5 | 5.5 | 1 | 2 |
| [0.8, 0.805) | 4.67 | 4.67 | 1 | 3 |
| ... | | | | ... |
| ... | | | | ... |
| [2.11, 2.115) | 7 | 7 | 1 | 102 |
| [2.12, 2.125) | 7 | 7 | 1 | 103 |
| [2.15, 2.155) | 8 | 8 | 1 | 104 |
| All | 5.7531 | 5.67 | 300 | |

| | | | Cumulative | |
|---|---|---|---|---|
| Value | Count | Percent | Count | No. |
| 3.835 | 4 | 1.33 | 4 | 1 |
| 4 | 8 | 2.67 | 12 | 2 |
| 4.33 | 4 | 1.33 | 16 | 3 |
| ... | | | | ... |
| ... | | | | ... |
| 7.33 | 1 | 0.33 | 295 | 30 |
| 7.5 | 2 | 0.67 | 297 | 31 |
| 7.75 | 2 | 0.67 | 299 | 32 |
| 8 | 1 | 0.33 | 300 | 33 |
| Total | 300 | 100 | 300 | |

**Figure 1.25** A part of the descriptive statistics of *MedY1_X1* by *X1* and its one-way tabulation, using 300 as the *Max # Bins*.

Figure 1.25 presents a part of the descriptive statistics of *MedY1_X1* by *X1* and its one-way tabulation, using 300 as the *Max # Bins*. Based on these results, the findings and notes are as follows:

1) A *Max # of Bins* of 300 is used to obtain the maximum number of classifications for the values of *X1*, because the data has 300 observations.
2) Like previous results, these findings also are unpredictable results, and they are highly dependent on the data and the maximum number of bins used.

    2.1 The descriptive statistics present 104 closed-open intervals of the values of *X1*, with the length of 0.005, such as [0.78, 0.785). Because the values of *X1* have only two decimal points, each interval has a single observation. As a result, *X1* has 104 distinct values for the 300 scores of *X1*.

    2.2 On the other hand, the one-way tabulation of *MedY1_X1* presents only 33 distinct medians for the 300 values of *Y1*.

    2.3 As an additional analysis, Figure 1.26a presents the graphs of quartiles of normal on the quartiles of *MedY1_X1*. This shows that the distribution of *MedY1_X1* is close to a normal distribution. For a comparison, see the following example.

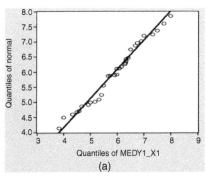
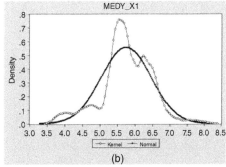

**Figure 1.26** The Quantiles – Quantiles graph of *MedY1_X1* and its kernel density and theoretical density functions.

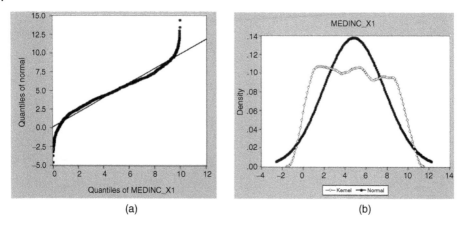

(a)                                                    (b)

**Figure 1.27**   The Quartiles – Quartiles graph of *MedINC_X1* in MLogit.wf1 and its kernel and theoretical density functions.

2.4 Figure 1.26b presents the kernel and theoretical density functions of *MedY1_X1*. Note that it is known that all numerical variables have theoretical normal density functions, which are supported by the central limit theorem, as presented in Agung (2011a). And the kernel density function is a special non-parametric graph.

**Example 1.17**   *Application of @Mediansby(INC,X1) in MLogit.wf1*
As a comparison with the graphs in Figure 1.26, Figure 1.27 presents the Quartiles – Quartiles graph of *MedINC_X1* in MLogit.wf1 and its kernel and theoretical density functions.

Based on the graphs in Figure 1.27, the following findings and notes are presented.

1) The Quartiles – Quartiles graph shows a systematic pattern. The points are below and above the line. Hence, the distribution of the values of *MedINC_X1* is far from a normal distribution.
2) Similarly, its kernel density also shows that the density of *MedINC_X1* is far from a normal density. However, its theoretical density is a normal density.

| Descriptive statistics for MEDINC_X1 | | | |
|---|---|---|---|
| Categorized by values of X1 | | | |
| Included observations: 1000 | | | |
| X1 | Median | Obs. | No. |
| [0.002, 0.003) | 0.61959 | 1 | 1 |
| [0.003, 0.004) | 2.169228 | 1 | 2 |
| ... | ... | ... | ... |
| [0.994, 0.995) | 6.489855 | 2 | 647 |
| [0.995, 0.996) | 0.43968 | 1 | 648 |
| [0.997, 0.998) | 8.769523 | 1 | 649 |
| All | 4.840846 | 1000 | |

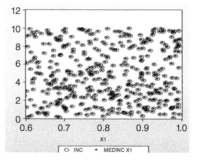

**Figure 1.28**   The descriptive statistics for *MedINC_X1* categorized by values of *X1*, using 1000 as the *Max # of Bins*, and the scatter graphs of *INC* and *MedINC_X1* on *X1*, only for *X1* > 0.6.

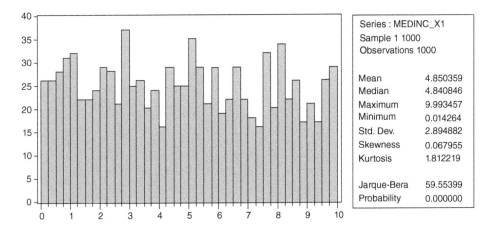

**Figure 1.29** Histogram and descriptive statistics for *MedINC_X1*.

3) Figure 1.28 presents a part of the descriptive statistics for *MedINC_X1* categorized by values of *X1*, using 1000 as the *Max # of Bins*, which shows that *MedINC_X1* has 649 different values out of the 1,000 observations. Note that 1000 is used to obtain the maximum number of distinct values of the medians of *INC* by *X1*.

The scatter graphs of *INC* and *MedINC_X1* on *X1* are presented only for $X1 > 0.6$. That is to better illustrate that the observed scores of *INC* are the same as the values of *MedINC_X1* for many scores of *X1*.

4) In addition, Figure 1.29 presents the histogram and descriptive statistics for *MedINC_X1*. Based on these results, the findings and notes are as follows:

4.1 The histogram clearly shows that the distribution of the scores (observed values) of *MedINC_X1* is very far from normal.

4.2 The Jarque-Bera (JB) test statistic also does not support the normality of the *MedINC_X1* scores. In other words, the normality of the *MedINC_X1* scores is rejected based on the JB statistic of 59.554 with a *p*-value $= 0.000\,000$.

# 2

# One- and Two-way ANOVA Quantile Regressions

## 2.1 Introduction

The equation specifications (ESs) of the *analysis of variance* (ANOVA) models presented in Agung (2011a), as well as other books, can be used directly to apply the *quantile regressions* (QRs) with categorical predictors, namely ANOVA-QR. Similarly, the specification equations of other mean regressions (MRs) also can be used directly to obtain the outputs of QRs. With EViews, the output of the MR is obtained using the LS – Least Squares (NLS and ARMA) estimation setting method. Another estimation setting method, QREG – Quantile Regression (including LAD), is used for the quantile regression analysis. The basic estimation process of the QR and its more advanced statistical analysis are presented in Appendix A, Section A.3, for the readers who never do the regression analysis using EViews.

In addition, it is well known that the error terms of MR are assumed to have *independent identical normal distribution* (IID) $\sim N(0, \sigma^2)$. But, QR, its error terms are assumed to have IID of zero means only. So, they don't need the normal distribution assumption. The ANOVA-QR, in fact, is *nonparametric quantile regression* (NP-QR), and the QR with numerical predictors of independent variables (IVs) is called *semiparametric quantile regression* (SP-QR). Agung (2011a) has presented alternative ANOVA models based on selected data sets, including Data_Faad.wf1, which is presented in Figure 1.2 with additional ordinal variables that will be used to present various examples of *one-way* ANOVA-QRs and selected $I \times J$ Factorial QRs, starting with the simplest one, namely the $2 \times 2$ Factorial ANOVA-QR.

## 2.2 One-way ANOVA Quantile Regression

We can run a one-way ANOVA QR of $Y$ on a single factor *F1* using the following alternative equation specifications, without and with an intercept.

$$Y @ \text{Expand}(F1) \tag{2.1a}$$

$$Y \ C \ @ \text{Expand}(F1, @ \text{Dropfirst}) \tag{2.1b}$$

**Example 2.1**  *An Application of the QR (2.1b)*
With Data_Faad.wf1 on-screen, we can obtain the output of the ANOVA-QR(0.5) by selecting *Quick/Estimation Equation*, inserting the following ES (2.2), and clicking *OK*. Similarly,

*Quantile Regression: Applications on Experimental and Cross Section Data Using EViews*, First Edition.
I Gusti Ngurah Agung.

**Table 2.1** Statistical results summary of the QR (2.2) for tau = 0.1, 0.5, and 0.9.

| Variable | QR(0.1) | | | QR(0.5) | | | QR(0.9) | | |
|---|---|---|---|---|---|---|---|---|---|
| | Coef. | t-stat. | Prob. | Coef. | t-stat. | Prob. | Coef. | t-stat. | Prob. |
| C | 3.670 | 18.38 | 0.000 | 5.330 | 33.36 | 0.000 | 6.330 | 30.15 | 0.000 |
| G4 = 2 | 1.000 | 3.877 | 0.000 | 0.340 | 1.580 | 0.115 | 0.340 | 1.400 | 0.163 |
| G4 = 3 | 1.000 | 3.641 | 0.000 | 0.340 | 1.745 | 0.082 | 0.670 | 2.425 | 0.016 |
| G4 = 4 | 1.660 | 7.212 | 0.000 | 1.170 | 6.028 | 0.000 | 1.000 | 3.888 | 0.000 |
| Pseudo $R$-sq | 0.123 | | | 0.092 | | | 0.071 | | |
| Adj. $R$-sq | 0.114 | | | 0.083 | | | 0.061 | | |
| S.E. of reg | 1.456 | | | 0.869 | | | 1.393 | | |
| Quantile DV | 4.330 | | | 5.670 | | | 7.000 | | |
| Sparsity | 4.595 | | | 2.211 | | | 4.603 | | |
| Quasi LR-stat | 29.43 | | | 37.12 | | | 16.10 | | |
| Prob. | 0.000 | | | 0.000 | | | 0.001 | | |

we can obtain the output of ANOVA-QR($\tau$) for any $\tau \varepsilon (0, 1)$. Table 2.1 presents the statistical summary of the one-way QR($\tau$) of *Y1* for $\tau = 0.1, 0.5$, and 0.9, using the following ES:

$$Y1 \; C @ Expand\,(G4, @dropfirst) \tag{2.2}$$

Based on this summary, the findings and notes are as follows:

1) The factor *G4* has significant effect on *Y1* for each QR, based on the quasi-likelihood ratio (QLR) statistics, at the 1% level of significance. In other words, the population quantiles 0.1, 0.5, and 0.9 of *Y1* for the four levels of the factor *G4* have significant differences for each QR.
2) The *t*-statistic can be used to test the one- or two-sided hypothesis for each pair of quantiles, with the level *G4* = 1 as the referent group. Then, the statistical hypothesis can be defined based on each of the parameters C(2), C(3), and C(4), as the coefficient of the dummy variables *(G4 = 2), (G4 = 3)*, and *(G4 = 4)*, respectively. Do the testing hypotheses as exercises.
3) The other hypothesis should be tested using the *redundant variables test* (RVT), or the *Wald test*, based on each ANOVA-QR, such as follows:
   3.1 Figure 2.1 presents the output of the Wald test based on the QR(0.5), for the statistical hypothesis $H_0$: C(2) = C(3) = C(4) vs. $H_1$: Otherwise.
   At the 1% level, we can conclude the medians of *Y1* in the three levels of *G4* = 2, 3, and 4 have significant differences, based on the *Chi-square* statistic of $\chi_0^2 = 34.56401$ with $df = 2$ and p-value = 0.0000. In other words, the three dummy variables *(G4 = 2), (G4 = 3)*, and *(G4 = 4)* have significant joint effects on the median of *Y1*.
   3.2 As another test, at the 5% level, based on the QR(0.9), the null hypothesis $H_0$; C(2) = C(3) = 0 is rejected based on the Wald test (*Chi-square*) of 5.947037

Wald test:
Equation: untitled

| Test statistic | Value | df | Probability |
|----------------|-----------|---------|-------------|
| F-statistic | 17.28200 | (2,296) | 0.0000 |
| Chi-square | 34.56401 | 2 | 0.0000 |

Null hypothesis: $C(2) = C(3) = C(4)$

**Figure 2.1** Output of the RVT for the IVs, @Expand(*G4*,@Dropfirst).

with $df = 2$ and $p$-value $= 0.011$. So, we can conclude that the QR(0.9) of *Y1* has significant differences between the first three levels of *G4*.

3.3 As a comparison, the null hypothesis $H_0$: $C(2) = C(3)$ used for testing the quantiles difference between the two levels $G4 = 2$ and $G4 = 3$, which can be tested only using the Wald test. Do this as an exercise.

3.4 In contrast with the ES (2.1b), the ES (2.1a) is important only to present a set of quantiles of *Y1* in the form of a table or graph, as presented in the Figures 2.5 and 2.6, for a two-way ANOVA-QR(Median).

## 2.3 Alternative Two-way ANOVA Quantile Regressions

This section presents five alternative ESs of two-way ANOVA-QRs, or $I \times J$ Factorial ANOVA-QRs, starting with the simplest ES.

### 2.3.1 Applications of the Simplest Equation Specification

Based on the four categorical factors *A, B, (G2 or G4),* and *(H2 or H3)* in Data_Faad.wf1, as presented in Figure 1.2, we can easily run various $I \times J$-Factorial ANOVA-QRs, using the simplest ES as follows:

$$Y @ \text{Expand}(F1, F2) \tag{2.3}$$

where *Y* is a numerical independent or objective variable, and the two factors *F1* and *F2* can be selected from the previous four factors. So, we would have various $I \times J$ Factorial ANOVA-QRs.

All types of hypotheses on the quantile differences can be tested using the Wald test, such as the following:

1. The hypothesis on the joint effects of both factors on a quantile of *Y*.
2. Specific for a level of *F1*, say *F1* = *i*, the hypotheses on the conditional quantile differences between the levels of *F2*, or the effect of *F2* on the quantile of *Y*, conditional for *F1* = *i*.
3. Specific for a level of *F2*, say *F2* = *j*, the hypotheses on the conditional quantile differences between the levels of *F1*, or the effect of *F1* on the quantile of *Y*, conditional for *F2* = *j*.

4. The hypothesis on the quantile difference-in-differences (DID) of $Y$ by the two factors. In other words, the hypothesis on the interaction $F1*F2$ on the quantile of $Y$. See the following examples.

**Example 2.2   ANOVAQR**

Figure 2.2 presents the statistical results of an ANOVA-QR(Median) of $Y = Y1$, with $F1 = A$ and $F2 = B$ in ES (2.3). Based on these results, the following findings and notes are presented:

1) The steps of the analysis are as follows:
   1.1 With Data_Faad.wf1 on-screen, click *Quick/Estimation Equation* and then insert the ES: *Y1 @expand(A,B)*.
   1.2 Then, by selecting the QREG method and clicking *OK,* we obtain the results in Figure 2.1a, , which show 0.5 (median) as the quantile to estimate. To estimate the other quantile, we can insert that quantile.
   1.3 Finally, by clicking *View/Representations*, we obtain the results in Figure 2.1b.
2) Based on the estimation function, we can develop a table of the model parameters, as presented in Table 2.2, where $C(1)$, $C(2)$, $C(3)$, and $C(4)$ are the median parameters of $Y1$ by the factors $A$ and $B$.
3) In addition, Table 2.2 presents four *conditional median differences* $A(1-2|B = 1)$, $A(1-2|B = 2)$, $B(1-2|A = 1)$, and $B(1-2|A = 2)$. And the *median DID* or the *interaction effect* of $A*B$ is defined as

$$IE\,(A*B) = A\,(1-2\mid B = 1) - A\,(1-2\mid B = 2)$$
$$= B\,(1-2\mid A = 1) - B\,(1-2\mid A = 2)$$
$$= C\,(1) - C\,(2) - C\,(3) + C\,(4)$$

4) Then all the preceding statistical hypotheses can be written using these parameters, and they then can be tested using the Wald test. For instance, Figure 2.3 presents the

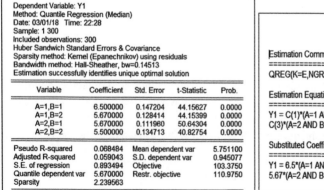

**Figure 2.2**   The statistical results of the $2 \times 2$ factorial median regression.

**Table 2.2** The median parameters of the QR in Figure 2.1.

|  | B = 1 | B = 2 | B(1 − 2) |
|---|---|---|---|
| A = 1 | C(1) | C(2) | C(1) − C(2) |
| A = 2 | C(3) | C(4) | C(3) − C(4) |
| A(1 − 2) | C(1) − C(3) | C(2) − C(4) | C(1) − C(2) − C(3) + C(4) |

| Wald Test: Equation: EQ01 | | | |
|---|---|---|---|
| Test Statistic | Value | df | Probability |
| t-statistic | 2.515458 | 296 | 0.0124 |
| F-statistic | 6.327531 | (1, 296) | 0.0124 |
| Chi-square | 6.327531 | 1 | 0.0119 |

| Null Hypothesis: C(1)-C(3)=C(2)-C(4) Null Hypothesis Summary: | | |
|---|---|---|
| Normalized Restriction (= 0) | Value | Std. Err. |
| C(1) - C(2) - C(3) + C(4) | 0.660000 | 0.262378 |
| Restrictions are linear in coefficients. | | |

**Figure 2.3** The Wald test for testing the DID of the medians of *Y1* by *A* and *B*.

statistical results for testing the interaction effect *A\*B* on the endogenous variable *Y1*, which shows it has an significant effect on the median of *Y1* at the 5% level of significance. In other words, at the 5% level, the effect of the factor *A* on *Y1* is significantly dependent on the factor *B*. Or the median DID of *Y1* is significant.

### 2.3.2 Application of the Quantile Process

With the statistical results on-screen, we can select *View/Quantile Process* to get three alternative options for doing more advanced analysis: *Process Coefficients, Slope Equality Test,* and *Symmetric Quantiles Test*. However, the *Slope Equality Test* is only applicable for the models with an intercept, including the model (2.5). See the alternative model presented in the Example 2.4.

**Example 2.3** *Application of the Process Coefficients*
By selecting *View/Quantile Process/Process Coefficient* and clicking *OK*, we obtain the output in Figure 2.4, which presents the default options of the *Output = Table, Quantiles specification = 4,* and *Coefficient specification = All coefficients.* However, Figure 2.5 shows the statistical results of using *Quantiles specification = 10.* Based on this table, the following findings and notes are presented:

1. Within each cell (*A = i, B = j*), the table presents nine quantiles of *Y1* from $\tau = 0.1$ up to 0.9, with their values presented in the Coefficient column . The quantile specification can easily be modified. For instance, by using 20 quantiles, we obtain the quantiles of 0.05 up to 0.95.
2. Each of the quantiles is significantly greater than zero, based on the *t*-statistic, with a *p*-value = 0.000. However, we can't conduct the testing hypothesis on quantile differences of *Y1* between the cells generated by the two factors *A* and *B*.

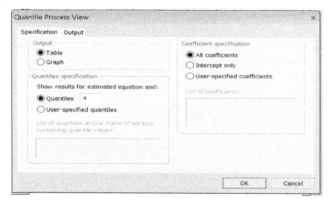

**Figure 2.4** The default options for the process coefficients.

| | Quantile | Coef. | t-Stat. | Prob. | | Quantile | Coef. | t-Stat. | Prob. |
|---|---|---|---|---|---|---|---|---|---|
| A=1,B=1 | 0.1 | 5.000 | 22.551 | 0.000 | A=2,B=1 | 0.1 | 4.330 | 26.890 | 0.000 |
| | 0.2 | 5.500 | 38.435 | 0.000 | | 0.2 | 4.670 | 22.350 | 0.000 |
| | 0.3 | 5.670 | 36.213 | 0.000 | | 0.3 | 5.330 | 47.746 | 0.000 |
| | 0.4 | 6.300 | 35.719 | 0.000 | | 0.4 | 5.500 | 49.582 | 0.000 |
| | 0.5 | 6.500 | 44.156 | 0.000 | | 0.5 | 5.670 | 50.643 | 0.000 |
| | 0.6 | 6.670 | 48.700 | 0.000 | | 0.6 | 5.670 | 52.450 | 0.000 |
| | 0.7 | 7.000 | 49.933 | 0.000 | | 0.7 | 6.000 | 50.628 | 0.000 |
| | 0.8 | 7.000 | 55.652 | 0.000 | | 0.8 | 6.300 | 36.779 | 0.000 |
| | 0.9 | 7.500 | 50.549 | 0.000 | | 0.9 | 6.500 | 42.095 | 0.000 |
| A=1,B=2 | 0.1 | 4.330 | 33.274 | 0.000 | A=2,B=2 | 0.1 | 4.330 | 27.526 | 0.000 |
| | 0.2 | 4.670 | 18.423 | 0.000 | | 0.2 | 4.670 | 25.362 | 0.000 |
| | 0.3 | 5.330 | 38.468 | 0.000 | | 0.3 | 5.000 | 18.353 | 0.000 |
| | 0.4 | 5.670 | 45.866 | 0.000 | | 0.4 | 5.500 | 42.543 | 0.000 |
| | 0.5 | 5.670 | 44.154 | 0.000 | | 0.5 | 5.500 | 40.828 | 0.000 |
| | 0.6 | 6.000 | 47.616 | 0.000 | | 0.6 | 5.670 | 43.182 | 0.000 |
| | 0.7 | 6.330 | 45.573 | 0.000 | | 0.7 | 6.000 | 46.325 | 0.000 |
| | 0.8 | 6.500 | 52.840 | 0.000 | | 0.8 | 6.330 | 51.424 | 0.000 |
| | 0.9 | 6.670 | 61.761 | 0.000 | | 0.9 | 6.500 | 63.758 | 0.000 |

**Figure 2.5** The table of statistics using *Quantiles* = 10.

3. In addition, by selecting *Output = Graph*, we obtain the four graphs of the quantile by the two factors *A* and *B*, as presented in Figure 2.6.

**Example 2.4** **Application of the Symmetric Quantiles Test**
Selecting *View/Quantile Process/Symmetric Quantiles Test* displays the dialog shown in Figure 2.7, which presents the default options of *Test Quantiles = 4,* and *All coefficients*. Then by inserting *Test Quantiles = 10* and clicking *OK,* we obtain the output in Figure 2.8. Based on this output, the following findings and notes are presented:

1. The output presents two types of *symmetric quantiles tests* (SQTs). The first test is to compare all coefficients, namely 16 pairs of quantiles, with the conclusion that the null hypothesis is rejected, based on the *Chi-square* statistic of 16.554 593 with $df = 16$ and a $p$-value = 0.0002. So, the 16 pairs of quantiles are jointly significant.

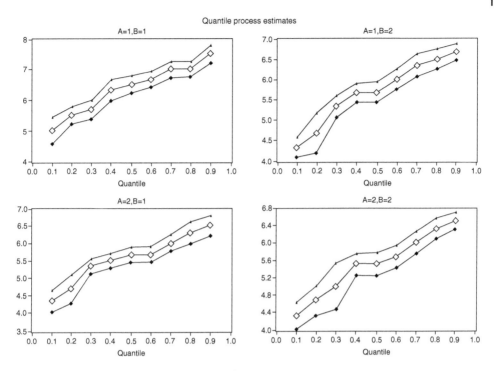

**Figure 2.6** The graphs of statistics in Figure 2.4.

2. The second type of test is to test each of the 16 pairs of coefficients. For instance, four pairs of the coefficients of the quantiles 0.1, 0.9, within the four cells generated by the two factors $A$ and $B$, has the following statistical hypothesis:

$$H_0: b(0.1) + b(0.9) - 2*b(0.5) = 0 \text{ vs.} H_1: \text{Otherwise}$$

At the 10% level of significance, the null hypothesis is rejected only for the cell ($A = 2$, $B = 1$), based on the *restriction value* of −0.51 with a *p*-value = *Prob.* = 0.0347.

### 2.3.3 Applications of the Models with Intercepts

Based on the four categorical factors $A$, $B$, ($G2$ or $G4$), and ($H2$ or $H3$), we can easily present various types of factorial quantile regression analysis, using a model with an intercept as a referent group, with the following alternative ESs:

(i) The simplest model with the ES is as follows:

$$Y \ C@ \text{Expand} (F1, F2, @ \text{Drop} (i, j)) \tag{2.4}$$

where $Y$ is a numerical independent or objective variable, the two factors $F1$ and $F2$ can be selected from the previous four factors, and the cell ($F1 = i$ $F2 = j$) is selected as a reference group. As other alternative ESs, the function @Droplast or @Dropfirst can be used.

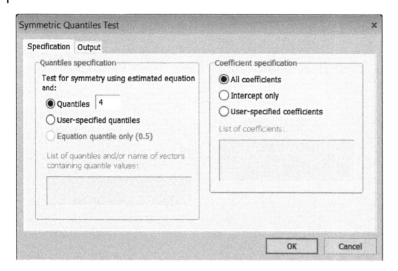

**Figure 2.7** The default options of the symmetric quantiles test.

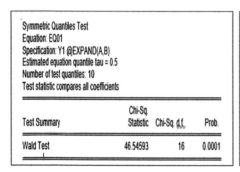

| Symmetric Quantiles Test |
| --- |
| Equation: EQ01 |
| Specification: Y1 @EXPAND(A,B) |
| Estimated equation quantile tau = 0.5 |
| Number of test quantiles: 10 |
| Test statistic compares all coefficients |

| Test Summary | Chi-Sq. Statistic | Chi-Sq. d.f. | Prob. |
| --- | --- | --- | --- |
| Wald Test | 46.54593 | 16 | 0.0001 |

Restriction Detail: b(tau) + b(1-tau) - 2*b( 5) = 0

| Quantiles | Variable | Restr. Value | Std. Error | Prob. |
| --- | --- | --- | --- | --- |
| 0.1, 0.9 | A=1,B=1 | -0.500000 | 0.304173 | 0.1002 |
| | A=1,B=2 | -0.340000 | 0.238557 | 0.1541 |
| | A=2,B=1 | -0.510000 | 0.241558 | 0.0347 |
| | A=2,B=2 | -0.170000 | 0.254414 | 0.5040 |
| 0.2, 0.8 | A=1,B=1 | -0.500000 | 0.229812 | 0.0296 |
| | A=1,B=2 | -0.170000 | 0.253461 | 0.5024 |
| | A=2,B=1 | -0.370000 | 0.236418 | 0.1176 |
| | A=2,B=2 | 0.000000 | 0.224065 | 1.0000 |
| 0.3, 0.7 | A=1,B=1 | -0.330000 | 0.187791 | 0.0789 |
| | A=1,B=2 | 0.320000 | 0.166278 | 0.0543 |
| | A=2,B=1 | -0.010000 | 0.143224 | 0.9443 |
| | A=2,B=2 | 0.000000 | 0.228115 | 1.0000 |
| 0.4, 0.6 | A=1,B=1 | -0.030000 | 0.134573 | 0.8236 |
| | A=1,B=2 | 0.330000 | 0.114876 | 0.0041 |
| | A=2,B=1 | -0.170000 | 0.100147 | 0.0896 |
| | A=2,B=2 | 0.170000 | 0.120534 | 0.1584 |

**Figure 2.8** The results of a symmetric quantile test based on the QR in Figure 2.2.

(ii) The full-factorial model has the following ES:

$$Y\ C\ @\,\mathrm{Expand}\,(F1, @\,\mathrm{Drop}\,(i))\ @\,\mathrm{Expand}\,(F2, @\,\mathrm{Drop}\,(j))$$
$$@\,\mathrm{Expand}\,(F1, @\,\mathrm{Drop}\,(i))\,{}^{*}@\,\mathrm{Expand}\,(F2, @\,\mathrm{Drop}\,(j)) \tag{2.5}$$

Specific for the dichotomous factors with 1 and 2 as their levels, and @Drop(*) = @Droplast, the ES of the model can be presented as follows:

$$Y\ C\ (F1 = 1)\ (F2 = 1)\ (F1 = 1)\,{}^{*}(F2 = 1) \tag{2.6}$$

(iii) A special model with the ES is as follows:

$$Y\ C@\,\mathrm{Expand}\,(F1, @\,\mathrm{Drop}\,(i))\ @\,\mathrm{Expand}\,(F1)\,{}^{*}@\,\mathrm{Expand}\,(F2, @\,\mathrm{Drop}\,(j)) \tag{2.7}$$

Specific for the dichotomous factors with 1 and 2 as their levels, and @Drop(*) = @Droplast, the ES of the model can be presented as follows:

$$Y\ C\ (F1 = 1)\ (F1 = 1)\,{}^{*}(F2 = 1)\ (F1 = 2)\,{}^{*}(F2 = 1) \tag{2.8}$$

In contrast with the output of the model (2.5), we can directly test several hypotheses based on the outputs of the models (2.6)–(2.8), using the *Quasi-LR* statistic or the *t*-statistic presented in the first stage of the statistical results. See the following example.

### Example 2.5  *A 2×2 Factorial ANOVA QR with an Intercept*

As an alternative ES of the model in the Example 2.1, Figure 2.9 presents the statistical results of a 2 × 2 Factorial QR in ES (2.6). So based on this model, the following findings and notes are presented:

1.  Based on the estimation equation, we can write the equation of the model and develop a table of the model parameters as presented in Table 2.3. Then, we can easily write the statistical hypotheses of all hypotheses on the QR(Median) differences between selected cells and its DID . Based on this model, we can directly present the conclusion of several testing hypotheses by using the *Quasi-LR* statistic or the *t*-statistic in Figure 2.9, as follows:

    2.1  The joint effects of the factors $A$ and $B$ on $Y1$ is significant, based on the *Quasi-LR* statistic of 27.148 15 with a *Prob.* = 0.000 005. In other words, the four medians of $Y1$ have significant differences.

    2.2  Specific for $B = 2$, the factor $A$ has insignificant effect on the median of $Y1$, based on the *t*-statistic $t_0 = 0.913\,427$ with a *p*-value = *Prob.* = 0.3618.

    2.3  Specific for $A = 2$, the factor $B$ has insignificant effect on the median of $Y1$, based on the *t*-statistic $t_0 = 0.970\,515$ with a *p*-value = *Prob.* = 0.3326.

    2.4  The other hypotheses should be tested using the Wald test. Do these as exercises.

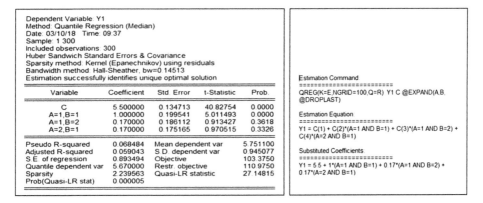

**Figure 2.9**  Statistical results of a 2 × 2 Factorial QR (2.6a).

**Table 2.3**  The parameters of the QR in Figure 2.1.

|           | B = 1           | B = 2           | B(1 – 2)                          |
|-----------|-----------------|-----------------|-----------------------------------|
| A = 1     | C(1) + C(2)     | C(1) + C(3)     | C(2) – C(3)                       |
| A = 2     | C(1) + C(4)     | C(1)            | C(4)                              |
| A(1 – 2)  | C(2) – C(4)     | C(3)            | DID = C(2) – C(3) + C(4)          |

```
Quantile Slope Equality Test
Equation EQ03
Specification Y1 C @EXPAND(A,B,@DROPLAST)
Estimated equation quantile tau = 0.5
Number of test quantiles  4
Test statistic compares all coefficients

Test Summary          Chi-Sq Statistic  Chi-Sq d f      Prob

Wald Test                  21 68551          6          0 0014
```

```
Restriction Detail  b(tau_h) - b(tau_k) = 0

Quantiles       Variable    Restr Value    Std Error     Prob

0 25, 0 5       A=1,B=1      0 000000      0 220759     1 0000
                A=1,B=2      0 490000      0 208212     0 0186
                A=2,B=1     -0 170000      0 254603     0 5043
0 5, 0 75       A=1,B=1      0 300000      0 180110     0 0958
                A=1,B=2      0 140000      0 172100     0 4159
                A=2,B=1      0 470000      0 161949     0 0037
```

**Figure 2.10**  Quantile slope equality test using the default options.

2. Based on this model, we can conduct the *quantile slope equality test* (QSET), with the results presented in Figure 2.10, by using the default option, *Quantiles* = 4. Based on the results, the findings and notes are as follows:

   2.1 Figure 2.10 shows there are six pairs of quantiles (slopes) that should be tested. At the 1% level, the six slopes have significant differences, based on the *Chi-square* statistic of $\chi 2 = 22.685\,51$ with $df = 6$ and $p$-value = 0.0014.

   2.2 Three of the six pairs of quantiles have significant differences, at the 1%, 5%, or 10% level.

For instance, at the 5% level, the pair of quantiles (0.25, 0.5) in the cell $(A = 1, B = 2)$ is significantly different, with $p$-value = *Prob.* = 0.0186.

### Example 2.6   *A Special 2 × 3 Factorial QR (2.7)*

Figure 2.11 presents the statistical results from a special 2 × 3 Factorial QR (2.7). Based on these results, the following findings and notes are presented:

1. Based on the estimation equation, we can develop a table of the model parameters as presented in Table 2.4. Then, we can easily write the statistical hypotheses of all hypotheses on the QR(Median) differences between selected cells and its DID. Based on this model, we can directly present the conclusion of several testing hypotheses by using the *Quasi-LR* statistic or the $t$-statistic in Figure 2.11, as follows:

   1.1 The null hypothesis $H_0$: C(2) = C(3) = … = C(6) = 0 is rejected based the *Quasi-LR* statistic of 44.534 33 with a *Prob.* = 0.000 000. Hence, we can conclude the joint

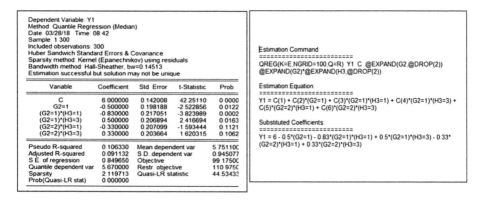

**Figure 2.11**  Statistical results of a 2 × 3 Factorial QR (2.8).

**Table 2.4** The parameters of the QR in Figure 2.11.

| | H = 1 | H = 2 | H = 3 | H(1 − 2) | B(3 − 3) |
|---|---|---|---|---|---|
| G = 1 | C(1) + C(2) + C(3) | C(1) + C(2) | C(1) + C(2) + C(4) | **C(3)** | **C(4)** |
| G = 2 | C(1) + C(5) | C(1) | C(1) + C(6) | **C(5)** | **C(6)** |
| G(1 − 2) | C(2) + C(3) − C(5) | **C(2)** | C(2) + C(4) − C(6) | **C(3) − C(5)** | **C(4) − C(6)** |

      effects of the factors *G* and *H* on *Y1* are significant. In other words, the six medians of *Y1* have significant differences.

1.2  Based on each of the parameters from C(2) to C(5), a two- or one-sided hypothesis on the median difference between two specific cells can be defined and then easily tested using the *t*-statistic presented in Figure 2.11. Do this for each of the parameters as exercises. Refer to the Appendix B for testing the hypotheses using the *t*-statistic.

1.3  The other hypotheses can be tested using the Wald test, for instance the following:

    a.  Specific for $G2 = 1$, the effect of the factor *H3* on *Y1*, with the following statistical hypothesis. It is found at the level of 1% that $H_0$ is rejected based on the *F*-statistic of $F_0 = 17.298$ with $df = (2294)$ and a *p*-value = 0.0000.

$$H_0: \text{C}(3) = \text{C}(4) = 0 \text{ vs. } H_1: \text{Otherwise}$$

    b.  Specific for $G2 = 2$, the effect of the factor *H3* on *Y1*, with the following statistical hypothesis. It is found at the level of 1% that $H_0$ is rejected based on the *F*-statistic of $F_0 = 4.948$ with $df = (2294)$ and a *p*-value = 0.0077.

$$H_0: \text{C}(5) = \text{C}(6) = 0 \text{ vs. } H_1: \text{Otherwise}$$

    c.  The bivariate quantile DID, or the interaction effect of *G2\*H3* on *Y1*, has the following statistical hypothesis. We find at the 10% level that $H_0$ is rejected based on the *F*-statistic of $F_0 = 2.511$ with $df = (2294)$ and a *p*-value = 0.0829.

$$H_0: \text{C}(3) - \text{C}(5) = \text{C}(4) - \text{C}(6) = 0 \text{ vs. } H_1: \text{Otherwise}$$

**Example 2.7**   *A Special 4 × 3 Factorial QR 2.7 (2.7)*

Figure 2.12 presents the statistical results of a special 4 × 3 Factorial ANOVA-QR (2.7) with the output of its representations. Based on these results, the following findings and notes are presented:

1.  The following equation command, which shows the level $G4 = 2$ is used as the referent group of the factor *G4*. Similarly, for the factor $H3 = 2$, we can easily modify the model by using other levels for each of the factors.

$$Y1\ C @\text{EXPAND}\,(G4, @\text{DROP}\,(2))$$

$$@\text{EXPAND}\,(G4)\,{}^*@\text{EXPAND}\,(H3, @\text{DROP}\,(2)) \tag{2.9a}$$

2.  The model has the following equation, which can be obtained using the block copy-paste method of the estimation equation in Figure 2.12, but with an additional error term $\varepsilon$.

Dependent Variable: Y1
Method: Quantile Regression (Median)
Date: 03/28/18   Time: 15:32
Sample: 1 300
Included observations: 300
Huber Sandwich Standard Errors & Covariance
Sparsity method: Kernel (Epanechnikov) using residuals
Bandwidth method: Hall-Sheather, bw=0.14513
Estimation successful but solution may not be unique

| Variable | Coefficient | Std. Error | t-Statistic | Prob. |
|---|---|---|---|---|
| C | 5.670000 | 0.194135 | 29.20642 | 0.0000 |
| G4=1 | -0.170000 | 0.269113 | -0.631705 | 0.5281 |
| G4=3 | 0.000000 | 0.242165 | 0.000000 | 1.0000 |
| G4=4 | 0.660000 | 0.253067 | 2.608009 | 0.0096 |
| (G4=1)*(H3=1) | -0.830000 | 0.273178 | -3.038314 | 0.0026 |
| (G4=1)*(H3=3) | 0.000000 | 0.375133 | 0.000000 | 1.0000 |
| (G4=2)*(H3=1) | -0.340000 | 0.334551 | -1.016287 | 0.3103 |
| (G4=2)*(H3=3) | 0.660000 | 0.244258 | 2.702066 | 0.0073 |
| (G4=3)*(H3=1) | 0.000000 | 0.251429 | 0.000000 | 1.0000 |
| (G4=3)*(H3=3) | 0.630000 | 0.321570 | 1.959140 | 0.0511 |
| (G4=4)*(H3=1) | -0.660000 | 0.262350 | -2.515725 | 0.0124 |
| (G4=4)*(H3=3) | 0.170000 | 0.217293 | 0.782354 | 0.4346 |

| | | | |
|---|---|---|---|
| Pseudo R-squared | 0.167921 | Mean dependent var | 5.751100 |
| Adjusted R-squared | 0.136140 | S.D. dependent var | 0.945077 |
| S.E. of regression | 0.833787 | Objective | 92.34000 |
| Quantile dependent var | 5.670000 | Restr. objective | 110.9750 |
| Sparsity | 1.901121 | Quasi-LR statistic | 78.41688 |
| Prob(Quasi-LR stat) | 0.000000 | | |

Estimation Command:
=========================
CREG(K=E,NGRID=100,Q=R) Y1 C @EXPAND(G4,@DROP(2))
@EXPAND(G4)*@EXPAND(H3,@DROP(2))

Estimation Equation:
=========================
Y1 = C(1) + C(2)*(G4=1) + C(3)*(G4=3) + C(4)*(G4=4) + C(5)*(G4=1)*(H3=1) +
C(6)*(G4=1)*(H3=3) + C(7)*(G4=2)*(H3=1) + C(8)*(G4=2)*(H3=3) +
C(9)*(G4=3)*(H3=1) + C(10)*(G4=3)*(H3=3) + C(11)*(G4=4)*(H3=1) +|
C(12)*(G4=4)*(H3=3)

Substituted Coefficients:
=========================
Y1 = 5.67 - 0.17*(G4=1) + 0*(G4=3) + 0.66*(G4=4) - 0.83*(G4=1)*(H3=1) +
0*(G4=1)*(H3=3) - 0.34*(G4=2)*(H3=1) + 0.66*(G4=2)*(H3=3) + 0*(G4=3)*(H3=1)
+ 0.63*(G4=3)*(H3=3) - 0.66*(G4=4)*(H3=1) + 0.17*(G4=4)*(H3=3)

**Figure 2.12**   Statistical results of a $4 \times 3$ Factorial QR (2.8).

Then we can develop a table of the model parameters as presented in Table 2.5.

$$Y1 = C(1) + C(2)*(G4 = 1) + C(3)*(G4 = 3) + C(4)*(G4 = 4)$$
$$+ C(5)*(G4 = 1)*(H3 = 1) + C(6)*(G4 = 1)*(H3 = 3)$$
$$+ C(7)*(G4 = 2)*(H3 = 1) + C(8)*(G4 = 2)*(H3 = 3) \qquad (2.9b)$$
$$+ C(9)*(G4 = 3)*(H3 = 1) + C(10)*(G4 = 3)*(H3 = 3)$$
$$+ C(11)*(G4 = 4)*(H3 = 1) + C(12)*(G4 = 4)*(H3 = 3) + \varepsilon$$

3. The steps to develop Table 2.5 are as follows:

    3.1 As the intercept, C(1) should be inserted in the 12 cells.

    3.2 The parameter C(2) as the coefficient of $(G4 = 1)$ should be added in all levels of H3.

    3.3 The parameter C(3) as the coefficient of $(G4 = 3)$ should be added in all levels of H3.

    3.4 The parameter C(4) as the coefficient of $(G4 = 4)$ should be added in all levels of H3.

**Table 2.5**   The parameters of the QR in Figure 2.11.

| | H3 = 1 | H3 = 2 | H3 = 3 | H3(1 − 2) | H3(3 − 2) |
|---|---|---|---|---|---|
| G = 1 | C(1) + C(2) + C(5) | C(1) + C(2) | C(1) + C(2) + C(6) | *C(5)* | *C(6)* |
| G = 2 | C(1) + C(7) | C(1) | C(1) + C(8) | *C(7)* | *C(8)* |
| G = 3 | C(1) + C(3) + C(9) | C(1) + C(3) | C(1) + C(3) + C(10) | *C(9)* | *C(10)* |
| G = 4 | C(1) + C(4) + C(11) | C(1) + C(4) | C(1) + C(4) + C(12) | *C(11)* | *C(12)* |
| G(1 − 2) | C(2) + C(5) − C(7) | C(2) | C(2) + C(6) − C(8) | C(5) − C(7) | C(6) − C(8) |
| G(3 − 2) | C(3) + C(9) − C(7) | C(3) | C(3) + C(10) − C(8) | C(9) − C(7) | C(10) − C(8) |
| G(4 − 2) | C(4) + C(11) − C(7) | C(4) | C(4) + C(12) − C(8) | C(11) − C(7) | C(12) − C(8) |

3.5 The parameter C(5) as the coefficient of *(G4 = 4)\*(H3 = 1)* should be added in the cell *(G4 = 4)\*(H3 = 1)* only. Similarly, each of the parameters from C(6) to C(12) should be added in the corresponding cells.

3.6 The table can be completed by computing the differences: *G4*(1 − 2), *G4*(3 − 2), and *G4*(4 − 2) for each level of *H3*; *H3*(1 − 2) and *H3*(3 − 2) for each level of *G4*; and the six DIDs, (C(5) − C(7)) to (C(12) − C(8)).

4. Based on this table, we can identify all statistical hypotheses, which can be tested using the test statistics presented in the Figure 2.12, as follows:

4.1 The $H_0$: $C(k) = 0, \forall\, k > 1$ is rejected based on the *Quasi-LR* statistic of 78.418 88 with a *Prob.* = 0.000 000. Hence, the factors *G4* and *H3* have a joint significant effect on *Y1*. That is the 12 medians of *Y1* have significant differences.

4.2 Based on each of the parameters C(k), for $k = 2, 3$, and 4, a one- or two-sided statistical hypothesis on quartiles (medians) difference between two levels of *G4*, specific for *H3 = 2*, can be tested directly using the *t*-statistic in Figure 1.12.

4.3 Based on each of the parameters C(k), for $k > 4$, a one- or two-sided statistical hypothesis on quartiles (medians) difference between two levels of *H3*, specific for each level of *G4*, can be tested directly using the *t*-statistic in Figure 1.12.

5. The other hypotheses, such as the following, can be tested using the Wald test. Do the other tests as exercises.

5.1 The median DID or the interaction effect of *G4\*H3* on *Y1*, with the following statistical hypothesis:

$$H_0: C(5) - C(7) = C(9) - C(7) = C(11) - C(7) =$$
$$C(6) - C(8) = C(10) - C(8) = C(12) - C(8) = 0$$

$$\text{or } H_0: \ C(5) = C(7) = C(9) = C(11), C(6) = C(8) = C(10)$$
$$= C(12) \text{ vs. } H_1: \text{ Otherwise}$$

5.2 Specific for *G4 = 1*, the effect of the factor *H3* on *Y1*, with the following statistical hypothesis:

$$H_0: \ C(5) = C(6) = 0 \text{ vs. } H_1: \text{ Otherwise}$$

5.3 Specific for *H3 = 1*, the effect of the factor *G4* on *Y1*, with the following statistical hypothesis:

$$H_0: \ C(2) + C(5) = C(7) = C(3) + C(9) = C(4) + C(11) = 0, \text{ vs. } H_1: \text{ Otherwise}$$

5.4 Specific for *H3 = 3*, the effect of the factor *G4* on *Y1*, with the following statistical hypothesis:

$$H_0: \ C(2) = C(3) = C(4) = 0, \text{ vs. } H_1: \text{ Otherwise}$$

## 2.4 Forecasting

It is well known that forecasting is applicable for the time-series models, which has been presented in various books, such as Agung (2019), Wilson and Keating (1994), and Hankel and Reitsch (1992). However, it has been found that the *Forecast* button in EViews also can be applied for cross-section models, specifically the binary choice models. The results are *predicted probabilities,* namely $P(Y = 1)$, of the zero-one independent variables $Y$.

Now, I want to explore the application of the *Forecast* for the QR in the following example.

### Example 2.8    *Forecasting Based on the $2 \times 2$ Factorial QR in Example 2.1*

Figure 2.13a shows the estimate of the QR. With the output on-screen, click the *Forecast* button to display the Forecast dialog shown in Figure 2.13b. Then click *OK* to obtain the results in Figure 2.14.

Based on these results, the following findings and notes are presented:

1. The forecast variable of *Y1* is presented as *Y1F* in Figure 2.13b, which is inserted directly as a new variable in the data file. The characteristics of *Y1F* are as follows:

   1.1  The one-way tabulation of *Y1F* in Figure 2.15 shows that it has only three distinct values, which are exactly the same as the three different medians of *Y1*, as presented in Figure 2.13a, and its graph is a two-step function, as presented in Figure 2.14.

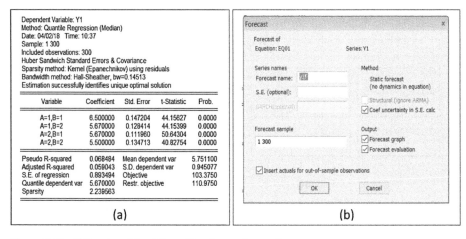

**Figure 2.13**  Estimates of the QR and the forecasting options.

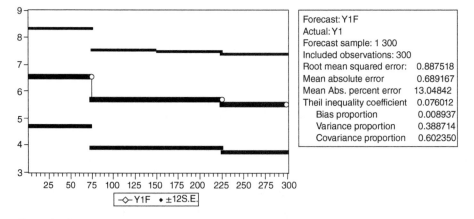

**Figure 2.14**  The forecast graph of *Y1F*, and its forecast evaluation.

Tabulation of Y1F
Date: 04/02/18  Time: 10:48
Sample: 1 300
Included observations: 300
Number of categories: 3

| Value | Count | Percent | Cumulative Count | Cumulative Percent |
|---|---|---|---|---|
| 5.500000 | 75 | 25.00 | 75 | 25.00 |
| 5.670000 | 150 | 50.00 | 225 | 75.00 |
| 6.500000 | 75 | 25.00 | 300 | 100.00 |
| Total | 300 | 100.00 | 300 | 100.00 |

**Figure 2.15**  One-way tabulation of Y1F.

Descriptive statistics for Y1F
Categorized by values of A and B
Date: 04/03/18   Time: 06:13
**Sample: 1 300**
Included observations: 300

| **Mean** **Std. Dev.** **Obs.** | | 1 | B 2 | All |
|---|---|---|---|---|
| | 1 | 6.500000 | 5.670000 | 6.085000 |
| | | 0.000000 | 0.000000 | 0.416390 |
| | | 75 | 75 | 150 |
| A | 2 | 5.670000 | 5.500000 | 5.585000 |
| | | 0.000000 | 0.000000 | 0.085285 |
| | | 75 | 75 | 150 |
| | All | 6.085000 | 5.585000 | 5.835000 |
| | | 0.416390 | 0.085285 | 0.390812 |
| | | 150 | 150 | 300 |

**Figure 2.16**  Descriptive statistic of Y1F by the factors A and B.

1.2 The descriptive statistics of Y1F by the factors A and B, in Figure 2.16, shows that its four means are exactly the same as the four different medians of Y1, as shown in Figure 2.13a. Note that the standard deviation of zero indicates that Y1F has a single value within each of the cells $(A = i, B = j)$.

1.3 Hence, I would say the forecast variable Y1F is acceptable in the statistical sense, but it does not have any benefit in practice, since Y1F presents the same medians as the variable Y1 by the two factors.

2. It is important to note the *Theil inequality coefficient* (TIC) = 0.076012, shown in Figure 2.14, where TIC = 0 indicates the perfect forecast., and TIC = 1 indicates the worst forecast.

**Table 2.6** The parameters of the QR in (2.11b).

|  | F2 = 1 | F2 = 2 | F2 = 3 | F2(1 − 3) | F2(2 − 3) |
|---|---|---|---|---|---|
| F1 = 1 | C(1) + C(2) + C(3) | C(1) + C(2) + C(4) | C(1) + C(2) | **C(3)** | **C(4)** |
| F1 = 2 | C(1) + C(3) | C(1) + C(4) | C(1) | **C(3)** | **C(4)** |
| F1(1 − 2) | **C(2)** | **C(2)** | **C(2)** | **0** | **0** |

## 2.5 Additive Two-way ANOVA Quantile Regressions

An additive QR of $Y$ with two categorical predictors, namely $F1$ and $F2$, is an acceptable QR in the statistical sense, but it has a specific assumption which never is observed in practice. Referring to the interaction QR with an intercept in (2.7), the additive QR has the following ES:

$$Y \; C \; @\text{Expand}\,(F1, @\text{Droplast}) \; @\text{Expand}\,(F2, @\text{Droplast}) \tag{2.10}$$

Note that for a $I \times J$ factorial model, this model has only $\{1 + (I - 1) + (J - 1)\} < I \times J$. So, this additive model has a hidden criterion or assumption, which never occurs in reality. See the following example.

**Example 2.9** *An Application of a $2 \times 3$ Factorial QR*
As an illustration, for a $2 \times 3$ Factorial QR, based on ES (2.10), we have the following ES with only four parameters, which is less that the number of cells:

$$Y \; C \; (F1 = 1) \; (F2 = 1) \; (F2 = 2) \tag{2.11a}$$

or

$$Y1 = C\,(1) + C\,(2)^*\,(F1 = 1) + C\,(3)^*\,(F2 = 1) + C\,(4)^*\,(F2 = 2) \tag{2.11b}$$

Table 2.6 presents the parameters of the QR (2.11b) by the factors $F1$ and $F2$, which shows that $F1(1 - 2) = C(2)$ for all levels of $F2$, and $F2(1 - 3) = C(3)$ and $F2(2 - 3) = C(4)$ for all levels of $F1$. So, this is a case that never occurs in reality and has never been observed, based on sample data, and for higher factorial QRs. Hence, I would say the additive QR with categorical predictors is the worst ANOVA-QR.

## 2.6 Testing the Quantiles of *Y1* Categorized by *X1*

Referring to the medians of $Y1$ categorized by $X1$, presented in Section 1.3, this section presents the QRs of $Y1$ categorized by $X1$. So, we have to generate an ordinal variable based on $X1$ or use the rank variable $RX1 = @Rank(X1,a)$ as presented in Section 1.4. As an illustration, I generated an ordinal categorical variable based on $X1$ having 10 levels, namely $O10X1$, using the following equation:

$$O10X1 = 1 + 1^*\,(X1 >= @\text{Quantile}\,(X1, 0.1)) + 1^*\,(X1 >= @\text{Quantile}\,(X1, 0.2))$$

$$+ 1^* (X1 >= @\text{Quantile}\,(X1, 0.3)) + 1^* (X1 >= @\text{Quantile}\,(X1, 0.4))$$
$$+ 1^* (X1 >= @\text{Quantile}\,(X1, 0.5)) + 1^* (X1 >= @\text{Quantile}\,(X1, 0.6))$$
$$+ 1^* (X1 >= @\text{Quantile}\,(X1, 0.7)) + 1^* (X1 >= @\text{Quantile}\,(X1, 0.8))$$
$$+ 1^* (X1 >= @\text{Quantile}\,(X1, 0.9)) \tag{2.12}$$

### Example 2.10 *Quantile Regression of Y1 Categorized by X1*

We can apply the QR using the following ES, with the statistical results summary presented in Figure 2.17:

$$Y1\ C@\text{Expand}\,(O10X1, @\text{Dropfirst}) \tag{2.13}$$

Based on this summary, the findings and notes are as follows:

1) Based on each QR, the 10 quantiles have significant differences, based on the QLR statistic, at the 1% level. For instance, with the QR(0.1), the 10 quantiles have significant differences based on the $QRL$ statistic of $QLR_0 = 73.33$ with $df = 290\ (300 - 10)$ and a $p$-value $= 0.000$.
2) The difference of quantiles between the levels $O10X1 = 1$ and each of the other levels can be tested directly using the $t$-statistic in the output. The hypothesis can be either a

| | QR(0.1) | | | QR(0.5) | | | QR(0.9) | | |
|---|---|---|---|---|---|---|---|---|---|
| Variable | Coef. | t-stat. | Prob. | Coef. | t-stat. | Prob. | Coef. | t-stat. | Prob. |
| C | 3.670 | 27.31 | 0.000 | 4.330 | 22.61 | 0.000 | 5.670 | 38.85 | 0.000 |
| O10X1 = 2 | 0.660 | 3.476 | 0.001 | 1.170 | 4.000 | 0.000 | 0.830 | 2.183 | 0.030 |
| O10X1 = 3 | 0.660 | 2.981 | 0.003 | 1.170 | 4.480 | 0.000 | 0.830 | 3.291 | 0.001 |
| O10X1 = 4 | 1.000 | 1.436 | **0.152** | 1.670 | 6.527 | 0.000 | 1.330 | 3.190 | 0.002 |
| O10X1 = 5 | 1.000 | 3.197 | 0.002 | 1.340 | 4.927 | 0.000 | 0.830 | 4.011 | 0.000 |
| O10X1 = 6 | 0.660 | 2.836 | 0.005 | 1.170 | 4.860 | 0.000 | 0.660 | 3.219 | 0.001 |
| O10X1 = 7 | 1.660 | 8.369 | 0.000 | 1.670 | 6.233 | 0.000 | 1.330 | 5.623 | 0.000 |
| O10X1 = 8 | 1.660 | 9.361 | 0.000 | 1.340 | 5.025 | 0.000 | 1.330 | 4.800 | 0.000 |
| O10X1 = 9 | 1.660 | 8.703 | 0.000 | 1.970 | 6.611 | 0.000 | 1.330 | 4.324 | 0.000 |
| O10X1 = 10 | 1.830 | 7.251 | 0.000 | 2.170 | 8.454 | 0.000 | 1.830 | 6.674 | 0.000 |
| Pse. R-sq | 0.208 | | | 0.154 | | | 0.147 | | |
| Adj. R-sq | 0.183 | | | 0.127 | | | 0.121 | | |
| S.E. of reg | 1.312 | | | 0.818 | | | 1.261 | | |
| Quantile DV | 4.330 | | | 5.670 | | | 7.000 | | |
| Sparsity | 3.122 | | | 2.011 | | | 3.749 | | |
| QLR stat | 73.33 | | | 67.80 | | | 41.19 | | |
| Prob | 0.000 | | | 0.000 | | | 0.000 | | |

**Figure 2.17** Statistical results summary of the QR in (2.13) for tau = 0.1, 0.5, and 0.9.

one- or two-sided hypothesis. Note that all of the $t$-statistics have positive values, which indicates that the quantile in the level $O10X1 = 1$ is the smallest.

3) Note the difference between the following two hypotheses, which can be tested using the Wald test. Do them as exercises.

$$H_{10} : C(2) = C(3) = C(4) = 0 \text{ vs } H_{11} : \text{Otherwise} \tag{2.14}$$

$$H_{20} : C(2) = C(3) = C(4) \text{ vs } H_{21} : \text{Otherwise} \tag{2.15}$$

The first hypothesis is for testing the quantile differences between the first four levels of $O10X1 = 1, 2, 3,$ and 4; but the second hypothesis is for testing the quantile differences between the levels $O10X1 = 2, 3,$ and 4 only.

4) As an additional analysis, Figure 2.18 presents the QSET for the pair of quantiles (0.1, 0.9), based on the QR(0.9) in (2.13), which shows the differences of the slopes of nine IVs. In the ANOVA-QRs, the slopes in fact are the quantiles within each level of $O10X1$. Note that the *restriction values* are the differences of the coefficient of the IVs of the QR(0.1) and QR(0.9) in Figure 2.17. For instance, the *restriction value* of $O10X1 = -0.17$ equals the difference of the coefficients of the IV $O10X1$, in Figure 2.17, that is $(0.660 - 0.830) = -0.17$.

Quantile slope equality test
Equation: EQ03_O10X1
Specification: Y1 C @EXPAND(O10X1,@DROPFIRST)
Estimated equation quantile tau = 0.9
User-specified test quantiles: 0.1
Test statistic compares all coefficients

| Test summary | Chi-Sq. statistic | Chi-Sq. d.f. | Prob. |
|---|---|---|---|
| Wald test | 5.181649 | 9 | 0.8182 |

Restriction detail: $b(tau\_h) - b(tau\_k) = 0$

| Quantiles | Variable | Restr. value | Std. error | Prob. |
|---|---|---|---|---|
| 0.1, 0.9 | O10X1 = 2 | −0.170000 | 0.407215 | 0.6763 |
| | O10X1 = 3 | −0.170000 | 0.316534 | 0.5912 |
| | O10X1 = 4 | −0.330000 | 0.771476 | 0.6688 |
| | O10X1 = 5 | 0.170000 | 0.356501 | 0.6335 |
| | O10X1 = 6 | 0.000000 | 0.292831 | 1.0000 |
| | O10X1 = 7 | 0.330000 | 0.291353 | 0.2574 |
| | O10X1 = 8 | 0.330000 | 0.312750 | 0.2914 |
| | O10X1 = 9 | 0.330000 | 0.344237 | 0.3377 |
| | O10X1 = 10 | 4.44E-16 | 0.351414 | 1.0000 |

**Figure 2.18** A quantile slope equality test based on QR(0.9) in (2.13).

## 2.7    Applications of QR on Population Data

In contrast to the QRs of birthweight in the National Health of Statistics presented by Koenker and Hallock (2001), this section presents alternative ANOVA-QRs based on the Children Weekly Working Hours (*CWWH*) in the Indonesian National Social Economic Survey 2013 (*SUSENAS* 2013). In this section, the variables considered are *CWWH; AGE* in years, within the closed–open interval [10, 20); *SEX:* 1 (Male) and 2 (Female); UR: 1 (Urban) and 2 (Rural); *PROV* (provinces); *KAB* (Kabupaten); *SCHL* (schooling level): 1 (Elementary), 2 (Junior High), 3 (Senior High), and 4 (College/University); and *DST* (dummy student): 1 (Student) and 0 (Not Student).

In addition, the numerical variables *INC* (income), *EXP_CAP* (expenditure capital), and *FOOD_CAP* can be transformed to ordinal variables, and the lowest 5 of the 10 or 20 generated groups also can be used as the predictors of *CWWH*.

### 2.7.1    One-way-ANOVA-QRs

Various one-way ANOVA-QRs of *CWWH*, with each of the preceding variables, as well as other numerical objective or dependent variables (DVs) in SUSENAS, can be presented as one-way QRs. The following example presents only a single QR of *CWWH* on *AGE*, conditional for $10 \leq AGE < 20$; $DST = 1$ (the children are still going to school); and $CWWH < 98$, because $CWWH = 98$ indicates an NA score.

**Example 2.11**    *One-way ANOVA-QRs of CWWH by AGE*
Figure 2.19 presents the statistical results of two QRs, with the following ESs. Based on these results, the following findings and notes are presented:

$$CWWH @ \text{Expand} (AGE) \tag{2.16a}$$

and

$$CWWH \ C \ @ \text{Expand} (AGE, @ \text{Dropfirst}) \tag{2.16b}$$

1) The first QR in (2.16a) should be applied, if and only if we want to identify directly the medians of *CWWH* for the *AGE* variable from 10 to 19 years. For instance, the median = 11 for $AGE = 10$, which indicates that 50% of the children $AGE = 10$ have weekly working hours below 11 hours. This QR is the simplest QR of *CWWH* by *AGE*, for the sample $\{10 \leq AGE < 20$ and $DST = 1$ (Student) and $CWWH < 98\}$.
2) It is important to note the use of the ESs without the intercept, such (2.1a), (2.3), and (2.16a) in order to compute the Quantile Process Estimates (QPEs) for various sets of quantiles, in the form of a table and graph, as presented in Figure 2.20 The figure shows the outputs of four selected $AGE = 10, 12, 16$, and 19, for 10 quantiles. It could easily be extended to more quantiles, such as 20 quantiles from 0.05 to 0.95.
   Note that the table only presents the coefficient (Median) and the standard error of *CWWH*. In this case, we don't need to test the hypothesis of each median of *CWWH* because the medians should always be greater than zero.

| | QR(median) in (2.16a) | | | | QR(median) in (2.16b) | | |
|---|---|---|---|---|---|---|---|
| Variable | Coef. | *t*-stat | Prob. | Variable | Coef. | *t*-stat | Prob. |
| AGE = 10 | 10 | 12.45 | 0.000 | C | 10 | 12.45 | 0.0000 |
| AGE = 11 | 12 | 18.17 | 0.000 | AGE = 11 | 2 | 1.92 | 0.0545 |
| AGE = 12 | 14 | 25.07 | 0.000 | AGE = 12 | 4 | 4.09 | 0.0000 |
| AGE = 13 | 14 | 30.42 | 0.000 | AGE = 13 | 4 | 4.32 | 0.0000 |
| AGE = 14 | 14 | 27.53 | 0.000 | AGE = 14 | 4 | 4.21 | 0.0000 |
| AGE = 15 | 14 | 31.39 | 0.000 | AGE = 15 | 4 | 4.35 | 0.0000 |
| AGE = 16 | 15 | 26.40 | 0.000 | AGE = 16 | 5 | 5.08 | 0.0000 |
| AGE = 17 | 14 | 24.44 | 0.000 | AGE = 17 | 4 | 4.05 | 0.0001 |
| AGE = 18 | 16 | 21.95 | 0.000 | AGE = 18 | 6 | 5.53 | 0.0000 |
| AGE = 19 | 41 | 60.47 | 0.000 | AGE = 19 | 31 | 29.49 | 0.0000 |
| Pse $R$^2 | 0.1996 | | | Pse $R$^2 | 0.1996 | QLR stat | 1576.3 |
| Adj. $R$^2 | 0.1975 | | | Adj. $R$^2 | 0.1975 | Prob. | 0.0000 |
| S.E. of reg | 14.446 | | | S.E. of reg | 14.446 | | |
| Quant DV | 16 | | | Quant DV | 16 | | |
| Sparsity | 22.184 | | | Sparsity | 22.184 | | |

**Figure 2.19** Outputs of the QR(Median)s in (2.16a) and (2.16b) for $10 \leq AGE < 20$ and $DST = 1$.

3) It is an unbelievable finding that for the 19-year-old students, 50% have weekly working hours above 41 hours, as presented in Figure 2.19. In addition, we find the maximum number of hours is 96. I would say that the value of 96 indicates a missing value in the data set. For these reasons, for a more advanced analysis, I am using the subsample of the $AGE = 10$ to 18 years only.

4) Figure 2.21 presents the output of the QSET of three selected *AGE* groups in the QR (2.16). Based on this output, the following findings and notes are presented:

4.1 I try to insert C(1), C(3), C(7), and C(10) as the user-specified coefficient, in order to obtain the same four age groups ($AGE = 10, 12, 16,$ and 19) presented in Figure 2.20. But the output shows only three. Why? Because C(1) is an intercept and not a slope.

4.2 In order to obtain the output of the four age groups, we have to run the QR, using the following ES, for $i \neq$ the four age groups 10, 12, 16, and 19.

$$CWWH\ C@\,\text{Expand}\,(AGE, @\,\text{Drop}\,(i)) \tag{2.17}$$

5) As an illustration, Figure 2.22 presents the statistical results of QR(Median) in (2.17) for @Drop($i$) = @Drop(13). Based on these results, the findings and notes are as follows:

5.1 In this QR, the intercept C represents $AGE = 13$. Note that the three QRs in (2.16a), (2.16b), and (2.17) are exactly the same model.

5.2 Furthermore, I have to insert the four parameters C(2), C(4), C(7), and C(10) as the user-specified quantiles, in order to obtain the QSET for $AGE = 10, 12, 16,$ and 19.

| | Quantile | Coef. | Std. err | | Quantile | Coef. | Std. err |
|---|---|---|---|---|---|---|---|
| Age=10 | 0.1 | 5 | 0.5929 | Age=16 | 0.1 | 6 | 0.3683 |
| | 0.2 | 7 | 0.6388 | | 0.2 | 8 | 0.5245 |
| | 0.3 | 8 | 0.7024 | | 0.3 | 12 | 0.4902 |
| | 0.4 | 10 | 0.7830 | | 0.4 | 14 | 0.4829 |
| | 0.5 | 10 | 0.8031 | | 0.5 | 15 | 0.5681 |
| | 0.6 | 12 | 0.8501 | | 0.6 | 20 | 0.9204 |
| | 0.7 | 14 | 1.1764 | | 0.7 | 25 | 1.5383 |
| | 0.8 | 18 | 2.0578 | | 0.8 | 35 | 1.7294 |
| | 0.9 | 22 | 1.5269 | | 0.9 | 48 | 1.1201 |
| Age=12 | 0.1 | 5 | 0.4480 | Age=19 | 0.1 | 14 | 0.6478 |
| | 0.2 | 7 | 0.5760 | | 0.2 | 20 | 1.1021 |
| | 0.3 | 10 | 0.6027 | | 0.3 | 28 | 1.2388 |
| | 0.4 | 12 | 0.5424 | | 0.4 | 35 | 0.8612 |
| | 0.5 | 14 | 0.5584 | | 0.5 | 41 | 0.6780 |
| | 0.6 | 15 | 0.7003 | | 0.6 | 48 | 0.3745 |
| | 0.7 | 18 | 1.3548 | | 0.7 | 48 | 0.3102 |
| | 0.8 | 21 | 0.7767 | | 0.8 | 54 | 0.7934 |
| | 0.9 | 28 | 1.8663 | | 0.9 | 60 | 2.0134 |

Quantile process estimates

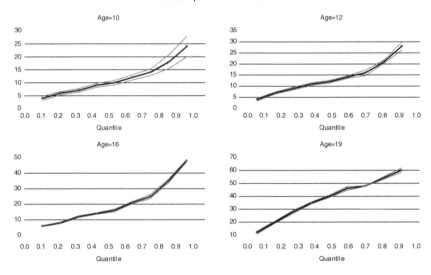

**Figure 2.20** Table and graph of the Quantile Process Estimates of selected *AGE* groups in the QR (2.16a).

Redundant variables test

Null hypothesis: @EXPAND(AGE)*(UR = 1)

Equation: UNTITLED

Specification: CWWH @EXPAND(AGE) @EXPAND(AGE)*(UR = 1)

CWWH

| | Value | df | Probability |
|---|---|---|---|
| QLR L-statistic | 229.0754 | 10 | 0.0000 |
| QLR lambda-statistic | 225.2160 | 10 | 0.0000 |

Wald test:

Equation: untitled

| Test statistic | Value | df | Probability |
|---|---|---|---|
| F-statistic | 20.41569 | (10, 3472) | 0.0000 |
| Chi-square | 204.1569 | 10 | 0.0000 |

Null hypothesis: C(11) = C(12) = C(13) = C(14) = C(15) = C(16)

= C(17) = C(18) = C(19) = C(20) = 0

**Figure 2.21** Two alternative tests for testing the medians differences between the levels of *UR* for all *AGE* levels.

5.3 However, compare the QSET outputs in Figure 2.21 to those of the ages 12, 16, and 19 in Figure 2.23 We have different values. Why?

For instance, note that for the pairs of quantiles (0.1, 0.5), *Restr. Value* = -1.00 for *AGE* = 12 in this figure 2.23. But Figure 2.21 presents the value of −4.00. Why? It is because the coefficient of the dummy variable *(AGE = 12)* in (2.16b) is the deviation of the medians of *CWWH* between the two groups *AGE* = 12 and *AGE* = 10, that is (12 − 10) = 2, based on the output of ES (2.16a) in Figure 2.19. But the coefficient of the dummy variable *(AGE = 12)* in Figure 2.22 is the deviation of the medians of *CWWH* between the two groups *AGE* = 12 and *AGE* = 13, that is (12 − 14) = −2. Similarly the dummy variable *(AGE=12)* also has different coefficients based on the QR(0.1) in (2.16b) and (2.17), also based on the output of (2.16a).

### 2.7.2 Application of the Forecasting

In response to a query about how to obtain the fitted values for *CWWH* based on the QR(Median) of *CWWH* on *AGE* in (2.16a), using a subset of the original observations, a staff member of the EViews provider sent me an email (December 20, 2018): "You can get the fitted values, by using the forecast command from the equation. This will compute the fitted values and put them in the appropriate places in the work-file so that they line up with the correct values for *AGE*."

Dependent Variable: CWWH

Method: Quantile Regression (Median)

Date: 12/18/18 Time: 11:46

Sample: 1 177852 IF AGE>=10 AND AGE<20 AND DST=1 AND

CWWH<98

Included observations: 3492

Huber Sandwich Standard Errors & Covariance

Sparsity method: Kernel (Epanechnikov) using residuals

Bandwidth method: Hall-Sheather, bw = 0.064039

Estimation successful but solution may not be unique

| Variable | Coefficient | Std. Error | t-Statistic | Prob. |
|----------|-------------|------------|-------------|-------|
| C | 14.00000 | 0.460264 | 30.41732 | 0.0000 |
| AGE=10 | −4.000000 | 0.925684 | −4.321128 | 0.0000 |
| AGE=11 | −2.000000 | 0.805092 | −2.484188 | 0.0130 |
| AGE=12 | 0.000000 | 0.723610 | 0.000000 | 1.0000 |
| AGE=14 | 0.000000 | 0.685882 | 0.000000 | 1.0000 |
| AGE=15 | 0.000000 | 0.640946 | 0.000000 | 1.0000 |
| AGE=16 | 1.000000 | 0.731162 | 1.367686 | 0.1715 |
| AGE=17 | 0.000000 | 0.734891 | 0.000000 | 1.0000 |
| AGE=18 | 2.000000 | 0.861978 | 2.320245 | 0.0204 |
| AGE=19 | 27.00000 | 0.819473 | 32.94802 | 0.0000 |

| | | | |
|----------|-------------|------------|-------------|
| Pseudo R-squared | 0.199603 | Mean dependent var | 22.42354 |
| Adjusted R-squared | 0.197534 | S.D. dependent var | 16.88172 |
| S.E. of regression | 14.44585 | Objective | 17527.50 |
| Quantile dependent var | 16.00000 | Restr. objective | 21898.50 |
| Sparsity | 22.18425 | Quasi-LR statistic | 1576.253 |
| Prob(Quasi-LR stat) | 0.000000 | | |

**Figure 2.22** The output of the QR(Median) in (2.17).

Even though I am using the original observations, I also will use the *Forecast* option to compute the values of the fitted variable because it is easier to do it that way. With the output of the QR of any numerical dependent variable, say $Y$, on-screen, select *Forecast* and click *OK*. The forecast variable $YF$ is inserted directly as an additional variable in the data file.

**Example 2.12** *Two Alternative Scatter Graphs of CWWHF on AGE*

Referring to the note presented in point 3, Example 2.7, Figure 2.24 presents two alternative scatter graphs of *CWWHF* on *AGE*, based on the QR(Median) in (2.16a) for $10 \leq AGE < 20$,

Quantile slope equality test
Equation: EQ01_C
Specification: CWWH C @EXPAND(AGE,@DROP(13))
Estimated equation quantile tau = 0.5
User-specified test quantiles: 0.1, 0.9
Test statistic compares user-specified coefficients: **C(2) C(4) C(7) C(10)**

| Test summary | Chi-Sq. statistic | Chi-Sq. d.f. | Prob. |
|---|---|---|---|
| Wald test | 767.0022 | 8 | 0.0000 |

Restriction detail: $b(tau\_h) - b(tau\_k) = 0$

| Quantiles | Variable | Restr. value | Std. error | Prob. |
|---|---|---|---|---|
| 0.1, 0.5 | AGE = 10 | 3.000000 | 0.961347 | 0.0018 |
| | AGE = 12 | −1.000000 | 0.768648 | 0.1933 |
| | AGE = 16 | −1.000000 | 0.751004 | 0.1830 |
| | AGE = 19 | −19.00000 | 0.911949 | 0.0000 |
| 0.5, 0.9 | AGE = 10 | 2.000000 | 1.900294 | 0.2926 |
| | AGE = 12 | 0.000000 | 2.133620 | 1.0000 |
| | AGE = 16 | −19.00000 | 1.614119 | 0.0000 |
| | AGE = 19 | −5.000000 | 2.248465 | 0.0262 |

**Figure 2.23** The QSET based on the QR(Median) in Figure 2.22 for AGE = 10, 12, 16, and 19.

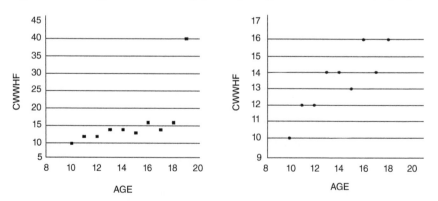

**Figure 2.24** The scatter graphs of *CWWHF* < 80 on *AGE* based on the QR(Median) in (2.16a), for $10 \le AGE < 20$ (a), and $10 \le AGE < 19$ (b).

and $10 \leq AGE < 19$, respectively. Based on these graphs, the following findings and notes are presented:

1) The scatter graph in Figure 2.24a shows the 10 points of the *(AGE,CWWHF)*, where the point for $AGE = 19$ can be considered as an outlier.
2) Because $AGE = 19$ presents an outlier, the linear regression fit is inappropriate for presenting the relationship of *CWWHF* on *AGE;* however, a *linear polynomial fit*, covered in Chapter 4, will work.
3) On the other hand, the scatter graph in Figure 2.24b shows that the linear regression fit would be a good fit for presenting the relationship of *CWWHF* on *AGE*.

### 2.7.3  Two-way ANOVA-QRs

Referring to the three alternative ESs of the two-way ANOVA-QRs in (2.3), (2.4), and (2.7), let's consider the following three alternative ESs of any numerical $Y$, on two categorical factors *F1* and *F2*, for the population data:

$$Y @ \text{Epand}\,(F1, F2) \tag{2.18a}$$

$$Y\ C\ @ \text{Epand}\,(F1, F2, @\text{Droplast}) \tag{2.18b}$$

$$Y @ \text{Epand}\,(F1)\,@\text{Expand}\,(F1)\,{}^*@\text{Expand}\,(F2, @\text{Droplast}) \tag{2.18c}$$

ES (2.18a) is the simplest ES. Its main objective is to present the set of quantiles for each cell generated by the two factors, which are the output of the *QPE*. The main objectives of ES (2.18b) are to conduct the *slope equality test* and test the joint effects of the two factors, or to test the quantile differences between all cells generated by the two factors. Finally, the main objective of ES (2.18c) is to test the quantile differences between the level of the factor *F2* for each level of *F1*.

**Example 2.13**  *Application of the Simplest ES*
As an illustration, Figure 2.25 presents the output of QR(Median) of *CWWH* on *SCHL* and *SEX*, using the simplest ES, as follows:

$$CWWH @ \text{Expand}\,(SEX, SCHL) \tag{2.19}$$

With the output in Figure 2.25 on-screen, select *View/Quantile Process/Quantile Coefficient* and click *OK* to obtain a set of eight graphs. However, the output in Figure 2.26 is obtained by inserting "C(1) C(2) C(3) C(4)" as the *User-specified coefficients*.

In addition, the parameters tabulation of the model (2.19) can easily be developed, as presented in Table 2.7. You can use it to define various hypotheses on the *CWWH*'s quantile differences between cells/groups generated by *SCHL* and *SEX*, which could be tested using the Wald test, as presented before. Do these as exercises. Specific for the DID, we have the following statistical hypothesis:

$$H_0: \text{C}\,(1) - \text{C}\,(5) = \text{C}\,(2) - \text{C}(6) = \text{C}\,(3) - \text{C}\,(7)$$
$$= \text{C}\,(4) - \text{C}\,(8) \text{ vs } H_1: \text{Otherwise}$$

Dependent variable: CWWH
Method: quantile regression (Median)
Date: 12/20/18 Time: 16:19
Sample: 1 177852 IF AGE > =10 AND AGE < 20 AND DST = 1
AND CWWH < 98
Included observations: 3492
Huber sandwich standard errors and covariance
Sparsity method: Kernel (Epanechnikov) using residuals
Bandwidth method: Hall-Sheather, bw = 0.064039
Estimation successful but solution may not be unique

| Variable | Coefficient | Std. error | $t$-statistic | Prob. |
|---|---|---|---|---|
| SEX = 1,SCHL = 1 | 12.00000 | 0.457464 | 26.23160 | 0.0000 |
| SEX = 1,SCHL = 2 | 14.00000 | 0.342970 | 40.81994 | 0.0000 |
| SEX = 1,SCHL = 3 | 15.00000 | 0.483190 | 31.04366 | 0.0000 |
| SEX = 1,SCHL = 4 | 42.00000 | 0.781014 | 53.77622 | 0.0000 |
| SEX = 2,SCHL = 1 | 12.00000 | 0.599912 | 20.00293 | 0.0000 |
| SEX = 2,SCHL = 2 | 14.00000 | 0.453044 | 30.90209 | 0.0000 |
| SEX = 2,SCHL = 3 | 15.00000 | 0.631344 | 23.75885 | 0.0000 |
| SEX = 2,SCHL = 4 | 40.00000 | 1.509304 | 26.50228 | 0.0000 |
| Pseudo $R$-squared | 0.205790 | Mean dependent var | | 22.42354 |
| Adjusted $R$-squared | 0.204195 | S.D. dependent var | | 16.88172 |
| S.E. of regression | 14.44972 | Objective | | 17 392.00 |
| Quantile dependent var | 16.00000 | Restr. objective | | 21 898.50 |
| Sparsity | 22.67141 | | | |

**Figure 2.25** The output of the QR(Median) in (2.19).

**Example 2.14** *A Special Two-way ANOVA-QR of CWWH by SCHL and SEX*
Figure 2.27 shows the output of the QR(Median) of *CWWH* on *SCHL* (schooling level) and
*SEX*, using the following ES. Based on this output, the following findings and notes are
presented:

$$CWWH @\text{Expand}(SCHL) @\text{Expand}(SCHL) * (SEX = 1) \qquad (2.20)$$

1) Based on the output in Figure 2.27, the findings are as follows:
   1.1. The coefficients of the first four IVs are the medians of *CWWH* by *SCHL* for the Girl.
   For instance, the coefficient of $(SCHL = 1)$, $C(1) = 12$, is the *CWWH*-median for the
   Girl in $SCHL = 1$.aejust
   1.2. And the coefficients of the last four IVs, are the medians deviations of *CWWH* of the
   Boy and Girl by *SCH*, namely *Dev(Bot-Girl)*
   1.3. Then we can develop the output summary as presented in Figure 2.28, which shows
   the medians of *CWWH* by *SCHL*, specific for $SEX = 2$, and the *Dev(Male-Female)* of
   the medians by *SCHL*.

Quantile process estimates

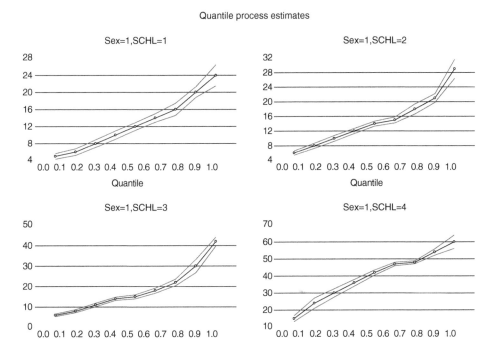

**Figure 2.26** The graphs of the first four IVs of the QR(Median) in Figure 2.25.

**Table 2.7** The model parameters of the QR (2.19).

| SEX | SCHL | | | |
|-----|---|---|---|---|
| | 1 | 2 | 3 | 4 |
| Male | C(1) | C(2) | C(3) | C(4) |
| Female | C(5) | C(6) | C(7) | C(8) |
| (M − F) | C(1) − C(5) | C(2) − C(6) | C(3) − C(7) | C(4) − C(8) |

2) The main objective of this model is to identify directly the medians differences of *CWWH* between *SEX* by *SCHL*, and it can be tested using the *t*-statistic in the column of *Dev(Male-Female)*. Do this as an exercise.

3) The medians differences of *CWWH* between *SEX* by *SCHL* can be tested using the Wald test, with the output presented in Figure 2.29. We can conclude they have insignificant differences. Note that the parameters C(5), C(6), C(7), and C(8) are the coefficients of the model's last four IVs.

4) I have found that the QPE based on this QR presents the output for each of its IVs, but the output of the QPE using ES (2.16a) presents the equations of QRs within the eight cells/groups generated by *SCHL* and *SEX*. So if you want to present the QPE based on any QR, I recommend applying the simplest ES, such as in (2.16a). Do this as an exercise.

Dependent variable: CWWH
Method: quantile regression (Median)
Sample: IF AGE > =10 AND AGE < 20 AND DST = 1 AND CWWH < 98
Included observations: 3492

| Variable | Coefficient | Std. error | $t$-statistic | Prob. |
|---|---|---|---|---|
| SCHL = 1 | 12.00000 | 0.599912 | 20.00293 | 0.0000 |
| SCHL = 2 | 14.00000 | 0.453044 | 30.90209 | 0.0000 |
| SCHL = 3 | 15.00000 | 0.631344 | 23.75885 | 0.0000 |
| SCHL = 4 | 40.00000 | 1.509304 | 26.50228 | 0.0000 |
| (SCHL = 1)*(SEX = 1) | 0.000000 | 0.754432 | 0.000000 | 1.0000 |
| (SCHL = 2)*(SEX = 1) | 0.000000 | 0.568223 | 0.000000 | 1.0000 |
| (SCHL = 3)*(SEX = 1) | 0.000000 | 0.795027 | 0.000000 | 1.0000 |
| (SCHL = 4)*(SEX = 1) | 2.000000 | 1.699407 | 1.176881 | 0.2393 |

| | | | |
|---|---|---|---|
| Pseudo $R$-squared | 0.205790 | Mean dependent var | 22.42354 |
| Adjusted $R$-squared | 0.204195 | S.D. dependent var | 16.88172 |
| S.E. of regression | 14.44972 | Objective | 17 392.00 |
| Quantile dependent var | 16.00000 | Restr. objective | 21 898.50 |
| Sparsity | 22.67141 | | |

**Figure 2.27** The output of the QR(Median) in (2.20).

| Variable | GILR (SEX = 2) | | | Dev(Boy-Girl) | | |
|---|---|---|---|---|---|---|
| | Coeff. | $t$-stat | Prob | Coeff. | $t$-stat | Prob. |
| SCHL = 1 | 12 | 20.0029 | 0.0000 | 0 | 0.0000 | 1.0000 |
| SCHL = 2 | 14 | 30.9021 | 0.0000 | 0 | 0.0000 | 1.0000 |
| SCHL = 3 | 15 | 23.7589 | 0.0000 | 0 | 0.0000 | 1.0000 |
| SCHL = 4 | 40 | 26.5023 | 0.0000 | 2 | 1.1769 | 0.2393 |

**Figure 2.28** The summary of the output of the QR(Median) in Figure 2.27.

Wald Test:

Equation: Untitled

| Test Statistic | Value | df | Probability |
|---|---|---|---|
| F-statistic | 0.346262 | (4, 3484) | 0.8468 |
| Chi-square | 1.385050 | 4 | 0.8468 |

Null Hypothesis: C(5)=C(6)=C(7)=C(8)=0

**Figure 2.29** Wald Test for testing the medians differences between SEX for all SCHL levels.

**Example 2.15   *Two-way ANOVA-QR of CWWH on AGE and UR***
Figure 2.30 shows a summary of the output of the two-way ANOVA-QR of *CWWH* on *AGE* and *UR* (urban-rural regions), with the following ES.

$$CWWH @\text{Expand}(AGE, UR) @\text{Expand}(AGE) * (UR = 1) \tag{2.21}$$

Based on this summary, the following findings and notes are presented:

1) Refer to the Table 2.7 in Example 2.13 for how to develop this summary based on the output of the QR in (2.20) or the general ES in (2.18c).
2) The hypothesis on the medians difference, namely *Dev(Urban-Rural),* for each level of *AGE*, can easily be tested using the *t*-statistic, with either one- or two-sided hypotheses.
3) Figure 2.31 presents the outputs of two alternative tests, the RVT and the Wald test, for testing the *Dev(Urban-Rural)* for all levels of *AGE*. They show the null hypothesis is rejected at the 1% level. Note that the RVT is easier to do because its null hypothesis can be inserted using the block copy-paste method.

**Example 2.16   *Two-way ANOVA-QR of CWWH on AGE and SCHL***
As a special illustration, Figure 2.32 presents a part of the output, using the following ES, with the notes presented in italic. Why? Many of the cells/groups generated by the two factors have no observations, which can be checked using the two-way tabulation of *AGE* and *SCHL*. In other words, the table is an incomplete table.

$$CWWH\ C\ @\text{Expand}(AGE, \text{Dropfirst})$$
$$@\text{Expand}(AGE) *@\text{Expand}(SCHL, @\text{Droplast}) \tag{2.22}$$

However, the results are obtained by using the following ES, with a part of the output presented in Figure 2.33. This is because the estimation method directly selects 26 non-empty

| | Rural(UR = 2) | | | Dev(Urban-Rural) | | |
|---|---|---|---|---|---|---|
| Variable | Coef. | t-stat. | Prob. | Coef. | t-stat. | Prob. |
| AGE = 10 | 10 | 11.8180 | 0.0000 | 2 | 0.9661 | 0.3340 |
| AGE = 11 | 11 | 15.8061 | 0.0000 | 3 | 1.8061 | 0.0710 |
| AGE = 12 | 12 | 18.4602 | 0.0000 | 3 | 2.0708 | 0.0384 |
| AGE = 13 | 14 | 23.1113 | 0.0000 | 0 | 0.0000 | 1.0000 |
| AGE = 14 | 12 | 21.1868 | 0.0000 | 2 | 1.8777 | 0.0605 |
| AGE = 15 | 12 | 23.7884 | 0.0000 | 2 | 1.9609 | 0.0500 |
| AGE = 16 | 14 | 27.1254 | 0.0000 | 10 | 3.0776 | 0.0021 |
| AGE = 17 | 14 | 19.2156 | 0.0000 | 1 | 0.6517 | 0.5146 |
| AGE = 18 | 15 | 20.6116 | 0.0000 | 6 | 2.3213 | 0.0203 |
| AGE = 19 | 35 | 39.0750 | 0.0000 | 13 | 13.1536 | 0.0000 |
| Pse. *R*-sq. | 0.2263 | Mean DV var | | 22.424 | | |
| Ad. *R*-sq | 0.2220 | S.D. DV | | 16.882 | | |
| S.E. of reg | 14.075 | Objective | | 16 943.5 | | |
| Quant DV | 16 | Restr. objective | | 21 898.5 | | |
| Sparsity | 20.395 | | | | | |

**Figure 2.30** The summary of the output of the QR(Median) in (2.21).

Redundant Variables Test
Null hypothesis: @EXPAND(AGE)*(UR=1)
Equation: UNTITLED
Specification: CWWH @EXPAND(AGE) @EXPAND(AGE)*(UR=1)
CWWH

| | Value | df | Probability |
|---|---|---|---|
| QLR L-statistic | 229.0754 | 10 | 0.0000 |
| QLR Lambda-statistic | 225.2160 | 10 | 0.0000 |

Wald Test:
Equation: Untitled

| Test Statistic | Value | df | Probability |
|---|---|---|---|
| F-statistic | 20.41569 | (10, 3472) | 0.0000 |
| Chi-square | 204.1569 | 10 | 0.0000 |

Null Hypothesis: C(11)=C(12)=C(13)=C(14)=C(15)=C(16)
=C(17)=C(18)=C(19)=C(20)=0

**Figure 2.31** Two alternative test for testing the medians differences between the levels of *UR* for all *AGE* levels.

Dependent variable: CWWH
Method: quantile regression (Median)
Date: 12/20/18   Time: 16:43
Sample: 1 177852 IF AGE >=10 AND AGE < 20 AND DST = 1 AND CWWH < 98
Included observations: 3492
WARNING: linear dependencies found in regressors (some coefficients set to initial values)
*Singularity in sandwich covariance estimation*

| Variable | Coefficient | Std. error | t-statistic | Prob. |
|---|---|---|---|---|
| C | 0.000000 | NA | NA | NA |
| AGE = 11 | 12.00000 | NA | NA | NA |
| AGE = 12 | 0.000000 | NA | NA | NA |
| AGE = 13 | 0.000000 | NA | NA | NA |
| ... | | | | |
| .... | | | | |
| (AGE = 19)*(SCHL = 1) | 0.000000 | NA | NA | NA |
| (AGE = 19)*(SCHL = 2) | −24.00000 | NA | NA | NA |
| (AGE = 19)*(SCHL = 3) | −24.00000 | NA | NA | NA |
| Pseudo R-squared | 0.218805 | Mean dependent var | | 22.42354 |
| Adjusted R-squared | 0.213397 | S.D. dependent var | | 16.88172 |
| S.E. of regression | 14.18635 | Objective | | 17 107.00 |
| Quantile dependent var | 16.00000 | Restr. objective | | 21 898.50 |
| Sparsity | 22.64793 | Quasi-LR statistic | | 1692.516 |
| Prob(quasi-LR stat) | 0.000000 | | | |

**Figure 2.32**   A part of the output of the QR(Median) in (2.22).

cells out of the 40 cells (*AGE* and *SCHL* have 10 and 4 levels, respectively), which can be identified in the output of its *representations*.

$$CWWH \; C @ Expand \, (AGE, SCHL, @Dropfirst) \tag{2.23}$$

### Example 2.17   *A Special QR of log(Food_CAP) on PROV and UR*

Figure 2.34 presents a summary of the output of a QR(Median) of *log(Food_CAP)* on *PROV* (province) and *UR*, using the following ES (for *PROV* <> 31 because *PROV* = 31 is Jakarta, which does not have rural area):

$$log \, (Food\_CAP) \; @ExpandPROV) @ Expand \, (PROV) * (UR = 2) \tag{2.24}$$

Based on this output, the findings and notes are as follows:

1) Figure 2.34 can be developed based on the output of the QR in (2.23), using a method similar to what we saw in Figures 2.28 and 2.30.
2) The main objective of this analysis is to study and test the hypotheses of the median differences of *log(Food_CAP)* between the urban and rural areas for each of the provinces in

Dependent variable: CWWH

Method: quantile regression (Median)

Bandwidth method: Hall-Sheather, bw = 0.064039

Estimation successful but solution may not be unique

| Variable | Coefficient | Std. error | *t*-statistic | Prob. |
|---|---|---|---|---|
| C | 10.00000 | 0.803149 | 12.45099 | 0.0000 |
| AGE = 11,SCHL = 1 | 2.000000 | 1.041196 | 1.920867 | 0.0548 |
| AGE = 11,SCHL = 2 | −1.000000 | 3.033774 | −0.329622 | 0.7417 |
| AGE = 19,SCHL = 2 | 4.000000 | 4.247428 | 0.941746 | 0.3464 |
| ... | | | | |
| AGE = 19,SCHL = 3 | 8.000000 | 1.993237 | 4.013572 | 0.0001 |
| AGE = 19,SCHL = 4 | 32.00000 | 1.054914 | 30.33424 | 0.0000 |

| | | | |
|---|---|---|---|
| Pseudo *R*-squared | 0.218805 | Mean dependent var | 22.42354 |
| Adjusted *R*-squared | 0.213397 | S.D. dependent var | 16.88172 |
| S.E. of regression | 14.16083 | Objective | 17 107.00 |
| Quantile dependent var | 16.00000 | Restr. objective | 21 898.50 |
| Sparsity | 22.56572 | Quasi-LR statistic | 1698.683 |
| Prob(quasi-LR stat) | 0.000000 | | |

**Figure 2.33** A part of the output of the QR(Median) in (2.23).

Indonesia, accept Jakarta, which are represented as the *Dev(Rural-Urban)* in Figure 2.34 The output shows that 9 of 31 provinces have insignificant differences at the 10% level of significance.

3) Based on the output of the RVT in Figure 2.35, at the 1% level, it can be concluded that the 31 provinces in *Dev(Rural_Urban)* have significant differences, based on the *QLR L*-statistic of 1040.585 with $df = 31$ and *p*-value = 0.0000.

**Example 2.18** *Another Special QR of log(Food_CAP) by PROV and UR*

The following QR has three main objectives: testing the conditional main effect of *UR*, the conditional main effect of *PROV*, and the interaction effect of *PROV\*UR*, which can easily be tested using the Wald test. As an illustration, Figure 2.36 shows the output of the model for only PROV < 16. See the codes of the provinces in Figure 3.34. Based on this output, we can develop the model parameters, as presented in Table 2.8. Based on the parameters C(2) through C(10), presented in bold, we can observe the following characteristics:

$$\log(Food\_CAP) \; C \; (UR = 1) @ \text{Expand}(PROV, @\text{Droplast})$$

$$@\text{Expand}(PROV, @\text{Droplast}) * (UR = 1) \qquad (2.25)$$

| Variable | Urban(UR = 1) | | | Dev(Rural-Urban) | | |
|---|---|---|---|---|---|---|
| | Coef. | t-stat. | Prob. | Coef. | t-stat. | Prob. |
| PROV = 11 | 12.6233 | 1645.19 | 0.0000 | 0.0148 | 1.5612 | **0.1185** |
| PROV = 12 | 12.6669 | 2204.32 | 0.0000 | 0.0498 | 5.9399 | 0.0000 |
| PROV = 13 | 12.6609 | 1950.84 | 0.0000 | −0.0074 | −0.7641 | **0.4448** |
| PROV = 14 | 12.8489 | 2080.19 | 0.0000 | −0.0054 | −0.4962 | **0.6198** |
| PROV = 15 | 13.1331 | 662.44 | 0.0000 | 0.0067 | 0.2985 | **0.7653** |
| PROV = 16 | 12.5740 | 1309.72 | 0.0000 | 0.1511 | 12.0195 | 0.0000 |
| PROV = 17 | 12.5241 | 593.91 | 0.0000 | −0.1106 | −4.4320 | 0.0000 |
| PROV = 18 | 12.6466 | 1613.80 | 0.0000 | 0.0710 | 6.9523 | 0.0000 |
| PROV = 19 | 12.6485 | 711.19 | 0.0000 | 0.0036 | 0.1142 | **0.9091** |
| PROV = 21 | 12.6900 | 936.21 | 0.0000 | −0.0493 | −2.1471 | 0.0318 |
| PROV = 32 | 12.7552 | 1755.73 | 0.0000 | −0.0746 | −5.6755 | 0.0000 |
| PROV = 33 | 12.6224 | 2562.26 | 0.0000 | −0.0639 | −8.9435 | 0.0000 |
| PROV = 34 | 12.7513 | 790.87 | 0.0000 | 0.0790 | 3.0276 | 0.0025 |
| PROV = 35 | 12.5992 | 2285.03 | 0.0000 | 0.1080 | 13.0777 | 0.0000 |
| PROV = 36 | 12.5500 | 1282.03 | 0.0000 | 0.1468 | 8.2136 | 0.0000 |
| PROV = 51 | 12.7799 | 1073.50 | 0.0000 | −0.0573 | −2.9117 | 0.0036 |
| PROV = 52 | 12.7491 | 738.11 | 0.0000 | 0.0839 | 3.8429 | 0.0001 |
| PROV = 53 | 12.7236 | 1229.26 | 0.0000 | −0.0918 | −7.5370 | 0.0000 |
| PROV = 61 | 12.9066 | 1382.69 | 0.0000 | −0.0660 | −5.3656 | 0.0000 |
| PROV = 62 | 12.9204 | 821.97 | 0.0000 | −0.0721 | −3.9749 | 0.0001 |
| PROV = 63 | 12.8024 | 1103.50 | 0.0000 | 0.0131 | 0.7733 | **0.4393** |
| PROV = 64 | 12.6872 | 1856.74 | 0.0000 | 0.0898 | 7.1219 | 0.0000 |
| PROV = 71 | 12.9066 | 864.42 | 0.0000 | −0.0495 | −2.7220 | 0.0065 |
| PROV = 72 | 12.8623 | 965.99 | 0.0000 | −0.0699 | −4.6540 | 0.0000 |
| PROV = 73 | 12.9161 | 1606.35 | 0.0000 | 0.1035 | 10.1640 | 0.0000 |
| PROV = 74 | 12.8388 | 916.27 | 0.0000 | 0.1496 | 9.2293 | 0.0000 |
| PROV = 75 | 12.7405 | 859.90 | 0.0000 | 0.0193 | 0.8300 | **0.4065** |
| PROV = 76 | 12.6459 | 546.60 | 0.0000 | 0.0373 | 1.3459 | **0.1783** |
| PROV = 81 | 12.8296 | 735.03 | 0.0000 | 0.1544 | 7.0855 | 0.0000 |
| PROV = 82 | 12.7667 | 661.18 | 0.0000 | 0.0456 | 1.9102 | 0.0561 |
| PROV = 91 | 12.7718 | 1172.40 | 0.0000 | 0.0111 | 0.5695 | **0.5690** |
| PROV = 94 | 12.8168 | 1032.64 | 0.0000 | −0.1016 | −7.1568 | 0.0000 |
| Pse R-sq | 0.04764 | Mean DV | | 12.76928 | | |
| Adj. R-sq | 0.047298 | S.D. DV | | 0.436576 | | |
| S.E. of reg | 0.417727 | Objective | | 28 639.41 | | |
| Quant DV | 12.73628 | Restr. objective | | 30 072.05 | | |
| Sparsity | 0.963641 | | | | | |

**Figure 2.34** The output of the QR(Median) in (2.24).

Redundant variables test

Equation: EQ09_FOOD_CAP

Specification: LOG(FOOD_CAP) @EXPAND(PROV)

      @EXPAND(PROV)*(UR = 2)

K

| | Value | df | Probability |
|---|---|---|---|
| QLR L-statistic | 1040.585 | 31 | 0.0000 |
| QLR lambda-statistic | 1038.315 | 31 | 0.0000 |

**Figure 2.35** The output of the RVT of @Expand(*PROV*)*(*UR* = 2).

1) The parameter C(2) represents the main effect of *UR* on *log(Food_CAP)*, conditional for *PROV* = 15, which is the median difference of *log(Food_CAP)* between the levels of *UR*, conditional for *PROV* = 15. Based on the *t*-statistic, we can conclude they have an insignificant difference with a *p*-value = 0.7636.

2) The parameters C(3), C(4), C(5), and C(6) represent the main effect of *PROV*, conditional for *UR* = 2, which can be tested using the Wald test with the null hypothesis C(3) = C(4) = C(5) = C(6) = 0. The null hypothesis is rejected based on the Chi-square statistic of $\chi_0^2$ = 2194.028 with *df* = 4 and *p*-value = 0.0000.

3) The parameters C(7), C(8), C(9), and C(10) represent the effect of the interaction *PROV*UR*, which can be tested using the Wald test with the null hypothesis C(7) = C(8) = C(9) = C(10) = 0. The null hypothesis is rejected based on the Chi-square statistic of $\chi_0^2$ = 25.484 23 with *df* = 4 and *p*-value = 0.0000. Hence, we can conclude that median differences of *log(Food_CAP)* between the five provinces are significantly dependent on the *UR*-regions.

4) If using all provinces, we would obtain output with NAs similar to the output in Figure 2.30, since Jakarta (*PROV* = 31) does not have a rural area (*UR* = 2).

## 2.8 Special Notes and Comments on Alternative Options

Figure 2.37 presents a set of possible combination options, which can be applied for each QR(*τ*), ∀ *τ* *ε*(0.1).

Note that by using the *Bootstrap* coefficient covariance method, as presented in this figure, each of the other options is available for selection, such as *Coefficient Covariance* with three options, *Weight* with five options, *Bootstrap Setting Method* with four options, and *Random Generator* with five alternative options. However, using *Coefficient Covariance: Ordinary (IID)*, or *Huber Sandwich,* would result in a smaller number of options. So, we have a lot of possible option combinations. We would accept the output of each QR using all option combinations to estimate the only parameter standard errors, in the statistical sense. Hence,

Dependent variable: LOG(FOOD_CAP)
Method: quantile regression (Median)
Date: 12/21/18   Time: 18:59
**Sample: 1 177852 IF PROV < 16**
Included observations: 38 027
Huber sandwich standard errors and covariance
Sparsity method: Kernel (Epanechnikov) using residuals
Bandwidth method: Hall-Sheather, bw = 0.028892
Estimation successful but solution may not be unique

| Variable | Coefficient | Std. error | *t*-statistic | Prob. |
|---|---|---|---|---|
| C | 13.13981 | 0.010133 | 1296.747 | 0.0000 |
| UR = 1 | −0.006679 | 0.022204 | −0.300798 | 0.7636 |
| PROV = 11 | −0.501772 | 0.011599 | −43.25821 | 0.0000 |
| PROV = 12 | −0.423087 | 0.011766 | −35.95723 | 0.0000 |
| PROV = 13 | −0.486294 | 0.012293 | −39.55909 | 0.0000 |
| PROV = 14 | −0.296313 | 0.013331 | −22.22687 | 0.0000 |
| (PROV = 11)*(UR = 1) | −0.008103 | 0.024296 | −0.333518 | 0.7387 |
| (PROV = 12)*(UR = 1) | −0.043132 | 0.023798 | −1.812406 | 0.0699 |
| (PROV = 13)*(UR = 1) | 0.014083 | 0.024103 | 0.584280 | 0.5590 |
| (PROV = 14)*(UR = 1) | 0.012045 | 0.024989 | 0.482030 | 0.6298 |

| | | | | |
|---|---|---|---|---|
| Pseudo *R*-squared | 0.057275 | Mean dependent var | | 12.75982 |
| Adjusted *R*-squared | 0.057052 | S.D. dependent var | | 0.415844 |
| S.E. of regression | 0.391314 | Objective | | 5797.191 |
| Quantile dependent var | 12.72900 | Restr. objective | | 6149.396 |
| Sparsity | 0.905902 | Quasi-LR statistic | | 3110.312 |
| Prob(Quasi-LR stat) | 0.000000 | | | |

**Figure 2.36** The output of the QR(Median) in (2.25) for *PROV* < 16.

the estimates of the parameter coefficients of the QRs are invariant or constant for all option combinations . Since the parameters' estimates are invariant, the residuals of the QRs are constant for all combinations of options, as well as the outputs of all residual tests.

On the other hand, note that we never know parameter standard error would be the true population standard error—refer to special notes presented in Agung (2009a). For these reasons, if we don't have a good reason to select a specific combination of options, it is better to apply the default combination, which I have presented in previous examples. I also will apply the default combination in examples in the following chapters, when using the *Coefficient Covariance: Ordinary (IID) Standard Error and Covariance,* which assumes the error terms of the QR have identical and independent distribution.

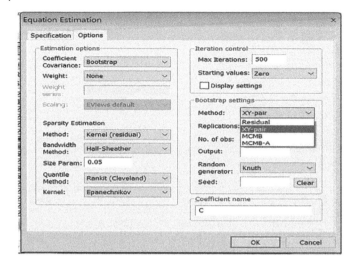

**Figure 2.37** Alternative estimation options for the estimation equation of QRs.

**Table 2.8** Model parameters of the QR in Figure 2.34.

| PROV | UR = 1 | UR = 2 | (1 − 2) |
|------|--------|--------|---------|
| 11 | $C(1) + C(2) + C(3) + C(7)$ | $C(1) + C(3)$ | $C(2) + C(7)$ |
| 12 | $C(1) + C(2) + c(4) + C(8)$ | $C(1) + C(4)$ | $C(2) + C(8)$ |
| 13 | $C(1) + C(2) + C(5) + C(9)$ | $C(1) + C(5)$ | $C(2) + C(9)$ |
| 14 | $C(1) + C(2) + C(6) + C(10)$ | $C(1) + C(6)$ | $C(2) + C(10)$ |
| 15 | $C(1) + C(2)$ | $C(1)$ | **C(2)** |
| (11 − 15) | $C(3) + C(7)$ | **C(3)** | **C(7)** |
| (12 − 15) | $C(4) + C(8)$ | **C(4)** | **C(8)** |
| (13 − 15) | $C(5) + C(9)$ | **C(5)** | **C(9)** |
| (14 − 15) | $C(6) + C(10)$ | **C(6)** | **C(10)** |

However, we can easily learn the differential outputs between selected combinations of options. The parameters' estimates of a specific QR are invariant or constant for all combinations of the options. See the following examples.

Specific for the Bootstrap method, it is one of the simulation estimation methods that uses a number of samples of size *n,* where each sample is *randomly selected with replacement* from the observed sample data (Hao and Naiman 2007). Note that the Bootstrap method is used only to estimate the parameter standard errors. EViews provides a default option using $n = 100$.

On the other hand, Davino, Furno, and Vistocco (2014, p. 122) present resampling methods, which can be used to estimate the parameter standard errors, without using the assumption on the error distribution. And Koenker (2005, p. 108) employs another

| Variable | (a) IID Std. Err. | | | (b) HS Std. Err. | | | (c) Bootstrap Std. Err. | | |
|---|---|---|---|---|---|---|---|---|---|
| | Coef. | t-stat. | Prob. | Coef. | t-stat. | Prob. | Coef. | t-stat. | Prob. |
| C | 5.500 | 41.646 | 0.000 | 5.500 | 40.828 | 0.000 | 5.500 | 105.75 | 0.000 |
| $A = 1, B = 1$ | 1.000 | 5.354 | 0.000 | 1.000 | 5.011 | 0.000 | 1.000 | 8.325 | 0.000 |
| $A = 1, B = 2$ | 0.170 | 0.910 | 0.364 | 0.170 | 0.913 | 0.362 | 0.170 | 1.773 | 0.077 |
| $A = 2, B = 1$ | 0.170 | 0.910 | 0.364 | 0.170 | 0.971 | 0.333 | 0.170 | 1.620 | 0.106 |
| Pse. R-sq | 0.068 | | | 0.068 | | | 0.068 | | |
| Adj. R-sq | 0.059 | | | 0.059 | | | 0.059 | | |
| S.E. of reg. | 0.893 | | | 0.893 | | | 0.893 | | |
| Quantile DV | 5.670 | | | 5.670 | | | 5.670 | | |
| Sparsity | 2.287 | | | 2.240 | | | 2.287 | | |
| QLR stat | 26.580 | | | 27.148 | | | 26.580 | | |
| Prob | 0.000 | | | 0.000 | | | 0.000 | | |
| (a) Ordinary (IID) standard errors and covariance | | | | | | | | | |
| (b) Huber sandwich standard errors and covariance | | | | | | | | | |
| (c) Bootstrap standard errors and covariance with Bootstrap method XY-pair | | | | | | | | | |

**Figure 2.38** The outputs summary of the QR(Median) in (2.26) using three alternative standard errors and covariances.

approach to refinement of inference for QRs, called the *saddlepoint methods*, which were introduced by Daniel (1954).

**Example 2.19    *The First Illustrative Statistical Results***
Figure 2.38 presents the summary of the outputs of a $2 \times 2$ Factorial ANOVA-QR(Median) with the following ES, using the default options and the three alternative options of the *Coefficient Covariance* method. Based on these outputs, the following findings and notes are presented:

$$Y1\ C@\text{Expand}\ (A, B, @\text{Droplast}) \tag{2.26}$$

1) The parameters' estimates of the QR are invariant or constant for the three options of the *Coefficient Covariance* method. So, the medians of Y1 by the factors A and B are invariant for the three options. I am very sure that the quantile($\tau$), for each $\tau \varepsilon (0, 1)$ also are invariant.
2) The three outputs present different values of the *Quasi-LR* statistics. Even though, at the 1% level, they give the same conclusion, the medians of Y1 have significant differences between the four cells generated by the factors A and B.
3) Note that the three options give exactly the same parameters' estimates. However, the t-statistic for testing the medians differences between pairs of cells gives different

conclusions because the *t*-statistic is using different assumptions of coefficient covariance. And each is acceptable, in the statistical sense. However, we never can tell which output is the best in representing the population standard error and covariance.

4) I am very confident that all combinations of options would give the same parameters' estimates of all QRs. But they present different standard errors, as well as values of the *t*-test statistic and the *Quasi LR* statistic.

### Example 2.20  *Additional Bootstrap Selection Settings*

In addition to the result of the Bootstrap method presented in Figure 2.38, Figure 2.39 shows the outputs of the QR (2.26), using three additional Bootstrap selection settings.

Based on this summary, the following findings and notes are presented:

1) The four Bootstrap selection settings present the same parameters' estimates of the QR (2.26). This is similarly the case for the parameters' estimates of any other QR. Note that the statistical results of the Bootstrap selection method use 100 random samples selected from the data file as the default option.

2) However, each of the outputs presents different estimates of parameter standard errors. Thus, they present different values of the *t*-statistics. Even by doing reanalyzed, I have found different values of the *t*-statistic are obtained because it is using different replication samples.

| Variable | (a) BM residual | | | (b) MCMB | | | (c) MCMB-A | | |
|---|---|---|---|---|---|---|---|---|---|
| | Coef. | *t*-stat. | Prob. | Coef. | *t*-stat. | Prob. | Coef. | *t*-stat. | Prob. |
| C | 5.500 | 107.9 | 0.000 | 5.500 | NA | 0.000 | 5.500 | 76.593 | 0.000 |
| $A = 1, B = 1$ | 1.000 | 12.72 | 0.000 | 1.000 | 8.367 | 0.000 | 1.000 | 13.206 | 0.000 |
| $A = 1, B = 2$ | 0.170 | 2.500 | 0.013 | 0.170 | 2.036 | 0.043 | 0.170 | 1.985 | 0.048 |
| $A = 2, B = 1$ | 0.170 | 2.673 | 0.008 | 0.170 | 2.280 | 0.023 | 0.170 | 1.273 | 0.204 |
| Pse. *R*-sq | 0.068 | | | 0.068 | | | 0.068 | | |
| Adj. *R*-sq | 0.059 | | | 0.059 | | | 0.059 | | |
| S.E. of reg. | 0.893 | | | 0.893 | | | 0.893 | | |
| Quantile DV | 5.670 | | | 5.670 | | | 5.670 | | |
| Sparsity | 2.287 | | | 2.287 | | | 2.287 | | |
| QLR stat | 26.58 | | | 26.58 | | | 26.58 | | |
| Prob | 0.000 | | | 0.000 | | | 0.000 | | |

(a) Bootstrap method: residual, reps = 100, rng = kn, seed = 473 136 867

(b) Bootstrap method: MCMB, reps = 100, rng = kn, seed = 1 202 951 143

(c) Bootstrap method: MCMB-A, reps = 100, rng = kn, seed = 571 009 815

**Figure 2.39** The outputs summary of the QR(Median) in (2.26) using three alternative Bootstrap selection settings.

3) The MCMB and MCMB-A, respectively, indicate the Markov Change Marginal Bootstrap and its modification. For additional details, see He and Hu (2002).

**Example 2.21**   *The Bootstrap Method: Residual, with Multiple Options*

Figure 2.40 presents the outputs of two QRs (2.26) using the *Bootstrap method: Residual*, using multiple options, with different weighting series. Based on this output, the following findings and notes are presented:

1) I have found that we can use any variable as a weighting series, even using the interaction factors or variables, as presented in the outputs. In this example, I am using $X1$ and $A*B*X1$, respectively, as weighting series, because they are the cause factors/variables of the DV, $Y1$. For the weight type, inverse standard deviation is used to show that the larger standard deviation has a smaller contribution on the estimates of the parameter standard error.

2) Both QRs use the *Bootstrap method: Residual*, which is applied with replication samples, *reps* = 200, random generator, *rng* = *mt* (Mersenne), *seed* = *, the Sparsity method using fitted quantile, and the Bandwidth method Bofinger, which gives unique optimal solutions. So, the statistical results in Figure 2.40 are acceptable.

| Dependent Variable: Y1 | | | | |
|---|---|---|---|---|
| Method: Quantile Regression (Median) | | | | |
| Date: 02/21/19  Time: 07:47 | | | | |
| Sample: 1 300 | | | | |
| Included observations: 300 | | | | |
| **Weighting series: X1** | | | | |
| Weight type: Inverse standard (EViews default scaling) | | | | |
| Bootstrap Standard Errors & Covariance | | | | |
| Bootstrap method: XY-pair, reps=200, rng=mt, seed=1273810964 | | | | |
| Sparsity method: Siddiqui using fitted quantiles | | | | |
| Bandwidth method: Bofinger, bw=0.20699 | | | | |
| Estimation successfully identifies unique optimal solution | | | | |
| Variable | Coef. | Std. Err. | t-Stat. | Prob. |
| C | 5.5 | 0.0999 | 55.0301 | 0.0000 |
| A=1,B=1 | 1 | 0.1518 | 6.5893 | 0.0000 |
| A=1,B=2 | 0.17 | 0.2011 | 0.8454 | 0.3986 |
| A=2,B=1 | 0.17 | 0.1107 | 1.5355 | 0.1257 |
| **Weighted Statistics** | | | | |
| Pse. R^2 | 0.0901 | Mean dependent var | | 5.8537 |
| Adj. R^2 | 0.0809 | S.D. dependent var | | 1.8657 |
| S.E. of reg | 0.8717 | Weighted mean dep. | | 5.9317 |
| Objective | 100.4004 | Quantile dep var | | 5.6700 |
| Restr. Obj. | 110.3389 | Sparsity | | 2.3301 |
| QLR stat. | 34.1219 | Prob(QLR stat) | | 0.0000 |
| **Unweighted Statistics** | | | | |
| Mean DV | 5.7511 | S.D. dependent var | | 0.9451 |
| Quant DV | 5.6700 | | | |

| Dependent Variable: Y1 | | | | |
|---|---|---|---|---|
| Method: Quantile Regression (Median) | | | | |
| Date: 02/21/19  Time: 07:50 | | | | |
| Sample: 1 300 | | | | |
| Included observations: 300 | | | | |
| **Weighting series: A*B*X1** | | | | |
| Weight type: Inverse standard deviation (EViews default scaling) | | | | |
| Bootstrap Standard Errors & Covariance | | | | |
| Bootstrap method: XY-pair, reps=200, rng=mt, seed=1273810964 | | | | |
| Sparsity method: Siddiqui using fitted quantiles | | | | |
| Bandwidth method: Bofinger, bw=0.20699 | | | | |
| Estimation successfully identifies unique optimal solution | | | | |
| Variable | Coef. | Std. Err. | t-Stat. | Prob. |
| C | 5.5 | 0.0999 | 55.0301 | 0.0000 |
| A=1,B=1 | 1 | 0.1518 | 6.5893 | 0.0000 |
| A=1,B=2 | 0.17 | 0.2011 | 0.8454 | 0.3986 |
| A=2,B=1 | 0.17 | 0.1107 | 1.5355 | 0.1257 |
| **Weighted Statistics** | | | | |
| Pse. R^2 | 0.0480 | Mean dependent var | | 5.7213 |
| Adj. R^2 | 0.0383 | S.D. dependent var | | 2.9373 |
| S.E. of reg | 0.9176 | Weighted mean dep. | | 5.3884 |
| Objective | 98.5160 | Quantile dep var | | 5.6700 |
| Restr. Obj. | 103.4813 | Sparsity | | 2.2187 |
| QLR stat. | 17.9037 | Prob(QLR stat) | | 0.0005 |
| **Unweighted Statistics** | | | | |
| Mean DV | 5.7511 | S.D. dependent var | | 0.9451 |
| Quant DV | 5.6700 | | | |

**Figure 2.40**   The outputs of two QRs in (2.26) using multiple options with different weighting series.

3) It is important to note that the estimates of the parameter coefficients are the same as the outputs of six estimation methods presented in Figures 2.38 and 2.39. So, the weighted series does not have any effect on the estimates of the parameter coefficients and their standard errors. However, both outputs present different sets of weighted statistics, except the quantile dependent variable. Based on these findings, we don't need to use the weight in doing the QR analysis because the weighted statistics are not important for testing hypotheses on the parameters of the QR model.

# 3

# *N*-Way ANOVA Quantile Regressions

## 3.1  Introduction

As the extension of the two-way ANOVA-QR, this chapter presents $I \times J \times K$ *Factorial QRs* of any objective or problem variables on three factors *F1, F2,* and *F3*, which can have various equation specifications (ESs). This chapter presents several such ESs, which are extensions of the ESs (2.3), (2.4), and (2.7).

Note that conditional for a level of *F3*, say *F3 = k,* all types of hypotheses on the effects of the two factors, *F1* and *F2* on *Y,* can be tested using either the Wald test, the *Quasi-LR* statistic, or the *t*-statistic, by using alternative models. Refer to all examples presented in the previous chapter. In addition, based on three-way a ANOVA-QR, we have a hypothesis of a three-way interaction effect: *F11\*F2\*F3* on *Y*. Hence, in the following examples, I consider and present only the conditional two-way interaction effects and the three-way interaction effect, based on several selected $I \times J \times K$ Factorial QRs.

## 3.2  The Models Without an Intercept

Based on the four sets of categorical factors, *A, B, (G2 or G4)* and *(H2 or H3)*, in Data_Faad.wf1, presented in Chapter 1, we can easily conduct various three-way QR analyses, using the simplest ES as follows:

$$Y @ Expand (F1, F2, F3) \tag{3.1}$$

where *Y* is a numerical dependent, objective, or predicted variable, and the three factors, *F1, F2,* and *F3,* can be selected from the four sets of the categorical factors. So, we could have various $I \times J \times K$ Factorial QRs. However, all hypotheses have to be tested using the Wald test.

If various hypotheses will be tested, I recommend conducting the analysis using the models with intercepts, mainly the full factorial and specific QR models, which will be presented in the following section.

### Example 3.1  *A 2×2×2 Factorial QR*
Figure 3.1 present the estimates of a $2 \times 2 \times 2$ Factorial QR of *Y1* on the factors *A, B,* and *G2,* and its forecast graph *Y1F* and forecast evaluation in Figure 3.2. Based on these results,

*Quantile Regression: Applications on Experimental and Cross Section Data Using EViews,* First Edition.
I Gusti Ngurah Agung.
© 2021 John Wiley & Sons Ltd. Published 2021 by John Wiley & Sons Ltd.

Dependent variable: Y1
Method: quantile regression (median)
Included observations: 300
Huber sandwich standard errors and covariance
Sparsity method: Kernel (Epanechnikov) using residuals
Bandwidth method: Hall-Sheather, bw = 0.14513
Estimation successful but solution may not be unique

| Variable | Coefficient | Std. error | t-Statistic | Prob. |
|---|---|---|---|---|
| A = 1,B = 1,G2 = 1 | 6.000000 | 0.304510 | 19.70378 | 0.0000 |
| A = 1,B = 1,G2 = 2 | 6.500000 | 0.159605 | 40.72564 | 0.0000 |
| A = 1,B = 2,G2 = 1 | 5.670000 | 0.155507 | 36.46131 | 0.0000 |
| A = 1,B = 2,G2 = 2 | 6.300000 | 0.199800 | 31.53151 | 0.0000 |
| A = 2,B = 1,G2 = 1 | 5.330000 | 0.257409 | 20.70633 | 0.0000 |
| A = 2,B = 1,G2 = 2 | 5.670000 | 0.124634 | 45.49323 | 0.0000 |
| A = 2,B = 2,G2 = 1 | 5.500000 | 0.172004 | 31.97600 | 0.0000 |
| A = 2,B = 2,G2 = 2 | 5.670000 | 0.210686 | 26.91214 | 0.0000 |

| | | | | |
|---|---|---|---|---|
| Pseudo R-squared | 0.095382 | Mean dependent var | | 5.751100 |
| Adjusted R-squared | 0.073696 | S.D. dependent var | | 0.945077 |
| S.E. of regression | 0.868783 | Objective | | 100.3900 |
| Quantile dependent var | 5.670000 | Restr. objective | | 110.9750 |
| Sparsity | 3.207000 | | | |

**Figure 3.1** Estimates of a $2 \times 2 \times 2$ Factorial QR(Median).

only the following selected notes are presented, which are the extension of the two-way QR presented in the previous chapter:

1) Based on the model parameters in Table 3.1, we have the following conditional two-way interaction effects, which can easily be tested using the Wald test:

1.1 Specific for the level $A = i$, we have the following conditional two-way interaction effect (IE):

$$IE\,(B^*G2|A = 1) = C(1) - C(2) - C(3) + C(4)$$

$$IE\,(B^*G2|A = 2) = C(5) - C(6) - C(7) + C(8)$$

Which of these indicates the effect of G2 on Y1 depends on the factor B, conditional for the level of $A = i$.

1.2 Specific for the level $B = j$, we have the following conditional two-way interaction effect:

$$IE\,(A^*G2|B = 1) = C(1) - C(2) - C(5) + C(6)$$

$$IE\,(A^*G2|B = 2) = C(3) - C(4) - C(7) + C(8)$$

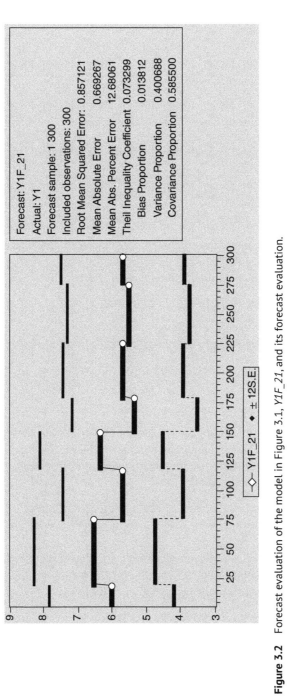

**Figure 3.2** Forecast evaluation of the model in Figure 3.1, *Y1F_21*, and its forecast evaluation.

**Table 3.1** The model parameters by the factors *A, B* and *G2*.

| | B = 1 | | | B = 2 | | |
|---|---|---|---|---|---|---|
| | G2 = 1 | G2 = 2 | G2(1–2) | G2 = 1 | G2 = 2 | G2(1–2) |
| A = 1 | C(1) | C(2) | C(1)–C(2) | C(3) | C(4) | C(3)–C(4) |
| A = 2 | C(5) | C(6) | C(5)–C(6) | C(7) | C(8) | C(7)–C(8) |
| A(1–2) | C(1)–C(5) | C(2)–C(6) | | C(1)–C(5) | C(1)–C(5) | |

Which of these indicates the effect of *G2* on *Y1* depends on the factor *A*, conditional for the level of *B* = *j*.

1.3 Specific for the level *G2* = *k*, we have the following conditional two-way interaction effect:

$$IE\,(A^*B|G2 = 1) = C(1) - C(3) - C(5) + C(7)$$

$$IE\,(A^*B|G2 = 2) = C(2) - C(4) - C(6) + C(8)$$

Which of these indicates the effect of *A* on *Y1* depends on the factor *B*, conditional for the level of *G2* = *k*.

2) Then we have the following three-way interaction effect:

$$IE\,(A^*B^*G2) = IE\,(B^*G2|A = 1) - IE\,(B^*G2|A = 2)$$

$$= IE\,(A^*G2|B = 1) - IE\,(A^*G2|B = 2)$$

$$= IE\,(A^*B|G2 = 1) - IE\,(A^*B|G2 = 2)$$

$$IE\,(A^*B^*G2) = (C(1) - C(2) - C(3) + C(4)) - (C(5) - C(6) - C(7) + C(8))$$

**Example 3.2  *A* 2×2×3 *Factorial QR***

Figure 3.3 presents the estimates of a 2 × 2 × 3 Factorial QR of *Y1* on the factors *A, B*, and *H3*, and its forecast graph *Y1F_23* and forecast evaluation in Figure 3.4. Based on these results, only the following selected notes are presented, which are the extension of the 2×2×2 Factorial QR in the preceding example:

1) Following from the two-way interaction effects presented in Tables 2.2 through 2.5, and based on the model parameters in Table 3.2, we have the following conditional two-way interaction effects:

1.1 Specific for the level *A* = 1, we have a 2 × 3 tabulation of the model parameters *C*(1) to *C*(6), by the factors *B* and *H3* with their pairwise differences, such as the four conditional differences in the last two rows of Table 3.2:

$$H3\,(1 - 3|A = 1, B = 1) = C(1) - C(3)$$

$$H3\,(1 - 3|A = 1, B = 2) = C(4) - C(6)$$

$$H3\,(2 - 3|A = 1, B = 1) = C(2) - C(3)$$

$$H3\,(2 - 3|A = 1, B = 2) = C(5) - C(6)$$

Dependent variable: Y1
Method: quantile regression (Median)
Date: 04/04/18   Time: 21:05
Sample: 1 300
Included observations: 300
Huber sandwich standard errors & covariance
Sparsity method: Kernel (Epanechnikov) using residuals
Bandwidth method: Hall-Sheather, bw = 0.14513
Estimation successful but solution may not be unique

| Variable | Coefficient | Std. error | t-Statistic | Prob. |
|---|---|---|---|---|
| A = 1,B = 1,H3 = 1 | 5.500000 | 0.334024 | 16.46589 | 0.0000 |
| A = 1,B = 1,H3 = 2 | 6.500000 | 0.184543 | 35.22215 | 0.0000 |
| A = 1,B = 1,H3 = 3 | 6.500000 | 0.199770 | 33.53749 | 0.0000 |
| A = 1,B = 2,H3 = 1 | 5.330000 | 0.235812 | 23.60279 | 0.0000 |
| A = 1,B = 2,H3 = 2 | 5.670000 | 0.177343 | 31.97197 | 0.0000 |
| A = 1,B = 2,H3 = 3 | 6.300000 | 0.149028 | 43.27402 | 0.0000 |
| A = 2,B = 1,H3 = 1 | 5.500000 | 0.204042 | 26.95519 | 0.0000 |
| A = 2,B = 1,H3 = 2 | 5.670000 | 0.159726 | 35.49831 | 0.0000 |
| A = 2,B = 1,H3 = 3 | 5.500000 | 0.217464 | 25.29150 | 0.0000 |
| A = 2,B = 2,H3 = 1 | 5.500000 | 0.222488 | 24.72043 | 0.0000 |
| A = 2,B = 2,H3 = 2 | 5.500000 | 0.159942 | 34.38740 | 0.0000 |
| A = 2,B = 2,H3 = 3 | 6.300000 | 0.176574 | 35.67900 | 0.0000 |

| | | | | |
|---|---|---|---|---|
| Pseudo R-squared | 0.146249 | Mean dependent var | | 5.751100 |
| Adjusted R-squared | 0.113641 | S.D. dependent var | | 0.945077 |
| S.E. of regression | 0.858501 | Objective | | 94.74500 |
| Quantile dependent var | 5.670000 | Restr. objective | | 110.9750 |
| Sparsity | 1.914722 | | | |

**Figure 3.3** Statistical results of a $2 \times 2 \times 3$ Factorial QR(Median).

Then we have the following conditional two-way bivariate interaction effect:

$$IE\,(B^*H3|A = 1) = H3\,(1 - 3|A = 1, B = 1) - H3\,(1 - 3|A = 1, B = 2),$$
$$H3\,(2 - 3|A = 1, B = 1) - H3\,(2 - 3|A = 1, B = 2)$$

$$IE\,(B^*H3|A = 1) = [C(1) - C(3) - C(4) + C(6), C(2) - C(3) - C(5) + C(6)]$$

Similarly, for the interaction effect of $B^*H3$, we have a $2 \times 3$ tabulation of the model parameters $C(7)$ to $C(12)$, by the factors $B$ and $H3$, conditional for $A = 2$. Then we have the following conditional two-way bivariate interaction effect:

$$IE(B^*H3\,|\,A = 2) = [C(7) - C(9) - C(10) + C(12),$$
$$C(8) - C(9) - C(11) + C(12)]$$

Wald test:

Equation: EQ07

| Test statistic | Value | df | Probability |
|---|---|---|---|
| F-statistic | 0.868928 | (2, 288) | 0.4205 |
| Chi-square | 1.737857 | 2 | 0.4194 |

Null hypothesis: $(C(1) - C(3) - C(4) + C(6) - C(7) + C(9) + C(10) - C(12)) = 0$,
$(C(2) - C(3) - C(5) + C(6) - C(8) + C(9) + C(11) - C(12)) = 0$

Null hypothesis summary:

| Normalized restriction (= 0) | | Value | Std. err. |
|---|---|---|---|
| $C(1) - C(3) - C(4) + C(6) - C(7) + C(9) + C(10) - C(12)$ | | −0.830000 | 0.631585 |
| $C(2) - C(3) - C(5) + C(6) - C(8) + C(9) + C(11) - C(12)$ | | −0.340000 | 0.507136 |

Restrictions are linear in coefficients.

**Figure 3.4** A Wald test of the three-way interaction effect, based on the model in Figure 3.3.

**Table 3.2** The model parameters by the factors *A*, *B*, and *H3*.

| | A = 1 | | | A = 2 | | |
|---|---|---|---|---|---|---|
| | B = 1 | B = 2 | B(1−2) | B = 1 | B = 2 | B(1−2) |
| H3 = 1 | C(1) | C(4) | C(1)–C(4) | C(7) | C(10) | C(7)–C(10) |
| H3 = 2 | C(2) | C(5) | C(2)–C(5) | C(8) | C(11) | C(8)–C(11) |
| H3 = 3 | C(3) | C(6) | c(3)–c(6) | C(9) | C(12) | C(9)–c(12) |
| H3(1−3) | C(1)–C(3) | C(4)–C(6) | | C(7)–C(9) | C(10)–C(12) | |
| H3(2−3) | C(2)–C(3) | C(5)–C(6) | | C(8)–C(9) | C(11)–C(12) | |

1.2 Specific for the level $B = 1$, we have a $2 \times 3$ tabulation by the factors *A* and *H3*. with the model parameters $C(1)$, $C(2)$, $C(3)$, $C(7)$, $C(8)$, and $C(9)$ and their pairwise differences, such as the four conditional differences in the last two rows of the Table 3.2:

$$H3 (1 - 3 | A = 1, B = 1) = C(1) - C(3)$$
$$H3 (1 - 3 | A = 2, B = 1) = C(7) - C(9),$$

$$H3 (2 - 3 | A = 1, B = 1) = C(2) - C(3)$$
$$H3 (2 - 3 | A = 2, B = 1) = C(8) - C(9)$$

Then we have a conditional two-way bivariate interaction effect, as follows:

$$IE\,(A^*H3|B = 1) = H3\,(1 - 3|A = 1, B = 1) - H3\,(1 - 3|A = 2, B = 1),$$
$$H3\,(2 - 3|A = 1, B = 1) - H3\,(2 - 3|A = 2, B = 1)$$

$$IE\,(A^*H3|B = 1) = [C(1) - C(3) - C(7) + C(9), C(2) - C(3) - C(8) + C(9)]$$

Similarly, for the interaction effect of *A\*H3*, conditional for $B = 2$, we have a $2 \times 3$ tabulation, by the factors *A* and *H3*, with the model parameters $C(4)$, $C(5)$, $C(6)$, $C(10)$, $C(11)$, and $C(12)$ and their pairwise differences, which are the four conditional differences in the last two rows of the Table 3.2:

$$H3\,(1 - 3|A = 1, B = 2) = C(4) - C(6)$$
$$H3\,(1 - 3|A = 2, B = 2) = C(10) - C(12)$$

$$H3\,(2 - 3|A = 1, B = 2) = C(5) - C(6)$$
$$H3\,(2 - 3|A = 2, B = 2) = C(11) - C(12)$$

Then we obtain the following conditional two-way bivariate interaction effect:

$$IE\,(A^*H3|B = 2) = H3\,(1 - 3|A = 1, B = 2) - H3\,(1 - 3|A = 2, B = 2),$$
$$H3\,(2 - 3|A = 1, B = 2) - H3\,(2 - 3|A = 2, B = 2)$$

$$IE\,(A^*H3|B = 2) = [C(4) - C(6) - C(10) + C(12),$$
$$C(5) - C(6) - C(11) + C(12)]$$

1.3 Specific for the level $H3 = k$, we have three $2 \times 2$ tabulations, by the factors *A* and *B*, for $k = 1$, 2 and 3. Referring to the difference in differences (DIDs) presented in Table 2.2, we have three conditional DIDs or two-way interaction effects of *A\*B*, as follows:

$$IE\,(A^*B|H3 = 1) = B\,(1 - 2|A = 1, H3 = 1) - B\,(1 - 2|A = 2, H3 = 1)$$
$$= C(1) - C(4) - C(7) + C(10)$$

$$IE\,(A^*B|H3 = 2) = B\,(1 - 2|A = 1, H3 = 2) - B\,(1 - 2|A = 2, H3 = 2)$$
$$= C(2) - C(5) - C(8) + C(11)$$

$$IE\,(A^*B|H3 = 3) = B\,(1 - 2|A = 1, H3 =) - B\,(1 - 2|A = 2, H3 = 3)$$
$$= C(3) - C(6) - C(9) + C(12)$$

Which of these indicates the effect of *A* on *Y1* depends on the factor *B*, conditional for the level of $H3 = k$.

2) Then we have the following three-way bivariate interaction effect:

$$IE\,(A^*B^*H3) = IE\,(B^*H3|A = 1) - IE\,(B^*H3|A = 2)$$
$$= [(C(1) - C(3) - C(4) + C(6), C(2) - C(3) - C(5) + C(6)) -$$
$$(C(7) - C(9) - C(10) + C(12), C(8) - C(9) - C(11) + C(12))]$$

$$= [(C(1) - C(3) - C(4) + C(6)) - (C(7) - C(9) - C(10) + C(12)),$$
$$(C(2) - C(3) - C(5) + C(6)) - (C(8) - C(9) - C(11) + C(12)))]$$

$$IE(A^*B^*H3) = [C(1) - C(3) - C(4) + C(6) - C(7) + C(9) + (10) - C(12),$$
$$C(2) - C(3) - C(5) + C(6) - C(8) + C(9) + C(11) - C(12)]$$

3) In addition, note that the three-way interaction effect also can be computed as follows:

$$IE(A^*B^*H3) = IE(A^*H3|B = 1) - IE(A^*H3|B = 2) = (IE(A^*B|H3 = 1) -$$
$$IE(A^*B|H3 = 3)), (IE(A^*B|H3 = 2) - IE(A^*B|H3 = 3))$$

For example, based on the output of the Wald test in Figure 3.4, we can conclude that the three-way interaction $A^*B^*H3$ has an insignificant effect on $Y1$, based on the $F$-statistic of $F_0 = 0.868\,928$ with $df = 2$ and $p$-value $= 0.4205$.

## 3.3 Models with Intercepts

Based on the four sets categorical factors, $A$, $B$, $(G2 \text{ or } G4)$ and $(H2 \text{ or } H3)$ in Datat_Faad.wf1, we can easily present various types of three-way quantile regression analysis, using a model with an intercept as a referent group, with the following alternative ESs:

1) The simplest model with the following ES:

$$Y \ C@ Expand(F1, F2, F3, @Drop(i, j, k)) \tag{3.2a}$$

where $Y$ is a numerical dependent or objective variable; the three factors, $F1$, $F2$ and $F3$, can be selected from the four categorical factors; and the cell $(F1 = i, F2 = j, F3 = k)$ is selected as a reference group. As with other alternative ESs, we can use the function @Droplast or @Dropfirst. Specific for the dichotomous factors with 1 and 2 as their levels, and @Drop(*) = @Droplast, the ES of the model can be presented as follows:

$$Y \ C \ (F1 = 1)^*(F2 = 1)^* \ (F3 = 1) \ (F1 = 1)^*(F2 = 1)^* \ (F3 = 2)$$
$$(F1 = 1)^*(F2 = 2)^* \ (F3 = 1) \ (F1 = 1)^*(F2 = 2)^* \ (F3 = 2)$$
$$(F1 = 2)^*(F2 = 1)^* \ (F3 = 1) \ (F1 = 2)^*(F2 = 2)^* \ (F3 = 1) \tag{3.2b}$$

2) The full-factorial model with the following ES:

$$Y \ C@ Expand(F1, @Drop(i)) @Expand(F2, @Drop(j)) @Expand(F3, @Drop(k))$$
$$@Expand(F1, @Drop(i))^* @Expand(F2, @Drop(j))$$
$$@Expand(F1, @Drop(i))^* @Expand(F3, @Drop(k))$$
$$@Expand(F2, @Drop(j))^* @Expand(F3, @Drop(k))$$
$$@Expand(F1, @Drop(i))^* @Expand(F2, @Drop(j))^* @Expand(F3, @Drop(k)) \tag{3.3a}$$

Specific for the dichotomous factors with 1 and 2 as their levels, and @Drop(*) = @Droplast, the ES can be presented as follows:

$$Y \ C \ (F1 = 1) \ (F2 = 1) \ (F3 = 1) \ (F1 = 1)^* \ (F2 = 1) \ (F1 = 1)^* \ (F3 = 1)$$
$$(F2 = 1)^* \ (F3 = 1) \ (F1 = 1)^*(F2 = 1)^* \ (F3 = 1) \tag{3.3b}$$

3) A special model with the following ES (Agung 2011a, p. 181):

$$Y\ C@Expand\ (F1, F2, @Drop\ (i, j))\ @Expand(F1, F2)^*\ @Expand\ (F3, @Drop\ (k))$$

(3.4a)

Specific for the dichotomous factors with 1 and 2 as their levels, and @Drop(*) = @Droplast, the ES can be presented as follows:

$$Y\ C\ (F1 = 1)^*\ (F2 = 1)\ (F1 = 1)^*\ (F2 = 2)\ (F1 = 2)^*\ (F2 = 1)$$
$$(F1 = 1)^*(F2 = 1)^*\ (F3 = 1)\ (F1 = 1)^*(F2 = 1)^*\ (F3 = 2)$$
$$(F1 = 1)^*(F2 = 2)^*\ (F3 = 1)\ (F1 = 1)^*(F2 = 2)^*\ (F3 = 2)$$
$$(F1 = 2)^*(F2 = 1)^*\ (F3 = 1)\ (F1 = 2)^*(F2 = 1)^*\ (F3 = 2)$$
$$(F1 = 2)^*(F2 = 2)^*\ (F3 = 1)\ (F1 = 2)^*(F2 = 2)^*\ (F3 = 2)$$

(3.4b)

**Example 3.3  A 2×2×2 Factorial QR in (3.3b)**

Figure 3.5 presents the estimate of a $2 \times 2 \times 2$ Factorial QR $Y1$ on the factors $A$, $B$, and $G2$ in (3.3b). First, this estimate shows that the $H_0$: $C(m) = 0$, $\forall m > 1$, is rejected based on the $Quasi\text{-}LR$ statistic of 38.368 83 with a $Prob. = 0.000\,003$. Hence, we can conclude that the factors $A$, $B$, and $G2$ have significant joints effects on $Y1$.

In other words, the medians of $Y1$ have significant differences between the eight groups of students, generated by the three factors $A$, $B$, and $G2=O2X1$ (the dichotomous factor of $X1$ = motivation).

Furthermore, based on this estimate, we can develop the model parameters by the three factors as presented in Table 3.3. Then, based on this table, we have the following statistics:

1) The following conditional main effects (MEs), which can be tested using the $t$-statistic in the output:

   1.1 The main effect $A$, conditional for $B = 2$ and $G2 = 2$, or the median difference between the cells ($A = 1$, $B = 2$, $G2 = 2$) and ($A = 2$, $B = 2$, $G2 = 2$), as follows:

   $$ME\ (A|B = 2, G2 = 2) = C(1) + C(2) - C(1) = C(2)$$

   1.2 The main effect $B$, conditional for $A = 2$ and $G2 = 2$, or the median difference between the cells ($A = 2$, $B = 1$, $G2 = 2$) and ($A = 2$, $B = 2$, $G2 = 2$), as follows:

   $$ME\ (B|A = 2, G2 = 2) = C(1) + C(3) - C(1) = C(3)$$

   1.3 The main effect $G2$, conditional for $A = 2$ and $B = 2$, or the median difference between the cells ($A = 2$, $B = 2$, $G2 = 1$) and ($A = 2$, $B = 2$, $G2 = 2$), as follows:

   $$ME\ (G2|A = 2, B = 2) = C(1) + C(4) - C(1) = C(4)$$

2) In addition, we have the following conditional two-way interaction effects, which also can be tested using the $t$-statistic in the output:

   2.1 The interaction effect $A*B$, conditional $G2 = 2$, as follows:

   $$IE\ (A^*B|G2 = 2) = C(3) + C(5) - C(3) = C(5)$$

   2.2 The interaction effect $A*G2$, conditional $B = 2$, as follows:

   $$IE\ (A^*G2|B = 2) = C(4) + C(6) - C(4) = C(6)$$

Dependent variable: Y1
Method: quantile regression (Median)
Date: 04/08/18   Time: 07:39
Sample: 1 300
Included observations: 300
Huber sandwich standard errors & covariance
Sparsity method: Kernel (Epanechnikov) using residuals
Bandwidth method: Hall-Sheather, bw = 0.14513
Estimation successful but solution may not be unique

| Variable | Coefficient | Std. error | t-Statistic | Prob. |
|---|---|---|---|---|
| C | 5.670000 | 0.210686 | 26.91214 | 0.0000 |
| A = 1 | 0.630000 | 0.290359 | 3.169726 | 0.0308 |
| B = 1 | 0.000000 | 0.244790 | 0.000000 | 1.0000 |
| G2 = 1 | −0.170000 | 0.271981 | −0.625043 | 0.5324 |
| (A = 1)*(B = 1) | 0.200000 | 0.354000 | 0.564972 | 0.5725 |
| (A = 1)*(G2 = 1) | −0.460000 | 0.371586 | −1.237936 | 0.2167 |
| (B = 1)*(G2 = 1) | −0.170000 | 0.394673 | −0.430736 | 0.6670 |
| (A = 1)*(B = 1)*(G2 = 1) | 0.300000 | 0.581438 | 0.515963 | 0.6063 |

| | | | | |
|---|---|---|---|---|
| Pseudo R-squared | 0.095382 | Mean dependent var | | 5.751100 |
| Adjusted R-squared | 0.073696 | S.D. dependent var | | 0.945077 |
| S.E. of regression | 0.868783 | Objective | | 100.3900 |
| Quantile dependent var | 5.670000 | Restr. objective | | 110.9750 |
| Sparsity | 3.207000 | Quasi-LR statistic | | 38.36883 |
| Prob(Quasi-LR stat) | 0.000003 | | | |

**Figure 3.5**   The results of the full $2 \times 2 \times 2$ Factorial ANOVA QR(Median) in (3.3b).

2.3 The interaction effect $B^*G2$, conditional $A = 2$, as follows:

$$IE\,(B^*G2|A = 2) = C(4) + C(7) - C(4) = C(7)$$

3) Finally, we have the following three-way interaction effect, which also can be tested using the $t$-statistic in the output:

$$IE\,(A^*B^*G2) = IE\,(B^*G2|A = 1) - IE\,(B^*G2|A = 2)$$
$$= C(7) + C(8) - C(7) = C(8)$$

**Example 3.4   *A $2 \times 2 \times 2$ Factorial QR in (3.4b)***
Figure 3.6 presents the estimate of a $2 \times 2 \times 2$ Factorial QR in (3.4b). With this model, the $H_0$: $C(m) = 0$, $\forall m > 1$, is rejected based on the *Quasi-LR* statistic of 38.368 83 with a *Prob.* = 0.000 003. Hence, we can conclude that the factors *A*, *B*, and *G2* have significant joint effects on *Y1*. In other words, the eight medians *Y1* have significant differences,

**Table 3.3** The model parameters of the QR in Figure 3.5

| | A = 1 | | | A = 2 | | |
|---|---|---|---|---|---|---|
| | B = 1 | B = 2 | B(1−2) | B = 1 | B = 2 | B(1−2) |
| G2 = 1 | C(1) + C(2) + C(3) + C(4) + C(5) + C(6) + C(7) + C(8) | C(1) + C(2) + C(4) + C(6) | C(3) + C(5) + C(7) + C(8) | C(1) + C(3) + C(4) + C(7) | C(1) + **C(4)** | C(3) + C(7) |
| G2 = 2 | C(1) + C(2) + C(3) + C(5) | C(1) + **C(2)** | C(3) + **C(5)** | C(1) + C(3) | C(1) | **C(3)** |
| G2(1−2) | C(4) + C(6) + C(7) + C(8) | C(4) + **C(6)** | C(7) + **C(8)** | C(4) + C(7) | **C(4)** | **C(7)** |

---

Dependent variable: Y1
Method: quantile regression (Median)
Date: 04/05/18   Time: 21:13
Sample: 1 300
Included observations: 300
Huber sandwich standard errors & covariance
Sparsity method: Kernel (Epanechnikov) using residuals
Bandwidth method: Hall-Sheather, bw = 0.14513
Estimation successful but solution may not be unique

| Variable | Coefficient | Std. error | t-Statistic | Prob. |
|---|---|---|---|---|
| C | 5.670000 | 0.210686 | 26.91214 | 0.0000 |
| A = 1,B = 1 | 0.830000 | 0.264314 | 3.140201 | 0.0019 |
| A = 1,B = 2 | 0.630000 | 0.290359 | 3.169726 | 0.0308 |
| A = 2,B = 1 | 0.000000 | 0.244790 | 0.000000 | 1.0000 |
| (A = 1 AND B = 1)*(G2 = 1) | −0.500000 | 0.343802 | −1.454324 | 0.1469 |
| (A = 1 AND B = 2)*(G2 = 1) | −0.630000 | 0.253185 | −3.488300 | 0.0134 |
| (A = 2 AND B = 1)*(G2 = 1) | −0.340000 | 0.285995 | −1.188832 | 0.2355 |
| (A = 2 AND B = 2)*(G2 = 1) | −0.170000 | 0.271981 | −0.625043 | 0.5324 |

| | | | |
|---|---|---|---|
| Pseudo R-squared | 0.095382 | Mean dependent var | 5.751100 |
| Adjusted R-squared | 0.073696 | S.D. dependent var | 0.945077 |
| S.E. of regression | 0.868783 | Objective | 100.3900 |
| Quantile dependent var | 5.670000 | Restr. objective | 110.9750 |
| Sparsity | 3.207000 | Quasi-LR statistic | 38.36883 |
| Prob(Quasi-LR stat) | 0.000003 | | |

**Figure 3.6** The results of the $2 \times 2 \times 2$ Factorial ANOVA QR(Median) in (3.4b).

**Table 3.4** The model parameters of the QR in Figure 3.6

|  | A = 1 | | A = 2 | | AB: 11–22 | AB: 12–22 | AB: 21–22 |
|---|---|---|---|---|---|---|---|
|  | B = 1 | B = 2 | B = 1 | B = 2 |  |  |  |
| G2 = 1 | C(1) + C(2) + C(5) | C(1) + C(3) + C(6) | C(1) + C(4) + C(7) | C(1) + c(8) | C(2) + C(5) − C(8) | C(3) + C(6) − C(8) | C(4) + C(7) − C(8) |
| G2 = 2 | C(1) + C(2) | C(1) + C(3) | C(1) + C(4) | C(1) | **C(2)** | **C(3)** | **C(4)** |
| G2(1–2) | **C(5)** | **C(6)** | **C(7)** | **C(8)** |  |  |  |

Wald test:
Equation: EQ08

| Test statistic | Value | df | Probability |
|---|---|---|---|
| t-statistic | 0.515963 | 292 | 0.6063 |
| F-statistic | 0.266217 | (1,292) | 0.6063 |
| Chi-square | 0.266217 | 1 | 0.6059 |

Null hypothesis: $C(5) - C(6) - C(7) + C(8) = 0$
Null hypothesis Summary:

| Normalized restriction (= 0) | Value | Std. err. |
|---|---|---|
| $C(5) - C(6) - C(7) + C(8)$ | 0.300000 | 0.581438 |

Restrictions are linear in coefficients.

**Figure 3.7** The Wald test of the three-way interaction effect.

which could be compared to a test hypothesis of the eight means of *Y1*, based on the LS or mean regression (MR) of *Y1* by the three factors.

Furthermore, based on this estimate, we can develop the model parameters by the three factors, as presented in Table 3.4. This table shows that based on each of the parameters $C(2)$ to $C(8)$, we can write a one-side or two-sided hypothesis on the medians difference between specific pairs of cells ($A = i$, $B = j$, $G2 = k$) and then test it using the *t*-statistic in the output. For instance:

(i) For the right-sided hypothesis $H_0$: $C(3) \leq 0$ vs. $H_1$: $C(3) > 0$, the null hypothesis is rejected based on the *t*-statistic of $t_0 = 3.169\,726$ with a *p*-value $= 0.0308/2 = 0.0154$. Hence, at the 5% level, the cell ($A = 1$, $B = 2$, $G2 = 2$) has a significantly greater median than the cell ($A = 2$, $B = 2$, $G2 = 2$).

(ii) For the right-sided hypothesis $H_0$: $C(5) \leq 0$ vs. $H_1$: $C(5) > 0$, the null hypothesis is accepted directly, since the *t*-statistic of $t_0 = -1.454\,324 < 0$. In other words, $H_0$ is accepted based on the *t*-statistic with a *p*-value $= 1 - 0.1469/2 = 0.926\,55$. Hence, the data does not support the right-sided hypothesis $C(5) > 0$.

Compare this to the model (3.3a), based on this model, the following three-way interaction effect $A*B*G2$ on $Y1$ should be tested using the Wald test. The result is shown in Figure 3.7, which shows exactly the same $t$-statistic as presented in Figure 3.5 for the parameter $C(8)$:

$$IE\,(A^*B^*G2) = IE\,(B^*G2|A = 1) - IE\,(B^*G2|A = 2)$$

$$= (C(5) - C(6)) - (C(7) - C(8))$$

$$= C(5) - C(6) - C(7) + C(8)$$

**Example 3.5  *A 2×2×3 Factorial QR in (3.4a)***

Figure 3.8 presents the estimate of a $2 \times 2 \times 3$ Factorial QR in (3.4a). With this model, the $H_0$: $C(m) = 0$, $\forall m > 1$ is rejected based on the *Quasi-LR* statistic of 67.683 16 with a *Prob.* = 0.000 000. Hence, we can conclude that the factors $A$, $B$, and $H3$ have significant joint effects on $Y1$. In other words, the medians of $Y1$ have significant differences between the 12 cells/groups, generated by the factors of $A$, $B$, and $H3=O3X2$, which is presented in Chapter 1.

Furthermore, based on this estimate, we can develop the model parameters by the three factors $A$, $B$, and $H3$, as presented in Table 3.5. Similar to the Table 3.4, this table also shows that based on each of the parameters $C(2)$ to $C(12)$, we can write a one-side or two-sided hypothesis on the medians difference between specific pairs of cells ($A = i$, $B = j$, $G2 = k$) and then test it using the $t$-statistic in the output.

In this case, it is important to note the following bivariate interactions effects:

1) Specific for the levels $A = 1$ and 2, we have the following conditional bivariate interaction effects:

$$IE\,(B^*H3|A = 1) = (C(5) - C(6), C(9) - C(10))$$

$$IE\,(B^*H3|A = 2) = (C(7) - C(8), C(11) - C(12))$$

2) Then the following bivariate three-way interaction effect $A*B*H3$, with its Wald test, is presented in Figure 3.9, which shows that the interaction has an insignificant effect:

$$IE\,(A^*B^*H3) = IE\,(B^*H3|A = 1) - IE\,(B^*H3|A = 2)$$

$$= (C(5) - C(6), C(9) - C(10)) - (C(7) - C(8), C(11) - C(12))$$

$$= (C(5) - C(6) - C(7) + C(8), C(9) - C(10) - C(11) + C(12))$$

**Example 3.6  *A 2×2×4 Factorial QR in (3.4a)***

Figure 3.10 presents the estimate of a $2 \times 2 \times 4$ Factorial QR in (3.4a). With this model, the $H_0$: $C(m) = 0$, $\forall m > 1$, is rejected based on the *Quasi-LR* statistic of 69.877 32 with a *Prob.* = 0.000 000. Hence, we can conclude that the factors $A$, $B$, and $G4$ have significant joint effects on $Y1$, or the 16 medians of $Y1$ have significant differences. As a comparison, based on the MR of $Y1$ by the three factors, we would have the $F$-statistic for testing the means differences of $Y1$.

Dependent variable: Y1
Method: quantile regression (median)
Date: 04/05/18 Time: 21:19
Sample: 1 300
Included observations: 300
Huber sandwich standard errors & covariance
Sparsity method: Kernel (Epanechnikov) using residuals
Bandwidth method: Hall-Sheather, bw = 0.14513
Estimation successful but solution may not be unique

| Variable | Coefficient | Std. error | t-Statistic | Prob. |
|---|---|---|---|---|
| C | 6.300000 | 0.176223 | 35.75023 | 0.0000 |
| A = 1,B = 1 | 0.200000 | 0.266317 | 0.750986 | 0.4533 |
| A = 1,B = 2 | 0.000000 | 0.230645 | 0.000000 | 1.0000 |
| A = 2,B = 1 | −0.630000 | 0.290223 | −3.170741 | 0.0308 |
| (A = 1 AND B = 1)*(H3 = 1) | −1.000000 | 0.388461 | −3.574258 | 0.0105 |
| (A = 1 AND B = 1)*(H3 = 2) | 0.000000 | 0.271704 | 0.000000 | 1.0000 |
| (A = 1 AND B = 2)*(H3 = 1) | −0.970000 | 0.278577 | −3.481981 | 0.0006 |
| (A = 1 AND B = 2)*(H3 = 2) | −0.630000 | 0.231237 | −3.724481 | 0.0068 |
| (A = 2 AND B = 1)*(H3 = 1) | −0.170000 | 0.307732 | −0.552429 | 0.5811 |
| (A = 2 AND B = 1)*(H3 = 2) | 0.000000 | 0.280360 | 0.000000 | 1.0000 |
| (A = 2 AND B = 2)*(H3 = 1) | −0.800000 | 0.283468 | −3.822192 | 0.0051 |
| (A = 2 AND B = 2)*(H3 = 2) | −0.800000 | 0.237826 | −3.363807 | 0.0009 |

| | | | |
|---|---|---|---|
| Pseudo R-squared | 0.146249 | Mean dependent var | 5.751100 |
| Adjusted R-squared | 0.113641 | S.D. dependent var | 0.945077 |
| S.E. of regression | 0.857628 | Objective | 94.74500 |
| Quantile dependent var | 5.670000 | Restr. objective | 110.9750 |
| Sparsity | 1.918350 | Quasi-LR statistic | 67.68316 |
| Prob(Quasi-LR stat) | 0.000000 | | |

**Figure 3.8** The results of the $2 \times 2 \times 3$ Factorial ANOVA QR(Median) in (3.4b).

Similar to Table 3.4, this table also shows that, based on each of the parameters $C(2)$ to $C(12)$, we can write a one-side or two-sided hypothesis on the medians between specific pairs of cells $(A = i, B = j, G2 = k)$ and then test it using the $t$-statistic in the output.

Furthermore, based on this estimate, we can develop the model parameters by the three factors $A$, $B$, and $G4$, as presented in Table 3.6. Similar to Table 3.4, this table also shows that based on each of the parameters $C(2)$ to $C(16)$, we can write a one-side or two-sided hypothesis on the medians difference between specific pairs of cells $(A = i, B = j, G2 = k)$ and then test it using the $t$-statistic in the output.

**Table 3.5** The model parameters of the QR in Figure 3.8.

| | A = 1 | | | A = 2 | | |
|---|---|---|---|---|---|---|
| | B = 1 | B = 2 | B(1–2) | B = 1 | B = 2 | B(1–2) |
| H3 = 1 | C(1)+C(2) +C(5) | C(1)+C(3) +C(6) | | C(1)+C(4) +C(7) | C(1)+c(8) | |
| H3 = 2 | C(1)+C(2) +C(9) | C(1)+C(3) +C(10) | | C(1)+C(4) C(11) | C(1)+C(12) | |
| H3 = 3 | C(1)+**C(2)** | C(1)+**C(3)** | | C(1)+**C(4)** | C(1) | |
| H3(1–3) | **C(5)** | **C(6)** | **C(5)–C(6)** | **C(7)** | **C(8)** | **C(7)–C(8)** |
| H3(2–3) | **C(9)** | **C(10)** | **C(9)–C(10)** | **C(11)** | **C(12)** | **C(11)–C(12)** |

Wald test:

Equation: EQ09

| Test statistic | Value | df | Probability |
|---|---|---|---|
| F-statistic | 1.041637 | (2,288) | 0.3542 |
| Chi-square | 3.083275 | 2 | 0.3529 |

Null hypothesis: $C(5) - C(6) - C(7) + C(8) = 0$, $C(9) - C(10) - C(11) + C(12) = 0$
Null hypothesis summary:

| Normalized restriction (= 0) | Value | Std. err. |
|---|---|---|
| $C(5) - C(6) - C(7) + C(8)$ | −0.660000 | 0.481431 |
| $C(9) - C(10) - C(11) + C(12)$ | −0.170000 | 0.376505 |

Restrictions are linear in coefficients.

**Figure 3.9** The Wald test of the effect of $A^*B^*H3$ on $Y1$.

In this case, it is important to note the following tri-variate interaction effects:

1) Specific for the levels $A = 1$ and 2, we have the following conditional two-way interaction effects:

$$IE\,(B^*G4|A = 1) = (C(5) - C(6), C(9) - C(10), C(13) - C(14))$$

$$IE\,(B^*G4|A = 2) = (C(7) - C(8), C(11) - C\,(12, C(15) - C(16))$$

Which of these indicates the effect of $G4 = O4X1$ as presented in Chapter 1 depends on the factor $B$, conditional for $A = i$. Then we have the tri-variate interaction effect:
$IE(A * B * G4) = IE(B * G4|A = 1) - IE(B * G4|A = 2) = (C(5) - C(6) - C(7) + C(8), C(9) - C(10) - C(11) + C(12), C(13) - C(14) - C(15) + C(16))$

Dependent variable: Y1
Method: quantile regression (median)
Date: 04/09/18   Time: 16:27
Sample: 1 300
Included observations: 300
Huber sandwich standard errors & covariance
Sparsity method: Kernel (Epanechnikov) using residuals
Bandwidth method: Hall-Sheather, bw = 0.14513
Estimation successful but solution may not be unique

| Variable | Coefficient | Std. error | t-Statistic | Prob. |
|---|---|---|---|---|
| C | 6.330000 | 0.304997 | 20.75428 | 0.0000 |
| A = 1,B = 1 | 0.170000 | 0.337195 | 0.504159 | 0.6145 |
| A = 1,B = 2 | 0.170000 | 0.366995 | 0.463222 | 0.6436 |
| A = 2,B = 1 | −0.660000 | 0.348450 | −1.894103 | 0.0592 |
| (A = 1 AND B = 1)*(G4 = 1) | −1.170000 | 0.476756 | −3.454086 | 0.0147 |
| (A = 1 AND B = 1)*(G4 = 2) | 8.59E−17 | 0.321513 | 3.67E−16 | 1.0000 |
| (A = 1 AND B = 1)*(G4 = 3) | −0.830000 | 0.284897 | −3.913335 | 0.0039 |
| (A = 1 AND B = 2)*(G4 = 1) | −1.170000 | 0.283577 | −4.125860 | 0.0000 |
| (A = 1 AND B = 2)*(G4 = 2) | −0.500000 | 0.276970 | −1.805252 | 0.0721 |
| (A = 1 AND B = 2)*(G4 = 3) | −0.830000 | 0.317660 | −3.612860 | 0.0095 |
| (A = 2 AND B = 1)*(G4 = 1) | −0.340000 | 0.368332 | −0.923081 | 0.3567 |
| (A = 2 AND B = 1)*(G4 = 2) | 0.000000 | 0.343941 | 0.000000 | 1.0000 |
| (A = 2 AND B = 1)*(G4 = 3) | 0.000000 | 0.239309 | 0.000000 | 1.0000 |
| (A = 2 AND B = 2)*(G4 = 1) | −1.660000 | 0.444741 | −3.732508 | 0.0002 |
| (A = 2 AND B = 2)*(G4 = 2) | −0.830000 | 0.356473 | −3.328369 | 0.0206 |
| (A = 2 AND B = 2)*(G4 = 3) | −0.660000 | 0.389794 | −1.693203 | 0.0915 |

| | | | | |
|---|---|---|---|---|
| Pseudo R-squared | 0.152332 | Mean dependent var | | 5.751100 |
| Adjusted R-squared | 0.107560 | S.D. dependent var | | 0.945077 |
| S.E. of regression | 0.849489 | Objective | | 94.07000 |
| Quantile dependent var | 5.670000 | Restr. objective | | 110.9750 |
| Sparsity | 1.935392 | Quasi-LR statistic | | 69.87732 |
| Prob(Quasi-LR stat) | 0.000000 | | | |

**Figure 3.10** The results of the $2 \times 2 \times 4$ Factorial ANOVA QR(Median) in (3.4b).

2) As an illustration, the three-way interaction effect $A*B*G4$ has a significant effect, based on the Wald test, as presented in Figure 3.11. In other words, the factor $G4$, as an ordinal variable of $X1 =$ motivation, has a significant effect on $Y1$, conditional for the interaction $A*B$. So, we can say motivation has a significant effect on $Y1$, conditional for the interaction $A*B$, or the effect of the two-way interaction $A*B$ on $Y1$ significantly depends on the ordinal variable of motivation ($G4$).

**Table 3.6** The model parameters of the QR in Figure 3.10

| | A = 1 | | | A = 2 | | |
|---|---|---|---|---|---|---|
| | B = 1 | B = 2 | B(1−2) | B = 1 | B = 2 | B(1−2) |
| G4 = 1 | C(1) + C(2) + C(5) | C(1) + C(3) + C(6) | | C(1) + C(4) + C(7) | C(1) + c(8) | |
| G4 = 2 | C(1) + C(2) + C(9) | C(1) + C(3) + C(10) | | C(1) + C(4) C(11) | C(1) + C(12) | |
| G4 = 3 | C(1) + C(2) + C(13) | C(1) + C(3) + C(14) | | C(1) + C(4) + C(15) | C(1) + C(16) | |
| G4 = 4 | C(1) + **C(2)** | C(1) + **C(3)** | | C(1) + **C(4)** | C(1) | |
| G4(1−4) | **C(5)** | **C(6)** | **C(5)−C(6)** | **C(7)** | **C(8)** | **C(7)−C(8)** |
| G4(2−4) | **C(9)** | **C(10)** | **C(9)−C(10)** | **C(11)** | **C(12)** | **C(11)−C(12)** |
| G4(3−4) | **C(13)** | **C(14)** | **C(13)−C(14)** | **C(15)** | **C(16)** | **C(15)−C(16)** |

Wald test:
Equation: EQ12

| Test statistic | Value | df | Probability |
|---|---|---|---|
| F-statistic | 5.406478 | (3,284) | 0.0012 |
| Chi-square | 16.21943 | 3 | 0.0010 |

Null Hypothesis: C(5) − C(6) − C(7) + C(8) = 0,
C(9) − C(10) − C(11) + C(12) = 0, C(13) − C(14) − C(15) + C(16) = 0
Null hypothesis summary:

| Normalized restriction (= 0) | Value | Std. err. |
|---|---|---|
| C(5)−C(6)−C(7)+C(8) | −1.510000 | 0.671487 |
| C(9)−C(10)−C(11)+C(12) | 0.670000 | 0.539881 |
| C(13)−C(14)−C(15)+C(16) | 1.830000 | 0.589940 |

Restrictions are linear in coefficients.

**Figure 3.11** The Wald test of the effect of A*B*G4 on Y1.

## 3.4 *I × J × K* Factorial QRs Based on susenas.wf1

As the extensions of the two-way or $I \times J$ Factorial QRs of the *children weekly working hours* (*CWWH*), presented in the previous chapter, the following examples present the statistical results of selected $I \times J \times K$ factorial QRs of *CWWH*, with a specific main objective and testing hypothesis. For the testing hypotheses, such as the conditional two-way interaction effects and three-way interaction effects, refer to the alternative types of factor effects presented in the previous sections.

### 3.4.1 Alternative ESs of *CWWH* on *F1, F2,* and *F3*

As the ESs of the *CWWH* on three selected factors, I present three alternative ESs with the following specific objectives:

$$CWWH \ @Expand(F1, F2, F3) \tag{3.5a}$$

$$CWWH \ C \ @Expand(F1, F2, F3, @Dropfirst) \tag{3.5b}$$

$$CWWH \ C \ @Expand(F1, F2, @Dropfirst) \ @Expand(F1, F2)^* \ @Expand(F3, @Droplast) \tag{3.5c}$$

The main objective of ES (3.5a) is to present the quantile of *CWWH* for each cell or group generated by the three factors. No testing hypothesis would be conducted using ES (3.5a). If the testing hypotheses were done, I recommend doing the analysis using ES (3.5b) as the simplest ES, or ES (3.5c) as the more complex ES. The *t*-statistic presented in the output can be used directly to test the quantiles difference of *CWWH* between certain pairs of cells, and the *Quasi-likelihood ratio* (*QLR*) statistic can be used to test the hypothesis of the quantile of *CWWH* between all cells generated by the three factors. In other words, use it to test the joint effects of the three factors on the quantile of *CWWH*.

Finally, the main objective of ES (3.5c) is to test the hypotheses of the quantile differences of *CWWH* between pairs of *F3*'s levels, with its last level used as the reference cell or group, for each cell or group generated by the two factors *F1* and *F2*.

#### 3.4.1.1 Applications of the Simplest ES in (3.5a)

**Example 3.7** *Application of a Specific QR*
Figure 3.12 presents a part of the *Quantile Process Estimates* (QPEs) of a QR of *CWWH* on *AGE, UR,* and *SEX*, for tau = 0.1 to 0.9, using the following QR:

$$CWWH \ @Expand(AGE, UR, SEX) \tag{3.6}$$

Based on the full output, the following findings and notes are presented:

1) The full output presents $360 = 9 \times 10 \times 2 \times 2$ quantiles. So, it is too many to be presented in a book or paper.
2) Even though we are doing the analysis of a three-way ANOVA-QR, we could present conditional statistical results, such as presented previously in 3.6, specific for $AGE = 13$ and $UR = 1$. So, we can compare the quantiles differences between the *BOY* and *GIRL*, but we can't test their differences.
3) Note that by running the QR (3.6), based on the sample ($AGE = 13$ and $UR = 1$), we will obtain the output of the QR by *SEX* only. Then, by selecting *View/Quantile Process/ Process Coefficient* and clicking *OK*, we obtain the output presented in Figure 3.12. Do this as an exercise. The following example presents other conditional statistical results, from using the QR (3.6).

Quantile process estimates
Equation: EQ_(3.6)
Specification: CWWH @EXPAND(AGE,UR,SEX)
Estimated equation quantile tau = 0.5
Number of process quantiles: 10
Display all coefficients

|  | Quantile | Coefficient | Std. error | t-Statistic | Prob. |
|---|---|---|---|---|---|
| AGE = 13,UR = 1,SEX = 1 | 0.100 | 5.000000 | 0.804271 | 6.216814 | 0.0000 |
|  | 0.200 | 7.000000 | 0.789862 | 8.862308 | 0.0000 |
|  | 0.300 | 10.00000 | 2.292017 | 4.362970 | 0.0000 |
|  | 0.400 | 12.00000 | 1.066131 | 11.25565 | 0.0000 |
|  | 0.500 | 14.00000 | 0.497193 | 28.15807 | 0.0000 |
|  | 0.600 | 15.00000 | 0.540914 | 27.73083 | 0.0000 |
|  | 0.700 | 18.00000 | 6.584980 | 2.733494 | 0.0063 |
|  | 0.800 | 24.00000 | 5.332384 | 4.500801 | 0.0000 |
|  | 0.900 | 35.00000 | 2.225422 | 15.72735 | 0.0000 |
| AGE = 13,UR = 1,SEX = 2 | 0.100 | 6.000000 | 0.999761 | 6.001437 | 0.0000 |
|  | 0.200 | 9.000000 | 1.119965 | 8.035966 | 0.0000 |
|  | 0.300 | 14.00000 | 0.377200 | 37.11559 | 0.0000 |
|  | 0.400 | 14.00000 | 0.416446 | 33.61781 | 0.0000 |
|  | 0.500 | 14.00000 | 0.430941 | 32.48704 | 0.0000 |
|  | 0.600 | 16.00000 | 1.193061 | 13.41088 | 0.0000 |
|  | 0.700 | 20.00000 | 0.899459 | 22.23558 | 0.0000 |
|  | 0.800 | 21.00000 | 0.620890 | 33.82240 | 0.0000 |
|  | 0.900 | 30.00000 | 8.600723 | 3.488079 | 0.0005 |

**Figure 3.12** A part of the QPE of the QR (3.6) for tau = 0.1 to tau = 0.9.

**Example 3.8** *The Application of the QR (3.6), Conditional for AGE = 13 and SEX = 2*
Figure 3.13 presents the output of QR (3.6), conditional for $AGE = 13$ and $SEX = 2$ (Girl). Based on this output, findings and notes are as follows:

1) The inequality $CWWH < 98$ is used to delete the 12 "NA" observations of $CWWH$.
2) The three variables, $AGE, UR,$ and $SEX$, in the output indicate the unconditional results of 40 cells or groups generated by the three factors.
3) The output shows a girl in an urban area has greater median weekly working hours than in a rural area. And at the 5% level, the median in an urban area is significantly greater than in a rural area, based on the Wald test $t$-statistic of $t_0 = 2.212812$ with $df = 517$ and a $p$-value $= 0.0274/2 = 0.0137$.
4) Figure 3.14 presents the output of the QR's QPE, for tau = 0.1 to 0.9. However, we can't test their differences.

Dependent variable: CWWH
Method: quantile regression (median)
Date: 12/22/18   Time: 09:56
**Sample: 1 177852 IF CWWH > 0 AND CWWH < 98 AND AGE = 13 AND SEX = 2**
Included observations: 509
Huber sandwich standard errors and covariance
Sparsity method: Kernel (Epanechnikov) using residuals
Bandwidth method: Hall-Sheather, bw = 0.12168
Estimation successful but solution may not be unique

| Variable | Coefficient | Std. error | t-Statistic | Prob. |
|---|---|---|---|---|
| AGE = 13,UR = 1,SEX = 2 | 14.00000 | 0.751431 | 18.63113 | 0.0000 |
| AGE = 13,UR = 2,SEX = 2 | 12.00000 | 0.502251 | 23.89242 | 0.0000 |

| | | | | |
|---|---|---|---|---|
| Pseudo R-squared | 0.002874 | Mean dependent var | | 15.19843 |
| Adjusted R-squared | 0.000907 | S.D. dependent var | | 9.975063 |
| S.E. of regression | 10.28375 | Objective | | 1735.000 |
| Quantile dependent var | 14.00000 | Restr. objective | | 1740.000 |
| Sparsity | 18.88272 | | | |

**Figure 3.13** The output of the QR(Median) (3.6), conditional for $AGE = 13$ and $SEX = 2$.

**Example 3.9**   *The Application of the QR in (3.6), Conditional for AGE = 13*
As an additional illustration, Figure 3.15 presents the summary of the outputs of the QR(0.9) (3.6), conditional for $AGE = 13$ (junior high school children). Based on these outputs, the following findings and notes are presented:

1) Specific for the children $AGE = 13$, 10% (=100 − 90) of the boys in an urban area are working over 35 hours per week, and the girls are working over 30 hours per week. However, their quantiles (0.9) have an insignificant difference, based on the Wald test $t$-statistic of $t_0 = 1.075\,241$ with $df = 1218$ and a $p$-value $= 0.2825$.
2) However, in the rural area, the boys have a significant greater quantile(0.9) than the girls, based on the Wald test $t$-statistic of $t_0 = 2.630\,740$ with $df = 1218$ and a $p$-value $= 0.0056/2 = 0.0028$.

### 3.4.1.2   Applications of the ES in (3.5b)
As mentioned earlier, the main objectives of this model are to test the joint effects of the three factors, and to present the slope equality test, which is in fact to test the quantiles differences between the cells generated by the three factors. See the following examples.

**Example 3.10**   *A 2×2×2 Factorial QR of CWWH*
As the simplest three-way ANOVA-QR with an intercept, Figure 3.16 presents the output of a 2×2×2 Factorial QR of *CWWH*, with the following ES:

$$CWWH\ C\ @Expand\ (UR, DST, SEX, @Dropfirst) \qquad (3.7)$$

Quantile process estimates
Equation: UNTITLED
Specification: CWWH @EXPAND(AGE,UR,SEX)
Estimated equation quantile tau = 0.5
Number of process quantiles: 10
Display all coefficients

| | Quantile | Coefficient | Std. error | t-Statistic | Prob. |
|---|---|---|---|---|---|
| AGE = 13,UR = 1,SEX = 2 | 0.100 | 6.000000 | 1.036340 | 5.789608 | 0.0000 |
| | 0.200 | 9.000000 | 1.191823 | 7.551456 | 0.0000 |
| | 0.300 | 14.00000 | 0.652168 | 21.46684 | 0.0000 |
| | 0.400 | 14.00000 | 0.724769 | 19.31651 | 0.0000 |
| | 0.500 | 14.00000 | 0.751431 | 18.63113 | 0.0000 |
| | 0.600 | 16.00000 | 0.797807 | 20.05497 | 0.0000 |
| | 0.700 | 20.00000 | 1.248681 | 16.01690 | 0.0000 |
| | 0.800 | 21.00000 | 1.158102 | 18.13312 | 0.0000 |
| | 0.900 | 30.00000 | 3.378798 | 8.878898 | 0.0000 |
| AGE = 13,UR = 2,SEX = 2 | 0.100 | 4.000000 | 0.422026 | 9.478096 | 0.0000 |
| | 0.200 | 6.000000 | 0.437490 | 13.71461 | 0.0000 |
| | 0.300 | 9.000000 | 0.505192 | 17.81502 | 0.0000 |
| | 0.400 | 12.00000 | 0.488858 | 24.54699 | 0.0000 |
| | 0.500 | 12.00000 | 0.502251 | 23.89242 | 0.0000 |
| | 0.600 | 14.00000 | 0.503632 | 27.79809 | 0.0000 |
| | 0.700 | 18.00000 | 0.688242 | 26.15359 | 0.0000 |
| | 0.800 | 21.00000 | 0.774877 | 27.10106 | 0.0000 |
| | 0.900 | 25.00000 | 1.255223 | 19.91678 | 0.0000 |

**Figure 3.14** The output of the Quantile Process Estimates based on the QR in Figure 3.13.

Based on this output, the following findings and notes are presented:

1) We can conclude that the medians of *CWWH* have significant differences between the cells generated by *UR, DIST* and *SEX,* based on the *QLR* statistic of $QLR_0 = 766.9478$ with $df = 7$ and *p*-value = 0.000 000.
2) Compared with the median of *CWWH* in the first cell ($UR = 1$, $DST = 0$, $SEX = 1$), each of the other cells has a significantly smaller median, at the 1% level, except the cell ($UR = 1$, $DST = 1$, $SEX = 1$).
3) In fact, at the 10% level, the median in the cell ($UR = 1$, $DST = 1$, $SEX = 1$) is significantly smaller than the median in the first cell, based on the *t*-statistic of $t_0 = -1.417\,227$ with $df = (1432 - 8) = 1428$ and a *p*-value = 0.1564/2 = 0.0782.
4) Based on this model, it is important to note the difference between the following two hypotheses:

$\qquad$ (i) $H_0$: $C(2) = C(3) = C(4) = C(5) = C(6) = C(7) = C(8) = 0$ vs $H_1$: *Otherwise*

Dependent variable: CWWH
Method: quantile regression (tau = 0.9)
Date: 12/22/18   Time: 10:23
**Sample: 1 177852 IF CWWH > 0 AND CWWH < 98 AND AGE = 13**
Included observations: 1222
Huber sandwich standard errors and covariance
Sparsity method: Kernel (Epanechnikov) using residuals
Bandwidth method: Hall-Sheather, bw = 0.032363
Estimation successfully identifies unique optimal solution

| Variable | Coefficient | Std. error | t-Statistic | Prob. |
|----------|-------------|------------|-------------|-------|
| AGE = 13,UR = 1,SEX = 1 | 35.00000 | 3.463420 | 10.10562 | 0.0000 |
| AGE = 13,UR = 1,SEX = 2 | 30.00000 | 3.102954 | 9.668205 | 0.0000 |
| AGE = 13,UR = 2,SEX = 1 | 30.00000 | 0.950622 | 31.55829 | 0.0000 |
| AGE = 13,UR = 2,SEX = 2 | 25.00000 | 1.645789 | 15.19029 | 0.0000 |

| | | | | |
|----------|-------------|------------|-------------|-------|
| Pseudo R-squared | 0.005751 | Mean dependent var | | 15.65466 |
| Adjusted R-squared | 0.003302 | S.D. dependent var | | 10.53267 |
| S.E. of regression | 17.08469 | Objective | | 2870.000 |
| Quantile dependent var | 28.00000 | Restr. objective | | 2886.600 |
| Sparsity | 91.94142 | | | |

**Figure 3.15**   The output of the QR(0.9) (3.8), conditional for *AGE* = 13.

(ii) $H_0$: $C(2) = C(3) = C(4) = C(5) = C(6) = C(7) = C(8)$ *vs* $H_1$: *Otherwise*

The hypothesis (i) is for testing quantiles differences between the eight cells generated by the three factors, and the hypothesis (ii) is for testing the quantiles differences between the last seven cells.

5) Figure 3.17 presents the output of a quantile slope equality test (QSET), based on the QR(Median), with "0.1 0.9" as the user-specified test quantiles. Note that the test presents two pairs of quantiles: (0.1, 0.5) and (0.5, 0.9). To have the test for the pair of quantiles (0.1, 0.9), the test should be conducted based on QR(0.1). Do this as an exercise.

**Example 3.11**   *The Application of a 10 × 2 × 2 Factorial QR of CWWH*
Figure 3.18 presents a part of the output of a 10 × 2 × 2 Factorial QR of *CWWH* by *AGE, UR,* and *SEX,* using the following ES:

$$CWWH\ C@Expand\ (AGE, UR, SEX, @Dropfirst).$$   (3.8)

Based on this output, the findings and notes are as follows:

1) Similar to the output in the previous example, the quantile(0.1) of *CWWH* have significant differences between the 40 cells generated by the three factors, based on the *QLR* statistic of $QLR_0 = 827.9191$ with $df = 39$ and *p*-value = 0.000 000.

Dependent variable: CWWH
Method: quantile regression (median)
**Sample: 1 177852 IF CWWH > 0 AND CWWH < 98**
Included observations: 11432
Bandwidth method: Hall-Sheather, bw = 0.043128
Estimation successful but solution may not be unique

| Variable | Coefficient | Std. error | t-Statistic | Prob. |
|---|---|---|---|---|
| C | 28.00000 | 0.967641 | 28.93636 | 0.0000 |
| UR = 1,DST = 0,SEX = 2 | −6.000000 | 1.124069 | −5.337753 | 0.0000 |
| UR = 1,DST = 1,SEX = 1 | −3.000000 | 2.116810 | −1.417227 | 0.1564 |
| UR = 1,DST = 1,SEX = 2 | −7.000000 | 1.259136 | −5.559366 | 0.0000 |
| UR = 2,DST = 0,SEX = 1 | −13.00000 | 0.995035 | −13.06487 | 0.0000 |
| UR = 2,DST = 0,SEX = 2 | −14.00000 | 0.991613 | −14.11840 | 0.0000 |
| UR = 2,DST = 1,SEX = 1 | −13.00000 | 1.023965 | −12.69574 | 0.0000 |
| UR = 2,DST = 1,SEX = 2 | −13.00000 | 1.073232 | −12.11295 | 0.0000 |

| | | | | |
|---|---|---|---|---|
| Pseudo R-squared | 0.039388 | Mean dependent var | | 22.31009 |
| Adjusted R-squared | 0.038799 | S.D. dependent var | | 16.74588 |
| S.E. of regression | 16.70982 | Objective | | 68336.50 |
| Quantile dependent var | 16.00000 | Restr. objective | | 71138.50 |
| Sparsity | 29.22754 | Quasi-LR statistic | | 766.9478 |
| Prob(Quasi-LR stat) | 0.000000 | | | |

**Figure 3.16** The output of the QR(Median) in (3.7).

2) The $t$-statistic can be used to test the quantile(0.1) of $CWWH$ between pairs of cells, with $(AGE = 10, UR = 1, SEX = 1)$ as the reference group.
3) Based on this QR(0.1), we can conduct the QSET between pairs of tau = 0.1 and another tau. Figure 3.19 presents the QSET between pairs of (0.1, 0.9) for the slopes of the 39 independent variables, which is significant, based on the *Chi-square* statistic of $\chi_0^2 = 2845.937$ with $df = 3$, and $p$-value = 0.0000.
4) But, for the pairs of the quantiles of each slope, some of them have insignificant differences. Two of them are presented in Figure 3.19
5) Based on the output in Figure 3.18, we can test various conditional quantile(0.1) differences using the Wald test, as follows:
   5.1 The quantiles difference between Boy and Girl, conditional for $AGE = 10$, and $UR = 1$, with the null hypothesis $H_0$: $C(3) = C(4)$.
   5.2 The quantile differences between the four cells generated by the factors $UR$ and $SEX$, conditional for $AGE = 19$, with the null hypothesis $H_0$: $C(37) = C(38) = C(39) = C(40)$. And the null hypothesis is rejected, based on the *Chi-square* statistic of $\chi_0^2 = 15.1537$ with $df = 3$, and $p$-value = 0.0017.

Quantile slope equality test
Specification: CWWH C @EXPAND(UR,DST,SEX,@DROPFIRST)
Estimated equation quantile tau = 0.5
User-specified test quantiles: 0.1 0.9
Test statistic compares all coefficients

| Test summary | Chi-Sq. statistic | Chi-Sq. d.f. | Prob. |
|---|---|---|---|
| Wald test | 1054.810 | 14 | 0.0000 |

Restriction detail: b(tau_h) − b(tau_k) = 0

| Quantiles | Variable | Restr. value | Std. error | Prob. |
|---|---|---|---|---|
| 0.1, 0.5 | UR = 1,DST = 0,SEX = 2 | 6.000000 | 1.073827 | 0.0000 |
| | UR = 1,DST = 1,SEX = 1 | 3.000000 | 1.998825 | 0.1334 |
| | UR = 1,DST = 1,SEX = 2 | 7.000000 | 1.240100 | 0.0000 |
| | UR = 2,DST = 0,SEX = 1 | 12.00000 | 0.942849 | 0.0000 |
| | UR = 2,DST = 0,SEX = 2 | 13.00000 | 0.944131 | 0.0000 |
| | UR = 2,DST = 1,SEX = 1 | 12.00000 | 0.971203 | 0.0000 |
| | UR = 2,DST = 1,SEX = 2 | 12.00000 | 1.019252 | 0.0000 |
| 0.5, 0.9 | UR = 1,DST = 0,SEX = 2 | −6.000000 | 1.202622 | 0.0000 |
| | UR = 1,DST = 1,SEX = 1 | −3.000000 | 2.029964 | 0.1394 |
| | UR = 1,DST = 1,SEX = 2 | −7.000000 | 1.600610 | 0.0000 |
| | UR = 2,DST = 0,SEX = 1 | 1.000000 | 1.069128 | 0.3496 |
| | UR = 2,DST = 0,SEX = 2 | 6.000000 | 1.105945 | 0.0000 |
| | UR = 2,DST = 1,SEX = 1 | 1.000000 | 1.147200 | 0.3834 |
| | UR = 2,DST = 1,SEX = 2 | 7.000000 | 1.246958 | 0.0000 |

**Figure 3.17** The output of a QSET based on the QR(Median) (3.7).

5.3 The model parameters could be identified in the estimation equation in the full output of the representations of the QR(0.1) in Figure 3.18.

6) As an additional analysis, Figure 3.20 presents the graphs of the QPE of the parameters $C(10)$, $C(20)$, $C(30)$, and $C(40)$, based on the QR(0.1) in Figure 3.18. These show the *quantile deviations* of four selected cells from the quantiles(0.1) in the first cell, namely ($AGE = 10$, $UR = 1$, $SEX = 1$). Note that they present two pairs of ages differences, conditional for ($UR = 1$, $SEX = 2$) and ($UR = 2$, $SEX = 2$), respectively.

7) As a comparison, Figure 3.21 presents the graphs of the QPE within the same four selected cells, based on the QR(0.1) without the intercept, that is, the following ES:

$$CWWH @Expand (AGE, UR, SEX) \tag{3.9}$$

Dependent variable: CWWH
Method: Quantile regression (tau = 0.1)
**Sample: 1 177852 IF CWWH > 0 AND CWWH < 98**
Included observations: 12574
Bandwidth method: Hall-Sheather, bw = 0.014879
Estimation successful but solution may not be unique

| Variable | Coefficient | Std. error | t-Statistic | Prob. |
|---|---|---|---|---|
| C | 3.000000 | 0.938441 | 3.196792 | 0.0014 |
| AGE = 10,UR = 1,SEX = 2 | 1.000000 | 1.443130 | 0.692938 | 0.4884 |
| AGE = 10,UR = 2,SEX = 1 | 1.000000 | 1.052570 | 0.950056 | 0.3421 |
| AGE = 10,UR = 2,SEX = 2 | 3.000000 | 1.065052 | 2.816763 | 0.0049 |
| ... | | | | |
| .... | | | | |
| AGE = 19,UR = 1,SEX = 1 | 13.00000 | 1.512678 | 8.594030 | 0.0000 |
| AGE = 19,UR = 1,SEX = 2 | 15.00000 | 1.769524 | 8.476857 | 0.0000 |
| AGE = 19,UR = 2,SEX = 1 | 11.00000 | 1.100631 | 9.994271 | 0.0000 |
| AGE = 19,UR = 2,SEX = 2 | 9.000000 | 1.281121 | 7.025096 | 0.0000 |
| Pseudo R-squared | 0.049886 | Mean dependent var | | 22.35271 |
| Adjusted R-squared | 0.046930 | S.D. dependent var | | 16.80109 |
| S.E. of regression | 20.73169 | Objective | | 21673.90 |
| Quantile dependent var | 6.000000 | Restr. objective | | 22811.90 |
| Sparsity | 30.54512 | Quasi-LR statistic | | 827.9191 |
| Prob(Quasi-LR stat) | 0.000000 | | | |

**Figure 3.18** The output of the QER(0.1) in (3.8).

### 3.4.1.3 Applications of the ES in (3.5c)

**Example 3.12** *Application of a 2 × 2 × 2 Factorial QR of CWWH*
As a complex three-way ANOVA-QR in (3.5c), Figure 3.22 presents the output of a $2 \times 2 \times 2$ Factorial QR(Median) of *CWWH* on *UR, DST*, and *SEX*:

$$CWWH\ C@Expand\ (UR, DST, @Dropfirst)\ @Expand(UR, DST)^*(SEX = 1) \qquad (3.10)$$

Based on this output, the following findings and notes are presented:

1) Based on the output in Figure 3.22, we can develop the model parameters, as presented in Table 3.7, with a special location of the variables *UR, DST,* and *SEX*. (These will be applied for the $2 \times 10 \times 2$ Factorial ANOVA-QR, in the following example.) The steps are as follows:

   1.1 The parameter $C(1)$, as the intercept of the model, should be inserted in the eight cells generated by the three factors.

Quantile slope equality test
Equation: EQ09
Specification: CWWH C @EXPAND(AGE,UR,SEX,@DROPFIRST)
Estimated equation quantile tau = 0.1
User-specified test quantiles: 0.1 0.9
Test statistic compares all coefficients

| Test summary | Chi-Sq. statistic | Chi-Sq. d.f. | Prob. |
|---|---|---|---|
| Wald test | 2845.937 | 39 | 0.0000 |

Restriction detail: b(tau_h) − b(tau_k) = 0

| Quantiles | Variable | Restr. value | Std. error | Prob. |
|---|---|---|---|---|
| 0.1, 0.9 | AGE = 10,UR = 1,SEX = 2 | −13.00000 | 4.142740 | 0.0017 |
| | AGE = 11,UR = 1,SEX = 1 | 6.000000 | 3.025743 | 0.0474 |
| | AGE = 11,UR = 1,SEX = 2 | 2.000000 | 2.135122 | 0.3489 |
| | ... | | | |
| | AGE = 18,UR = 2,SEX = 1 | −5.000000 | 2.438284 | 0.0403 |
| | AGE = 18,UR = 2,SEX = 2 | −4.000000 | 2.521877 | 0.1127 |
| | AGE = 19,UR = 1,SEX = 1 | −22.00000 | 2.206380 | 0.0000 |
| | AGE = 19,UR = 1,SEX = 2 | −27.00000 | 2.297187 | 0.0000 |

**Figure 3.19** **A** part of the output of a QSET based on the QR(0.1) in (3.8).

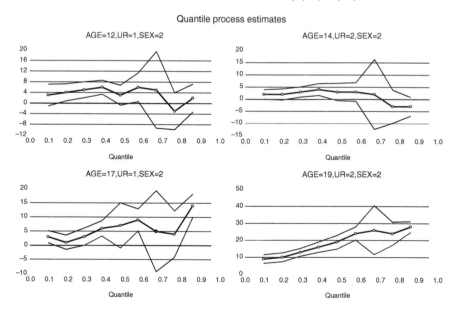

**Figure 3.20** The graphs of the quantiles deviations of the selected cells from the quantile(0.1) in cell $(AGE = 10, UR = 1, SEX = 1)$.

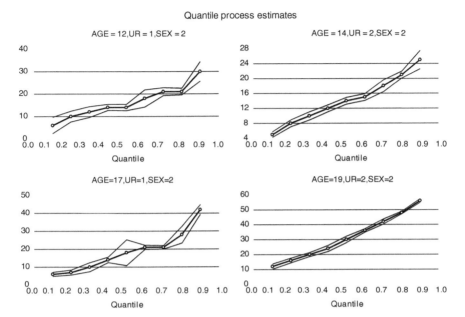

**Figure 3.21** The graphs of the QPE of QR(0.1) in (3.9) for the same four selected cells.

1.2 The parameter $C(2)$, as the coefficient of $UR = 1$, $DST = 1$, should be added to both values of *SEX*. Similarly, the parameter $C(3)$, as the coefficient of $UR = 2$, $DST = 0$, and $C(4)$, as the coefficient of $UR = 2$, $DST = 1$, should be added to both values of *SEX*.

1.3 Finally, the last four parameters, $C(5)$, $C(6)$, $C(7)$, and $C(8)$, should be added to the four cells generated by $UR$ and $DST$, specific for $SEX = 1$ (Boy).

2) Furthermore, based on this table, two important hypotheses can easily be defined for all QR(tau), in addition to the quantile differences between all cells generated by the three factors, which can be tested using the *QLR* statistic in the output. The two hypotheses are as follows:

2.1 First, the hypothesis on the quantile(0.1) difference of *CWWH* between boys and girls, for each cell or group generated by the factors $UR$ and $DST$, can be defined based on each of the parameters, $C(5)$, $C(6)$, $C(7)$, and $C(8)$, as either a one- or two-sided hypothesis. Then, it can be tested easily using the $t$-statistic in Figure 3.22.

2.2 And the hypothesis on the quantile(0.1) differences of *CWWH,* between the cells or groups generated by the factors $UR$ and $DST$, depend on *SEX*. In other words, the two-way interaction $(UR,DST)*SEX$ has an effect on *CWWH*, with the following statistical hypothesis, which can be tested using the Wald test or the *redundant variables test* (RVT).

$$H_0: C(5) = C(6) = C(7) = C(8) = 0 \text{ vs. } H_1: \text{Otherwise} \tag{3.11}$$

The outputs presented in Figure 3.23 show that the null hypothesis is rejected.

Dependent variable: CWWH
Method: quantile regression (median)
Date: 12/22/18   Time: 22:05
**Sample: 1 177852 IF CWWH > 0 AND CWWH < 98**
Included observations: 11432
Huber sandwich standard errors and covariance
Sparsity method: Kernel (Epanechnikov) using residuals
Bandwidth method: Hall-Sheather, bw = 0.043128
Estimation successful but solution may not be unique

| Variable | Coefficient | Std. error | t-Statistic | Prob. |
|---|---|---|---|---|
| C | 22.00000 | 0.572015 | 38.46050 | 0.0000 |
| UR = 1,DST = 1 | −1.000000 | 0.988078 | −1.012066 | 0.3115 |
| UR = 2,DST = 0 | −8.000000 | 0.611695 | −13.07842 | 0.0000 |
| UR = 2,DST = 1 | −7.000000 | 0.736682 | −9.502069 | 0.0000 |
| (UR = 1,DST = 0)*(SEX = 1) | 6.000000 | 1.124069 | 5.337753 | 0.0000 |
| (UR = 1,DST = 1)*(SEX = 1) | 4.000000 | 2.047841 | 1.953277 | 0.0508 |
| (UR = 2,DST = 0)*(SEX = 1) | 1.000000 | 0.317387 | 3.150727 | 0.0016 |
| (UR = 2,DST = 1)*(SEX = 1) | 0.000000 | 0.572429 | 0.000000 | 1.0000 |

| | | | | |
|---|---|---|---|---|
| Pseudo R-squared | 0.039388 | Mean dependent var | 22.31009 |
| Adjusted R-squared | 0.038799 | S.D. dependent var | 16.74588 |
| S.E. of regression | 16.70982 | Objective | 68336.50 |
| Quantile dependent var | 16.00000 | Restr. objective | 71138.50 |
| Sparsity | 29.22754 | Quasi-LR statistic | 766.9478 |
| Prob(Quasi-LR stat) | 0.000000 | | |

**Figure 3.22**   The output of the QR(Median) in (3.10).

**Table 3.7**   The parameters of the model (3.10) by *UR*, *DST*, and *SEX*.

| | Urban(UR = 1) | | | Rural(UR = 2) | | |
|---|---|---|---|---|---|---|
| | Boy(SEX = 1) | Girl(SEX = 2) | Boy–Girl | Boy(SEX = 1) | Girl(SEX = 2) | Boy–Girl |
| DST = 0 | C(1) + C(5) | C(1) | C(5) | C(1) + C(3) + C(7) | C(1) + C(3) | C(7) |
| DST = 1 | C(1) + C(2) + C(6) | C(1) + C(2) | C(6) | C(1) + C(4) + C(8) | C(1) + C(4) | C(8) |

### Example 3.13   *Application of a 2 × 10 × 2 Factorial QR of CWWH*

As an extension of the model parameters presented in Table 3.7, Figure 3.24 presents a part of three outputs of the following QR(Median). Table 3.8 presents the parameters of the model.

$$CWWH \ C @ Expand \ (UR, AGE, @Dropfirst) \ @Expand(UR, AGE)^*(SEX = 1) \qquad (3.12)$$

Wald test:
Equation: EQ12

| Test statistic | Value | df | Probability |
|---|---|---|---|
| F-statistic | 10.55849 | (4, 11424) | 0.0000 |
| Chi-square | 42.23397 | 4 | 0.0000 |

Null hypothesis: $C(5) = C(6) = C(7) = C(8) = 0$

Redundant variables test
Null hypothesis: @EXPAND(UR,DST)*(SEX = 1)
Equation: EQ12
Specification: CWWH C @EXPAND(UR,DST,@DROPFIRST)
    @EXPAND(UR,DST)*(SEX = 1)

| | Value | df | Probability |
|---|---|---|---|
| QLR L-statistic | 36.81459 | 4 | 0.0000 |
| QLR Lambda-statistic | 36.77841 | 4 | 0.0000 |

**Figure 3.23** Alternative outputs for testing the hypothesis (3.11).

**Table 3.8** The parameters of the model in (3.12).

| AGE | Urban | | | Rural | | |
|---|---|---|---|---|---|---|
| | Boy | Girl | Boy–Girl | Boy | Girl | Boy–Girl |
| 10 | C(1) + C(21) | C(1) | **C(21)** | C(1) + C(11) + C(31) | C(1) + C(11) | **C(31)** |
| 11 | C(1) + C(2) + C(22) | C(1) + C(2) | **C(22)** | C(1) + C(12) + C(32) | C(1) + C(12) | **C(32)** |
| 12 | C(1) + C(3) + C(23) | C(1) + C(3) | **C(23)** | C(1) + C(13) + C(33) | C(1) + C(13) | **C(33)** |
| 13 | C(1) + C(4) + C(24) | C(1) + C(4) | **C(24)** | C(1) + C(14) + C(34) | C(1) + C(14) | **C(34)** |
| 14 | C(1) + C(5) + C(25) | C(1) + C(5) | **C(25)** | C(1) + C(15) + C(35) | C(1) + C(15) | **C(35)** |
| 15 | C(1) + C(6) + C(26) | C(1) + C(6) | **C(26)** | C(1) + C(16) + C(36) | C(1) + C(16) | **C(36)** |
| 16 | C(1) + C(7) + C(27) | C(1) + C(7) | **C(27)** | C(1) + C(17) + C(37) | C(1) + C(17) | **C(37)** |
| 17 | C(1) + C(8) + C(28) | C(1) + C(8) | **C(28)** | C(1) + C(18) + C(38) | C(1) + C(18) | **C(38)** |
| 18 | C(1) + C(9) + C(29) | C(1) + C(9) | **C(29)** | C(1) + C(19) + C(39) | C(1) + C(19) | **C(39)** |
| 19 | C(1) + C(10) + C(30) | C(1) + C(10) | **C(30)** | C(1) + C(20) + C(40) | C(1) + C(20) | **C(40)** |

Based on the output of any QR(tau), we can test the following two hypotheses:

1) The hypothesis of the quantile differences of *CWWH* between all cells generated by the three factors can be tested using the QLR statistic in the output. In general, they have significant differences. Based on the output in Figure 3.24, we can conclude that the median of *CWWH* have significant differences between the 40 cells generated by the three factors.

Dependent variable: CWWH
Method: quantile regression (median)
Date: 12/22/18   Time: 22:07
**Sample: 1 177852 IF CWWH > 0 AND CWWH < 98**
Included observations: 12574
Huber sandwich standard errors and covariance
Sparsity method: Kernel (Epanechnikov) using residuals
Bandwidth method: Hall-Sheather, bw = 0.041781
Estimation successful but solution may not be unique

| Variable | Coefficient | Std. error | t-Statistic | Prob. |
|---|---|---|---|---|
| C | 12.00000 | 3.057996 | 3.924139 | 0.0001 |
| UR = 1,AGE = 11 | 2.000000 | 3.151673 | 0.634584 | 0.5257 |
| UR = 1,AGE = 12 | 2.000000 | 3.138791 | 0.637188 | 0.5240 |
| ***... | | | | |
| UR = 2,AGE = 18 | 2.000000 | 3.152277 | 0.634462 | 0.5258 |
| UR = 2,AGE = 19 | 18.00000 | 3.259950 | 5.521556 | 0.0000 |
| (UR = 1,AGE = 10)*(SEX = 1) | −1.000000 | 3.498164 | −0.285864 | 0.7750 |
| (UR = 1,AGE = 11)*(SEX = 1) | −4.31E−16 | 1.073983 | −4.01E−16 | 1.0000 |
| ...*** | | | | |
| (UR = 2,AGE = 18)*(SEX = 1) | 0.000000 | 0.996863 | 0.000000 | 1.0000 |
| (UR = 2,AGE = 19)*(SEX = 1) | 6.000000 | 1.225358 | 4.896527 | 0.0000 |

| | | | |
|---|---|---|---|
| Pseudo R-squared | 0.237602 | Mean dependent var | 22.35271 |
| Adjusted R-squared | 0.235230 | S.D. dependent var | 16.80109 |
| S.E. of regression | 13.89617 | Objective | 59996.50 |
| Quantile dependent var | 16.00000 | Restr. objective | 78694.50 |
| Sparsity | 18.40059 | Quasi-LR statistic | 8129.306 |
| Prob(Quasi-LR stat) | 0.000000 | | |

**Figure 3.24**   A part of the output of the QR(Median) (3.12).

2) One- or two-sided hypotheses on the median differences between boys and girls, for each cell generated by UR and AGE, can be easily tested using the $t$-statistic of the parameters $C(21)$ to $C(40)$ in Figure 3.24.

3) The $t$-statistic of the parameters $C(2)$ to $C(10)$ can be used to test the quantiles difference between the pairs of the AGE levels, conditional for the girls in an urban area. For instance, based on the $t$-statistic of $t_0 = 0.634\,584$ with $p$-value $= 0.5257$ for the parameter $C(2)$, we can conclude the median of CWWH has an insignificant difference between $AGE = 10$ and $AGE = 11$, conditional for girls in an urban area.

4) And the hypothesis on the median differences of CWWH, between the cells or groups generated by the factors UR and AGE, depends on SEX. In other words, the two-way interaction, (UR,AGE)*SEX, has a significant effect on CWWH, with the following

Redundant variables test

Null hypothesis: @EXPAND(UR,AGE)*(SEX = 1)

Equation: EQ13

Specification: CWWH C @EXPAND(UR,AGE,@DROPFIRST)
@EXPAND(UR,AGE)*(SEX = 1)

|  | Value | df | Probability |
|---|---|---|---|
| QLR L-statistic | 40.65088 | 20 | 0.0041 |
| QLR Lambda-statistic | 40.61923 | 20 | 0.0042 |

**Figure 3.25** The RVT's output for testing the hypothesis (3.13).

statistical hypothesis, which can be tested using the RVT. The output presented in Figure 3.25 shows the null hypothesis is rejected, at the 1% level.

$$H_0 : @Expand(UR, AGE)^* (SEX = 1) \text{ vs.} H_1 : Otherwise \tag{3.13}$$

**Example 3.14** *An Alternative ES of the 2 × 10 × 2 Factorial QR in (3.12)*
The following ES is introduced with the main objective that the quantile differences of *CWWH* between the *AGE* levels depend on both factors, *UR* and *SEX,* or the quantile differences between the four cells generated by *UR* and *SEX* depend on *AGE*.

$$CWWH C@Expand (AGE, @Dropfirst) @Expand(UR, SEX, @Droplast)^* @Expand (AGE) \tag{3.14}$$

In this case, I present only the output of the RVT of the hypothesis, based on a QR(0.25), as presented in Figure 3.26, which also presents the null hypothesis. Note the ordering of the interaction @Expand(UR,SEX,@Droplast)*@Expand(AGE) is made to ensure the levels of *AGE* are ordered from 10 to 19 in the output of the QR. As an exercise, develop the table of the model parameters using the same format as in Table 3.8.

Redundant variables test

Null hypothesis:
@EXPAND(UR,SEX,@DROPLAST)*@EXPAND(AGE)

Equation: UNTITLED

Specification: CWWH C @EXPAND(AGE,@DROPFIRST)
@EXPAND(UR,SEX,@DROPLAST)*@EXPAND(AGE) QUANT

|  | Value | df | Probability |
|---|---|---|---|
| QLR L-statistic | 469.2504 | 30 | 0.0000 |
| QLR Lambda-statistic | 463.1481 | 30 | 0.0000 |

**Figure 3.26** The output of the RVT based on the QR(0.25) (3.14).

---

Redundant variables test

Null hypothesis: @EXPAND(SEX,AGE)*(UR = 1)

Equation: EQ16

Specification: CWWH C @EXPAND(SEX,AGE,@DROPFIRST)
  @EXPAND(SEX,AGE)*(UR = 1)

|  | Value | df | Probability |
|---|---|---|---|
| QLR L-statistic | 512.6126 | 20 | 0.0000 |
| QLR Lambda-statistic | 504.9232 | 20 | 0.0000 |

---

**Figure 3.27**  The output of the RVT based on the QR(0.75) (3.15).

Another alternative ES in (3.15) is applied for testing the hypothesis that the quantile difference of *CWWH* between the urban and rural areas depend on *SEX* and *AGE*:

$$CWWH\ C\ @Expand\ (SEX, AGE, @Dropfirst)\ @Expand(SEX, AGE)^*\ (UR = 1)\quad (3.15)$$

with the output presented in Figure 3.27. Based on this output, we can conclude that the quantile(0.75) of *CWWH* between the urban and rural areas is significantly dependent on *AGE* and *SEX*.

## 3.5  Applications of the *N*-Way ANOVA-QRs

### 3.5.1  Four-Way ANOVA-QRs

For the analysis of a four-way ANOVA-QR of *Y* on the four factors *F1, F2, F3*, and *F4*, I recommend applying one of the following alternative ESs, where each QR has a specific main objective:

$$Y @Expand\ (F1, F2, F3, F4)\quad (3.16a)$$

$$Y\ C\ @Expand\ (F1, F2, F3, F4, @Dropfirst)\quad (3.16b)$$

$$Y\ C\ @Expand\ (F1, F2, F3, @Dropfirst)$$
$$@Expand(F4, @Droplast)^*\ @Expand\ (F1, F2, F3)\quad (3.16c)$$

$$Y\ C\ @Expand\ (F1, F2, @Dropfirst)\ @Expand(F3, F4, @Droplast)^*\ @Expand\ (F1, F2)$$
$$(3.16d)$$

ES (3.16a) should be applied if and only if we want to present the QPE of *Y* for all cells generated by the four factors, or a subset of the cells. Otherwise, we should apply either of the other ESs.

ES (3.16b) has to be applied if and only if we want to present the QSET for all cells generated by the four factors, or a subset of the cells, as presented in Figure 3.28. These, in fact, are the quantiles deviations.

ES (3.16c) has two specific objectives The first is to test the quantile differences of $Y$ between the two levels of $F4$, with the last level as the reference cell, for each cell generated by the three factors $F1$, $F2$, and $F3$. The second is to test whether the quantile differences between the levels of $F4$ are significantly dependent on the other three factors, $F1$, $F2$, and $F3$, or the quantiles differences between all cells generated by the factors $F1$, $F2$, and $F3$, are significantly dependent on the factor $F4$.

Similar to ES (3.16c), ES (3.16d) also has two specific objectives The first is to test the quantile differences of $Y$ between the two cells generated by $F3$ and $F4$, with the last cell as the reference group. The second is to test whether the quantile differences between all cells generated by $F3$ and $F4$ are significantly dependent on the factors $F1$ and $F2$.

**Example 3.15   *The Application of QR (3.16c)***
As an illustration, Figure 3.28 presents a part of the output of the following $2 \times 2 \times 2 \times 10$ Factorial QR(0.5). Note that $@Expand(SEX, @Droplast) = (SEX = 1)$.

$$CWWH \ C \ @Expand \ (UR, DST, AGE, \ @Dropfirst) \ (SEX = 1)^*$$

$$@Expand \ (UR, DST, AGE) \tag{3.17}$$

Dependent variable: CWWH
Method: quantile regression (median)
**Sample: 1 177852 IF CWWH> 0 AND CWWH < 98**
**Included observations: 11432**
Estimation successful but solution may not be unique

| Variable | Coefficient | Std. error | t-Statistic | Prob. |
|---|---|---|---|---|
| C | 12.00000 | 3.411162 | 3.517863 | 0.0004 |
| UR = 1,DST = 0,AGE = 11 | 2.000000 | 3.571401 | 0.560004 | 0.5755 |
| UR = 1,DST = 0,AGE = 12 | 2.000000 | 3.518992 | 0.568345 | 0.5698 |
| ... | | | | |
| (SEX = 1)*(UR = 1 AND DST = 0 AND AGE = 10) | −2.000000 | 4.414000 | −0.453104 | 0.6505 |
| (SEX = 1)*(UR = 1 AND DST = 0 AND AGE = 11) | 0.000000 | 1.365421 | 0.000000 | 1.0000 |
| (SEX = 1)*(UR = 1 AND DST = 0 AND AGE = 12) | 0.000000 | 1.121723 | 0.000000 | 1.0000 |
| ... | | | | |
| (SEX = 1)*(UR = 2 AND DST = 1 AND AGE = 18) | 0.000000 | 1.205994 | 0.000000 | 1.0000 |
| (SEX = 1)*(UR = 2 AND DST = 1 AND AGE = 19) | 6.000000 | 1.815680 | 3.304548 | 0.0010 |

| | | | |
|---|---|---|---|
| Pseudo R-squared | 0.236644 | Mean dependent var | 22.31009 |
| Adjusted R-squared | 0.231332 | S.D. dependent var | 16.74588 |
| S.E. of regression | 13.84398 | Objective | 54304.00 |
| Quantile dependent var | 16.00000 | Restr. objective | 71 138.50 |
| Sparsity | 17.82380 | Quasi-LR statistic | 7555.966 |
| Prob(Quasi-LR stat) | 0.000000 | | |

**Figure 3.28**   A part of the output of the QR(Median) (3.17).

Redundant variables test
Null hypothesis: (SEX = 1)*@EXPAND(UR,DST,AGE)
Equation: EQ17
Specification: CWWH C @EXPAND(UR,DST,AGE,@DROPFIRST)
   (SEX + 1)*@EXPAND(UR,DST,AGE)

|  | Value | df | Probability |
| --- | --- | --- | --- |
| QLR L-statistic | 57.00245 | 40 | 0.0396 |
| QLR Lambda-statistic | 56.93590 | 40 | 0.0401 |

**Figure 3.29** The output an RVT based on the QR(Median) (3.17).

Based on this output, the findings and notes are as follows:

1) Based on the $QRL$ statistic of $QLR_0 = 755.633$ with $Prob. = 0.000\,000$, we can conclude that the median of $CWWH$ have significant differences between the 80 cells generated by the four factors.

2) The $t$-statistic for each of the parameters $C(2)$ of the cell $(UR = 1, DST = 0, AGE = 11)$ up to $C(40)$ of the cell $(UR=2,DST=1,AGE=19)$ can be used the test either one- or two sided hypotheses on the median difference between two cells generated by $UR, DST,$ and $AGE$, specific for the girls, with $(UR = 1, DST = 0, AGE = 10)$ as the referent cell. See the output of the *representations*.

3) And the $t$-statistic for each parameters $C(41)$ of the cell $(SEX = 1)*(UR = 1\ AND\ DST = 0\ AND\ AGE = 10)$, up to $C(80)$ of the cell $(SEX = 2)*(UR = 1\ AND\ DIST = 1\ AND\ AGE = 19)$, can be used to test either one- or two-sided hypotheses on the median difference between boys and girls, for each cell generated by $UR,\ DST$ and $AGE$.

4) Figure 3.29 presents the output of the RVT for the hypothesis on the effect of $SEX$ on $CWWH$(Median) depending on the other three factors, which shows the null hypothesis is rejected at the 5% level. So, we can conclude that the medians of $CWWH$ between boys and girls are significantly dependent on the factors $UR, DST$, and $AGE$, based on the $QRL$ statistic of $QLR_0 = 57.002\,45$ with $df = 40$ and $Prob. = 0.0396$.

**Example 3.16   RVTs Based on the QR(Median) Alternatives**
As an additional illustration, Figures 3.30 and 3.31 present two outputs of the RVTs based on alternative QRs of $CWWH$ on $AGE, UR, DST$, and $SEX$. Note that each output already presents the equation of the QR(Median) specification and its null hypothesis. In this case, we have the following conclusions:

1) Based on the output in Figure 3.30, we can conclude that the median of $CWWH$ between urban and rural areas are significantly dependent on the factors $UR, DST$, and $AGE$, based on the $QRL$ statistic of $QLR_0 = 955.3890$ with $df = 40$ and $Prob. = 0.0000$.

2) Based on the output in Figure 3.31, we can conclude that the median of $CWWH$ between the $AGE$ levels are significantly dependent on the factors $UR, DST$, and $SEX$, based on the $QRL$ statistic of $QLR_0 = 6251.563$ with $df = 72$ and $Prob. = 0.0000$.

Redundant variables test
Null hypothesis:
@EXPAND(DST,SEX,AGE)*@EXPAND(UR,@DROPLAST)
Equation: Alternative-1
Specification: CWWH C @EXPAND(DST,SEX,AGE,@DROPFIRST)
    @EXPAND(DST,SEX,AGE)*@EXPAND(UR,@DROPLAST)

|  | Value | df | Probability |
|---|---|---|---|
| QLR L-statistic | 955.3980 | 40 | 0.0000 |
| QLR Lambda-statistic | 937.1160 | 40 | 0.0000 |

**Figure 3.30** The output of the RVT for the Alternative-1 of the QR(Median).

Redundant variables test
Null hypothesis:
@EXPAND(UR,DST,SEX)*@EXPAND(AGE,@DROPLAST)
Equation: Alternative-2
Specification: CWWH C @EXPAND(UR,DST,SEX,@DROPFIRST)
    @EXPAND(UR,DST,SEX)*@EXPAND(AGE,@DROPLAST)

|  | Value | df | Probability |
|---|---|---|---|
| QLR L-statistic | 6251.563 | 72 | 0.0000 |
| QLR Lambda-statistic | 5560.612 | 72 | 0.0000 |

**Figure 3.31** The output of the RVT for the Alternative-2 of the QR(Median).

3) However, by using the following alternative QR, I obtain the output of the QR(Median), but for its RVT of *@Expand(DST,@Droplast)\*@Expand(UR,SEX,AGE)*, we get an unexpected error message, shown in Figure 3.32.

$$CWWH \; C @ Expand \, (UR, SEX, AGE, @ Dropfirst)$$

$$@ Expand(DST, @ Droplast)^* @ Expand \, (UR, SEX, AGE) \tag{3.18}$$

4) To find out what is the problem, I tried to apply the following simpler QR, with its RVT of *@Expand(UR,SEX)\*@Expand(DST,@Droplast)*. That also produces the same error message, but the Wald test gives the surprising output presented in Figure 3.34, based on the output of the QR(Median) in Figure 3.33.

$$CWWH \; C @ expand \, (UR, SEX, @ Dropfirst)$$

$$@ Expand(UR, SEX)^* @ Expand \, (DST, @ Droplast) \tag{3.19}$$

Finally, I made the list of the variables *CWWH, DST, UR, SEX* and *AGE* and found the factor *DST* has many NAs, as shown in Figure 3.35. I also found that *DST* has only

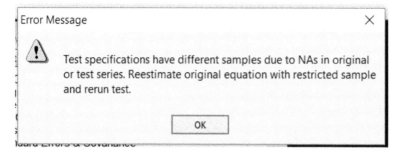

**Figure 3.32** An unexpected error message for the RVT based on QR (3.18).

Dependent variable: CWWH
Method: quantile regression (median)
Date: 12/24/18   Time: 12:27
**Sample: 1 177852 IF CWWH > 0 AND CWWH < 98**
**Included observations: 11 432**
Huber sandwich standard errors and covariance
Sparsity method: Kernel (Epanechnikov) using residuals
Bandwidth method: Hall-Sheather, bw = 0.043128
Estimation successful but solution may not be unique

| Variable | Coefficient | Std. error | t-Statistic | Prob. |
|---|---|---|---|---|
| C | 25.00000 | 1.882699 | 13.27881 | 0.0000 |
| UR = 1,SEX = 2 | −4.000000 | 2.047841 | −1.953277 | 0.0508 |
| UR = 2,SEX = 1 | −10.00000 | 1.912259 | −5.229418 | 0.0000 |
| UR = 2,SEX = 2 | −10.00000 | 1.939086 | −5.157069 | 0.0000 |
| (UR = 1 AND SEX = 1)*(DST = 0) | 3.000000 | 2.116810 | 1.417227 | 0.1564 |
| (UR = 1 AND SEX = 2)*(DST = 0) | 1.000000 | 0.988078 | 1.012066 | 0.3115 |
| (UR = 2 AND SEX = 1)*(DST = 0) | 0.000000 | 0.407360 | 0.000000 | 1.0000 |
| (UR = 2 AND SEX = 2)*(DST = 0) | −1.000000 | 0.512315 | −1.951923 | 0.0510 |

| | | | | |
|---|---|---|---|---|
| Pseudo R-squared | 0.039388 | Mean dependent var | | 22.31009 |
| Adjusted R-squared | 0.038799 | S.D. dependent var | | 16.74588 |
| S.E. of regression | 16.70982 | Objective | | 68336.50 |
| Quantile dependent var | 16.00000 | Restr. objective | | 71138.50 |
| Sparsity | 29.22754 | Quasi-LR statistic | | 766.9478 |
| Prob(Quasi-LR stat) | 0.000000 | | | |

**Figure 3.33** The output of the QR(Median) in (3.19).

Wald test:
Equation: untitled

| Test statistic | Value | df | Probability |
|---|---|---|---|
| F-statistic | 1.710703 | (4, 11424) | 0.1445 |
| Chi-square | 6.842813 | 4 | 0.1444 |

Null hypothesis: $C(5) = C(6) = C(7) = C(8) = 0$

**Figure 3.34** The output Wald test for testing the hypothesis, based on QR (3.19), which can't be obtained using the RVT.

G Group: UNTITLED  Workfile: SUBSAMPLE::Subsample\

View Proc Object Print Name Freeze Default ∨ Sort Edit+/- Smpl+/- Compare+/-

| | CWWH | DST | UR | SEX | AGE |
|---|---|---|---|---|---|
| 73 | 60 | 0 | 2 | 2 | 19 |
| 116 | 12 | 0 | 2 | 1 | 16 |
| 122 | 8 | NA | 2 | 1 | 15 |
| 139 | 56 | NA | 2 | 1 | 19 |
| 154 | 4 | 0 | 2 | 1 | 16 |
| 157 | 38 | 1 | 2 | 1 | 17 |
| 161 | 34 | 0 | 2 | 1 | 17 |
| 182 | 24 | 0 | 2 | 2 | 13 |
| 191 | 48 | 0 | 2 | 1 | 19 |
| 195 | 12 | NA | 2 | 1 | 17 |
| 207 | 36 | 0 | 2 | 1 | 19 |
| 223 | 6 | 1 | 2 | 2 | 19 |
| 224 | 8 | 1 | 2 | 1 | 19 |
| 226 | 10 | 0 | 2 | 1 | 19 |

**Figure 3.35** A part of the scores of the variables used in the models.

11 432 valid scores, compared to 12 574 for the others. Note that the outputs of the QRs in Figures 3.28 and 3.34 are using the valid scores of 11 432 for the factor *DST*. However, for the RVT, all the observations of 12 574 are used. Similarly, based on the QR in (3.18). For this reason, the error message is obtained. See the following example using only the scores of *DST* <> NA.

## Example 3.17  *An Application of QR (3.18) Using DST <> NA*

Figure 3.36 presents a part of the output of the QR(Median) in QR (3.18) using only the valid scores of *DST* <> NA and *CWWH* < 98. Figure 3.37 presents the output of the RVT for *@Expand(UR,SEX,AGE)\*@Expand(DST,@Droplast)*

Dependent variable: CWWH
Method: quantile regression (median)
Date: 12/24/18   Time: 14:22
**Sample: 1 177852 IF CWWH > 0 AND CWWH < 98 AND DST<>NA**
**Included observations: 11432**
Huber sandwich standard errors and covariance
Sparsity method: Kernel (Epanechnikov) using residuals
Bandwidth method: Hall-Sheather, bw = 0.043128
Estimation successful but solution may not be unique

| Variable | Coefficient | Std. error | t-Statistic | Prob. |
|---|---|---|---|---|
| C | 12.00000 | 2.569641 | 4.669913 | 0.0000 |
| UR = 1,SEX = 1,AGE = 11 | 2.000000 | 3.485810 | 0.573755 | 0.5661 |
| UR = 1,SEX = 1,AGE = 12 | 2.000000 | 2.768772 | 0.722342 | 0.4701 |
| UR = 1,SEX = 1,AGE = 13 | 2.000000 | 2.675057 | 0.747648 | 0.4547 |
| *** .... | | | | |
| ...***. | | | | |
| (UR = 2 AND SEX = 2 AND AGE = 16)*(DST = 0) | 4.64E-15 | 1.224663 | 3.79E-15 | 1.0000 |
| (UR = 2 AND SEX = 2 AND AGE = 17)*(DST = 0) | 5.01E-15 | 1.590192 | 3.15E-15 | 1.0000 |
| (UR = 2 AND SEX = 2 AND AGE = 18)*(DST = 0) | −1.000000 | 1.266416 | −0.789630 | 0.4298 |
| (UR = 2 AND SEX = 2 AND AGE = 19)*(DST = 0) | −4.94E-15 | 2.268194 | −2.18E-15 | 1.0000 |

| | | | |
|---|---|---|---|
| Pseudo R-squared | 0.236644 | Mean dependent var | 22.31009 |
| Adjusted R-squared | 0.231332 | S.D. dependent var | 16.74588 |
| S.E. of regression | 13.85191 | Objective | 54304.00 |
| Quantile dependent var | 16.00000 | Restr. objective | 71138.50 |
| Sparsity | 17.89148 | Quasi-LR statistic | 7527.384 |
| Prob(Quasi-LR stat) | 0.000000 | | |

**Figure 3.36** A part of the output of the QR(Median) in (3.18) using the scores of *DST* <> NA.

Redundant variables test
Null hypothesis:
@EXPAND(UR,SEX,AGE)*@EXPAND(DST,@DROPLAST)
Equation: EQ03_18
Specification: CWWH C @EXPAND(UR,SEX,AGE,@DROPFIRST)
     @EXPAND(UR,SEX,AGE)*@EXPAND(DST,@DROPLAST)

| | Value | df | Probability |
|---|---|---|---|
| QLR L-statistic | 34.87695 | 40 | 0.6998 |
| QLR Lambda-statistic | 34.85192 | 40 | 0.7009 |

**Figure 3.37** The output the RVT based on the QR(Median) in Figure 3.36

Dependent variable: CWWH
Method: quantile regression (tau = 0.25)

| Variable | Coefficient | Std. error | t-Statistic | Prob. |
|---|---|---|---|---|
| C | 8.000000 | 1.534513 | 5.213380 | 0.0000 |
| UR = 1,AGE = 11 | 1.32E-15 | 2.722030 | 4.86E-16 | 1.0000 |
| UR = 1,AGE = 12 | 4.000000 | 2.160836 | 1.851136 | 0.0642 |
| UR = 1,AGE = 13 | 4.000000 | 2.738929 | 1.460425 | 0.1442 |
| ... | | | | |
| UR = 2,AGE = 10 | −1.000000 | 1.720575 | −0.581201 | 0.5611 |
| UR = 2,AGE = 11 | −2.000000 | 1.851679 | −1.080101 | 0.2801 |
| UR = 2,AGE = 12 | 1.40E-15 | 1.885931 | 7.43E-16 | 1.0000 |
| ... | | | | |
| (SEX = 1 AND DST = 0)*(UR = 1,AGE = 10) | −1.000000 | 1.770580 | −0.564787 | 0.5722 |
| (SEX = 1 AND DST = 0)*(UR = 1,AGE = 11) | 0.000000 | 2.928526 | 0.000000 | 1.0000 |
| (SEX = 1 AND DST = 0)*(UR = 1,AGE = 12) | −5.000000 | 1.688800 | −2.960683 | 0.0031 |
| (SEX = 2 AND DST = 0)*(UR = 1,AGE = 10) | −2.000000 | 2.037508 | −0.981591 | 0.3263 |
| (SEX = 2 AND DST = 0)*(UR = 1,AGE = 11) | −1.000000 | 2.415131 | −0.414056 | 0.6788 |
| (SEX = 2 AND DST = 0)*(UR = 1,AGE = 12) | −2.000000 | 2.271965 | −0.880295 | 0.3787 |

| | | | |
|---|---|---|---|
| Pseudo R-squared | 0.121763 | Mean dependent var | 22.31009 |
| Adjusted R-squared | 0.115651 | S.D. dependent var | 16.74588 |
| S.E. of regression | 16.41330 | Objective | 39615.75 |
| Quantile dependent var | 10.00000 | Restr. objective | 45108.25 |
| Sparsity | 24.84668 | Quasi-LR statistic | 2357.928 |
| Prob(Quasi-LR stat) | 0.000000 | | |

**Figure 3.38**   A part of the output of the QR(0.25) (3.20).

Redundant variables test
Null hypothesis: @EXPAND(SEX,DST,@DROPLAST)*@EXPAND(UR,AGE)
Equation: EQ_(3.20)
Specification: CWWH C @EXPAND(UR,AGE,@DROPFIRST)
@EXPAND(SEX,DST,@DROPLAST)*@EXPAND(UR,AGE) QUANT
The groups (SEX = 1)

| | Value | df | Probability |
|---|---|---|---|
| QLR L-statistic | 97.12902 | 60 | 0.0017 |
| QLR Lambda-statistic | 96.85272 | 60 | 0.0018 |

**Figure 3.39**   The output of the RVT based on the QR(0.25) in Figure 3.38

**Table 3.9** Statistical result summary of the QR(0.25) in Figure 3.39

| Variatile | SEX = 2.DST = 1 | | | AGE | Dev(SEX = 1.DST = 0) | | | Dev(SEX = 1.DST = 1) | | | Dev(SEX = 2.DST = 0) | | |
|---|---|---|---|---|---|---|---|---|---|---|---|---|---|
| | Coef. | T-stat | Prob. | | Coef. | T-stat | Prob. | Coef. | T-stat | Prob. | Coef. | T-stat | Prob. |
| C | 8 | 5.213 | 0.000 | 10 | -1 | -0.565 | 0.572 | -2 | -0.596 | 0.551 | -2 | -0.982 | 0.326 |
| UR = 1,AGE = 11 | 0 | 0.000 | 1.000 | 11 | 0 | 0.000 | 1.000 | 0 | 0.000 | 1.000 | -1 | -0.414 | 0.679 |
| UR = 1,AGE: 12 | 4 | 1.851 | 0.064 | 12 | -5 | -2.961 | 0.003 | 0 | 0.000 | 1.000 | -2 | -0.880 | 0.379 |
| UR = 1,AGE = 13 | 4 | 1.460 | 0.144 | 13 | -2 | -0.672 | 0.501 | -3 | -0.862 | 0.389 | 0 | 0.000 | 1.000 |
| UR = 1,AGE = 14 | 2 | 0.535 | 0.593 | 14 | 2 | 0.552 | 0.581 | 0 | 0.000 | 1.000 | -2 | -0.577 | 0.564 |
| UR = 1,AGE = 15 | 4 | 2.113 | 0.035 | 15 | -3 | -2.119 | 0.034 | -6 | -2.831 | 0.005 | -2 | -0.976 | 0.329 |
| UR = 1,AGE = 16 | 4 | 1.804 | 0.071 | 16 | 2 | 1.077 | 0.282 | 2 | 0.929 | 0.353 | 0 | 0.000 | 1.000 |
| UR = 1,AGE = 17 | 2 | 0.875 | 0.382 | 17 | 3 | 1.628 | 0.104 | 2 | 1.092 | 0.275 | 0 | 0.000 | 1.000 |
| UR = 1,AGE = 18 | 4 | 1.748 | 0.081 | 18 | 2 | 1.030 | 0.303 | 0 | 0.000 | 1.000 | 4 | 0.895 | 0.371 |
| UR = 1,AGE = 19 | 28 | 11.798 | 0.000 | 19 | -1 | -0.512 | 0.609 | -6 | -2.026 | 0.043 | -1 | -0.435 | 0.664 |
| UR = 2,AGE = 10 | -1 | -0.581 | 0.561 | 10 | -1 | -0.980 | 0.327 | 0 | 0.000 | 1.000 | 1 | 1.055 | 0.292 |
| UR = 2,AGE = 11 | -2 | -1.080 | 0.280 | 11 | 0 | 0.000 | 1.000 | 1 | 0.842 | 0.400 | 3 | 2.393 | 0.017 |
| UR = 2,AGE = 12 | 0 | 0.000 | 1.000 | 12 | 0 | 0.000 | 1.000 | -1 | -0.798 | 0.425 | 0 | 0.000 | 1.000 |
| UR = 2,AGE = 13 | -1 | -0.582 | 0.561 | 13 | 1 | 1.086 | 0.278 | 2 | 1.707 | 0.088 | 1 | 1.009 | 0.313 |
| UR = 2,AGE = 14 | 0 | 0.000 | 1.000 | 14 | 1 | 1.134 | 0.257 | 0 | 0.000 | 1.000 | 1 | 1.051 | 0.293 |
| UR = 2,AGE = 15 | 1 | 0.560 | 0.576 | 15 | 0 | 0.000 | 1.000 | 1 | -0.900 | 0.368 | -1 | -0.898 | 0.369 |
| UR = 2,AGE = 16 | 2 | 0.879 | 0.379 | 16 | 0 | 0.000 | 1.000 | -2 | -1.116 | 0.264 | 0 | 0.000 | 1.000 |
| UR = 2,AGE = 17 | 1 | 0.569 | 0.570 | 17 | 0 | 0.000 | 1.000 | -1 | -0.867 | 0.386 | 1 | 0.942 | 0.346 |
| UR = 2,AGE = 18 | 4 | 2.040 | 0.041 | 18 | -4 | -3.007 | 0.003 | -2 | -1.244 | 0.214 | -3 | -1.950 | 0.051 |
| UR = 2,AGE = 19 | 10 | 4.779 | 0.000 | 19 | 4 | 1.964 | 0.050 | 3 | 1.336 | 0.182 | 0 | 0.000 | 1.000 |

| | | | | | |
|---|---|---|---|---|---|
| Pseudo R-squared | 0.121763 | | Mean dependent var | 22.31009 |
| Adjusted R-squared | 0.115651 | | S.D. dependent var | 16.74588 |
| S.E. of regression | 16.4133 | | Objective | 39 615.75 |
| Quantile dependent var | 10 | | Restr. objective | 45 108.25 |
| Sparsiti | 24.84668 | | Quasi-LR statistic | 2357.928 |
| Prob(Quasi-LR-star) | 0.00000 | | | |

**Example 3.18**  *An Application of the QR in (1.16d) for DST<>NA*

Figure 3.38 presents a part of the output of the QR(Median) of *CWWH* on *UR, AGE, DST*, and *SEX*, using the following ES. Figure 3.39 presents the output of its RVT.

$$CWWH\ C@Expand\ (UR, AGE, @Dropfirst)$$

$$@Expand(SEX, DST, @Droplast)^* @Expand\ (UR, AGE) \tag{3.20}$$

Note that the variable *AGE* is presented as the last variable in each of the two functions in (3.20), so that the output presents the ordered *AGE* levels, which could be easily used to develop the statistical result summary using Excel, as presented Table 3.9. Based on this summary, the following findings and notes are presented:

1) In block $(SEX = 2, DST = 1)$, each of parameters, $C(2)$ to $C(20)$, as the coefficients of the cells or groups $(UR = 1, AGE = 11)$ to $(UR = 2, AGE = 19)$, respectively, can be used to present a one- or two-sided hypothesis on the quantile(0.25) difference between each of these cells and the first cell $(UR = 1, AGE = 10)$, conditional for the group $(SEX = 2, DST = 1)$. And it can be tested using the *t*-statistic in the summary, conditional for $(SEX = 2, DST = 1)$. For instance, at the 10% level, the quantiles(0.25) in $(UR = 1, AGE = 12)$ and $(UR = 1, AGE = 10)$ have a significant difference, conditional for $(SEX = 2, DST = 1)$.

# 4

# Quantile Regressions Based on (X1,Y1)

## 4.1 Introduction

The equation specifications (ESs) of all univariate regression models of a numerical dependent variable presented in Agung (2011a) can also be used to conduct the analysis of the quantile regression models, based on cross-section and experimental data sets. However, this chapter presents only several selected models, based on Data_Faad.wfl, *motorcycle data from MASS*, MCYCLE.wfl and the National Social-Economics Survey of Indonesia, namely SUSENAS-2013.

## 4.2 The Simplest Quantile Regression

The simplest quantile regressions (QRs) based on *(X1,Y1)* can be presented using the equation specification as follows:

$$Y1\ C\ X1 \tag{4.1}$$

**Example 4.1**  *An Application of the QR (4.1)*
Figure 4.1 presents the statistical results of the QR (4.1) based on Data_Faad.wfl, and Figure 4.2 presents the forecast graph *Y1F* and its forecast evaluation. Based on these figures, the following findings and notes are presented:

1) We can easily obtain the forecast graph and evaluation in Figure 4.1 by clicking the *Forecast button and then OK,* without selecting any options. Then there is a question "Why the forecast graph of *Y1*, namely *Y1F*?" presents a four step function? The explanation is that Data_Faad.wfl is sorted by the two factors *A* and *B*, which makes four different groups or sets of the values or scores of the numerical variables *X1* and *Y1*.
2) In fact, the QR presents a linear regression function, with the following equation:

$$\widehat{Y1} = 3.336\,667 + 1.754\,39 * X1, or$$

$$\widehat{Med\,(Y1)} = 3.336\,667 + 1.754\,39 * X1$$

*Quantile Regression: Applications on Experimental and Cross Section Data Using EViews,* First Edition.
I Gusti Ngurah Agung.

Dependent Variable: Y1
Method: Quantile Regression (Median)
Date: 04/13/18   Time: 21:13
*Sample: 1 300*
Included observations: 300
Huber Sandwich Standard Errors and Covariance
Sparsity method: Kernel (Epanechnikov) using residuals
Bandwidth method: Hall-Sheather, bw = 0.14513
Estimation successfully identifies unique optimal solution

| Variable | Coefficient | Std. error | t-Statistic | Prob. |
|---|---|---|---|---|
| C | 3.336667 | 0.356698 | 9.354309 | 0.0000 |
| X1 | 1.754386 | 0.236059 | 7.431996 | 0.0000 |

| | | | |
|---|---|---|---|
| Pseudo R-squared | 0.112769 | Mean dependent var | 5.751100 |
| Adjusted R-squared | 0.109792 | S.D. dependent var | 0.945077 |
| S.E. of regression | 0.811773 | Objective | 98.46044 |
| Quantile dependent var | 5.670000 | Restr. objective | 110.9750 |
| Sparsity | 2.293867 | Quasi-LR statistic | 43.64529 |
| Prob(*Quasi-LR* stat) | 0.000000 | | |

**Figure 4.1**   Statistical results of the simplest QR (4.1).

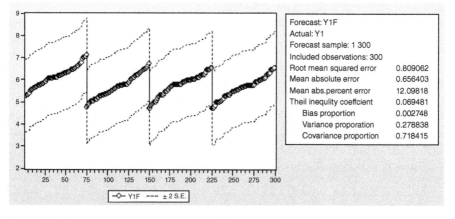

**Figure 4.2**   Forecast graph and evaluation of the simplest QR (4.1).

3) As an additional illustration, Figure 4.3 presents two scatter graphs of *Y1* and the fitted variable of *Y1*, namely *FVY1* = $\widehat{Y1}$, on *X1*. They include both full and partial values of *X1*, which show that *FVY1* is a straight line.

4) Referring to the variables *MedY1_X1* = *@Mediansby(X1,Y1)* presented in previous chapter, we have the following *Median Deviation from Linearity (MDFL)*, with its histogram and statistics presented in Figure 4.4:

$$MDFL = MedY1\_X1 - FVY1$$

Note that the histogram of *MDFL* does not look like it has a normal distribution, but it is unexpected that the data supports the normality of *MDFL* based on the Jarque-Bera test statistic of 0.829 351 with a *p*-value = 0.660 555. Compare this to the histograms of the variables presented in previous chapter.

5) As a comparison with the results in Figure 4.2, I developed Sort_DF.wf1, which is a sort of Data_Faad.wf1 by the numerical variable *X1*. As an alternative result, I found the forecast graph and evaluation of the QR (4.1) as shown in Figure 4.5, which presents exactly the same values of the evaluation statistics as in Figure 4.2.

## 4.3 Polynomial Quantile Regressions

This section presents an exploration of alternative polynomial QRs, starting with the quadratic QR based on *(X1,Y1)*, to find out what degree polynomial QR would give the best possible fit.

### 4.3.1 Quadratic Quantile Regression

The quadratic QR based on *(X1,Y1)* can be presented using the equation specification as follows:

$$Y1\ C\ X1\ X1\hat{}2 \tag{4.2}$$

**Example 4.2** *Application of the QR (4.2)*
Figure 4.6 presents the estimate of the QR (4.2) based on Data_Faad.wf1 and the scatter graphs of *Y1* and its fitted variable *FVPol2* on *X1*. Based on these results the following findings and notes are presented.

1) Even though $X1\hat{}2$ has a large Prob. > 0.30, the regression is an acceptable regression, in the statistical sense. Note that its graph is a part of parabolic curve having a maximum point out of the sample period, since $X1\hat{}2$ has a negative coefficient.

2) As a comparison, Figure 4.7 presents a reduced model of the model (4.2) which is obtained by deleting *X1,* and it is also an acceptable model, in the statistical sense. Note that $X1\hat{}2$ has a positive significant effect on *Y1,* and its graph is a part of parabolic curve having a minimum point.

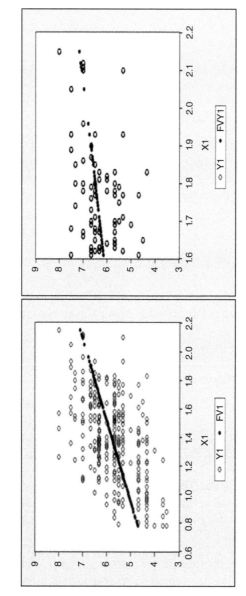

**Figure 4.3** The full scatter graphs of (*Y1*,*FVY1*) on *X1* and a part of them.

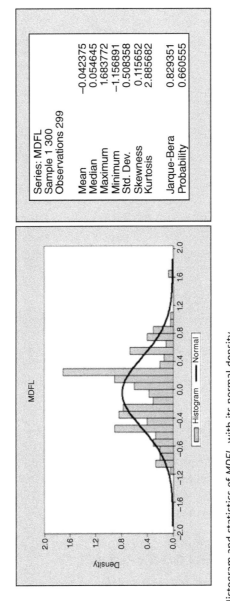

**Figure 4.4** Histogram and statistics of *MDFL*, with its normal density.

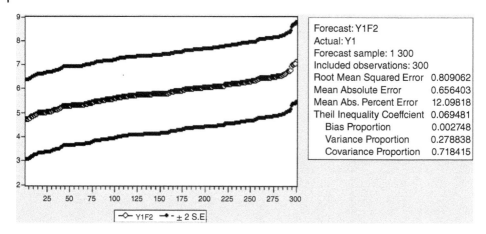

**Figure 4.5** Forecast graph and evaluation of the simplest QR (4.1), using Sort_DF.wf1.

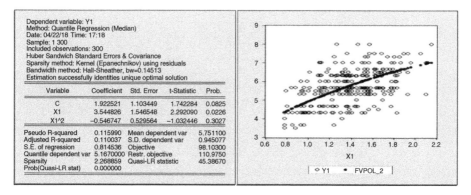

**Figure 4.6** Estimate of the QR (4.2) and the scatter graphs of *Y1* and its fitted variable, *FVPol2*, on *X1*.

3) Even though both models are acceptable models, we have to select which one is a better regression to represent the characteristics of the dependent variable *Y1* in the population. Note that the graph in Figure 2.6 is representing the growth with upper bound, and the graph in Figure 2.7 is representing the growth without upper bound! Since *Y1* is the scores of a test, which should have a maximum score, then I would select the regression in Figure 4.6 is a better regression.

### 4.3.2 Third Degree Polynomial Quantile Regression

The third degree polynomial QR based on *(X1,Y1)* can be presented using the equation specification (ES) as follows:

$$Y1\ C\ X1\ X1^2\ X1^3 \tag{4.3}$$

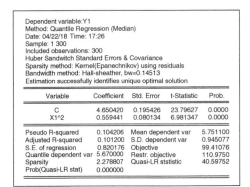
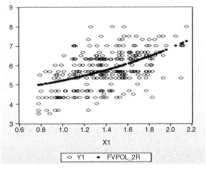

**Figure 4.7** Estimate of a reduced model of the QR (4.2), the scatter graphs of *Y1*, and its fitted variable, *FVPol2*, on *X1*.

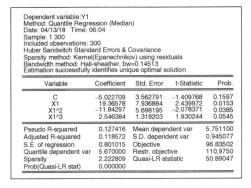
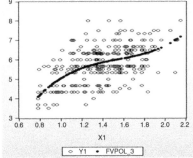

**Figure 4.8** Estimate of the QR (4.4) and the scatter graphs of *Y1* and its fitted variable, *FVPol_3*, on *X1*.

## Example 4.3    *An Application of the QR (4.3)*

Figure 4.8 presents the estimate of the QR (4.3) based on Data_Faad.wf1 and the scatter graphs of *Y1* and its fitted variable *FVPol_3* on *X1*. Based on these results, the following findings and notes are presented:

1) Compared with the regression in Figure 4.6, it is unexpected that *X, X^2,* and *X^3* each has an adjusted significant effect on *Y1,* and it has a greater adjusted R-squared. So, the third degree polynomial QR is a better regression than the quadratic QR.
2) However, note that its graph represents a curve without an upper bound. As an alternative, we will see models with upper and lower bounds.
3) The very small values of R-squared and adjusted R-squared correspond to a great spread of the *Y1*'s scores around its fitted regression. Hence, the IVs of the regression can explain only 12.7% of the *Y1*'s total variance.

### 4.3.3 Forth Degree Polynomial Quantile Regression

The fourth degree polynomial QR based on (X1,Y1) can be presented using the following ES:

$$Y1\ C\ X1\ X1\char`^2\ X1\char`^3\ X1\char`^4 \tag{4.4}$$

**Example 4.4** *An Application of the QR (4.4)*
Table 4.1 presents the statistical results summary of the QR (4.4) for tau = 0.1, 0.5, and 0.9, based on Data_Faad.wf1. Based on these results, the following findings and notes are presented:

1) Compared with the third degree polynomial QR(median), this QR(tau = 0.5) is a better fit model because it has a slightly greater Pseudo R-squared and adjusted R-squared.
2) For each of the IVs, $X^k$ has either a positive or negative significant effect on Y1 at the 1% or 5% level of significance. For instance, at the 5% level, $X^4$ has a negative significant adjusted effect on Y1, based on the $t$-statistic of $t_0 = -1.936$ with a $p$-value = 0.0538/2.
3) And all IVs have significant joint effects on Y1, based on the *Quasi-LR* statistic of 55.501 82 with the $p$-value = 0.000 000.
4) Specific for the QR(tau = 0.9), each of the IVs has a large $p$-value. So, in the statistical sense, it is not a good model. However, in this case, the model is kept as it is, as a comparison.
5) As an additional illustration, Figure 4.9 presents the scatter graphs of Y1, and its three fitted value variables, *Fitted_q10, Fitted_q50*, and *Fitted_q90*, of the three QRs in Table 4.1.

**Table 4.1**  Statistical results summary of the QR (4.4) for tau = 0.1, 0.5, and 0.9.

| Variable | QREG(tau = 0.1) | | | QREG(tau = 0.5) | | | QREG(tau = 0.9) | | |
|---|---|---|---|---|---|---|---|---|---|
| | Coef. | t-Stat. | Prob. | Coef. | t-Stat. | Prob. | Coef. | t-Stat. | Prob. |
| C | −23.86 | −1.451 | 0.148 | −34.20 | −2.421 | 0.016 | 3.551 | 0.206 | 0.837 |
| X1 | 82.421 | 1.657 | 0.099 | 107.16 | 2.497 | 0.013 | −0.371 | −0.007 | 0.994 |
| X1^2 | −90.49 | −1.654 | 0.099 | −107.2 | −2.269 | 0.024 | 7.000 | 0.126 | 0.900 |
| X1^3 | 43.781 | 1.685 | 0.093 | 46.900 | 2.095 | 0.037 | −5.236 | −0.203 | 0.839 |
| X1^4 | −7.728 | −1.714 | 0.088 | −7.484 | −1.936 | 0.054 | 1.166 | 0.268 | 0.789 |
| Pseudo R-sq | 0.1607 | | | 0.1340 | | | 0.1278 | | |
| Adj. R-sq | 0.1493 | | | 0.1222 | | | 0.1160 | | |
| S.E. of reg | 1.3507 | | | 0.8094 | | | 1.3121 | | |
| Quantile DV | 4.3300 | | | 5.6700 | | | 7.0000 | | |
| Sparsity | 3.9921 | | | 2.1428 | | | 4.5793 | | |
| Quasi-LR stat | 44.339 | | | 55.502 | | | 29.328 | | |
| Prob | 0.0000 | | | 0.0000 | | | 0.0000 | | |

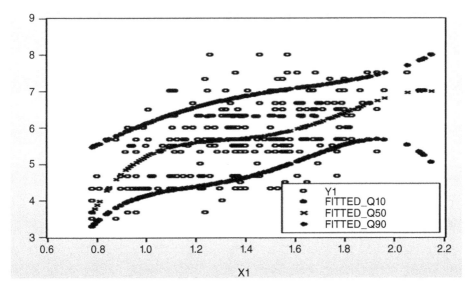

**Figure 4.9** The scatter graphs of *Y1*, *Fitted_Q10, _Q50*, and *_Q90* on *X1*.

### 4.3.4 Fifth Degree Polynomial Quantile Regression

The fifth degree polynomial QR based on *(X1,Y1)* can be presented using the following ES:

$$Y1\ C\ X1\ X1\char`\^2\ X1\char`\^3\ X\char`\^4\ X1\char`\^5 \tag{4.5}$$

**Example 4.5** *Application of the QR (4.5)*

Figure 4.10 presents the estimate of the QR (4.5) based on Data_Faad.wf1 and the scatter graphs of *Y1* and its fitted variable *FVPol5* on *X1*. Based on these results, the following findings and notes are presented:

1) Because, at the 10% level of significance, each of the independent variables has an insignificant adjusted effect on *Y1*, I would say that the fourth degree polynomial QR is the best fit polynomial QR.

2) Unexpectedly, deleting any of the four variables, $X1, X1\char`\^2, X1\char`\^3$, and $X1\char`\^4$, results in a reduced fifth degree polynomial QR, where each of their independent variables has significant adjusted effects at the 5% level. Two of the four reduced models with the smallest and the largest adjusted R-squared are presented in Figure 4.11, which are obtained by deleting $X1$ and $X1\char`\^4$, respectively.

3) Hence, it can be concluded that the reduced QR in Figure 4.11a is the best reduced fifth degree polynomial model of the model (4.5), based specifically on Data_Faad.wf1.

## 4.4 Logarithmic Quantile Regressions

For an extension of the polynomial QRs, we might have a nonlinear QR as the best possible QR. However, EViews does not provide an estimation method for the nonlinear QR. For this reason, I propose logarithmic QRs, as shown in this section.

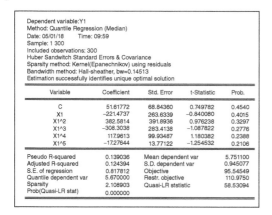
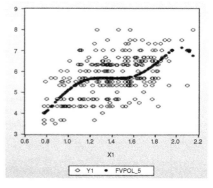

**Figure 4.10** Estimate of the QR (4.5) and the scatter graphs of Y1 and its fitted variable, *FVPol2*, on X1.

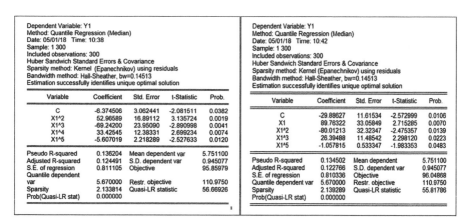

**Figure 4.11** Estimate of two reduced models of the QR (4.5).

### 4.4.1 The Simplest Semi-Logarithmic QR

The simplest semi-logarithmic QR based on $(X,Y)$ can be represented using the following ES for a positive variable $Y$:

$$\log(Y) \; C \; X \tag{4.6}$$

which represents the nonlinear model as follows:

$$Y = Exp(C1 + C(2) * X) \tag{4.7}$$

The model (4.6) can be extended to a bounded semi-logarithmic model for non-positive variable $Y$, with the following general ES:

$$Log((Y - L)/(U - Y)) \; C \; X \tag{4.8}$$

where $U$ and $L$ are subjectively selected fixed upper and lower bounds of the numerical variable $Y$.

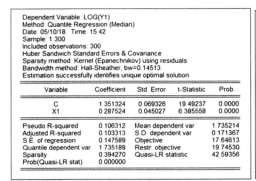

Figure 4.12 Statistical results of a semi-logarithmic QR in (4.6).

### Example 4.6 *An Application of the Model (4.6)*

Figure 4.12 presents the statistical results of a model (4.6) based on the bivariate *(X1,Y1)* in Data_Faad.wf1. Based on these results, the following findings and notes are presented:

1) The numerical variable *X1* has a significant linear effect on *log(Y1)* = *ln(Y1)*, based on the *Quasi-LR* statistic of 42.5935 with a *p*-value = 0.000 000. And based on the *t*-statistic of $t_0 = 8.386$ with a *p*-value = 0.000, we can conclude that *X1* has a positive significant effect on *log(Y1)*.

2) The semi-logarithmic regression function has a pseudo R-squared of 0.106 312, and it can be transformed to a nonlinear function as follows:

$$Y1 = Exp(C(1) + C(2) * X1) \quad Y1 = Exp(1.351\,32449925 + 0.287523874038 * X1)$$

$$(4.9)$$

### Example 4.7 *An Application of the Model (4.8)*

As a more complex or advanced model, Figure 4.13 presents the statistical results of a QR in (4.8) based on the bivariate *(X1,Y1)* in Data_Faad.wf1. This QR(Median) has the smallest adjusted R-squared compared with the other QRs. So, this QR is the worst fit model.

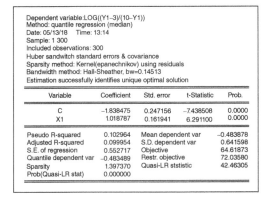
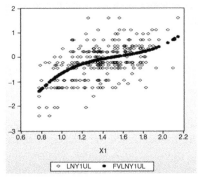

Figure 4.13 Statistical results of a bounded semi-logarithmic QR in (4.8).

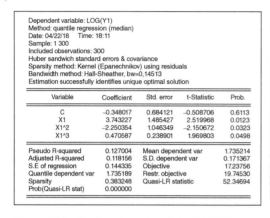

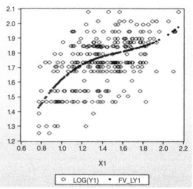

**Figure 4.14** Statistical results of the QR (4.10).

### 4.4.2 The Semi-Logarithmic Polynomial QR

As a comparison with the previous polynomial QR, this section presents only two semi-logarithmic third degree polynomial QRs.

#### 4.4.2.1 The Basic Semi-Logarithmic Third Degree Polynomial QR

Based on the bivariate $(X,Y)$, the model has the following general ES:

$$log(Y) \; C \; X \; X^{\wedge}2 \; X^{\wedge}3 \tag{4.10}$$

**Example 4.8   *An Application of the Model (4.10)***

Figure 4.14 presents the statistical results of a model in (4.10) based on the bivariate *(X1,Y1)* in Data_Faad.wf1. Based on these results, the following findings and notes are presented:

1) The independent variables *X1, X1^2,* and *X1^3* have joint significant effects on $log(Y1) = ln(Y1)$, based on the *Quasi-LR* statistic of 52.346 94 with a $p$-value $= 0.000\,000$.
2) At the 5% level of significance, each of the independent variables has a significant adjusted effect on *log(Y1)*.
3) The regression function has the pseudo R-squared of 0.127 004.

#### 4.4.2.2 The Bounded Semi-Logarithmic Third Degree Polynomial QR

As an extension of the model (4.10), based on the bivariate $(X,Y)$, this model has the following general ES:

$$LnYul \; C \; X \; X^{\wedge}2 \; X^{\wedge}3 \tag{4.11}$$

where $LnYul = log((Y - L)/(U - Y))$, where $U$ and $L$ are subjectively selected, fixed upper and lower bounds of the numerical variable $Y$.

**Example 4.9   *An Application of the Model (4.11)***

Figure 4.15 presents the statistical results of a model in (4.11) based on the bivariate *(X1,Y1)* in Data_Faad.wf1.

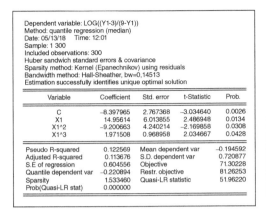
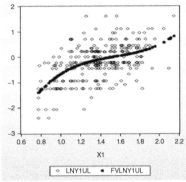

| Dependent variable: LOG((Y1-3)/(9-Y1)) |
| --- |
| Method: quantile regression (median) |
| Date: 05/13/18   Time: 12:01 |
| Sample: 1 300 |
| Included observations: 300 |
| Huber sandwich standard errors & covariance |
| Sparsity method: Kernel (Epanechnikov) using residuals |
| Bandwidth method: Hall-Sheather, bw=0,14513 |
| Estimation successfully identifies unique optimal solution |

| Variable | Coefficient | Std. error | t-Statistic | Prob. |
| --- | --- | --- | --- | --- |
| C | -8.397965 | 2.767368 | -3.034640 | 0.0026 |
| X1 | 14.95614 | 6.013855 | 2.486948 | 0.0134 |
| X1^2 | -9.200663 | 4.240214 | -2.169858 | 0.0308 |
| X1^3 | 1.971508 | 0.968958 | 2.034667 | 0.0428 |

| Pseudo R-squared | 0.122569 | Mean dependent var | -0.194592 |
| --- | --- | --- | --- |
| Adjusted R-squared | 0.113676 | S.D. dependent var | 0.720877 |
| S.E of regression | 0.604556 | Objective | 71.30228 |
| Quantile dependent var | -0.220894 | Restr. objective | 81.26253 |
| Sparsity | 1.533460 | Quasi-LR statistic | 51.96220 |
| Prob(Quasi-LR stat) | 0.000000 | | |

**Figure 4.15** Statistical results of the QR (4.11).

Compared with the simpler regression function in Figure 4.15, this regression has a smaller pseudo R-squared of 0.122 569. Similar to the previous findings, a simpler model is a better fit model compared with a more complex model.

## 4.5 QRs Based on MCYCLE.WF1

For *motorcycle data from MASS*, I thank Professor Koenker for helping me acquire the data, which contains only two variables *MILLI (times in milliseconds after impact)* and *ACCEL (Acceleration)*. However, I obtained slight different data than the data presented in Koenker (2005).

### 4.5.1 Scatter Graphs of *(MILL,ACCEL)* with Fitted Curves

As a preliminary analysis based on the data in MCYCLE.wf1, Figure 4.16 presents the scatter graphs of *(MILL,ACCEL)* with their *linear polynomial fit* with six degrees and *linear fit*. Based on these graphs, the following findings and notes are presented:

1) Even though a polynomial curve can be used as a good fit model for any sample data, these graphs clearly show that the polynomial curve never will be a good fit model of *ACCEL* on *MILLI*, because the observed scores of *ACCEL* are almost constant for small scores of *MILLI* < 13. So, I will try applying alternative piecewise QRs in the following sections.
2) The linear regression is not a good fit model either, but it always can be applied to present and test the trend of *ACCEL* with respect to *MILLI*.

**Example 4.10** *Advanced Fitted Curves*
As the extension of the fitted curves shown in Figure 4.16, Figure 4.17 presents the scatter graphs of *(MILLI,ACCEL)* with their *kernel fit* and the *LOWESS polynomial-6 fit*. Based on these graphs, the findings and notes are as follows:

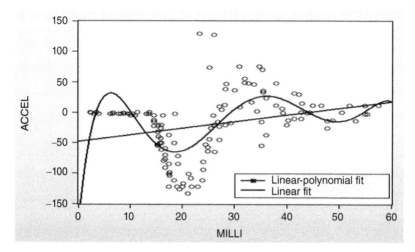

**Figure 4.16** Scatter graphs of *(MILLI,ACCEL)* with their linear polynomial fit and linear fit.

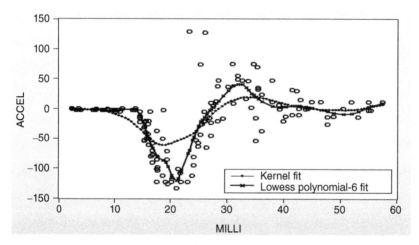

**Figure 4.17** Scatter graphs of *(MILLI,ACCEL)* with their kernel fit and the lowest polynomial-6 fit.

1) Compared with the graphs in Figure 4.16, we can conclude the LOWESS polynomial-6 fit is the best fit curve for the data among the four fit curves, which is nonparametric cure. So, it can't be presented using a common continuous least squares (LS) regression and QR. For this reason, I will try applying alternative piecewise QRs in following sections.
2) The graph of *ACCEL* on *MILLI* < 13 is a horizontal line because its scores are very small within the interval [−150, 150]. For this reason, I present two specific parts of the scatter graph of *(MILLI,ACCEL)*, with their LOWESS polynomial-6 fit only.
3) Specific for *Milli* < 13, the graph shows a regression line with a negative slope. Additionally, the observed scores of *ACCEL* have Mean = −1.316 667, Median = 2.7, Max. = 0, and Min. = −5.4, with 18 observations.

### 4.5.2 Applications of Piecewise Linear QRs

To obtain a piecewise linear QR using the scatter graph in Figure 4.16, at the first stage, I try to generate four intervals of the variables *MILLI* (Milliseconds) using the following equation, which could be modified after looking at the scatter graph of *ACCEL* and its fitted variable on *MILLI*. The four intervals are closed-open intervals, namely 1. [0, 13), 2. [13, 20), 3. [20, 32), and 4. (*Milli* > = 32).

$$INT4 = 1 + 1 * (Milli >= 13) + 1 * (Milli >= 20) + 1 * (Milli >= 32) \qquad (4.12)$$

**Example 4.11** *An Application of a Piecewise Linear QR*

Figure 4.18 presents the output of a piecewise linear *QR* of *ACCEL* on *MILLI* by *INT4*, using the following ES, the scatter graphs of *ACCEL*, and its *FITTED01* on *MILLI*:

$$ACCEL \ C \ @Expand\,(INT4, @Dropfirst)\ MILLI * \ @Expand\,(INT4) \qquad (4.13)$$

where the function *@Expand(INT4,@Dropfirst)* represents the three dummies of *INT4* = 2, *INT4* = 3, and *INT4* = 4, and the function *MILLI\*@Expand(INT4)* represents the linear effect of *MILL1* on *ACCEL* by *INT4*, or within each interval or level of *INT4*.

Based on the output of the QR(Median), the following findings and notes are presented:

1) The function *@Expand(INT4,@Dropfirst)* represents the three dummies of *INT4* = 2, *INT4* = 3, and *INT4* = 4, and the function *MILLI\*@Expand(INT4)* represents the linear effect of *MILL1* on *ACCEL* by *INT4*, or within each interval or level of *INT4*.
2) The results show that the IVs, *MILLI* and *INT4*, have significant joint effects on *ACCEL*, based on the *Quasi-LR* statistic of $QLR_0 = 175.6776$ with the *p*-value 0.000 000. And the QR has a Pseudo R-squared = 0.463 822, which indicates that the IVs, *MILLI* and *INT4*, can explain only 46.4% of the total variance of *ACCEL*. This is because of the great spread of the *ACCEL's* scores around its fitted variable, *FITTED01,* specifically for the large scores of *MILLI*.

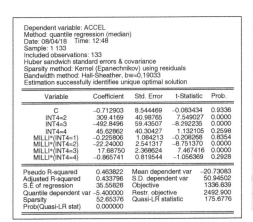
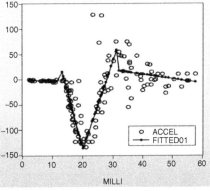

**Figure 4.18** Output of the QR(Median) in (4.13) and the scatter graphs of *ACCEL* and its fitted value variable, *FITTED01*, on *Milli*.

3) Specific for the level *INT4* = 1, the linear effect of *MILLI* on *ACCEL* has an insignificant effect on *ACCEL*, based on the *t*-statistic of $t_0 = -0.208\,268$ with $df = (133 - 8) = 125$ and a very large *p*-value = 0.8534. For a better picture of the QR(Median) within the level *INT4* = 1, Figure 4.19 presents two graphs of *ACCEL* and *FITTED01*, specific for the subsample {*MILLI* < 13}, namely the graphs with unsorted and sorted of *ACCEL*. Based on these graphs, the findings and notes are as follows:

3.1 The graphs clearly show the median-regression curve is not a straight line.

3.2 Note that Figure 4.19b shows that for most of the smaller scores of *ACCEL*, their predicted medians are greater than its observed scores; however, for most of the greater scores of *ACCEL*, their predicted medians are smaller than its observed scores. Similar findings also are found on the graphs of *ACCEL* and *FITTED01* for the whole sample. Do them as an exercise. These findings are unexpected, which should be the specific characteristic of the LAD (*least absolute deviation*) estimation method.

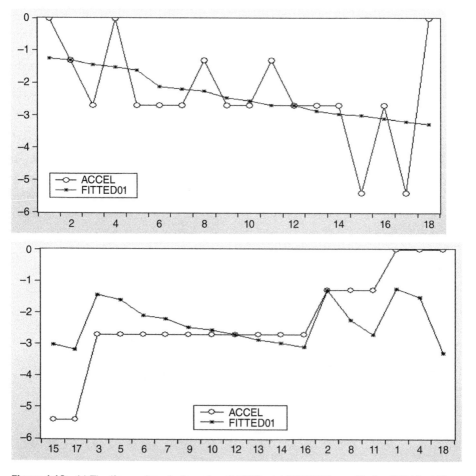

**Figure 4.19** (a) The line and symbol graphs of *ACCEL* and *FITTED01*, specific for {*MILLI* < 13}. (b) The graphs of sorted *ACCEL* and *FITTED01*, specific for {*MILLI* < 13}.

3.3 For additional characteristics of the median regression for *MILLI* < 13, we can run the QR(Median) based only on the subsample {*Milli* < 13}. We find the QR has a pseudo $R^2 = 0.002\,551$, but with a negative adjusted $R^2. = -0.059\,789$. So, the model is considered to be a poor fitted model (EViews 8, User Guide II, p. 13).

### 4.5.3 Applications of the Quantile Process

For a more advanced analysis, EViews provides the quantile process with three options: *Process Coefficients, Quantile Slope Equality Test (QSET)*, and *Symmetric Quantiles Test (SQT)*. See the following examples.

**Example 4.12** *Application Quantile Slope Equality Test*
Having the output of the QR(Median) in Figure 4.18, selecting *View/Quantile Process/Quantile Slope Equality Test and clicking OK* obtains the output in Figure 4.20. Based on this output, the following findings and notes are presented:

1) Based on the Wald test, the 14 slopes of the IVs, between the pairs (0.25, 0.5) and (0.5, 0.75), have jointly significant differences, based on the *Chi-square* statistic of 58.7832 with $df = 14$ and *p*-value = 0.0000.
   However, it is unexpected that only one of 14 pairs of slopes, namely the slopes of *MILLI\*(INT4 = 2)* between tau = 0.5 and tau = 0.75, has a significant difference with a *p*-value = 0.0007, based on the Chi-square test statistic; but the output does not show it. Note that the slopes difference is b(0.5) − b(0.75) = −8.871 579. Based on these findings, I can make the following statement. Refer also to the discussion on sampling in Agung (2011a).

   > Never believe the conclusion of a testing hypothesis, because it is only based on sample data, which happens to be selected by or available for us, and it might not represent the corresponding population characteristics.

2) It is important to mention that QSETs in fact represent the tests on the linear effect differences of *Milli* on *ACCEL* between pairs of quantiles (0.25, 0.5) and (0.5, 0.75) for each level of *INT4*.
3) Finally, note that to do the QSET, the QR with an intercept should be applied. Otherwise, the error message "Test requires equation estimated with explicit intercept" is shown on-screen.

**Example 4.13** *Application of the Process Coefficients*
To conduct the analysis using *Quantile Process/Process Coefficients*, I think it is better to apply the QR without an intercept using the following ES:

$$ACCEL\;@Expand\,(INT4)\;MILLI * @Expand\,(INT4) \tag{4.14}$$

With the output of the QR(Median) in (4.14) on-screen, selecting *View/Quantile Process/Process Coefficients* and clicking *OK* obtains the output in Figure 4.21. Based on this output, the following findings and notes are presented:

Quantile Slope Equality Test
Equation: EQ01
Specification: ACCEL C @EXPAND(INT4,@DROPFIRST) MILLI*@EXPAND(INT4)
Estimated equation quantile tau = 0.5
Number of test quantiles: 4
Test statistic compares all coefficients

| Test Summary | Chi-Sq. Statistic | Chi-Sq. d.f. | Prob. |
|---|---|---|---|
| Wald Test | 58.78320 | 14 | 0.0000 |

Restriction Detail: b(tau_h) − b(tau_k) = 0

| Quantiles | Variable | Restr. Value | Std. Error | Prob. |
|---|---|---|---|---|
| 0.25, 0.5 | INT4 = 2 | 2.735478 | 36.15466 | 0.9397 |
| | INT4 = 3 | 30.29960 | 57.20900 | 0.5964 |
| | INT4 = 4 | −37.16711 | 44.04732 | 0.3988 |
| | MILLI*(INT4 = 1) | 0.225806 | 0.942151 | 0.8106 |
| | MILLI*(INT4 = 2) | −0.712381 | 2.225039 | 0.7488 |
| | MILLI*(INT4 = 3) | −2.062500 | 2.256690 | 0.3607 |
| | MILLI*(INT4 = 4) | 0.481125 | 0.933645 | 0.6063 |
| 0.5, 0.75 | INT4 = 2 | 123.1598 | 40.80576 | 0.0025 |
| | INT4 = 3 | 1.561017 | 49.35063 | 0.9748 |
| | INT4 = 4 | −62.32278 | 33.24728 | 0.0609 |
| | MILLI*(INT4 = 1) | 0.103462 | 0.971671 | 0.9152 |
| | MILLI*(INT4 = 2) | −8.871579 | 2.621031 | 0.0007 |
| | MILLI*(INT4 = 3) | −0.655093 | 1.953488 | 0.7374 |
| | MILLI*(INT4 = 4) | 1.019676 | 0.680588 | 0.1341 |

**Figure 4.20** Output of the QSET of the QR (4.13), with four test quantiles.

1) The advantage of using the ES (4.14) is that the equation of each QR for each quantile, within each level of *INT4,* can easily be written based on the output. For instance, for the QR in *INT4* = 1 for tau = 0.25, we have the following equation:

$$\widehat{ACCEL} = -2.70 - (2.76E - 17) * Milli * (INT4 = 1)$$

Referring to the graph in Figure 4.19, this regression is represented as an horizontal line because the coefficient of *Milli*(INT4* = 1) = 2.76E-17 ≅ 0.

Quantile Process Estimates
Equation: EQ01A
Specification: ACCEL @EXPAND(INT4) MILLI*@EXPAND(INT4)
Estimated equation quantile tau = 0.5
Number of process quantiles: 4
Display all coefficients

| | Quantile | Coefficient | Std. Error | t-Statistic | Prob. |
|---|---|---|---|---|---|
| INT4=1 | 0.250 | -2.700000 | 7.483834 | -0.360778 | 0.7189 |
| | 0.500 | -0.712903 | 8.544469 | -0.083434 | 0.9336 |
| | 0.750 | 0.790244 | 8.097022 | 0.097597 | 0.9224 |
| INT4=2 | 0.250 | 309.4524 | 36.23064 | 8.541178 | 0.0000 |
| | 0.500 | 308.7040 | 40.08715 | 7.700822 | 0.0000 |
| | 0.750 | 187.0474 | 45.11377 | 4.146126 | 0.0001 |
| INT4=3 | 0.250 | -465.2500 | 63.79498 | -7.292894 | 0.0000 |
| | 0.500 | -493.5625 | 58.81769 | -8.391396 | 0.0000 |
| | 0.750 | -493.6204 | 40.81723 | -12.09343 | 0.0000 |
| INT4=4 | 0.250 | 5.761538 | 51.63021 | 0.111592 | 0.9113 |
| | 0.500 | 44.91574 | 39.38814 | 1.140337 | 0.2563 |
| | 0.750 | 108.7417 | 21.79284 | 4.989789 | 0.0000 |

| | Quantile | Coefficient | Std. Error | t-Statistic | Prob. |
|---|---|---|---|---|---|
| MILLI*(INT4=1) | 0.250 | -2.78E-17 | 0.948410 | -2.93E-17 | 1.0000 |
| | 0.500 | -0.225806 | 1.084213 | -0.208268 | 0.8354 |
| | 0.750 | -0.329268 | 1.026552 | -0.320752 | 0.7489 |
| MILLI*(INT4=2) | 0.250 | -22.95238 | 2.247573 | -10.21207 | 0.0000 |
| | 0.500 | -22.24000 | 2.541317 | -8.751370 | 0.0000 |
| | 0.750 | -13.36842 | 3.000779 | -4.454983 | 0.0000 |
| MILLI*(INT4=3) | 0.250 | 15.62500 | 2.518918 | 6.203060 | 0.0000 |
| | 0.500 | 17.68750 | 2.368624 | 7.467416 | 0.0000 |
| | 0.750 | 18.34259 | 1.609383 | 11.39728 | 0.0000 |
| MILLI*(INT4=4) | 0.250 | -0.384615 | 1.120709 | -0.343189 | 0.7320 |
| | 0.500 | -0.865741 | 0.819544 | -1.056369 | 0.2928 |
| | 0.750 | -1.885417 | 0.474424 | -3.974120 | 0.0001 |

**Figure 4.21** The output of the process coefficients of the QR (4.14).

2) The *t*-statistic in the output can be used to test the linear effect of *Milli* on *ACCEL* for each quantile, 0.25, 0.5, and 0.75, within each level of *INT4*. Note that if the QSET is conducted based on the model (4.14), then an error message, "Test requires equation estimated with explicit intercept," is shown on-screen.

### Example 4.14 *Application of the Symmetric Quantile Test*

As an additional analysis, with the output in Figure 4.21 on the screen, select *View/Quantile Process/Symmetric Quantiles Test* and click *OK* obtains the output in Figure 4.22. Based on this output, the following findings and notes are presented:

Symmetric Quantiles Test
Equation: EQ01A
Specification: ACCEL @EXPAND(INT4) MILLI*@EXPAND(INT4)
Estimated equation quantile tau = 0.5
Number of test quantiles: 4
Test statistic compares all coefficients

| Test Summary | Chi-Sq. Statistic | Chi-Sq. d.f. | Prob. |
|---|---|---|---|
| Wald Test | 6.880687 | 8 | 0.5496 |

Restriction Detail: b(tau) + b(1-tau) - 2*b(.5) = 0

| Quantiles | Variable | Restr. Value | Std. Error | Prob. |
|---|---|---|---|---|
| 0.25, 0.75 | INT4=1 | -0.483950 | 12.10572 | 0.9681 |
| | INT4=2 | -120.9083 | 58.30584 | 0.0381 |
| | INT4=3 | 28.25463 | 84.78355 | 0.7389 |
| | INT4=4 | 24.67172 | 58.92717 | 0.6754 |
| | MILLI*(INT4=1) | 0.122345 | 1.536020 | 0.9365 |
| | MILLI*(INT4=2) | 8.159198 | 3.720975 | 0.0283 |
| | MILLI*(INT4=3) | -1.407407 | 3.403685 | 0.6792 |
| | MILLI*(INT4=4) | -0.538551 | 1.240199 | 0.6641 |

**Figure 4.22** Output of the symmetric quantiles test based on the QR (4.14).

1) The output presents only the results of the testing statistical hypothesis as follows:

$$H_0 : b(0.25) + s\,b(0.75) - 2 * b(0.5) = 0 \text{ vs. } H_1 : Otherwise \tag{4.15}$$

2) Based on the Wald test, the combined total of eight $H_0$, for the eight IVs, is accepted based on the *Chi-square* statistic of 0.880 687 with $df = 8$ and $p$-value $= 5496$. The hypothesis can be considered as a multivariate hypothesis.
3) No test statistic is presented for any of the hypotheses. However, I am very sure the *Chi-square* test statistic is applied, because it is a nonparametric test.
   As an illustration, for the individual IVs, at the 5% level of significance, two of the null hypotheses, those for $INT4 = 2$, and $Milli*(INT4 = 2)$, are rejected, with the $p$-values of 0.0381 and 0.0283, respectively.

**Example 4.15** *A Special Quantile Process Estimate*
Unlike the previous examples, which use the QR(Median) and 4 (four) as the number of quantiles, this example presents a special quantile process estimate, using the QR(0.25) and 5 (five) as the number of quantiles.

With the output of QR(0.25) in (4.14) on-screen, select *View/Quantile Process/Process Coefficient*, insert 5 as the number of quantiles, and click *OK* to obtain the output in Figure 4.23. Based on this output, the findings and notes presented as follows:

1) The five quantiles presented have special spacing, specifically the first three quantiles, 0.20, 0.25, and 0.40, which I was not expecting at all.
2) For various statistical results, we can select any values $0 < \tau(\text{tau}) < 1$ as an estimated equation quantile and an integer for the number of quantile process. However, I recommend inserting only one of the integers 4, 5, 8, 10, or 20 as the number of quantiles. If 2 is inserted, for example, then an error message results; and, if 3 is inserted, then we obtain a value of quantile $= 0.333$, which is an uncertain decimal number. Do these as exercises.
3) In addition, we can apply the QR with an intercept as presented in (4.13), and obtain the output of the QSET easily.

Quantile Process Estimates
Equation: EQ01A
Specification: ACCEL @EXPAND(INT4) MILLI*@EXPAND(INT4)
Estimated equation quantile tau = 0.25
Number of process quantiles: 5
Display all coefficients

| | Quantile | Coefficient | Std. Error | t-Statistic | Prob. |
|---|---|---|---|---|---|
| INT4=1 | 0.200 | -1.157143 | 7.602574 | -0.152204 | 0.8793 |
| | 0.250 | -2.700000 | 7.483834 | -0.360778 | 0.7189 |
| | 0.400 | -2.700000 | 8.619069 | -0.313259 | 0.7546 |
| | 0.600 | -0.808108 | 8.916284 | -0.090633 | 0.9279 |
| | 0.800 | 0.709091 | 7.469966 | 0.094926 | 0.9245 |
| INT4=2 | 0.200 | 273.9261 | 37.15920 | 7.371690 | 0.0000 |
| | 0.250 | 309.4524 | 36.23064 | 8.541178 | 0.0000 |
| | 0.400 | 295.5435 | 36.65261 | 8.063367 | 0.0000 |
| | 0.600 | 271.4000 | 62.85215 | 4.318071 | 0.0000 |
| | 0.800 | 185.2000 | 41.73849 | 4.437152 | 0.0000 |
| INT4=3 | 0.200 | -437.3432 | 51.78960 | -8.444614 | 0.0000 |
| | 0.250 | -465.2500 | 63.79498 | -7.292894 | 0.0000 |
| | 0.400 | -508.5615 | 79.04290 | -6.433994 | 0.0000 |
| | 0.600 | -465.1533 | 46.44823 | -10.01445 | 0.0000 |
| | 0.800 | -500.1000 | 38.80474 | -12.88760 | 0.0000 |
| INT4=4 | 0.200 | 2.227273 | 49.63387 | 0.044874 | 0.9643 |
| | 0.250 | 5.761538 | 51.63021 | 0.111592 | 0.9113 |
| | 0.400 | 46.10769 | 44.42106 | 1.037969 | 0.3013 |
| | 0.600 | 101.3582 | 28.11998 | 3.604490 | 0.0005 |
| | 0.800 | 109.7333 | 20.16867 | 5.440782 | 0.0000 |

| | Quantile | Coefficient | Std. Error | t-Statistic | Prob. |
|---|---|---|---|---|---|
| MILLI*(INT4=1) | 0.200 | -0.385714 | 0.966591 | -0.399046 | 0.6905 |
| | 0.250 | -1.11E-16 | 0.948410 | -1.17E-16 | 1.0000 |
| | 0.400 | 0.000000 | 1.092248 | 0.000000 | 1.0000 |
| | 0.600 | -0.189189 | 1.131080 | -0.167264 | 0.8674 |
| | 0.800 | -0.295455 | 0.947102 | -0.311956 | 0.7556 |
| MILLI*(INT4=2) | 0.200 | -20.95652 | 2.258850 | -9.277516 | 0.0000 |
| | 0.250 | -22.95238 | 2.247573 | -10.21207 | 0.0000 |
| | 0.400 | -21.80435 | 2.287989 | -9.529917 | 0.0000 |
| | 0.600 | -19.66667 | 4.151018 | -4.737794 | 0.0000 |
| | 0.800 | -13.25000 | 2.778119 | -4.769415 | 0.0000 |
| MILLI*(INT4=3) | 0.200 | 14.28378 | 2.011932 | 7.099535 | 0.0000 |
| | 0.250 | 15.62500 | 2.518918 | 6.203060 | 0.0000 |
| | 0.400 | 18.03846 | 3.130279 | 5.762573 | 0.0000 |
| | 0.600 | 16.93333 | 1.862758 | 9.090464 | 0.0000 |
| | 0.800 | 18.73529 | 1.530384 | 12.24222 | 0.0000 |
| MILLI*(INT4=4) | 0.200 | -0.318182 | 1.074860 | -0.296022 | 0.7677 |
| | 0.250 | -0.384615 | 1.120709 | -0.343189 | 0.7320 |
| | 0.400 | -1.038462 | 0.965518 | -1.075549 | 0.2842 |
| | 0.600 | -1.945455 | 0.621967 | -3.127904 | 0.0022 |
| | 0.800 | -1.897436 | 0.438745 | -4.324686 | 0.0000 |

**Figure 4.23** The output of a special quantile estimates of the QR (4.14).

Dependent Variable: ACCEL
Method: Quantile Regression (tau = 0.25)
Date: 08/06/18   Time: 06:24
Sample: 1 133
Included observations: 133
Huber Sandwich Standard Errors & Covariance
Sparsity method: Kernel (Epanechnikov) using residuals
Bandwidth method: Hall-Sheather, bw=0.13182
Estimation successfully identifies unique optimal solution

| Variable | Coefficient | Std. Error | t-Statistic | Prob. |
|---|---|---|---|---|
| INT4=1 | -2.700000 | 7.483834 | -0.360778 | 0.7189 |
| INT4=2 | 309.4524 | 36.23064 | 8.541178 | 0.0000 |
| INT4=3 | -465.2500 | 63.79498 | -7.292894 | 0.0000 |
| INT4=4 | 5.761538 | 51.63021 | 0.111592 | 0.9113 |
| MILLI*(INT4=1) | -1.70E-16 | 0.948410 | -1.79E-16 | 1.0000 |
| MILLI*(INT4=2) | -22.95238 | 2.247573 | -10.21207 | 0.0000 |
| MILLI*(INT4=3) | 15.62500 | 2.518918 | 6.203060 | 0.0000 |
| MILLI*(INT4=4) | -0.384615 | 1.120709 | -0.343189 | 0.7320 |

| | | | |
|---|---|---|---|
| Pseudo R-squared | 0.581559 | Mean dependent var | -20.73083 |
| Adjusted R-squared | 0.558126 | S.D. dependent var | 50.94502 |
| S.E. of regression | 42.09610 | Objective | 972.6254 |
| Quantile dependent var | -50.80000 | Restr. objective | 2324.400 |
| Sparsity | 58.33377 | | |

Dependent Variable: ACCEL
Method: Quantile Regression (tau = 0.75)
Date: 08/06/18   Time: 06:30
Sample: 1 133
Included observations: 133
Huber Sandwich Standard Errors & Covariance
Sparsity method: Kernel (Epanechnikov) using residuals
Bandwidth method: Hall-Sheather, bw=0.13182
Estimation successfully identifies unique optimal solution

| Variable | Coefficient | Std. Error | t-Statistic | Prob. |
|---|---|---|---|---|
| INT4=1 | 0.790244 | 8.097022 | 0.097597 | 0.9224 |
| INT4=2 | 187.0474 | 45.11377 | 4.146126 | 0.0001 |
| INT4=3 | -493.6204 | 40.81723 | -12.09343 | 0.0000 |
| INT4=4 | 108.7417 | 21.79284 | 4.989789 | 0.0000 |
| MILLI*(INT4=1) | -0.329268 | 1.026552 | -0.320752 | 0.7489 |
| MILLI*(INT4=2) | -13.36842 | 3.000779 | -4.454983 | 0.0000 |
| MILLI*(INT4=3) | 18.34259 | 1.609383 | 11.39728 | 0.0000 |
| MILLI*(INT4=4) | -1.885417 | 0.474424 | -3.974120 | 0.0001 |

| | | | |
|---|---|---|---|
| Pseudo R-squared | 0.385350 | Mean dependent var | -20.73083 |
| Adjusted R-squared | 0.350930 | S.D. dependent var | 50.94502 |
| S.E. of regression | 38.52142 | Objective | 1158.984 |
| Quantile dependent var | 0.000000 | Restr. objective | 1885.600 |
| Sparsity | 61.56041 | | |

**Figure 4.24**   Outputs of the QR(0.25) and QR(0.75) based on the ES in (4.14).

**Example 4.16**   *Outputs of QR (4.14) for Selected Quantiles*
As a comparison with the output of the QR(Median) presented in Figure 4.18, Figure 4.24 presents its outputs of the QR(0.25) and QR(0.75). Based on these outputs, the findings and notes are as follows:

1) Each of these QRs should be applied to generate each of their fitted variables, namely *Fit01_Q25*, and *Fit01_Q75*, in addition to *Fit01_Q50*, all three variable can be generated using the equation specification (4.13).
2) The three QRs for tau = 0.25, 0.50, and 0.75, have adjusted Pseudo R-squared of 0.558, 0.434, and 0.351, respectively.
3) Figure 4.25 presents the scatter graphs of *ACCEL, Fit01_Q25, Fit01_Q50,* and *Fit01_Q75* on *Milli* by *INT4,* as a graphical comparison. Refer to the QSETs, which are presented in Figure 4.20. In fact, the QSETs are representing the tests on the linear effect differences of *Milli* on *ACCEL* between pair of quantiles (0.25, 0.5) and (0.5, 0.75) for each level of *INT4*.

### 4.5.4   Alterative Piecewise Linear QRs

Observing the graphs of the linear QRs in *INT4* = 4, I tried to modify the interval of *MILLI* by generating alternative intervals and observing each of the outcomes as presented in Figure 4.26. Finally, I found two of the best intervals, which are presented in the following examples.

**Example 4.17**   *The First Alternative Piecewise Linear QR*
As the first case, I propose a piecewise linear QR using the following ES:

$$ACCEL \ C \ @Expand\,(INT4a, Dropfirst)\,Milli * @Expand\,(INT4a) \qquad (4.16a)$$

and the *INT4a* of *Milli* having four levels generated using the following equation:

$$INT4a = 1 + 1 * (milli >= 13) + 1 * (milli >= 21) + 1 * (milli >= 28) \qquad (4.16b)$$

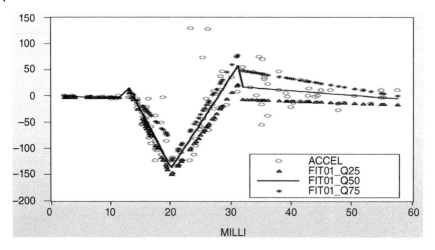

**Figure 4.25** Scatter graphs of *ACCEL*, *Fit01_q25*, *Fit01_q50*, and *Fit01_q75* on Milli.

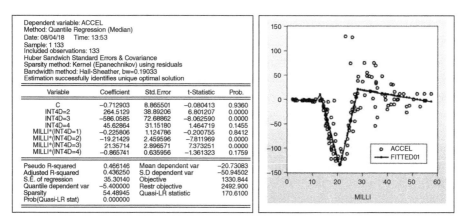

**Figure 4.26** The output of QR(Median) in (4.16a), with the scatter graphs of *ACCEL* and its fitted variable, *FITTED01*, on *MILLI*.

Figure 4.26 presents the statistical results of the QR(Median) in (4.16a) and the scatter graphs of *ACCEL* and its fitted variable, *FITTED02*, on *MILLI*. Compared with the QR in Figure 4.18, having the adjusted R-squared = 0.433, this QR has a slightly greater adjusted R-squared = 0.436. In addition, the scatter graphs of *ACCEL* and its fitted variable, *Fitted02*, on *Milli* do not have a step or break between the last two intervals. So, we can conclude this QR is a better fit model.

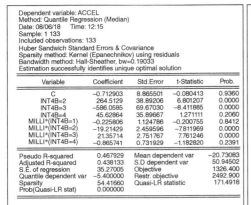
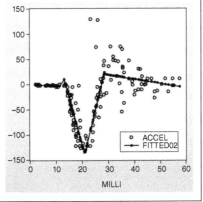

Dependent variable: ACCEL
Method: Quantile Regression (Median)
Date: 08/06/18    Time: 12:15
Sample: 1 133
Included observations: 133
Huber Sandwich Standard Errors & Covariance
Sparsity method: Kernel (Epanechnikov) using residuals
Bandwidth method: Hall-Sheather, bw=0.19033
Estimation successfully identifies unique optimal solution

| Variable | Coefficient | Std.Error | t-Statistic | Prob. |
|---|---|---|---|---|
| C | −0.712903 | 8.865501 | −0.080413 | 0.9360 |
| INT4B=2 | 264.5129 | 38.89206 | 6.801207 | 0.0000 |
| INT4B=3 | −586.0585 | 69.67030 | −8.411885 | 0.0000 |
| INT4B=4 | 45.62864 | 35.89667 | 1.271111 | 0.2060 |
| MILLI*(INT4B=1) | −0.225806 | 1.124786 | −0.200755 | 0.8412 |
| MILLI*(INT4B=2) | −19.21429 | 2.459596 | −7.811969 | 0.0000 |
| MILLI*(INT4B=3) | 21.35714 | 2.751767 | 7.761246 | 0.0000 |
| MILLI*(INT4B=4) | −0.865741 | 0.731929 | −1.182820 | 0.2391 |

| | | | |
|---|---|---|---|
| Pseudo R-squared | 0.467929 | Mean dependent var | −20.73083 |
| Adjusted R-squared | 0.438133 | S.D dependent var | 50.94502 |
| S.E. of regression | 35.27005 | Objective | 1326.400 |
| Quantile dependent var | −5.400000 | Restr. objective | 2492.900 |
| Sparsity | 54.41660 | Quasi-LR statistic | 171.4918 |
| Prob(Quasi-LR stat) | 0.000000 | | |

**Figure 4.27**   The output of QR(Median) in (4.17a), with the scatter graphs of *ACCEL* and its fitted variable, *FITTED02*, on *MILLI*.

### Example 4.18   *Another Alternative Piecewise Linear QR*

As a modification of the QR (4.16a), I propose another piecewise linear QR(Median) using the following ES:

$$ACCEL \; C \; @Expand \, (INT4b, @Dropfirst) \; Milli * @Expand \, (INT4b) \tag{4.17a}$$

$$INT4b = 1 + 1 * (milli >= 13) + 1 * (milli >= 21) + 1 * (milli >= 29) \tag{4.17b}$$

Figure 4.27 presents the statistical results of the QR(Median) in (4.17a) and the scatter graphs of *ACCEL* and its fitted variable, *FITTED02*, on *MILLI*. Compared with both QRs in Figures 4.18 and 4.25, this QR(Median) has the greatest adjusted R-squared = 0.4386. So, we can conclude this QR is the best fit model among the three QRs.

## 4.5.5   Applications of Piecewise Quadratic QRs

As an extension of the piecewise linear QRs, this section presents selected piecewise quadratic QRs of *ACCEL* on *MILLI*, by *INT3*. See the following examples.

### Example 4.19   *The Simplest Piecewise Quadratic QR*

As an extension of the QR (4.17a), Figure 4.28 presents the statistical results of the simplest piecewise quadratic *QR* of *ACCEL* on *MILLI* by *INT3*, using the following ES:

$$ACCEL \; C \; @Expand \, (Int3, @Dropfirst) \; Milli * @Expand \, (Int3) \; Milli \char`^2 * (Int3 = 2) \tag{4.18a}$$

$$Int3 = 1 + 1 * (Milli >= 13) + 1 * (Milli >= 29) \tag{4.18b}$$

Compared with both QRs (4.16a) and (4.17a), this QR(Median) has the smallest adjusted R-squared. So, in the statistical sense, the piecewise linear QR (4.17a) is a better fit model

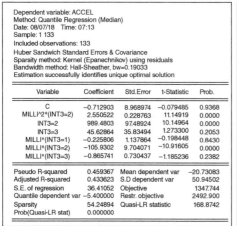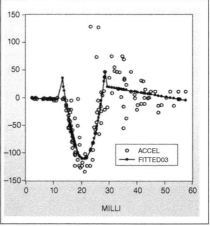

**Figure 4.28** The output of QR(Median) in (4.18a), with the scatter graphs of ACCEL and its fitted variable, *FITTED03*, on MILLI.

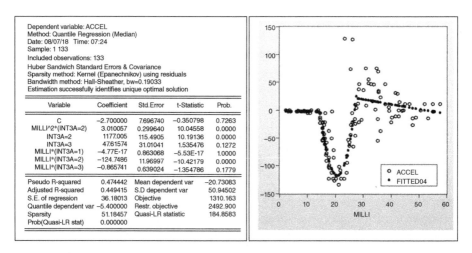

**Figure 4.29** The output of QR(Median) in (4.18a), with the scatter graphs of *ACCEL* and its fitted variable, *FITTED04,* on *MILLI.*

than this piecewise quadratic QR. By looking at its scatter graph, I am very confident a better piecewise can be obtained by modifying the interval of *INT3*.

By using the trial-and-error method to develop alternative intervals of *Milli*, I finally found the output of a quadratic QR(Median), presented in Figure 4.29, having a greater adjusted R-squared than the three QRs (4.16a), (4.17a), and (4.28). I used the following ES:

$$ACCEL\ C\ @Expand\,(Int3a, @Dropfirst)\ Milli * @Expand\,(Int3a)$$
$$Milli\hat{\ }2 * (Int3a = 2) \tag{4.19a}$$

$$Int3a = 1 + 1 * (milli >= 14) + 1 * (milli >= 28) \tag{4.19b}$$

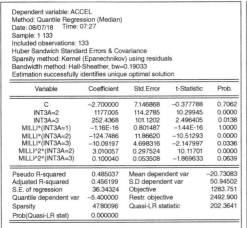
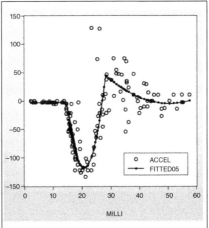

**Figure 4.30** The output of the QR (4.20), with the scatter graphs of *ACCEL* and its fitted variable, *FITTED05*, on *MILLI*.

Note that the functions or IVs, *Milli\*@Expand(Int3a)* and *Milli^2\*(Int3a = 2)*, indicate that the quadratic QR of *ACCEL* on *MILLI* is available only within the interval *Int3a = 2*, as presented in Figure 4.29.

Furthermore, note that the ESs (4.18a) and (4.19a) in fact are the same equations, but use difference intervals with three levels, namely *Int3* and *Int3a*, respectively. As exercises, do this for other intervals, such as by changing the lower bound of $1*(Milli >= 28)$.

### Example 4.20    *A More Advanced Piecewise Quadratic QR*

As an extension of the piecewise QR (4.19a), this example presents a more advanced piecewise quadratic QR(Median) having the following ES, with the statistical results presented in Figure 4.30:

$$ACCEL\ C\ @Expand\ (Int3a, @Dropfirst)\ Milli * @Expand\ (Int3a)$$

$$Milli^2 * @Expand\ (Int3a, @Dropfirst) \tag{4.20}$$

Based on these results, the findings and notes are as follows:

1) The function *Milli^2@Expand(Int3a,@Dropfirst)* in (4.20) indicates the quadratic QRs of *ACCEL* are available within the two intervals *Int3a = 2* and *Int3a = 3*.

2) Compared with all previous QRs of ACCEL, this QR has the greatest adjusted R-squared = 0.458 199. Hence, so far, this piecewise quadratic QR is the best. Nonetheless, I tried to extend this model, by using the interval *Int3a*, which is presented in the following example.

### Example 4.21    *An Extension of the QR (4.20)*

As an extension of the QR (4.20), I propose the following piecewise quadratic QR of *ACCEL* on *MILLI*, by *INT4*:

$$ACCEL\ C\ @Expand\ (INT4, @Dropfirst)\ Milli * @Expand\ (INT4)$$

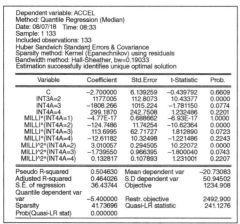
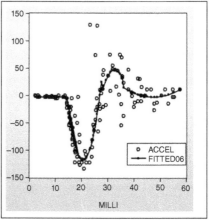

Dependent variable: ACCEL
Method: Quantile Regression (Median)
Date: 08/07/18   Time: 08:33
Sample: 1 133
Included observations: 133
Huber Sandwich Standard Errors & Covariance
Sparsity method: Kernel (Epanechnikov) using residuals
Bandwidth method: Hall-Sheather, bw=0.19033
Estimation successfully identifies unique optimal solution

| Variable | Coefficient | Std.Error | t-Statistic | Prob. |
|---|---|---|---|---|
| C | −2.700000 | 6.139259 | −0.439792 | 0.6609 |
| INT3A=2 | 1177.005 | 112.8073 | 10.43377 | 0.0000 |
| INT4A=3 | −1808.266 | 1015.224 | −1.781150 | 0.0774 |
| INT4A=4 | 299.1870 | 242.7508 | 1.232486 | 0.2201 |
| MILLI*(INT4A=1) | −4.77E-17 | 0.688662 | −6.93E-17 | 1.0000 |
| MILLI*(INT4A=2) | −124.7486 | 11.74254 | −10.62364 | 0.0000 |
| MILLI*(INT4A=3) | 113.6995 | 62.71727 | 1.812890 | 0.0723 |
| MILLI*(INT4A=4) | −12.61182 | 10.32498 | −1.221486 | 0.2243 |
| MILLI^2*(INT4A=2) | 3.010057 | 0.294505 | 10.22072 | 0.0000 |
| MILLI^2*(INT4A=3) | −1.739550 | 0.966395 | −1.800040 | 0.0743 |
| MILLI^2*(INT4A=4) | 0.132817 | 0.107893 | 1.231001 | 0.2207 |

| | | | |
|---|---|---|---|
| Pseudo R-squared | 0.504630 | Mean dependent var | −20.73083 |
| Adjusted R-squared | 0.464026 | S.D dependent var | 50.94502 |
| S.E. of regression | 36.43744 | Objective | 1234.908 |
| Quantile dependent var | | | |
| var | −5.400000 | Restr. objective | 2492.900 |
| Sparsity | 41.73696 | Quasi-LR statistic | 241.1276 |
| Prob(Quasi-LR stat) | 0.000000 | | |

**Figure 4.31** The output of the QR(Median) in (4.21a), with the scatter graphs of *ACCEL* and its fitted variable, *FITTED06*, on *MILLI*.

$$Milli^2 * @Expand\,(INT4, @Dropfirst) \tag{4.21a}$$

$$INT4 = 1 + 1 * (milli >= 14) + 1 * (milli >= 28) + 1 * (milli >= 36) \tag{4.21b}$$

Figure 4.31 presents the output of the QR(Median) of *ACCEL* in (4.21a) and the scatter graphs of *ACCEL* and its fitted variable, *FITTED06,* on *MILLI*. Based on these statistical results, the findings and notes are as follows:

1) The function *Milli^2*@Expand(INT4,@Dropfirst)* indicates the quadratic QRs of *ACCEL* are available within the three intervals *INT4* = 2, *INT4* = 3 and *INT4* = 4.
2) Compared with all previous QRs of ACCEL, this QR(Median) has the greatest adjusted R-squared = 0.464 026. So far, then, this piecewise quadratic QR is the best. However, I tried modifying the interval *INT4*, using the trial-and-error method, to obtain a better fit QR. Four of the findings are presented in the following examples.

**Example 4.22   *The First Modification of the QR (4.21a)***
As the first modification of the GR (4.21a), Figure 4.32 presents the statistical results based on the following ES:

$$ACCEL\ C\ @Expand\,(INT4a, @Dropfirst)\ Milli * @Expand\,(INT4a)$$

$$Milli^2 * @Expand\,(INT4a, @Dropfirst) \tag{4.22a}$$

$$INT4a = 1 + 1 * (milli >= 14) + 1 * (milli >= 28) + 1 * (milli >= 34) \tag{4.22b}$$

Unexpectedly, this QR(Median) of *ACCEL* has a slight smaller adjusted R-squared than the QR (4.21a). For this reason, I present the following modifications.

**Example 4.23   *The Second Modification of the QR (4.21a)***
As the second modification of the QR (4.21a), Figure 4.33 presents the statistical results of the QR(Median) in the following ES:

$$ACCEL\ C\ @Expand\,(INT4b, @Dropfirst)\ Milli * @Expand\,(INT4b)$$

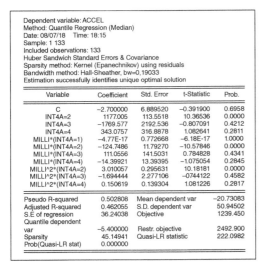
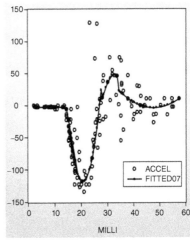

Dependent variable: ACCEL
Method: Quantile Regression (Median)
Date: 08/07/18   Time: 18:15
Sample: 1 133
Included observations: 133
Huber Sandwich Standard Errors & Covariance
Sparsity method: Kernel (Epanechnikov) using residuals
Bandwidth method: Hall-Sheather: bw=0,19033
Estimation successfully identifies unique optimal solution

| Variable | Coefficient | Std. Error | t-Statistic | Prob. |
|---|---|---|---|---|
| C | -2.700000 | 6.889520 | -0.391900 | 0.6958 |
| INT4A=2 | 1177.005 | 113.5518 | 10.36536 | 0.0000 |
| INT4A=3 | -1769.577 | 2192.536 | -0.807091 | 0.4212 |
| INT4A=4 | 343.0757 | 316.8878 | 1.082641 | 0.2811 |
| MILLI*(INT4A=1) | -4.77E-17 | 0.772668 | -6.18E-17 | 1.0000 |
| MILLI*(INT4A=2) | -124.7486 | 11.79270 | -10.57846 | 0.0000 |
| MILLI*(INT4A=3) | 111.0556 | 141.5031 | 0.784828 | 0.4341 |
| MILLI*(INT4A=4) | -14.39921 | 13.39395 | -1.075054 | 0.2845 |
| MILLI^2*(INT4A=2) | 3.010057 | 0.295631 | 10.18181 | 0.0000 |
| MILLI^2*(INT4A=3) | -1.694444 | 2.277106 | -0744122 | 0.4582 |
| MILLI^2*(INT4A=4) | 0.150619 | 0.139304 | 1.081226 | 0.2817 |

| | | | | |
|---|---|---|---|---|
| Pseudo R-squared | 0.502808 | Mean dependent var | -20.73083 | |
| Adjusted R-squared | 0.462055 | S.D. dependent var | 50.94502 | |
| S.E of regression | 36.24038 | Objective | 1239.450 | |
| Quantile dependent var | -5.400000 | Restr. objective | 2492.900 | |
| Sparsity | 45.14941 | Quasi-LR statistic | 222.0982 | |
| Prob(Quasi-LR stat) | 0.000000 | | | |

**Figure 4.32** The output of the QR (4.22a), with the scatter graphs of *ACCEL* and its fitted variable, *FITTED07*, on *MILLI*.

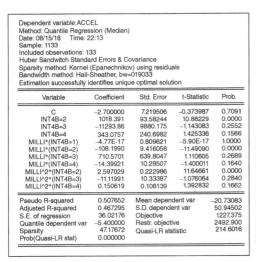
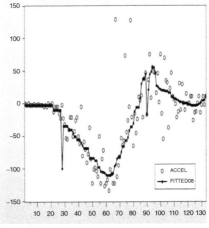

Dependent variable:ACCEL
Method: Quantile Regression (Median)
Date: 08/15/18   Time: 22:13
Sample: 1133
Included observations: 133
Huber Sandwitch Standard Errors & Covariance
Sparsity method: Kernel (Epanechnikov) using residuals
Bandwidth method: Hall-Sheather, bw=019033
Estimation successfully identifies unique optimal solution

| Variable | Coefficient | Std. Error | t-Statistic | Prob. |
|---|---|---|---|---|
| C | -2.700000 | 7.219506 | -0.373987 | 0.7091 |
| INT4B=2 | 1018.391 | 93.58244 | 10.88229 | 0.0000 |
| INT4B=3 | -11293.86 | 9880.175 | -1.143083 | 0.2552 |
| INT4B=4 | 343.0757 | 240.6982 | 1.425336 | 0.1566 |
| MILLI*(INT4B=1) | -4.77E-17 | 0.809621 | -5.90E-17 | 1.0000 |
| MILLI*(INT4B=2) | -108.1990 | 9.416058 | -11.49090 | 0.0000 |
| MILLI*(INT4B=3) | 710.5701 | 639.8047 | 1.110605 | 0.2689 |
| MILLI*(INT4B=4) | -14.39921 | 10.28507 | -1.400011 | 0.1640 |
| MILLI^2*(INT4B=2) | 2.597029 | 0.222986 | 11.64661 | 0.0000 |
| MILLI^2*(INT4B=3) | -11.91991 | 10.33387 | -1.076064 | 0.2840 |
| MILLI^2*(INT4B=4) | 0.150619 | 0.108139 | 1.392832 | 0.1662 |

| | | | | |
|---|---|---|---|---|
| Pseudo R-squared | 0.507652 | Mean dependent var | -20.73083 | |
| Adjusted R-squared | 0.467295 | S.D. dependent var | 50.94502 | |
| S.E. of regression | 36.02176 | Objective | 1227.375 | |
| Quantile dependent var | -5.400000 | Restr. objective | 2492.900 | |
| Sparsity | 47.17672 | Quasi-LR ststistic | 214.6016 | |
| Prob(Quasi-LR stat) | 0.000000 | | | |

**Figure 4.33** The output of the QR(Median) (4.23a), with the scatter graphs of *ACCEL* and its fitted variable, *FITTED08*, on *MILLI*.

$$Milli\^2 * @Expand\,(INT4b, @Dropfirst) \tag{4.23a}$$

$$INT4b = 1 + 1 * (milli >= 14) + 1 * (milli >= 29) + 1 * (milli >= 33) \tag{4.23b}$$

Based on these statistical results, the following findings and notes are presented:

1) Compared with the QR (4.21a) with the adjusted R-squared = 0.464 026, this QR has a slight greater adjusted R-squared = 0.467 286. Even so, in the statistical sense, this QR is a better fit model than the previous QRs.

2) However, unexpectedly, there are two medians far from the medians curve. The observations, numbers 28 and 118, with the values −113.2733 and −1.07E−14, respectively, have a very large difference, with the observed scores of −2.7 and −10.7 of *ACCEL*. I then tried a third modification, which follows.

**Example 4.24** *The Third Modification of the QR (4.21a)*
As the third modification of the GR in (4.21a), Figure 4.34 presents the statistical results of the QR(Median) using the following ES:

$$ACCEL \ C \ @Expand \ (INT4c, @Dropfirst) \ Milli * @Expand \ (INT4c)$$

$$Milli\char`^2 * @Expand \ (INT4c, @Dropfirst) \tag{4.24a}$$

$$INT4c = 1 + 1 * (milli >= 14) + 1 * (milli >= 26) + 1 * (milli >= 33) \tag{4.24b}$$

Based on these statistical results, the following findings and notes are presented:

1) Compared with the QR (4.23a) with adjusted R-squared $= 0.467\,286$, this QR has a slight greater adjusted R-squared $= 0.476\,677$. So, this QR is a better fit model.
2) However, the graph of its fitted value variable, *Fitted09*, in Figure 4.35, still presents a point far from the medians regression, which may be considered as a median outlier. I have tried to experiment using alternative intervals of the variable *MILLI*, but the median outlier can't be omitted. The following example presents the best median regression, with the adjusted R-squared $= 0.4772$, but it still has a median outlier.

```
Dependent Variable: ACCEL
Method: Quantile Regression (Median)
Date: 08/15/18   Time: 22:39
Sample: 1 133
Included observations: 133
Huber Sandwich Standard Errors & Covariance
Sparsity method: Kernel (Epanechnikov) using residuals
Bandwidth method: Hall-Sheather, bw=0.19033
Estimation successfully identifies unique optimal solution
```

| Variable | Coefficient | Std. Error | t-Statistic | Prob. |
|---|---|---|---|---|
| C | -2.700000 | 6.195336 | -0.435812 | 0.6637 |
| INT4C=2 | 1287.143 | 101.4244 | 12.69066 | 0.0000 |
| INT4C=3 | -1455.132 | 1729.467 | -0.841376 | 0.4018 |
| INT4C=4 | 343.0757 | 255.0752 | 1.344998 | 0.1811 |
| MILLI*(INT4C=1) | -4.77E-17 | 0.694940 | -6.87E-17 | 1.0000 |
| MILLI*(INT4C=2) | -137.0079 | 10.51426 | -13.03068 | 0.0000 |
| MILLI*(INT4C=3) | 89.86810 | 116.9619 | 0.768354 | 0.4438 |
| MILLI*(INT4C=4) | -14.39921 | 10.83327 | -1.329166 | 0.1863 |
| MILLI^2*(INT4C=2) | 3.333047 | 0.264343 | 12.60882 | 0.0000 |
| MILLI^2*(INT4C=3) | -1.337644 | 1.964666 | -0.680850 | 0.4973 |
| MILLI^2*(INT4C=4) | 0.150619 | 0.113129 | 1.331398 | 0.1855 |

| | | | | |
|---|---|---|---|---|
| Pseudo R-squared | 0.516323 | Mean dependent var | | -20.73083 |
| Adjusted R-squared | 0.476677 | S.D. dependent var | | 50.94502 |
| S.E. of regression | 35.90095 | Objective | | 1205.758 |
| Quantile dependent var | -5.400000 | Restr. objective | | 2492.900 |
| Sparsity | 41.01823 | Quasi-LR statistic | | 251.0380 |
| Prob(Quasi-LR stat) | 0.000000 | | | |

**Figure 4.34** The output of the QR(Median) (4.24a).

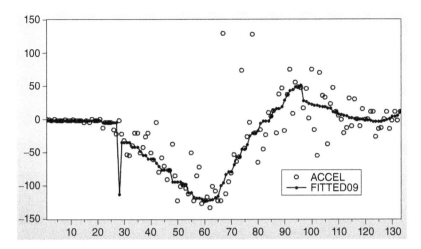

**Figure 4.35** The scatter graphs of *ACCEL* and its fitted variable, *FITTED09*, on *MILLI*.

**Example 4.25** *The Fourth Modification of the QR (4.21a)*

As the fourth modification of the QR in (4.21a), Table 4.2 presents the statistical summary of three QRs of ACCEL for tau = 0.1, 0.5, and 0.9, using the following ES:

$$ACCEL \ Ca \ @Expand \ (INT4d, @Dropfirst) \ Milli * @Expand \ (INT4d)$$

$$Milli\hat{\ }2 * @Expand \ (INT4d, @Dropfirst) \tag{4.25a}$$

$$INT4d = 1 + 1 * (milli >= 15) + 1 * (milli >= 26) + 1 * (milli >= 33) \tag{4.25b}$$

Based on this summary, the following findings and notes are presented:

1) Compared with all previous median regressions of *ACCEL*, this QR(Median) is the best, with an adjusted R-squared = 0.4772. But its fitted value variable still shows a median outlier, as you can see in Figure 4.36, with the value = −111.599, with an observed score −2.7 at the 28th observation.

2) Whatever the output of the estimation process of the QR (including LAD) is, it should be acceptable, in the statistical sense, under the assumption the model is a good model, in the theoretical sense. It is similar for the LS regression or the mean regression (MR). For comparison, do the MR as an exercise. You should obtain a similar form of graph as with the median regressions.

### 4.5.6 Alternative Piecewise Polynomial QRs

As alternative QRs, this section presents selected piecewise polynomial QRs.

**Example 4.26** *An Application of a Piecewise Third Degree Polynomial QR*

In this case, I propose a QR using the following explicit ES:

$$ACCEL = (C\,(10) + C\,(11) * MILLI) * (INT3 = 1) + (C\,(20) + C\,(21) * MILLI$$

$$+ C\,(22) * MILLI\hat{\ }2) * (INT3 = 2) + (C\,(30) + C\,(31) * MILLI +$$

$$C\,(32) * MILLI\hat{\ }2 + C\,(33) * MILLI\hat{\ }3) * (INT3 = 3) \tag{4.26a}$$

**Table 4.2** The statistical summary of three QRs, for tau = 0.1, 0.5, and 0.9.

| Variable | QREG (tau = 0.1) | | | QREG (tau = 0.5) | | | QREG (tau = 0.9) | | |
|---|---|---|---|---|---|---|---|---|---|
| | Coef. | t-Stat. | Prob. | Coef. | t-Stat. | Prob. | Coef. | t-Stat. | Prob. |
| C | 1.885 | 0.382 | 0.703 | 0.631 | 0.110 | 0.913 | 0.000 | 0.000 | 1.000 |
| INT4D = 2 | 962.47 | 3.328 | 0.001 | 1388.5 | 9.818 | 0.000 | 725.41 | 1.089 | 0.278 |
| INT4D = 3 | −798.88 | −0.412 | 0.681 | −1458.5 | −0.834 | 0.406 | 8103.4 | 0.923 | 0.358 |
| INT4D = 4 | −310.60 | −0.761 | 0.448 | 339.74 | 1.344 | 0.181 | 648.63 | 2.378 | 0.019 |
| MILLI*(INT4D = 1) | −1.225 | −2.473 | 0.015 | −0.406 | −0.689 | 0.492 | 0.000 | 0.000 | 1.000 |
| MILLI*(INT4D = 2) | −104.36 | −3.411 | 0.001 | −146.45 | −10.292 | 0.000 | −83.23 | −1.225 | 0.223 |
| MILLI*(INT4D = 3) | 38.15 | 0.288 | 0.774 | 89.868 | 0.760 | **0.449** | −552.57 | −0.889 | 0.376 |
| MILLI*(INT4D*4) | 11.68 | 0.658 | 0.512 | −14.399 | −1.340 | 0.183 | −24.95 | −2.142 | 0.034 |
| MILLI^2*(INT4D = 2) | 2.479 | 3.105 | 0.002 | 3.544 | 10.185 | 0.000 | 2.266 | 1.376 | 0.171 |
| MILLI^2*(INT4D = 3) | −0.396 | −0.176 | 0.861 | −1.338 | −0.673 | **0.502** | 9.470 | 0.863 | 0.390 |
| MILLI^2*(INT4D = 4) | −0.116 | −0.606 | 0.545 | 0.151 | 1.342 | 0.182 | 0.243 | 1.977 | 0.050 |
| Pseudo R-squared | 0.656 | | | 0.517 | | | 0.382 | | |
| Adj. R-squared | 0.628 | | | 0.4772 | | | 0.331 | | |
| S.E. of regression | 49.89 | | | 35.684 | | | 64.386 | | |
| Quantile DV | −101.9 | | | −5.400 | | | 36.200 | | |
| Sparsity | 70.315 | | | 40.899 | | | 113.45 | | |
| Quasi-LR Stat | 267.45 | | | 252.014 | | | 86.806 | | |
| Prob(QLR stat) | 0.0000 | | | 0.0000 | | | 0.0000 | | |

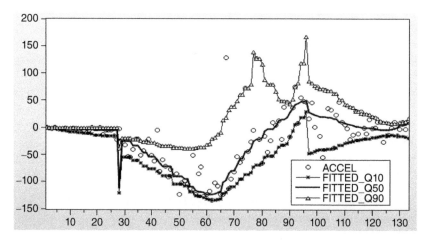

**Figure 4.36** The graphs of *ACCEL* and its fitted value variables, based on the ES (4.25a).

**Table 4.3** Statistical results summary of the QR(tau) (4.26a), for tau = 0.1, 0.5, and 0.9.

| Parameter | QREG(tau = 0.1) | | | QREG(tau = 0.5) | | | QREG(tau = 0.9) | | |
|---|---|---|---|---|---|---|---|---|---|
| | Coef. | t-Stat. | Prob. | Coef. | t-Stat. | Prob. | Coef. | t-Stat. | Prob. |
| C(10) | 1.885 | 0.398 | 0.691 | 0.631 | 0.101 | 0.920 | 0.000 | 0.000 | 1.000 |
| C(11) | −1.225 | −2.560 | 0.012 | −0.406 | −0.641 | 0.523 | 0.000 | 0.000 | 1.000 |
| C(20) | 947.37 | 6.155 | 0.000 | 1389.1 | 9.342 | 0.000 | 1035.2 | 1.413 | 0.160 |
| C(21) | −102.5 | −6.559 | 0.000 | −146.4 | −9.773 | 0.000 | −114.7 | −1.555 | 0.123 |
| C(22) | 2.428 | 6.203 | 0.000 | 3.544 | 9.667 | 0.000 | 3.057 | 1.744 | 0.084 |
| C(30) | −573.2 | −1.113 | 0.268 | −1120.7 | −2.205 | 0.029 | −2145.9 | −2.052 | 0.042 |
| C(31) | 44.031 | 1.119 | 0.265 | 87.084 | 2.272 | 0.025 | 170.36 | 2.095 | 0.038 |
| C(32) | −1.137 | −1.184 | 0.239 | −2.147 | −2.311 | 0.023 | −4.251 | −2.060 | 0.042 |
| C(33) | 0.010 | 1.263 | 0.209 | 0.017 | 2.330 | 0.021 | 0.034 | 1.989 | 0.049 |
| Pseudo R-sq | 0.636 | | | 0.494 | | | 0.389 | | |
| Adj. R-sq | 0.613 | | | 0.461 | | | 0.350 | | |
| S.E. of reg | 47.866 | | | 35.796 | | | 62.902 | | |
| Quantile DV | −101.9 | | | −5.400 | | | 36.200 | | |
| Sparsity | 67.915 | | | 45.592 | | | 113.95 | | |
| Mean DV | −20.731 | | | −20.731 | | | −20.731 | | |
| S.D. DV | 50.945 | | | 50.945 | | | 50.945 | | |

$$Int3 = 1 + 1 * (Milli >= 15) + 1 * (Milli >= 27) \qquad (4.26b)$$

Note the equation (4.26a) clearly shows the QR function within each of the three intervals of *INT3*. Table 4.3 presents the statistical results summary of the QRs in (4.26a) for tau = 0.10, 0.50, and 0.90. Based on these statistical results, the following findings and notes are presented:

1) The parameters $C(k0)$, $k = 1$, 2, and 3 are the intercept parameters of the QRs within each of the intervals *INT3*.
2) Each of the IVs of the QRs, within the three intervals *INT3,* has either positive or negative significant effects on *ACCEL* at the 1%, 5%, 10% or 15% level of significance, accept *MILLI* in *INT3* = 1, for the QR(0.5) and QR(0.9). For instance, consider the following:

   2.1 For the QR(0.1), at the 15% level, $MILLI^3$ has a positive significant adjusted effect on *ACCEL*, based on the *t*-statistic of $t_0 = 1.263$ with a *p*-value = 0.209/2 < 0.15, as presented in the row of C(33).

   2.2 For the QR(0.9), at the 5% level, $MILLI^2$ has a positive significant adjusted effect on *ACCEL*, based on the *t*-statistic of $t_0 = 1.774$ with a *p*-value = 0.084/2 < 0.05, as presented in the row of C(22).
3) Each of the QR functions can be written based on the output of the *Representations*. For the QR(0.50), we have the following three median regressions:

$$\widehat{ACCEL1} = (0.631\,25 - 0.406\,25 * MILLI)$$

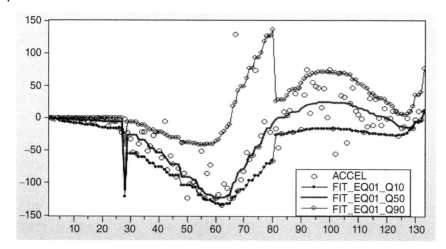

**Figure 4.37** The graphs of *ACCEL* and its three fitted value variables of the QRs.

$$\widehat{ACCEL2} = (1389.086 - 146.447 * MILLI + 3.544 * MILLI\widehat{\ }2)$$

$$\widehat{ACCEL3} = (-1120.682 + 87.084 * MILLI - 2.147 * MILLI\widehat{\ }2 + 0.017 * MILLI\widehat{\ }3)$$

4) As an additional illustration, Figure 4.37 presents the Line and Symbol graphs of *ACCEL*, and the three fitted value variables of the QRs for *tau* = 0.10, 0.50 and 0.90, namely *Fit_EQ01_Q10*, *Fit_EQ01_Q50*, and *Fit_EQ01_Q90*.

**Example 4.27**  *An Application of a Piecewise Fourth Degree Polynomial QR*
As another illustration, Figure 4.38 presents the statistical results of the QR(Median), using the following explicit ES:

$$ACCEL = (C(10) + C(11) * MILLI) * (INT3 = 1) + (C(20) + C(21) * MILLI$$

$$+ C(22) * MILLI\widehat{\ }2) * (INT3 = 2) + (C(30) + C(31) * MILLI +$$

$$C(32) * MILLI\widehat{\ }2 + C(33) * MILLI\widehat{\ }3 + C(34) * MILLI\widehat{\ }4) * (INT3 = 3) \quad (4.27)$$

Based on these statistical results, the following findings and notes are presented:

1) Compared with the previous QR (tau = 0.5) having an adjusted R-squared = 0.461, this QR has a slight greater adjusted R-squared = 0.4686. So, this QR is a better fit model.
2) Each of the IVs of the QR(Median), within each of the three intervals, has either positive or negative significant (adjusted) effect on ACCEL at the 1% level.
3) The QR in the interval *INT3* = 3 is a fourth degree polynomial QR(Median) with the equation as follows:

$$\widehat{ACCEL3} = (-7637.194 + 747.450 * MILLI - 26.623 * MILLI\widehat{\ }2 + 0.411 *$$

$$MILLI\widehat{\ }3 - 0.002 * MILLI\widehat{\ }4)$$

And its IVs have significant joint effects on *ACCEL* for *MILLI* ≥ 27, because the null hypothesis $H_0$: C(31) = C(32) = C(33) = C(34) = 0 is rejected based on the Wald test (Chi-square) of 28.8040 with *df* = 4 and *p*-value = 0.0000.

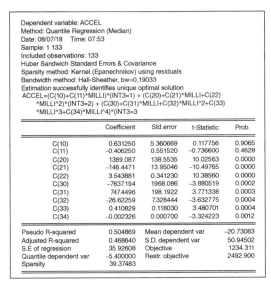

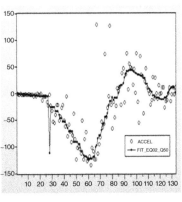

**Figure 4.38** An output of the QR(0.5) in (4.27) and the graphs of *ACCEL* and its fitted value variable.

**Example 4.28** *An Application of a Piecewise Fifth Degree Polynomial QR*

As an advanced QR example, Figure 4.39 presents the statistical results of a fifth degree polynomial QR(Median), using the following explicit ES:

$$ACCEL = (C(10) + C(11) * MILLI) * (INT3 = 1) + (C(20) + C(21) * MILLI$$

$$+ C(22) * MILLI^2) * (INT3 = 2) + (C(30) + C(31) * MILLI$$

$$+ C(32) * MILLI^2 + C(33) * MILLI^3 + C(34) * MILLI^4$$

$$+ C(35) * MILLI^5) * (INT3 = 3) \tag{4.28}$$

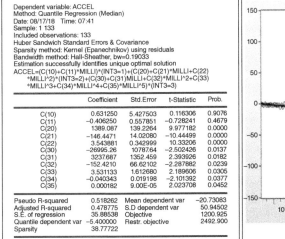

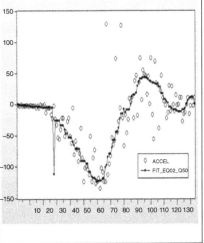

**Figure 4.39** Output of a QR(tau = 0.5) (4.28) with the graphs of *ACCEL* and its fitted value variable.

Based on these statistical results, the following findings and notes are presented:

1) Compared with the piecewise fourth degree polynomial QR(Median) adjusted R-squared = 0.4686, this QR(Median) has a greater adjusted R-squared = 0.4778. So, again, this QR is a better fit model.
2) Each of the IVs of the QR, within each of the three intervals, has either a positive or negative significant (adjusted) effect on *ACCEL* at the 1% or 5% level.
3) The QR in the interval *INT3* = 3 is a fifth degree polynomial QR(Median) with the equation as follows:

$$\widehat{ACCEL3} = (-26\,995.259 + 3237.687 * MILLI - 152.421 * MILLI^2 + 3.531 *$$
$$MILLI^3 - 0.040 * MILLI^4 + 0.0002 * MILLI^5)$$

And its IVs have significant joint effects on *ACCEL*, specific for *MILLI* ≥ 27, because the null hypothesis $H_0$: C(31) = C(32) = C(33) = C(34) = C(35) = 0 is rejected based on the Wald test (Chi-square) of 31.644 014 with $df = 5$ and $p$-value = 0.0000.

### 4.5.7 Applications of Continuous Polynomial QRs

Referring to the scatter graph of *ACCEL* on *MILLI* in Figure 4.17 with its polynomial MR, I tried to apply a high degree polynomial QR of *ACCEL* on *MILLI*. However, the trial-and-error method should be applied to obtain the best possible polynomial QR. The polynomial QR can be presented using the following general ES:

*ACCEL C Milli Milli^2..Milli^k*             (4.29)

**Example 4.29** *Applications of Polynomial QR for* **k = 5, 6, and 7**
As the first illustration, Table 4.4 presents the statistical results summary of three polynomial QR(Median)s in (4.29), for $k = 5, 6$, and 7. Based on these results, the following findings and notes are presented:

1) The coefficients of *Milli^5, Milli^6, and Milli^7* are equal to zero in three decimal points, but in fact they are not zero, as indicated by the values of their $t$-statistic. The cause is the scores of the three IVs are too large compared with the score of *ACCEL*. To increase their coefficients, the independent variable *Milli = Milli/10* can be used to replace either all IVs or only the three IVs, *Milli^5, Milli^6, and Milli^7*. See an illustration presented later in Figure 4.40.
2) Note that each of the IVs of the fifth and sixth degree polynomial QRs has either a positive or negative significant effect on *ACCEL* at the 1% or 5% level. However, it is unexpected that all of the IVs of the seventh degree polynomial QR have Prob. > 0.10. Because, at the 10% level, each of the IVs, *Milli^3* to *Milli^7*, of the seventh degree polynomial QR has either a negative or positive significant effect on *ACCEL* with a $p$-value = *Prob./2* < 0.10, we have a reason not to reduce the model.
3) Since the seventh degree polynomial model has the greatest pseudo R-squared and adjusted R-squared, it is the best fit model among the three models.

**Table 4.4** The outputs summary of the polynomial QR(Median) (4.29) for k = 5, 6, and 7.

| Variable | 5-th degree | | | 6-th degree | | | 7-th degree | | |
| --- | --- | --- | --- | --- | --- | --- | --- | --- | --- |
| | Coef. | t-Stat. | Prob. | Coef. | t-Stat. | Prob. | Coef. | t-Stat. | Prob. |
| C | −81.66 | −2.261 | 0.0255 | −150.5 | −2.564 | 0.0115 | 70.106 | 0.758 | 0.4500 |
| MILLI | 38.598 | 3.034 | 0.0029 | 70.705 | 3.116 | 0.0023 | −52.25 | −0.966 | 0.3358 |
| MILLI^2 | −5.078 | −3.628 | 0.0004 | −9.831 | −3.318 | 0.0012 | 12.570 | 1.186 | 0.2379 |
| MILLI^3 | 0.242 | 3.820 | 0.0002 | 0.552 | 3.158 | 0.0020 | −1.334 | −1.389 | 0.1673 |
| MILLI^4 | −0.005 | −3.814 | 0.0002 | −0.015 | −2.847 | 0.0052 | 0.068 | 1.508 | 0.1340 |
| MILLI^5 | 0.000 | 3.721 | 0.0003 | 0.000 | 2.494 | 0.0139 | −0.002 | −1.555 | 0.1224 |
| MILLI^6 | | | | 0.000 | −2.150 | 0.0334 | 0.000 | 1.554 | 0.1226 |
| MILLI^7 | | | | | | | 0.000 | −1.525 | 0.1299 |
| Pseudo R-sq | 0.2278 | | | 0.2429 | | | 0.3058 | | |
| Adj. R-sq | 0.1974 | | | 0.2069 | | | 0.2669 | | |
| S.E. of reg. | 41.212 | | | 40.201 | | | 40.331 | | |
| Quantile DV | −5.400 | | | −5.400 | | | −5.400 | | |
| Sparsity | 93.322 | | | 92.419 | | | 83.045 | | |
| Quasi-LR stat | 48.678 | | | 52.425 | | | 73.439 | | |
| Prob. | 0.0000 | | | 0.0000 | | | 0.0000 | | |

4) As a graphical presentation, Figure 4.41 presents the graphs of *ACCEL* and the three fitted variables, namely *FITTED_5DP, FITTED_6DP*, and *FITTED_7DP* (DP is degree polynomial). Based the deviation of the scores of *ACCEL* and its corresponding fitted values, I decided to explore higher degree polynomial models, which I present in the following example.

5) Referring to the seventh degree polynomial QR, with each of its IVs having an insignificant effect at the 10% level, I tried to explore a reduced model by deleting at least one $Milli^k$, for $k < 7$. By deleting the IV having the highest probability, namely *MILLI*, we get the output presented in Figure 4.40a. Based on this output, the findings and notes are as follows:

5.1 Each of the IVs of the reduced model has a significant effect at the 1% level. So, in the statistical sense, this reduced regression could be considered a better regression than its full model.

5.2 Compared with the fifth and sixth degree polynomial QRs, this reduced model is the best because it has the largest pseudo R-squared and adjusted R-squared.

5.3 As an additional illustration, Figure 4.40b presents the graphs *ACCEL,* the fitted value variable of the reduced model, namely *FITTED_RM,* and *FITTED_7DP* of the full model, which shows the graphs of *FITTED_RM* and *FITTED_7DP* are very closed. And they have a very large coefficient correlation of 0.975 930.

5.4 In fact, by deleting any one of the IVs, $Milli^k$, for $k < 7$, the reduced model obtained is a better fit model than the fifth and sixth degree polynomial QRs. Even by deleting the IV with the lowest probability namely $Milli^5$, each of the IVs of the reduced

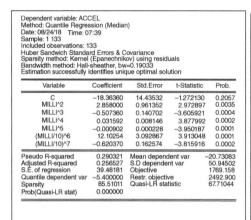

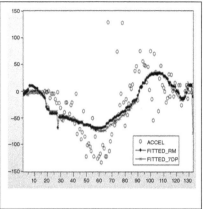

**Figure 4.40** Statistical results of a reduced model of the 7DP QR, and the graphs of *ACCEL*, its fitted value variables *FITTED_RM*, and *FITTED_7DP*.

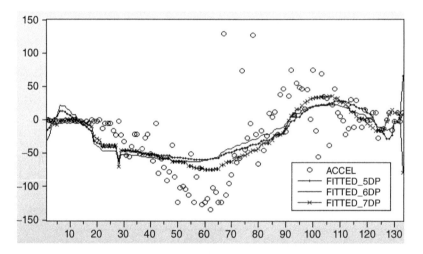

**Figure 4.41** The graphs of *ACCEL* and its fitted value variables in Table 4.4.

model has either a positive or negative significant effect at the 1% level, and *Milli^7* has a negative significant effect, based on the *t*-statistic of $t_0 = -2.519\,703$ with a *p*-value $= 0.0130/2 < 0.01$.

### Example 4.30 *Applications of Polynomial QR for* **k = 8, 9, and 10**

Compared with to the three previous models, Table 4.5 presents the statistical results summary of three polynomial QR(Median)s in (4.29), for *k* = 8, 9, and 10. Based on these results, the following findings and notes are presented:

1) Similar to the previous models, these models can be represented using the IVs *SMilli = Milli/10* to make the coefficient of *Milli^k*, for *k* > 5 not look like zeroes, as presented in Figure 4.40a.

**Table 4.5** Statistical results summary of the polynomial QR(Median) (4.29) for k = 8, 9, and 10.

| Variable | 8-th degree | | | 9-th degree | | | 10-th degree | | |
|---|---|---|---|---|---|---|---|---|---|
| | Coef. | t-Stat. | Prob. | Coef. | t-Stat. | Prob. | Coef. | t-Stat. | Prob. |
| C | 433.56 | 1.714 | 0.0890 | 119.20 | 0.945 | 0.346 | −314.2 | −2.214 | 0.0287 |
| MILLI | −278.0 | −2.139 | 0.0344 | −71.896 | −0.833 | 0.407 | 304.47 | 2.795 | 0.0060 |
| MILLI^2 | 61.650 | 2.478 | 0.0146 | 12.891 | 0.598 | 0.551 | −111.0 | −3.499 | 0.0007 |
| MILLI^3 | −6.479 | −2.724 | 0.0074 | −0.659 | −0.244 | 0.808 | 19.864 | 4.127 | 0.0001 |
| MILLI^4 | 0.364 | 2.862 | 0.0049 | −0.036 | −0.183 | 0.855 | −1.975 | −4.591 | 0.0000 |
| MILLI^5 | −0.012 | −2.916 | 0.0042 | 0.005 | 0.602 | 0.548 | 0.117 | 4.873 | 0.0000 |
| MILLI^6 | 0.000 | 2.911 | 0.0043 | 0.000 | −0.970 | 0.334 | −0.004 | −5.004 | 0.0000 |
| MILLI^7 | 0.000 | −2.870 | 0.0048 | 0.000 | 1.272 | 0.206 | 0.000 | 5.025 | 0.0000 |
| MILLI^8 | 0.000 | 2.808 | 0.0058 | 0.000 | −1.512 | 0.133 | 0.000 | −4.973 | 0.0000 |
| MILLI^9 | | | | 0.000 | 1.700 | 0.092 | 0.000 | 4.876 | 0.0000 |
| MILLI^10 | | | | | | | 0.000 | −4.751 | 0.0000 |
| Pseudo R-sq | 0.3546 | | | 0.361 | | | 0.4617 | | |
| Adj. R-sq | 0.3129 | | | 0.315 | | | 0.4175 | | |
| S.E. of reg. | 37.309 | | | 38.133 | | | 36.652 | | |
| Quantile DV | −5.400 | | | −5.400 | | | −5.400 | | |
| Sparsity | 75.901 | | | 73.769 | | | 55.086 | | |
| Quasi-LR stat | 93.168 | | | 97.675 | | | 167.14 | | |
| Prob. | 0.0000 | | | 0.0000 | | | 0.0000 | | |

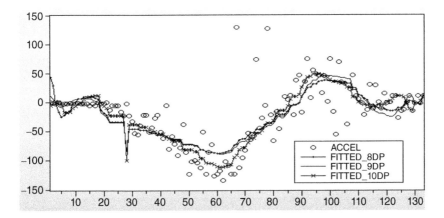

**Figure 4.42** The graphs of *ACCEL* and its fitted value variables in Table 4.5.

2) Among the six polynomial QRs presented, the tenth degree polynomial QR is the best because it has the largest pseudo R-squared and adjusted R-squared. As an exercise, you can try higher degree polynomial models.

3) Corresponding to the ninth degree polynomial QR, the notes presented are as follows:

   3.1 Since at the 5% level, $MillI^9$ has a positive significant adjusted effect on *ACCEL*, based on the *t*-statistic of $t_0 = 1.700$ with a *p*-value $= 0.092/2 = 0.046 < 0.05$, this regression can be accepted to present the ninth degree polynomial QR, even though the other IVs have large *p*-values.

   3.2 In fact, at the 10% level, $Milli^8$ has a negative significant effect, based on the *t*-statistic of $t_0 = -1.512$ with a *p*-value $= 0.132/2 = 0.076 < 0.10$.

   3.3 By deleting any one of the IVs, $Milli^k$, for $k < 8$, we would obtain a reduced model having a significant effect of each of its IVs. For instance, by deleting $Milli^4$, with the highest probability we obtain the output of the reduced QR in Figure 4.43, and by deleting $Milli^7$, with the lowest probability, we obtain the output in Figure 4.44. Note that for the reduced models, I am using $(Milli/10)^k$ for several IVs, in order to have sufficiently large coefficients of the IVs.

4) *Fitted_9DP* and *Fitted_10DP* have unexpected, very small negative values of $-82.1598$ and $-99.1250$, respectively, compared with the observed score $= -2.7000$ of *ACCEL*, at the observation number 28. Why? There is no explanation, which is an impact of the estimation process, the LAD estimation method, and the data which happen to be used. Whatever the results, they should be accepted, because the LAD estimation method has been accepted.

## 4.5.8 Special Notes and Comments

Referring to the four groups of QRs of *ACCEL* on *MILLI*, namely the groups of the piecewise linear QRs, piecewise quadratic QRs, and piecewise polynomial QRs, and the groups of continuous polynomial QRs, Table 4.6 presents their equation specifications, with their adjusted R-squared. Based on this summary, the following findings and notes are presented, specific for MCYCLE.wf1:

1) In general, I consider the group of piecewise polynomial QRs to be the best group; the second is the piecewise quadratic QRs; the third is the piecewise linear QRs; and the last is the continuous polynomial QRs.

2) Among all the models presented in Table 4.6, the model (4.28) is the best QR(Median) because it has the largest adjusted R-squared of 0.478 775. Note that this QR(Median) has specific characteristics, as follows:

   2.1 In the first interval [0,15), the QR(Median) is a simple linear QR of *ACCEL* on *MILLI*.

   2.2 In the second interval [15,27), the QR(Median) is a quadratic QR of *ACCEL* on *MILLI*.

   2.3 And in the third interval [27,last), the QR(Median) is a fifth degree polynomial-QR of *ACCEL* on *MILLI*.

Dependent Variable: ACCEL
Method: Quantile Regression (Median)
Included observations: 133
Huber Sandwich Standard Errors and Covariance
Sparsity method: Kernel (Epanechnikov) using residuals
Bandwidth method: Hall-Sheather, bw = 0.19033
Estimation successfully identifies unique optimal solution

| Variable | Coefficient | Std. error | t-Statistic | Prob. |
|---|---|---|---|---|
| C | 148.8339 | 59.39889 | 2.505669 | 0.0135 |
| MILLI | −90.93882 | 30.48013 | −2.983545 | 0.0034 |
| MILLI^2 | 17.29001 | 4.985060 | 3.468365 | 0.0007 |
| MILLI^3 | −1.178317 | 0.300897 | −3.916016 | 0.0001 |
| MILLI^5 | 0.003617 | 0.000827 | 4.375607 | 0.0000 |
| (MILLI/10)^6 | −181.9314 | 40.97801 | −4.439732 | 0.0000 |
| (MILLI/10)^7 | 40.52004 | 9.127613 | 4.439281 | 0.0000 |
| (MILLI/10)^8 | −4.373438 | 0.994512 | −4.397571 | 0.0000 |
| (MILLI/10)^9 | 0.186171 | 0.042994 | 4.330202 | 0.0000 |

| | | | | |
|---|---|---|---|---|
| Pseudo R-squared | 0.360962 | Mean dependent var | | −20.73083 |
| Adjusted R-squared | 0.319733 | S.D. dependent var | | 50.94502 |
| S.E. of regression | 37.87545 | Objective | | 1593.059 |
| Quantile dependent var | −5.400000 | Restr. objective | | 2492.900 |
| Sparsity | 74.30290 | Quasi-LR statistic | | 96.88358 |
| Prob(Quasi-LR stat) | 0.000000 | | | |

**Figure 4.43** The output of a reduced the ninth degree polynomial QR after deleting *Milli^4*.

2.4 By looking at the graphs of the polynomial QRs, in Figures 4.38, 4.39b and 4.40, I would say the fifth degree polynomial QR in the third interval can be extended to higher degree polynomial model. Do this as an exercise.

3) The second best QR(Median) is the model (4.24), with the following specific characteristics and the adjusted R-squared = 0.476 776.

3.1 In the first interval [0,14), the QR(Median) is a simple linear QR(Median).

3.2 And in the last three intervals [14,16), [16,33) and [33,last), the QRs are quadratic median regressions.

**Table 4.6** Summary of the equation specifications from (4.13) to (4.29) with their adjusted R-squared.

| ES | Quantile-Regression(median) | Adj.R^2 |
|---|---|---|
| | *Piecewise linear QR(median)* | |
| (4.13) | *ACCEL C @Expand(INT4,@Dropfirst)* *MILLI\*@Expand(INT4)* | 0.43796 |
| (4.16) | *ACCEL C @Expand(INT4a,Dropfirst)* *Milli\*@Expand(INT4a)* | 0.436250 |
| (4.17) | *ACCEL C @Expand(INT4b,@Dropfirst)* *Milli\*@Expand(INT4b)* | **0.438133** |
| | *Piecewise quadratic QR(median)* | |
| (4.18) | *ACCEL C @Expand(Int3,@Dropfirst) Milli\*@Expand(Int3)* *Milli^2\*(Int3 = 2)* | 0.433623 |
| (4.20) | *ACCEL C @Expand(Int3a,@Dropfirst)* *Milli\*@Expand(Int3a) Milli^2\*Expand(Int3a,@Dropfirst)* | 0.456199 |
| (4.21) | *ACCEL C @Expand(INT4,@Dropfirst)* *Milli\*@Expand(INT4) Milli^2\*@Expand(INT4,@Dropfirst)* | 0.464026 |
| (4.23) | *ACCEL C @Expand(INT4b,@Dropfirst)* *Milli\*@Expand(INT4b)* *Milli^2\*@Expand(INT4b,@Dropfirst)* | 0.467295 |
| (4.24) | *ACCEL C @Expand(INT4c,@Dropfirst)* *Milli\*@Expand(INT4c)* *Milli^2\*@Expand(INT4c,@Dropfirst)* | **0.476776** |
| | *Piecewise polynomial QR(median)* | |
| (4.26) | *ACCEL = (C(10) + C(11)\*MILLI)\*(INT3 = 1) + (C(20) +* *C(21)\*MILLI + C(22)\*MILLI^2)\*(INT3 = 2) + (C(30) +* *C(31)\*MILLI + C(32)\*MILLI^2 + C(33)\*MILLI^3)\*(INT3 = 3)* | 0.461 |
| (4.27) | *ACCEL = (C(10) + C(11)\*MILLI)\*(INT3 = 1) + (C(20) +* *C(21)\*MILLI + C(22)\*MILLI^2)\*(INT3 = 2) + (C(30) +* *C(31)\*MILLI + C(32)\*MILLI^2 + C(33)\*MILLI^3 +* *C(34)\*MILLI^4)\*(INT3 = 3)* | 0.468640 |
| (4.28) | *ACCEL = (C(10) + C(11)\*MILLI)\*(INT3 = 1) + (C(20) +* *C(21)\*MILLI + C(22)\*MILLI^2)\*(INT3 = 2) + (C(30) +* *C(31)\*MILLI + C(32)\*MILLI^2 + C(33)\*MILLI^3 +* *C(34)\*MILLI^4 + C(35)\*MILLI^5)\*(INT3 = 3)* | **0.478775** |
| | *Continuous polynomial QR(median)* | |
| (4.29) | *ACCEL C Milli Milli^2 ... Milli^k* | |
| 1 | Table 4.4 For k = 6 | 0.2069 |
| 2 | Table 4.4 For k = 7 | 0.2669 |
| 3 | Table 4.5 For k = 8 | 0.3129 |
| 4 | Table 4.5 For k = 9 | 0.3146 |
| 5 | Table 4.5 For k = 10 | **0.4175** |

Dependent variable: ACCEL
Method: quantile regression (Median)
Included observations: 133
Huber sandwich standard errors and covariance
Sparsity method: Kernel (Epanechnikov) using residuals
Bandwidth method: Hall-Sheather, bw = 0.19033
Estimation successfully identifies unique optimal solution

| Variable | Coefficient | Std. error | t-Statistic | Prob. |
|---|---|---|---|---|
| C | 357.3942 | 170.9353 | 2.090816 | 0.0386 |
| MILLI | −226.6939 | 86.93912 | −2.607502 | 0.0102 |
| MILLI^2 | 49.05294 | 16.20366 | 3.027275 | 0.0030 |
| MILLI^3 | −4.925446 | 1.471456 | −3.347327 | 0.0011 |
| MILLI^4 | 0.253937 | 0.071733 | 3.540042 | 0.0006 |
| MILLI^5 | −0.006842 | 0.001885 | −3.629609 | 0.0004 |
| (MILLI/10)^6 | 83.58342 | 22.91370 | 3.647749 | 0.0004 |
| (MILLI/10)^8 | −0.968999 | 0.271920 | −3.563549 | 0.0005 |
| (MILLI/10)^9 | 0.065252 | 0.018692 | 3.490860 | 0.0007 |

| | | | | |
|---|---|---|---|---|
| Pseudo R-squared | 0.357466 | Mean dependent var | | −20.73083 |
| Adjusted R-squared | 0.316012 | S.D. dependent var | | 50.94502 |
| S.E. of regression | 37.33146 | Objective | | 1601.772 |
| Quantile dependent var | −5.400000 | Restr. objective | | 2492.900 |
| Sparsity | 75.49753 | Quasi-LR statistic | | 94.42720 |
| Prob(Quasi-LR stat) | 0.000000 | | | |

**Figure 4.44** The output of a reduced the ninth degree polynomial QR after deleting *Milli^7*.

## 4.6 Quantile Regressions Based on SUSENAS-2013.wf1

As the extension of the ANOVA-QRs of *CWWH* (Children Weekly Working Hours) on *AGE* as a categorical predictor in Section 2.7, based on *SUSENAS-* 2013,wf1, the following examples present alternative QRs of *CWWH* and *LBIRTH* (Live Birth) on *AGE* as a numerical predictor.

### 4.6.1 Application of *CWWH* on *AGE*

As a preliminary analysis, refer to the scatter graphs in Figure 2.23 of *CWWF* on *AGE,* and note that Figures 4.45 and 4.46 present the same scatter graphs, for $10 \leq AGE < 20$, and

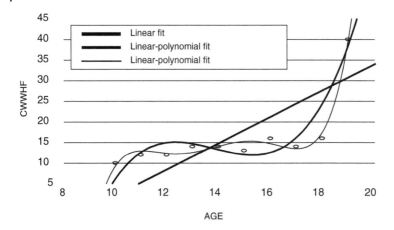

**Figure 4.45** The scatter graphs of *CWWH* < 80 on *AGE* for 10 ≤ *AGE* < 20, with the linear fit and linear polynomial fits having third and fifth degrees.

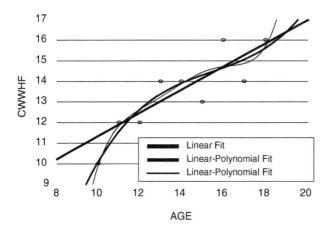

**Figure 4.46** The scatter graphs of *CWWH* < 80 on *AGE* for 10 ≤ *AGE* < 19, with the linear fit and linear polynomial fits having third and fifth degrees.

10 ≤ *AGE* < 19, respectively. However, their linear fits, and the third and fifth degree linear polynomial fits, are specific for *CWWH* < 80, which were obtained using the trial-and-error method. These graphs clearly show that alternative polynomial QR($\tau$)s are better fit QRs than the linear QR.

### 4.6.1.1 Quantile Regressions of CWWH on AGE

**Example 4.31** *Application of Three Alternative QRs of CWWH on AGE*

Figure 4.47 presents the statistical results summary of *CWWH* on *AGE*, as a numerical variable, using the following three ESs:

$$CWWH \ C \ AGE \tag{4.30}$$

| | 10 ≤ AGE < 20 | | | | 10 ≤ AGE < 19 | | |
|---|---|---|---|---|---|---|---|
| **Variable** | **Coef.** | **t-Stat.** | **Prob.** | **Variable** | **Coef.** | **t-Stat.** | **Prob.** |
| C | −36.00 | −59.229 | 0.000 | C | −28.286 | −35.583 | 0.0000 |
| AGE | 4.0000 | 100.77 | 0.000 | AGE | 3.4286 | 59.9150 | 0.0000 |
| Pse. R-sq | 0.1328 | | | Pse. R-sq | 0.0985 | | |
| Adj. R-sq | 0.1328 | | | Adj. R-sq | 0.0984 | | |
| QLR stat | 5601.7 | | | QLR stat | 3059.03 | | |
| Prob | 0.0000 | | | Prob | 0.0000 | | |

| | 10 ≤ AGE < 20 | | | | 10 ≤ AGE < 19 | | |
|---|---|---|---|---|---|---|---|
| **Variable** | **Coef.** | **t-Stat.** | **Prob.** | **Variable** | **Coef.** | **t-Stat.** | **Prob.** |
| C | 49.125 | 2.8166 | 0.0049 | C | −44.571 | −1.7916 | 0.0732 |
| AGE | −6.629 | −1.7868 | 0.0740 | AGE | 13.7333 | 2.4885 | 0.0128 |
| AGE^2 | 0.2500 | 0.9690 | 0.3325 | AGE^2 | −1.2000 | −2.9925 | 0.0028 |
| AGE^3 | 0.0042 | 0.7106 | 0.4774 | AGE^3 | 0.0381 | 3.9927 | 0.0001 |
| Pse. R-sq | 0.1462 | | | Pse. R-sq | 0.1124 | | |
| Adj. R-sq | 0.1461 | | | Adj. R-sq | 0.1123 | | |
| QLR stat | 7468.5 | | | QLR stat | 4145.36 | | |
| Prob | 0.0000 | | | Prob | 0.0000 | | |

| | 10 ≤ AGE < 20 | | | | 10 ≤ AGE < 19 | | |
|---|---|---|---|---|---|---|---|
| **Variable** | **Coef.** | **t-Stat.** | **Prob.** | **Variable** | **Coef.** | **t-Stat.** | **Prob.** |
| C | 981.75 | 1.3431 | 0.1793 | C | 925.00 | 0.7870 | 0.4313 |
| AGE | −374.36 | −1.4190 | 0.1559 | AGE | −352.46 | −0.8075 | 0.4194 |
| AGE^2 | 56.854 | 1.5121 | 0.1305 | AGE^2 | 53.51 | 0.8341 | 0.4042 |
| AGE^3 | −4.255 | −1.6082 | 0.1078 | AGE^3 | −4.0036 | −0.8579 | 0.3909 |
| AGE^4 | 0.1570 | 1.7069 | 0.0878 | AGE^4 | 0.1476 | 0.8786 | 0.3796 |
| AGE^5 | −0.002 | −1.7969 | 0.0724 | AGE^5 | −0.0021 | −0.8902 | 0.3734 |
| Pse. R-sq | 0.1467 | | | Pse. R-sq | 0.1124 | | |
| Adj. R-sq | 0.1466 | | | Adj. R-sq | 0.1122 | | |
| QLR stat | 7406.3 | | | QLR stat | 4125.09 | | |
| Prob | 0.0000 | | | Prob | 0.0000 | | |

**Figure 4.47** (a) The statistical results summary of the linear and third DP-QR(Median)s on *AGE* for *CWWH* < 80, by two age groups 10 ≤ *AGE* < 20 and 10 ≤ *AGE* < 19. (b) The statistical results summary of the fifth degree polynomial DP-QR(Median)s on *AGE*, for *CWWH* < 80, by two age groups 10 ≤ *AGE* < 20 and 10 ≤ *AGE* < 19.

$$CWWH \ C \ AGE \ AGE\hat{\ }2 \ AGE\hat{\ }3 \tag{4.31}$$

$$CWWH \ C \ AGE \ AGE\hat{\ }2 \ AGE\hat{\ }3 \ AGE\hat{\ }4 \ AGE\hat{\ }5 \tag{4.32}$$

Based on these results, the following findings and notes are presented:

1) Specific for the age group $10 \leq AGE < 20$, the QR(0.5) in (4.30) has the smallest adjusted R-squared of 0.1328. And the QR(0.5) in (4.31) and (4.32) have the adjusted R-squared of 0.1461 and 0.1466, respectively. Hence, the QR in (4.32) is the best fit QR, in the statistical sense, even though the three QRs have only small differences.

2) Specific for the age group $10 \leq AGE < 19$, the QR(0.5) in (4.30) has the smallest adjusted R-squared 0.0984. And the QR(0.5) in (4.31) and (4.32) have the adjusted R-squared of 0.1123 and 0.1122, respectively. So, for this age group, the QR(0.5) in (4.31) is the best fit QR, in the statistical sense, even though, again, they have very small differences.

**Example 4.32** *The Forecast Outputs Based on the Three QRs*
As an additional comparison, Figure 4.48 presents the forecast outputs based on the QR(0.5) in (4.30), (4.31), and (4.32). They are called *CWWHF-Lina, CWWHF-Pol3a, CWWHF-Pol5a* for the age group $10 \leq AGE < 20$, and *CWWHF-Linb, CWWHF-Pol3b, CWWHF-Pol5b* for the

| | CWWHF, $10 \leq AGE < 20$ | | | | CWWHF, $10 \leq AGE < 19$ | | | |
|---|---|---|---|---|---|---|---|---|
| AGE | Lina | Pol3a | Pol5a | Obs | Linb | Pol3b | Pol5b | Obs |
| 10 | 4 | 12 | 11 | 456 | 6 | 10.857 | 11 | 456 |
| 11 | 8 | 12 | 12 | 668 | 9.429 | 12 | 12 | 668 |
| 12 | 12 | 12.775 | 13.419 | 1079 | 12.857 | 13.257 | 13.412 | 1079 |
| 13 | 16 | 14.35 | 15 | 1589 | 16.286 | 14.857 | 15 | 1589 |
| 14 | 20 | 16.75 | 16.987 | 2044 | 19.714 | 17.029 | 17 | 2044 |
| 15 | 24 | 20 | 19.847 | 2654 | 23.143 | 20 | 19.863 | 2654 |
| 16 | 28 | 24.125 | 24 | 3461 | 26.571 | 24 | 24 | 3461 |
| 17 | 32 | 29.15 | 29.548 | 4638 | 30 | 29.257 | 29.525 | 4638 |
| 18 | 36 | 35.1 | 36 | 5916 | 33.429 | 36 | 36 | 5916 |
| 19 | 40 | 42 | 42 | 6466 | | | | |
| All | 32 | 29.15 | 29.548 | 28 971 | 26.571 | 24 | 24 | 22 505 |

**Figure 4.48** The scores of the forecast variables of *CWWH* based on the QR(0.5)s in Figure 4.47.

age group $10 \leq AGE < 19$. Based on these forecast outputs, the findings and notes are as follows:

1) Two IVs of the third polynomial QR(Median) for the first age group $10 \leq AGE < 20$ have p-values $> 0.30$, but each IV of the QR(Median) for the second age group $10 \leq AGE < 19$, has either a positive or negative significant effect at the 1% level. For instance, *AGE* has a positive significant effect, based on the *t*-statistic of $t_0 = 2.4885$ with $df = (25505 - 4)$ and p-value $= 0.0128/2 = 0.0059 < 0.01$, and *AGE*^2 has a negative significant effect, based on the *t*-statistic of $t_0 = -2.9925$ with $df = (25505 - 4)$ and p-value $= 0.0028/2$.

2) Each IV of the fifth degree polynomial QR(Median) for the first age group $10 \leq AGE < 20$ has either a positive or negative significant effect at the 5% or 10% level. For instance, *AGE* has a negative significant effect, based on the *t*-statistic of $t_0 = -1.4190$ with $df = (28971 - 6)$ and p-value $= 0.1559/2 = 0.077\,95 < 0.10$, and AGE^5 has a negative significant effect, based on the *t*-statistic of $t_0 = -1.7969$ with $df = (28971 - 6)$ and p-value $= 0.0724/2 = 0.0362 < 0.05$. But each IV of the QR(Median) for the second group has an insignificant effect at the 10% level of significance.

3) As additional outputs, Figure 4.48 presents the values of the forecast variables of *CWWH*, based on the six QRs in Figure 4.47, namely *CWWHF-Lina* and *b, CWWHF-Pol3a* and *b,* and *CWWHF-Pol5a* and *b,* by age group. Figure 4.47 presents their scatter graphs on *AGE*.

4) Note that the graphs clearly show that the values of forecast variables of the third and the fifth degree polynomial QR(Median)s coincide for each age group, with the coefficient correlations of $0.999\,139$ and $0.999\,953$, respectively, for the first and second age groups.

5) Hence, even though each IV of the third degree polynomial QR for the first age group and the fifth polynomial QR for the second age group has insignificant effects with large p-values, their IVs are good predictors for *CWWH*. This because their IVs have significant joint effects, based on the QRL statistic of 7468.5 (df $= 3$) and 4125.09 (df $= 5$) with p-value $= 0.0000$, respectively.

**Example 4.33** *Reduced QR(Median) of Two Selected QRs in Figure 4.47*
As an additional illustration, Figure 4.49 presents the output summary of the reduced models of the third and fifth degree polynomial QRs for the first and second age groups in Figure 4.47. Based on this summary, the following findings and notes are presented:

1) The reduced QR-Pol3a is obtained by deleting the IV, *AGE,* from the third degree polynomial QR for the first age group. In fact, by deleting the IV, *AGE*^2, we also obtain an acceptable third degree polynomial QR(Median) with a slight greater adjusted R-squared of 0.1461.

2) The reduced QR-Pol5b is obtained by using the trial-and-error method of deleting alternative IVs, *AGE, AGE*^2, *AGE*^3 *and AGE*^4, in order to have the fifth degree polynomial QR(Median). Note that *AGE*^5 has a very small p-value. In fact, another fifth degree polynomial QR(Median) also is available with a larger p-value for *AGE*^5.

3) The pair of scatter graphs presented in Figure 4.50a,b show they coincide.

| | Reduced QR-Pol3a | | | Reduced QR-Pol5b | | |
|---|---|---|---|---|---|---|
| Variable | Coef. | t-Stat. | Prob. | Coef. | t-Stat. | Prob. |
| C | 16.6618 | 14.4255 | 0.0000 | 3.4255 | 0.2195 | 0.8263 |
| AGE | | | | 1.3038 | 0.4525 | 0.6509 |
| AGE^2 | −0.1880 | −11.523 | 0.0000 | −0.0701 | −0.5024 | 0.6154 |
| AGE^3 | 0.0136 | 19.2040 | 0.0000 | | | |
| AGE^4 | | | | | | |
| AGE^5 | | | | 0.0000 | 3.4905 | 0.0005 |
| Pse. R-sq | 0.1461 | | | 0.1124 | | |
| Adj. R-sq | 0.1460 | | | 0.1123 | | |
| Quant DV | 28 | | | 24 | | |
| QLR stat | 7352.64 | | | 4147.58 | | |
| Prob | 0.0000 | | | 0.0000 | | |

**Figure 4.49** The outputs of the reduced QR of two selected QR(0.5)s in Figure 4.47.

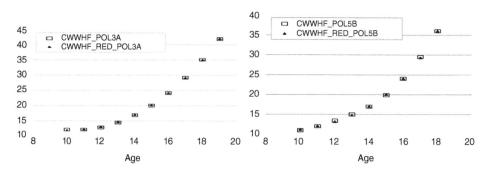

**Figure 4.50** (a) Scatter graph of *CWWHF_Pol3a* and its reduced model. (b) Scatter graph of *CWWHF_Pol5b* and its reduced model.

### Example 4.34  *Application of a Sixth Degree Polynomial QR of CWWH on AGE*

Through additional trial-and-error for a polynomial QR(Median) of *CWWH* on *AGE*, I obtained one of the best reduced sixth degree polynomial QRs, with the output presented in Figure 4.51. Based on this output, the findings and notes are as follows:

1) Each IV has either a positive or negative significant effect at the 5% level of significance. For instance, *AGE^6* has a negative significant effect, based on the *t*-statistic of $t_0 = -1.737\,434$ with $df = (25\,505 - 5)$ and p-value $= 0.0832/2 = 0.0416$. And the four IVs have significant joint effects, based on the *QRL*-statistic of $QRL_0 = 4150.772$ with $df = 4$ and p-value $= 0.0000$.

Dependent variable: CWWH
Method: quantile regression (Median)
**Sample: 1 1094179 IF CWWH < 80 AND AGE > =10 AND AGE < 19**
Included observations: 22505
Estimation successfully identifies unique optimal solution

| Variable | Coefficient | Std. error | t-Statistic | Prob. |
|----------|-------------|-----------|-------------|-------|
| C | −15.89620 | 11.45501 | −1.387707 | 0.1652 |
| AGE^3 | 0.180817 | 0.091603 | 1.973914 | 0.0484 |
| AGE^4 | −0.027945 | 0.014784 | −1.890238 | 0.0587 |
| AGE^5 | 0.001530 | 0.000838 | 1.826105 | 0.0678 |
| AGE^6 | −2.82E−05 | 1.63E−05 | −1.732434 | 0.0832 |

| | | | | |
|---|---|---|---|---|
| Pseudo R-squared | 0.112434 | Mean dependent var | 27.53042 |
| Adjusted R-squared | 0.112276 | S.D. dependent var | 18.00157 |
| S.E. of regression | 16.89092 | Objective | 149578.0 |
| Quantile dependent var | 24.00000 | Restr. objective | 168526.0 |
| Sparsity | 36.51954 | Quasi-LR statistic | 4150.772 |
| Prob(quasi-LR stat) | 0.000000 | | |

**Figure 4.51** The output of a reduced sixth degree polynomial QR for $10 \leq AGE < 19$.

2) Figure 4.52 shows that the five forecast variables of *CWWH* are highly or significantly correlated. We also have many more possible QRs with their forecast variables highly correlated, either some or all their independent variables have large *p*-values, such as the fifth degree polynomial QR for the second age group. However, we can never say which one gives the best prediction for the *population quantiles*, because we never observe the population, which is an abstract set of individuals or objects in all sample surveys.

3) In addition, all polynomial QRs of *CWWH* on *AGE* have demonstrated the unpredictable impacts of the multicollinearity between their IVs, which can never be removed or controlled, because $AGE^k$ and $AGE^n$ for all pairs of $k$ and $n$ are highly or significantly correlated.

Hence, the best possible QR is the QR which has the best fit to the data, that is, having the largest pseudo R-squared or adjusted R-squared, in the quantile regression sense, referring to the statement of Bezzecri (1973, in Gifi, 1990, p. 25): "*The model must follow the data, and not the other way around*". And referring to the statement "*In data analysis we must look to very emphasis on judgment*" (Tukey 1962, in Gifi, 1990, p. 22), I would say a researcher can use his or her best possible *expert's judgment* to select the best QR from among a set of defined alternative QRs.

| Covariance analysis: ordinary | | | | | |
| --- | --- | --- | --- | --- | --- |
| Included observations: 22505 | | | | | |
| Correlation | | | | | |
| t-Statistic | | | | | |

| Probability*_ | *_INB | *_POL3B | *_POL5B | *_RED_POL5B | *_POL6B |
| --- | --- | --- | --- | --- | --- |
| CWWHF_LINB | 1.000000 | | | | |
| | — | | | | |
| CWWHF_POL3B | 0.961831 | 1.000000 | | | |
| | 527.2683 | — | | | |
| | 0.0000 | — | | | |
| CWWHF_POL5B | 0.961220 | 0.999876 | 1.000000 | | |
| | 522.8493 | 9522.234 | — | | |
| | 0.0000 | 0.0000 | — | | |
| CWWHF_RED_ | | | | | |
| POL5B | 0.961232 | 0.999982 | 0.999846 | 1.000000 | |
| | 522.9338 | 25155.36 | 8533.765 | — | |
| | 0.0000 | 0.0000 | 0.0000 | — | |
| CWWHF_POL6B | 0.962851 | 0.999861 | 0.999903 | 0.999760 | 1.000000 |
| | 534.8877 | 8980.859 | 10779.32 | 6841.389 | — |
| | 0.0000 | 0.0000 | 0.0000 | 0.0000 | — |

**Figure 4.52** Bivariate correlation matrix based on five forecast variables of *CWWH* for $10 \leq AGE < 19$.

### 4.6.1.2 Application of Logarithmic QRs

**Example 4.35** *Application of* **log(CWWH)** *on AGE*
I am quite sure that all the QRs of *CWWH* on *AGE* presented in previous example can be applied for the QRs of *log(CWWH)*. However, in this example, I present only the fifth degree polynomial QR of *log(CWWH)* on *AGE*, using the following ES, which is derived from the ES in (4.32):

$$log\,(CWWH)\ C\,AGE\,AGE\char`^2\,AGE\char`^3\,AGE\char`^4\,AGE\char`^5 \qquad (4.33)$$

Figure 4.53 presents the output of the QR(Median) and its reduced model similar to the reduced model in Figure 4.54. Based on these outputs, the findings and notes are as follows:

1) Using the sample $\{CWWH < 98$ and $10 = < AGE < 19\}$, we obtain the error message "Log of non positive". For this reason, the statistical results presented based on the sample $\{CWWH > 0$ and $CWWH < 98$ and $10 = < AGE < 19\}$, with 22 135 observations, are compared to 22 505 observations for the previous example.

| Variable | 5th degree Pol | | | 4th degree Pol | | | 3rd degree Pol | | |
|---|---|---|---|---|---|---|---|---|---|
| | Coef. | t-Stat. | Prob. | Coef. | t-Stat. | Prob. | Coef. | t-Stat. | Prob. |
| C | 116.56 | 1.794 | 0.073 | -27.65 | -2.916 | 0.004 | 4.436 | 3.023 | 0.003 |
| AGE | -41.75 | -1.750 | 0.080 | 8.607 | 3.101 | 0.002 | -0.498 | -1.585 | 0.113 |
| AGE^2 | 6.049 | 1.745 | 0.081 | -0.923 | -3.072 | 0.002 | 0.036 | 1.610 | 0.107 |
| AGE^3 | -0.435 | -1.744 | 0.081 | 0.044 | 3.071 | 0.002 | -0.001 | -1.149 | 0.251 |
| AGE^4 | 0.016 | 1.750 | 0.080 | -0.001 | -3.044 | 0.002 | | | |
| AGE^5 | 0.000 | -1.759 | 0.079 | | | | | | |
| Pse. R-sq | 0.118 | | | 0.118 | | | 0.118 | | |
| Adj.R-sq | 0.118 | | | 0.118 | | | 0.118 | | |
| S.E. of reg. | 0.722 | | | 0.723 | | | 0.721 | | |
| Quant dv | 3.178 | | | 3.178 | | | 3.178 | | |
| Sparsity | 1.729 | | | 1.727 | | | 1.710 | | |
| QLR Stat | 3824.6 | | | 3823.3 | | | 3858.0 | | |
| Prob | 0.000 | | | 0.000 | | | 0.000 | | |

**Figure 4.53** The statistical results summary of the QR(0.5) in (4.33) and two of its reduced models.

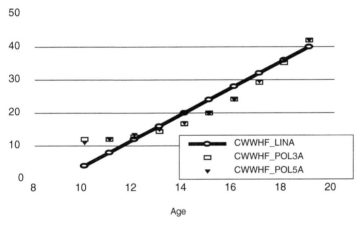

**Figure 4.54** Scatter graphs of the forecast variables of the three QR(Median)s in Figure 4.47.

2) In contrast to the fifth degree polynomial QR in Figure 4.32, each IV of this fifth degree polynomial QR(Median) unexpectedly has either a positive or negative significant effect at the 5% level. For instance, $AGE^5$ has a negative significant effect, based on the $t$-statistic of $t_0 = -1.759$ with $df = (22\,135 - 6)$ and $p$-value $= 0.079/2 = 0.0395 < 0.05$.

3) Then, in contrast to the third degree polynomial QR(0.5) of *CWWH* in Figure 4.47, this figure presents the output of the third degree polynomial QR(0.5)s of *log(CWWH)* beside the output of the fourth degree polynomial. Based on these outputs, the findings and notes are as follows:

    3.1 Again, it is unexpected that each IV of the third degree polynomial QR(0.5) of *log(CWWH)* has an insignificant effect at 10% level, compared with the QR of *CWWH* in Figure 4.47.

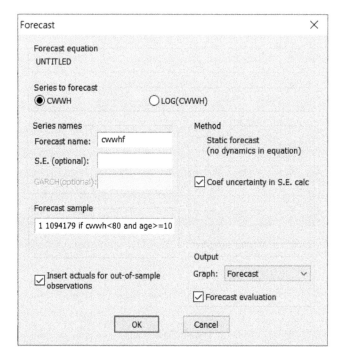

**Figure 4.55** The *Forecast* options .

| | CWWHF-L1 | CWWHF-LP3 | CWWHF-LP4 | CWWHF-LP5 | |
|---|---|---|---|---|---|
| AGE | Median | Median | Median | Median | Obs. |
| 10 | 9.684 | 11.353 | 10 | 12 | 442 |
| 11 | 11.148 | 12 | 12 | 12 | 659 |
| 12 | 12.834 | 13.102 | 13.496 | 13.243 | 1063 |
| 13 | 14.774 | 14.725 | 14.988 | 15 | 1570 |
| 14 | 17.008 | 16.975 | 16.999 | 17.141 | 2020 |
| 15 | 19.580 | 20 | 20 | 20 | 2610 |
| 16 | 22.540 | 24 | 24.340 | 24.177 | 3401 |
| 17 | 25.948 | 29.229 | 30 | 30 | 4546 |
| 18 | 29.871 | 36 | 36 | 36 | 5824 |
| All | 22.540 | 24 | 24.340 | 24.177 | 22 135 |

**Figure 4.56** Values of four forecast variables based on the QR(Median)s.

**Figure 4.57** The scatter graphs of *CWWHF_L1*, *CWWHF_LP3*, *CWWHF_LP4,* and *CWWHF_LP5* on *AGE*.

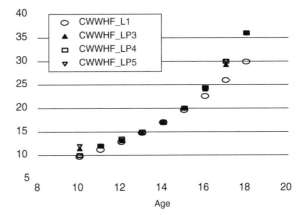

3.2 And, unexpectedly, the fourth degree polynomial QR of *log(CWWH)* is the best QR among the three QRs, since each of its IVs has either a positive or negative significant effect at the 1% level.

3.3 Note that the three QRs have the same values of pseudo-R-squared and adjusted R-squared in three decimal points.

4) Note that for the QR of *log(CWWH)*, the *Forecast* feature provides two options we can use to compute either the forecast values of *log(CWWH)* or *CWWH*, as presented in Figure 4.55. The figure shows *CWWHF* as the *Series to forecast*, with the method being the *Static forecast*, because we are using cross-section data. Refer to the "dynamic forecast" discussed in Agung (2019).

5) As an illustration, Figure 4.56 presents three forecast variables of QR(Median) in (4.33) of *CWWHF-LP5* and three forecast variables *CWWHF_LI*, *CWWHF_LP3*, and *CWWHF_LP4*, based on the following QR(Median):

$$\log(CWWH)\ C\ AGE \tag{4.34}$$

$$\log(CWWH)\ C\ AGE\ AGE\hat{}\ 2\ AGE\hat{}\ 3 \tag{4.35}$$

$$\log(CWWH)\ C\ AGE\ AGE\hat{}\ 2\ AGE\hat{}\ 3\ AG\hat{}\ 4 \tag{4.36}$$

6) Figure 4.57 presents the scatter graphs of the four forecast variables, namely *CWWHF_L1*, *CWWHF_L3*, *CWWHF_LP4*, and *CWWHF_LP5*, on *AGE*, which show the last three almost coincide. Note that Figure 4.53 shows that the three forecasts of the polynomial QR(Median)s have the same values of adjusted R-squared in three decimal points, but the third degree polynomial QR(Median) slightly has the smallest standard error of regression. Hence, in the statistical sense, we may consider it to be the best QR(Median) of *CWWH*.

**Example 4.36** *Unexpected Forecast Findings*

Figure 4.58 presents descriptive statistics of the forecast value variables based on the third, fourth, and fifth degree polynomial QR(0.1)s, using the ESs in (4.35), (4.36), and (4.33), respectively. Based on these descriptive statistics, the findings and notes are as follows:

1) The output of the fourth degree polynomial QR(0.1) of *log(CWWH)* is presented in Figure 4.59. However, it is very unexpected that running the *Forecast* option

| | LCWWHF_LP3 | LCWWHF_LP4 | LCWWHF_LP5 |
|---|---|---|---|
| Mean | 2.109373 | NA | 2.131285 |
| Median | 2.079442 | NA | 2.079442 |
| Maximum | 2.484907 | NA | 2.484907 |
| Minimum | 1.609438 | NA | 1.609438 |
| Std. Dev. | 0.299871 | NA | 0.286113 |
| Skewness | −0.173455 | NA | −0.165732 |
| Kurtosis | 1.694516 | NA | 1.676634 |
| Jarque-Bera | 1682.847 | NA | 1716.539 |
| Probability | 0.000000 | NA | 0.000000 |
| Sum | 46 690.98 | NA | 47176.00 |
| Sum Sq. Dev. | 1990.352 | NA | 1811.903 |
| Observations | 22135 | 0 | 22135 |

**Figure 4.58** Descriptive statistics of the forecast value variables based on the third, fourth, and fifth degree polynomial QR(0.1)s, using the ESs in (4.35), (4.36), and (4.33).

produces the message "Overflow," but the variable *LCWWHF_LP4* is inserted as a new variable in the data file. It is different from the third and fifth degree polynomial QR(0.1)s of *log(CWWH)*.

2) However, for tau = 0.4, 0.5, and 0.6, the forecast value variables are obtained, with their scatter graphs presented in Figure 4.60.

3) In addition, Figure 4.61 presents the output of a QSET, based on two selected pairs of quantiles. The QSET between the pairs of quantiles (0.4, 0.6) can easily be done. See the example 4.12 for how to conduct alternative QSETs.

4) In fact, I also obtained the overflow message based on the fourth degree polynomial QR(tau) for tau = 0.2, 0,3, 0,7, 0.8, and 0.9. So far, there is no explanation for this. See the following example.

**Example 4.37** *Correction of the Overflow Message*

To solve the overflow error, I followed the advice from my EViews provider and generated a new dependent variable, *LCWWH = log(CWWH)*, to replace the dependent variable *log(CWWH)* of the fourth degree polynomial QR(0.1) in (4.36), with the output presented in Figure 4.62. With the output still on the screen, clicking the *Forecast* button and then *OK* inserts the forecast value variable *LCWWHF* as a new variable in the data file. Then, we can generate the forecast value variable *CWWHF = Exp(LCWWHF)* and develop a list of scores, as presented in Figure 4.63.

In fact, even though there is no overflow message, all QR(tau) of *log(Y)* could be presented as QR(tau) of *LY*, where *LY* is a new generated variable *LY = log(Y)*.

Method: quantile regression (tau = 0.1)
Date: 01/02/19   Time: 05:37
**Sample: 1 1094179 IF CWWH < 80 AND AGE > = 10 AND AGE < 19 AND CWWH > 0**
Included observations: 22135
Huber sandwich standard errors and covariance
Sparsity method: Kernel (Epanechnikov) using residuals
Bandwidth method: Hall-Sheather, bw = 0.012323
Estimation successfully identifies unique optimal solution

| Variable | Coefficient | Std. error | t-Statistic | Prob. |
|---|---|---|---|---|
| C | 15.40181 | 19.37892 | 0.794771 | 0.4268 |
| AGE | −4.056272 | 5.597830 | −0.724615 | 0.4687 |
| AGE^2 | 0.438888 | 0.597885 | 0.734068 | 0.4629 |
| AGE^3 | −0.020919 | 0.028013 | −0.746764 | 0.4552 |
| AGE^4 | 0.000380 | 0.000486 | 0.781543 | 0.4345 |

| | | | | |
|---|---|---|---|---|
| Pseudo R-squared | 0.057886 | Mean dependent var | 3.081157 |
| Adjusted R-squared | 0.057716 | S.D. dependent var | 0.774629 |
| S.E. of regression | 1.201442 | Objective | 3185.499 |
| Quantile dependent var | 2.079442 | Restr. objective | 3381.224 |
| Sparsity | 4.630554 | Quasi-LR statistic | 939.2935 |
| Prob(Quasi-LR stat) | 0.000000 | | |

**Figure 4.59** The output of the fourth degree polynomial QR(0.1) in (4.36).

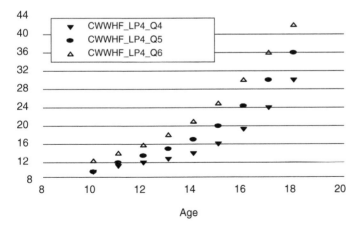

**Figure 4.60** The scatter graphs of three forecast value variables, based on the fourth degree polynomial QR(tau) of *log(CWWH)*, for tau = 0.4, 0.5, and 0.6.

Quantile slope equality test
Equation: EQ_(4.36)
Specification: LOG(CWWH) C AGE AGE^2 AGE^3 AGE^4
Estimated equation quantile tau = 0.4
User-specified test quantiles: 0.4 0.5 0.6
Test statistic compares all coefficients

| Test summary | Chi-Sq. statistic | Chi-Sq. d.f. | Prob. |
|---|---|---|---|
| Wald test | 297.4759 | 8 | 0.0000 |

Restriction detail: b(tau_h) − b(tau_k) = 0

| Quantiles | Variable | Restr. value | Std. error | Prob. |
|---|---|---|---|---|
| 0.4, 0.5 | AGE | −1.323576 | 1.763893 | 0.4530 |
| | AGE^2 | 0.140125 | 0.191364 | 0.4640 |
| | AGE^3 | −0.006953 | 0.009104 | 0.4450 |
| | AGE^4 | 0.000133 | 0.000160 | 0.4074 |
| 0.5, 0.6 | AGE | 5.303808 | 1.846713 | 0.0041 |
| | AGE^2 | −0.551409 | 0.198325 | 0.0054 |
| | AGE^3 | 0.024958 | 0.009345 | 0.0076 |
| | AGE^4 | −0.000415 | 0.000163 | 0.0109 |

**Figure 4.61** The output of the QSET of elected pairs of quantiles based on the QR in (4.36).

## 4.6.2 An Application of Life-Birth on *AGE* for Ever Married Women

As a preliminary analysis, Figure 4.64 presents the scatter graph of the forecast value variable of *LBIRTH* (Life Birth), namely *LBIRTHF*, on *AGE (from 15 to 49 years)*. It is based on a QR(Median) with the following ES, with its linear fit and third and fifth degree polynomial fits:

$$LBIRTH\ @Expand\ (AGE) \tag{4.37}$$

Based on these graphs, the following findings and notes are presented:

1) The linear (regression) fit can be considered a good fit regression.
2) It is very unexpected that the forecast value variable, *LBIRTHF*, based on the QR(Median) in (4.37), has only five values, namely 0, 1, 2, 3, and 4.

Dependent variable: LCWWH
Method: quantile regression (tau = 0.1)
Date: 01/05/19   Time: 05:02
*Sample: 1 1094179 IF CWWH < 80 AND AGE > =10 AND AGE < 19 AND CWWH > 0*
Included observations: 22135
Huber sandwich standard errors and covariance
Sparsity method: Kernel (Epanechnikov) using residuals
Bandwidth method: Hall-Sheather, bw = 0.012323
Estimation successfully identifies unique optimal solution

| Variable | Coefficient | Std. error | t-Statistic | Prob. |
|---|---|---|---|---|
| C | 15.40181 | 19.37892 | 0.794772 | 0.4268 |
| AGE | −4.056272 | 5.597830 | −0.724615 | 0.4687 |
| AGE^2 | 0.438888 | 0.597885 | 0.734068 | 0.4629 |
| AGE^3 | −0.020919 | 0.028013 | −0.746764 | 0.4552 |
| AGE^4 | 0.000380 | 0.000486 | 0.781543 | 0.4345 |

| | | | |
|---|---|---|---|
| Pseudo R-squared | 0.057886 | Mean dependent var | 3.081157 |
| Adjusted R-squared | 0.057716 | S.D. dependent var | 0.774629 |
| S.E. of regression | 1.201442 | Objective | 3185.499 |
| Quantile dependent var | 2.079442 | Restr. objective | 3381.224 |
| Sparsity | 4.630554 | Quasi-LR statistic | 939.2935 |
| Prob(Quasi-LR stat) | 0.000000 | | |

**Figure 4.62**  The output of the fourth degree polynomial QR(0.1) of *LCWWH*.

3) In addition, the third and fifth degree linear polynomial fits are presented as two of the best possible fits from among several alternative polynomial fits, which are obtained using the trial-and-error method.

### 4.6.2.1   QR(Median) of LBIRTH on AGE as a Numerical Predictor

Referring to the three linear fits in Figure 4.64, the following examples present the application of QRs of *LBIRTH on AGE* as a numerical predictor, using the following ESs:

$$LBIRTH \ C \ AGE \tag{4.38}$$

$$LBIRTH \ C \ AGE \ AGE\char`^2 \ AGE\char`^3 \tag{4.39}$$

$$LBIRTH \ C \ AGE \ AGE\char`^2 \ AGE\char`^3 \ AGE\char`^4 \ AGE\char`^5 \tag{4.40}$$

| | LCWWHF | Exp(CWWHF) | |
|---|---|---|---|
| **AGE** | **Median** | **Median** | **Obs.** |
| 10 | 1.6094 | 5 | 442 |
| 11 | 1.6094 | 5 | 659 |
| 12 | 1.6591 | 5.2548 | 1063 |
| 13 | 1.7379 | 5.6856 | 1570 |
| 14 | 1.8343 | 6.2609 | 2020 |
| 15 | 1.9459 | 7 | 2610 |
| 16 | 2.0794 | 8 | 3401 |
| 17 | 2.2508 | 9.4951 | 4546 |
| 18 | 2.4849 | 12 | 5824 |
| All | 2.0794 | 8 | 22 135 |

**Figure 4.63** The forecast value variables, based on the QR(0.1) in Figure 4.64.

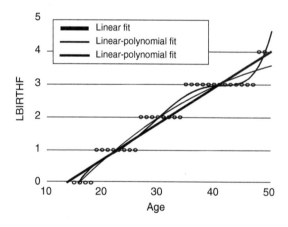

**Figure 4.64** Scatter graph of the forecast value variable of the QR(Median) in (4.37) on *AGE* with its linear fit and third and fifth linear polynomial fits.

**Example 4.38** *An Application of QR(Median) in (4.38) to (4.40)*

Figure 4.65 shows the statistical results summary of the QR(Median)s in (4.38) to (4.40). Based on this summary, the following findings and notes are presented:

1) Each IV of the three QR(Median)s has a significant effect on *LBIRTH* at the 1% level. It is similar for each dummy IV in QR(Median) (4.37), except the IV, *AGE* of the QR in (4.39), which has a significant effect at the 5% level. So, in the statistical sense, the QR(Median)s are acceptable or good fit models.
2) However, the QR(Median) of *LBIRTH* on *@Expand(AGE)* in (4.37) has the largest pseudo R-squared and adjusted R-squared of 0.174 235, and 0.171 373, respectively. So the QR in (4.37) should be considered the best fit QR among the four QR(Median)s. The second best is the fifth degree polynomial QR in (4.40).

| Variable | QR(Median) in (4.40) | | | QR(Median) in (4.39) | | | QR(Median) in (4.38) | | |
|---|---|---|---|---|---|---|---|---|---|
| | Coef. | t-Stat. | Prob. | Coef. | t-Stat. | Prob. | Coef. | t-Stat. | Prob. |
| C | −58.39 | −7.770 | 0.000 | 1.819 | 1.231 | 0.218 | −1.316 | −20.13 | 0.000 |
| AGE | 10.69 | 7.785 | 0.000 | −0.269 | −2.031 | 0.042 | 0.105 | 48.57 | 0.000 |
| AGE^2 | −0.751 | −7.737 | 0.000 | 0.014 | 3.627 | 0.000 | | | |
| AGE^3 | 0.026 | 7.687 | 0.000 | 0.000 | −4.508 | 0.000 | | | |
| AGE^4 | 0.000 | −7.562 | 0.000 | | | | | | |
| AGE^5 | 0.000 | 7.369 | 0.000 | | | | | | |
| Pse. R-sq | 0.158 | | | 0.152 | | | 0.146 | | |
| Adj. R-sq | 0.158 | | | 0.152 | | | 0.146 | | |
| S.E. of reg | 1.571 | | | 1.581 | | | 1.567 | | |
| Quant DV | 2.000 | | | 2.000 | | | 2.000 | | |
| Sparsity | 2.032 | | | 2.249 | | | 2.967 | | |
| QLR-stat | 4082.3 | | | 3552.0 | | | 2572.3 | | |
| Prob | 0.000 | | | 0.000 | | | 0.000 | | |

**Figure 4.65**  Statistical results summary of the QR(Median)s in (4.38) to (4.40).

**Figure 4.66**  Scatter graphs of the forecast variables of the QR(Median)s in (4.37) to (4.40) on AGE.

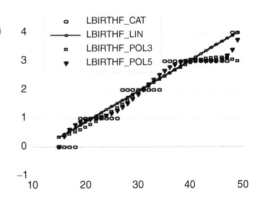

3) As an additional illustration, Figure 4.66 presents the scatter graphs of the forecast value variables, *LBIRTH_Cat, LBIRTH_Lin, LBIRTH_Pol3*, and *LBIRTH_Pol5*, of the QR(Median)s in (4.37) to (4.40), respectively.

4) In order to test the quantile differences between all age levels, I recommend doing the analysis using the following ES with an intercept, with *AGE* used as a categorical variable. When obtained, their medians have significant differences, based on the QLR-statistic of $QRL_0 = 6618.231$ with $df = 34$ and p-value = 0.000 00.

$$LBIRTH \ C \ @Expand \ (AGE, \ @Dropfirst) \tag{4.41}$$

# 5

# Quantile Regressions with Two Numerical Predictors

## 5.1 Introduction

As the extension of the quantile regressions (QRs) with a numerical predictor presented in previous chapter, this chapter presents examples of QRs with two numerical predictors. Figure 5.1 presents alternative up-and-down or causal relationships between three numerical variables *Y1, X1,* and *X2*.

   Figure 5.1a shows that both *X1* and *X2* have direct effects on *Y1*, and they do not have a causal relationship. This represents an additive model of *Y1* on *X1* and *X2*, which is the simplest model in a three-dimensional space. Figure 5.1b shows that *X1* has a direct effect on *Y1*, and *X2* has an indirect effect on *Y1* through *X1*. This represents a two-way interaction model. Even though there is no arrow from *X2* to *Y1*, *X2* is an upper or cause factor or variable of *Y2*. In other words, *X2* has a partial direct effect on *Y1*. And Figure 5.1c also represents a two-way interaction model, with the addition that *X2* is defined to have a direct effect on *Y1*.

## 5.2 Alternative QRs Based on Data_Faad.wf1

As an extension of the alternative models of *Y1* on a numerical variable *X1*, presented in previous chapter, this section will present selected QRs of *Y1* on two numerical predictors *X1* and *X2*.

### 5.2.1 Alternative QRs Based on (*X1,X2,Y1*)

#### 5.2.1.1 Additive QR
Based on any set of the numerical variables, *Y1, X1,* and *X2*, the additive linear regression, which is the simplest linear regression in a three-dimensional space, can be represented using the following general equation specification (ES):

$$Y1\ C\ X1\ X2 \tag{5.1}$$

*Quantile Regression: Applications on Experimental and Cross Section Data Using EViews,* First Edition.
I Gusti Ngurah Agung.
© 2021 John Wiley & Sons Ltd. Published 2021 by John Wiley & Sons Ltd.

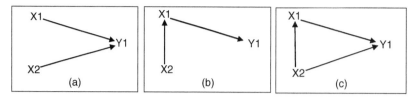

**Figure 5.1** Alternative up-and-down or causal relationships based on *(X1,X2,Y1)*.

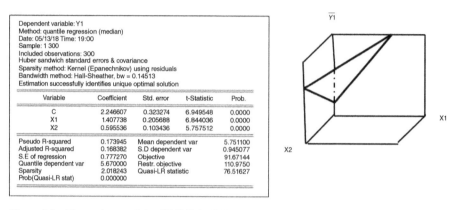

Dependent variable: Y1
Method: quantile regression (median)
Date: 05/13/18 Time: 19:00
Sample: 1 300
Included observations: 300
Huber sandwich standard errors & covariance
Sparsity method: Kernel (Epanechnikov) using residuals
Bandwidth method: Hall-Sheather, bw = 0.14513
Estimation successfully identifies unique optimal solution

| Variable | Coefficient | Std. error | t-Statistic | Prob. |
|---|---|---|---|---|
| C | 2.246607 | 0.323274 | 6.949548 | 0.0000 |
| X1 | 1.407738 | 0.205688 | 6.844036 | 0.0000 |
| X2 | 0.595536 | 0.103436 | 5.757512 | 0.0000 |

| | | | |
|---|---|---|---|
| Pseudo R-squared | 0.173945 | Mean dependent var | 5.751100 |
| Adjusted R-squared | 0.168382 | S.D dependent var | 0.945077 |
| S.E of regression | 0.777270 | Objective | 91.67144 |
| Quantile dependent var | 5.670000 | Restr. objective | 110.9750 |
| Sparsity | 2.018243 | Quasi-LR statistic | 76.51627 |
| Prob(Quasi-LR stat) | 0.000000 | | |

**Figure 5.2** The estimate of the QR(0.5) in (5.1) and the graph of its regression function.

### Example 5.1 *Application of QR (5.1)*

Figure 5.2 presents the estimate of the QR (5.1) and the graph of its regression function in a three orthogonal dimensional space. Based on this figure, the findings and notes are as follows:

1) The IVs (independent variables) *X1* and *X2* have significant joint effects, based on the *Quasi-LR* statistic, and each has a significant positive adjusted effect on *Y1*, based on the *t*-test statistic. In other words, one of the two IVs has a significant positive effect on *Y1*, adjusted for or by taking into account the other.
2) The regression function has the following fitted equation, and its graph is a plane in the three-dimensional space, with the axes $\widehat{Y1}$, *X1*, and *X2*.

$$\widehat{Y1} = 2.246\,607 + 1.407\,738*X1 + 0.595\,536*X2$$

3) Note that for *X2* = 0, we have the regression line $\widehat{Y1}$ = 2.246 607 + 1.407 738*X1, on the plane *(X1, $\widehat{Y1}$)*, and the regression line $\widehat{Y1}$ = 2.246 607 + 0.595 536*X2, on the plane *(X2, $\widehat{Y1}$)*, for *X1* = 0.

### 5.2.1.2 Semi-Logarithmic QR of log*(Y1)* on *X1* and *X2*

The semi-logarithmic linear regression of *log(Y1)*, for *Y1 > 0*, on *X1* and *X2*, can be represented using the following general ES:

$$log\,(Y1)\ C\,X1\,X2 \tag{5.2}$$

which can be extended to the bounded regression:

$$log((Y1 - L) / (U - Y1) \; C \; X1 \; X2 \tag{5.3}$$

where $U$ and $L$, respectively, are fixed upper and lower bounds of any variable $Y1$.

### Example 5.2 *Application of the QRs (5.2) and (5.3)*

Figure 5.3 presents the estimate of the QR(Median) (5.2) and a specific model in (5.3). Based on this figure, the findings and notes are as follows:

1) The observed scores of $Y1$ are within the closed interval [3.5, 8.0], and the substantial sense, the test scores of $Y1$ have a possible maximum of 10 and zero as the possible minimum scores. For this illustration, I chose $U = 10$ and $L = 3$, out of many possible choices.

2) The IVs $X1$ and $X2$ of both models have significant joint effects, based on the *Quasi-LR* statistic, and both have significant positive adjusted effects on their DVs (dependent variables), based on the $t$-test statistic.

3) Based on the model (5.2), the regression function has the following equation:

$$\widehat{log\,(Y1)} = 1.143\,7883 + 0.243\,010*X1 + 0.100\,796*X2$$

with the pseudo $R$-squared of 0.165 125. Hence, both IVs can explain 16.5% of the total variance of $\widehat{log\,(Y1)}$.

4) Based on the model (5.3), the regression function has the following equation:

$$log\left(\widehat{\frac{Y1 - 3}{10 - y1}}\right) = -2.618\,640 + 0.877\,531*X1 + 0.363\,700*X2$$

with the pseudo $R$-squared of 0.159 626. Thus, both IVs can explain 16.0% of the total variance of $log\left(\widehat{\frac{Y-3}{10-Y1}}\right)$, which is smaller than the regression function (5.2). So, in the statistical sense, the QR (5.2) is better than (5.3).

| Dependent Variable: LOG(Y1) | | | |
|---|---|---|---|
| Method: Quantile Regression (Median) | | | |
| Date: 05/18/18  Time: 19:50 | | | |
| Sample: 1 300 | | | |
| Included observations: 300 | | | |
| Huber Sandwich Standard Errors & Covariance | | | |
| Sparsity method: Kernel (Epanechnikov) using residuals | | | |
| Bandwidth method: Hall-Sheather, bw=0.14513 | | | |
| Estimation successfully identifies unique optimal solution | | | |

| Variable | Coefficient | Std. Error | t-Statistic | Prob. |
|---|---|---|---|---|
| C | 1.143788 | 0.068281 | 16.75109 | 0.0000 |
| X1 | 0.243010 | 0.038617 | 6.292779 | 0.0000 |
| X2 | 0.100796 | 0.019077 | 5.283559 | 0.0000 |

| | | | |
|---|---|---|---|
| Pseudo R-squared | 0.165125 | Mean dependent var | 1.735214 |
| Adjusted R-squared | 0.159503 | S.D. dependent var | 0.171367 |
| S.E. of regression | 0.140076 | Objective | 16.48485 |
| Quantile dependent var | 1.735189 | Restr. objective | 19.74530 |
| Sparsity | 0.361954 | Quasi-LR statistic | 72.06340 |
| Prob(Quasi-LR stat) | 0.000000 | | |

| Dependent Variable: LOG((Y1-3)/(10-Y1)) | | | |
|---|---|---|---|
| Method: Quantile Regression (Median) | | | |
| Date: 05/16/18  Time: 11:06 | | | |
| Sample: 1 300 | | | |
| Included observations: 300 | | | |
| Huber Sandwich Standard Errors & Covariance | | | |
| Sparsity method: Kernel (Epanechnikov) using residuals | | | |
| Bandwidth method: Hall-Sheather, bw=0.14513 | | | |
| Estimation successfully identifies unique optimal solution | | | |

| Variable | Coefficient | Std. Error | t-Statistic | Prob. |
|---|---|---|---|---|
| C | -2.618640 | 0.242004 | -10.82066 | 0.0000 |
| X1 | 0.877531 | 0.136831 | 6.413232 | 0.0000 |
| X2 | 0.363700 | 0.067548 | 5.384325 | 0.0000 |

| | | | |
|---|---|---|---|
| Pseudo R-squared | 0.159626 | Mean dependent var | -0.483878 |
| Adjusted R-squared | 0.153967 | S.D. dependent var | 0.641598 |
| S.E. of regression | 0.523455 | Objective | 60.53703 |
| Quantile dependent var | -0.483489 | Restr. objective | 72.03580 |
| Sparsity | 1.282176 | Quasi-LR statistic | 71.74536 |
| Prob(Quasi-LR stat) | 0.000000 | | |

**Figure 5.3**  Statistical results of the QR(Median) in (5.2) and a special model (5.3).

#### 5.2.1.3 Translog QR of *log(Y1)* on *log(X1)* and *log(X2)*

The translog (translogarithmic) linear regression based on positive variables $Y1, X1,$ and $X2$ can be represented using the following general ES:

$$\log(Y1) \; C \; \log(X1) \; \log(X2) \tag{5.4}$$

which can be extended to the bounded regression:

$$\log((Y1 - L)/(U - Y1) \; C \; \log(X1) \; \log(X2) \tag{5.5}$$

where $U$ and $L$, respectively, are fixed upper and lower bounds of the variable $Y1$.

**Example 5.3** *Applications of QR (5.4) and (5.5)*

Figure 5.4 presents the estimate of the QR(Median) (5.4) and a specific model in (5.5). Based on this figure, the findings and notes are as follows:

1) The IVs *log(X1)* and *log(X2)* of both models have significant joint effects, based on the *Quasi-LR* statistic, and each has a significant positive adjusted effect on the corresponding DV, based on the *t*-test statistic.
2) Based on the model (5.4), the regression function has the following equation:

$$\widehat{\log(Y1)} = 1.413 + 0.332 * \log(X1) + 0.251 * \log(X2)$$

with the pseudo $R$-squared of 0.180 501. Hence, both IVs can explain 18.1% of the total variance of $\widehat{\log(Y1)}$.
3) Based on the model (5.5), the regression function has the following equation:

$$\log\left(\widehat{\frac{Y1 - 3}{10 - y1}}\right) = -1.634\,208 + 1.179\,977 * \log(X1) + 0.899\,644 * \log(X2)$$

with the pseudo $R$-squared of 0.176 092. Thus, both IVs can explain 17.6% of the total variance of $\log\left(\frac{Y-3}{10-Y1}\right)$, which is smaller than the regression function (5.4). So, in the statistical sense, the QR (5.4) is better than (5.5).

### 5.2.2 Two-Way Interaction QRs

#### 5.2.2.1 Interaction QR of *Y1* on *X1* and *X2*

As an extension of the additive linear (5.1), based on any set of the numerical variables $Y1,$ $X1,$ and $X2$, with $X1$ and $X2$ being positive variables, the two-way interaction linear regression can be represented using the following general ES. The two-way interaction model has the main objective of testing the hypothesis that the effect of one of the main IVs on the DV depends on the other main IV.

In addition, note that the positive IVs of $X1$ and $X2$ are required because if both have negative scores, then the interaction will be misleading. That is because a (negative × negative) score is greater than 0, and thus it would be greater than all (negative × positive) scores and greater than some or many of the (positive × positive) scores. Whenever both $X1$ or $X2$ have negative scores, new positive IVs should be generated, such as $X1a = X1 + a > 0,$ and $X2b = X2 + b > 0$.

$$Y1 \; C \; X1 \; X2 \; X1 * X2 \tag{5.6a}$$

Dependent Variable: LOG(Y1)
Method: Quantile Regression (Median)
Date: 05/18/18   Time: 19:55
Sample: 1 300
Included observations: 300
Huber Sandwich Standard Errors & Covariance
Sparsity method: Kernel (Epanechnikov) using residuals
Bandwidth method: Hall-Sheather, bw=0.14513
Estimation successfully identifies unique optimal solution

| Variable | Coefficient | Std. Error | t-Statistic | Prob. |
|---|---|---|---|---|
| C | 1.412675 | 0.043334 | 32.59957 | 0.0000 |
| LOG(X1) | 0.332089 | 0.051313 | 6.471873 | 0.0000 |
| LOG(X2) | 0.251105 | 0.047638 | 5.271137 | 0.0000 |

| | | | | |
|---|---|---|---|---|
| Pseudo R-squared | 0.180501 | Mean dependent var | | 1.735214 |
| Adjusted R-squared | 0.174982 | S.D. dependent var | | 0.171367 |
| S.E. of regression | 0.138357 | Objective | | 16.18126 |
| Quantile dependent var | 1.735189 | Restr. objective | | 19.74530 |
| Sparsity | 0.348091 | Quasi-LR statistic | | 81.91047 |
| Prob(Quasi-LR stat) | 0.000000 | | | |

Dependent Variable: LOG((Y1-3)/(10-Y1))
Method: Quantile Regression (Median)
Date: 05/13/18   Time: 19:05
Sample: 1 300
Included observations: 300
Huber Sandwich Standard Errors & Covariance
Sparsity method: Kernel (Epanechnikov) using residuals
Bandwidth method: Hall-Sheather, bw=0.14513
Estimation successfully identifies unique optimal solution

| Variable | Coefficient | Std. Error | t-Statistic | Prob. |
|---|---|---|---|---|
| C | -1.634208 | 0.156401 | -10.44883 | 0.0000 |
| LOG(X1) | 1.179977 | 0.184794 | 6.385346 | 0.0000 |
| LOG(X2) | 0.899644 | 0.169500 | 5.307636 | 0.0000 |

| | | | | |
|---|---|---|---|---|
| Pseudo R-squared | 0.174092 | Mean dependent var | | -0.483878 |
| Adjusted R-squared | 0.168530 | S.D. dependent var | | 0.641598 |
| S.E. of regression | 0.515926 | Objective | | 59.49497 |
| Quantile dependent var | -0.483489 | Restr. objective | | 72.03580 |
| Sparsity | 1.237921 | Quasi-LR statistic | | 81.04444 |
| Prob(Quasi-LR stat) | 0.000000 | | | |

**Figure 5.4**  Statistical results of the QR(Median) in (5.4) and a special model (5.5).

In empirical studies or results, we might obtain a regression with an insignificant adjusted effect of $X1*X2$. In such cases, one of the following alternative reduced models (RMs) would be an acceptable model, in the statistical sense, to present a two-way interaction model. However, there is a possibility the interaction $X1*X2$ has an insignificant effect in the full model and in its interaction reduced models, which is highly dependent on the sample data used.

$$Y1\ C\ X1\ X1*X2 \tag{5.6b}$$

$$Y1\ C\ X2\ X1*X2 \tag{5.6c}$$

$$Y1\ C\ X1*X2 \tag{5.6d}$$

**Example 5.4**  *Application of the QRs (5.6a), (5.6b) and (5.6c)*
As an illustration Figure 5.5 presents the results summary of the QR(Median)s in (5.6a), (5.6b), and (5.6c). Based on this summary, the findings and notes are as follows:

1) Even though $X1*X2$ has a significant adjusted effect in the full model (5.6a), at the 10% level, I present the result of its reduced models (5.6b) and (5.6c) in order to show that the reduced models should always have smaller or much smaller $p$-values, because $X1$, $X2$, and $X1*X2$ are highly correlated. See the following additional example.
2) The set of IVs of each QR(Median) has significant joint effects on $Y1$, based on the *Quasi-LR* statistic with a $p$-value $= 0.0000$.
3) Note that at the 5% level of significance, the interaction $X1*X2$ has a negative significant effect on the median of $Y1$, based on the $t$-statistic of $t_0 = -1.8800$ with a $p$-value $= 0.0611/2 = 0.030\,55$, in the full model. However, in both reduced models (RMs), it has a positive significant effects.
4) Under the criterion that $X1$ is a more important cause factor or predictor than $X2$, based on each regression function, the following findings and notes are presented:
   4.1  For the full model, the median regression function can be written as follows:

$$\widehat{Y1} = (0.520 + 1.332*X2) + (2.730 - 0.536*X2)*X1$$

Note that in the two-dimensional space $(X1, \widehat{Y1})$, the graph of this function presents a set of heterogeneous lines where both their intercepts and slopes are the functions of $X2$.

| Variable | Full model (5.6a) | | | RM (5.6b) | | | RM (5.6c) | | |
|---|---|---|---|---|---|---|---|---|---|
| | Coef. | t-stat | Prob. | Coef. | t-stat | Prob. | Coef. | t-stat | Prob. |
| C | 0.5203 | 0.5395 | 0.5899 | 3.7931 | 11.030 | 0.0000 | 4.3791 | 14.101 | 0.0000 |
| X1 | 2.7300 | 3.6927 | 0.0003 | 0.5030 | 1.5034 | 0.1338 | | | |
| X2 | 1.3319 | 3.3851 | 0.0008 | | | | -0.1194 | -0.6548 | 0.5131 |
| X1*X2 | -0.5364 | -1.8800 | 0.0611 | 0.3484 | 4.8384 | 0.0000 | 0.4697 | 5.7926 | 0.0000 |
| Pseudo R-sq | 0.1821 | | | 0.1582 | | | 0.1530 | | |
| Adj. R-squared | 0.1738 | | | 0.1525 | | | 0.1473 | | |
| Quasi-LR stat | 82.083 | | | 66.862 | | | 64.766 | | |
| Prob(Q-LR stat) | 0.0000 | | | 0.0000 | | | 0.0000 | | |

**Figure 5.5**  The summary of the results of the QRs (5.6a), (5.6b), and (5.6c).

Since *X1* and *X2* are positive variables, the negative coefficient of *X1\*X2* indicates that the effect of *X1* on the median of *Y1* will decrease with increasing scores of *X2*, which is a special characteristic of this regression function. In other words, it can be said that the effect of *X2* is to decrease the effect of *X1* on *Y1*.

Furthermore, note that the effect of *X1* on the median of *Y1* would be zero or negative for $X2 \geq 2.730/0.536 = 5.093$. However, since the maximum observed scores of $X2 = 3.900 < 5.093$, the set of regression lines have only positive slopes and positive intercepts.

The regression function also can be presented as follows:

$$\widehat{Y1} = (0.520 + 2.730*X1) + (1.332 - 0.536*X1)*X2$$

which indicates the effect of *X2* on *Y1* is decreasing with the increasing scores of *X1*. I would say that either one of the previous special characteristics is the true characteristic, in many or most cases.

4.2 For the reduced model (5.6b), the median regression function can be written as follows:

$$\widehat{Y1} = 3.793 + (0.503 + 0.348*X2)*X1$$

Compared with the regression function (5.6a) in the two-dimensional space $(X1,\widehat{Y1})$, the graph of this function presents a set of heterogeneous lines with a single or constant intercept, and positive slopes represented by $(0.503 + 0.348*X2)$.

4.3 For the reduced model (5.6c), the median regression function can be written as follows:

$$\widehat{Y1} = (4.379 - 0.119*X2) + (0.470*X2)*X1$$

where the graphs are heterogeneous lines of $\widehat{Y1}$ on *X1*, with various positive slopes and intercepts, since $(4.379 - 0.119*X2) > (4.379 - 0.119*3.9) > 0$ for all observed scores of *X2*.

5) Since the full model has the largest pseudo *R*-squared, it is the best QR, in the statistical sense. Then there is a question which one of the three median regression functions is the best regression. Based on the theoretical concept, I would select the regression function (5.6a) as the best model. This is because the full two-way interaction model has a negative coefficient of the interaction is to represent the effect of one of the variable will be decreasing with the increasing scores of the other variable, which is the true condition in many or most cases in practice or reality. See also the following statistical results. However, based on only the statistical results, one would have selected the regression (5.6c) as the best, since the bivariate *(X1\*X2,Y1)* has a positive significant correlation of 0.818, based on the *t*-statistic of $t_0 = 24.532$ with a *p*-value $= 0.000$. The IV *X1\*X2* also has a positive significant adjusted effect on *Y1* in the regression (5.6c), and the regression (5.6d) is too specific.

## 5.2.2.2 Semi-Logarithmic Interaction QR Based on *(X1,X2,Y1)*

The semi-logarithmic linear regression of *log(Y1)*, for *Y1 > 0*, on positive variables *X1* and *X2*, can be represented using the following general ES:

$$log\,(Y1)\ C\ X1\ X2\ X1*X2 \tag{5.7}$$

which can be extended to the bounded regression:

$$log((Y1 - L) / (U - Y1)\ C\ X1\ X2\ X1*X2 \tag{5.8}$$

where $U$ and $L$, respectively, are fixed upper and lower bounds of any variable $Y1$.

### Example 5.5 *Applications of QRs (5.7) and (5.8)*

Figure 5.6 presents the estimate of the QR(Median) in (5.7) and a specific model in (5.8) based on Data_Faad.wf1. The findings and notes are as follows:

1) Similar to the full two-way interaction regression in previous example, the interaction $X1*X2$ in both of these regressions has a negative significant effect on the DV.
2) The IVs $X1$, $X2$, and $X1*X2$ of both QRs have significant joint effects on the DVs, based on the *Quasi-LR* statistic, and at the 1% level, each of the independent variables has either a positive or negative significant adjusted effect on its DV, based on the *t*-statistic. For instance, $X1*X2$ has a negative significant effect in both regressions, based on the *t*-statistic of $-2.666$ and $-2.628$, with the *p*-values of 0.0081/2 and 0.0090/2, respectively. Hence, referring to the notes in the previous example's point 4.1, both regressions are acceptable median regression functions, in both the theoretical and statistical senses.

#### 5.2.2.3 Translogarithmic Interaction QR Based on *(X1,X2,Y1)*

The translogarithmic interaction QR of $log(Y1)$, for $Y1 > 0$, on positive $log(X1)$ and $log(X2)$, can be represented using the following general ES:

$$log\,(Y1)\ C\ log\,(X1)\ log\,(X2)\ log\,(X1)*log\,(X2) \tag{5.9}$$

which can be extended to the bounded regression:

$$log\,((Y1 - L) / (U - Y1))\ C\,log\,(X1)\ log\,(X2)\ log\,(X1)*log\,(X2) \tag{5.10}$$

where $U$ and $L$, respectively, are fixed upper and lower bounds of any variable $Y1$.

### Example 5.6 *Applications of Translog Interaction QRs*

Figure 5.7 presents the estimates of the translog interaction QR(Median) in (5.9) and the following translog two-way interaction based on *(X1,X2,Y1)*. Since $log(X1)$ in Data-Faad.wf1

---

Dependent Variable: LOG(Y1)
Method: Quantile Regression (Median)
Date: 05/13/18   Time: 19:11
Sample: 1 300
Included observations: 300
Huber Sandwich Standard Errors & Covariance
Sparsity method: Kernel (Epanechnikov) using residuals
Bandwidth method: Hall-Sheather, bw=0.14513
Estimation successfully identifies unique optimal solution

| Variable | Coefficient | Std. Error | t-Statistic | Prob. |
|---|---|---|---|---|
| C | 0.618633 | 0.205164 | 3.015313 | 0.0028 |
| X1 | 0.641062 | 0.149350 | 4.292339 | 0.0000 |
| X2 | 0.305322 | 0.080055 | 3.813886 | 0.0002 |
| X1*X2 | -0.150184 | 0.056341 | -2.665614 | 0.0081 |

| | | | |
|---|---|---|---|
| Pseudo R-squared | 0.182838 | Mean dependent var | 1.735214 |
| Adjusted R-squared | 0.174556 | S.D. dependent var | 0.171367 |
| S.E. of regression | 0.139074 | Objective | 16.13511 |
| Quantile dependent var | 1.735189 | Restr. objective | 19.74530 |
| Sparsity | 0.349378 | Quasi-LR statistic | 82.66548 |
| Prob(Quasi-LR stat) | 0.000000 | | |

Dependent Variable: LOG((Y1-3)/(10-Y1))
Method: Quantile Regression (Median)
Date: 05/18/18   Time: 21:03
Sample: 1 300
Included observations: 300
Huber Sandwich Standard Errors & Covariance
Sparsity method: Kernel (Epanechnikov) using residuals
Bandwidth method: Hall-Sheather, bw=0.14513
Estimation successfully identifies unique optimal solution

| Variable | Coefficient | Std. Error | t-Statistic | Prob. |
|---|---|---|---|---|
| C | -4.540792 | 0.765838 | -5.929177 | 0.0000 |
| X1 | 2.358948 | 0.563854 | 4.183616 | 0.0000 |
| X2 | 1.116155 | 0.299131 | 3.731326 | 0.0002 |
| X1*X2 | -0.561790 | 0.213803 | -2.627608 | 0.0090 |

| | | | |
|---|---|---|---|
| Pseudo R-squared | 0.176020 | Mean dependent var | -0.483878 |
| Adjusted R-squared | 0.167669 | S.D. dependent var | 0.641598 |
| S.E. of regression | 0.517280 | Objective | 59.35609 |
| Quantile dependent var | -0.483489 | Restr. objective | 72.03580 |
| Sparsity | 1.243606 | Quasi-LR statistic | 81.56745 |
| Prob(Quasi-LR stat) | 0.000000 | | |

**Figure 5.6**   Estimates of the QR (5.7) and a specific QR in (5.8).

Dependent Variable: LOG(Y1)
Method: Quantile Regression (Median)
Date: 05/22/18   Time: 13:48
Sample: 1 300
Included observations: 300
Huber Sandwich Standard Errors & Covariance
Sparsity method: Kernel (Epanechikov) using residuals
Bandwidth method: Hall-Sheather, bw=0.14513
Estimation successfully identifies unique optimal solution

| Variable | Coefficient | Std. Error | t-Statistic | Prob. |
|---|---|---|---|---|
| C | 1.335428 | 0.047102 | 28.35155 | 0.0000 |
| LOG(X1) | 0.616447 | 0.128606 | 4.793311 | 0.0000 |
| LOG(X2) | 0.352909 | 0.060333 | 5.849378 | 0.0000 |
| LOG(X1)*LOG(X2) | -0.326779 | 0.147160 | -2.220560 | 0.0271 |

| Pseudo R-squared | 0.188384 | Mean dependent var | 1.735214 |
|---|---|---|---|
| Adjusted R-squared | 0.180158 | S.D. dependent var | 0.171367 |
| S.E. of regression | 0.138499 | Objective | 16.02560 |
| Quantile dependent var | 1.735189 | Restr. objective | 19.74530 |
| Sparsity | 0.338369 | Quasi-LR statistic | 87.94421 |
| Prob(Quasi-LR stat) | 0.000000 | | |

Dependent Variable: LOG(Y1)
Method: Quantile Regression (Median)
Date: 05/22/18   Time: 14:43
Sample: 1 300
Included observations: 300
Huber Sandwich Standard Errors & Covariance
Sparsity method: Kernel (Epanechikov) using residuals
Bandwidth method: Hall-Sheather, bw=0.14513
Estimation successfully identifies unique optimal solution

| Variable | Coefficient | Std. Error | t-Statistic | Prob. |
|---|---|---|---|---|
| C | 0.718982 | 0.158057 | 4.548870 | 0.0000 |
| LOG(X1)+1 | 0.616447 | 0.128606 | 4.793311 | 0.0000 |
| LOG(X2) | 0.679688 | 0.192960 | 3.522435 | 0.0005 |
| (LOG(X1)+1)*LOG(X2) | -0.326779 | 0.147160 | -2.220560 | 0.0271 |

| Pseudo R-squared | 0.188384 | Mean dependent var | 1.735214 |
|---|---|---|---|
| Adjusted R-squared | 0.180158 | S.D. dependent var | 0.171367 |
| S.E. of regression | 0.138499 | Objective | 16.02560 |
| Quantile dependent var | 1.735189 | Restr. objective | 19.74530 |
| Sparsity | 0.338369 | Quasi-LR statistic | 87.94421 |
| Prob(Quasi-LR stat) | 0.000000 | | |

**Figure 5.7**   Estimates of the QRs (5.9) and (5.10).

has 32 negative scores, I am using $(log(X1) + 1) > 0$ as an independent variable as a comparative study:

$$log(Y1)\ C\ (log(X1) + 1)\ log(X2)\ (log(X1) + 1) * log(X2) \tag{5.11}$$

Based on these results, the findings and notes are as follows:

1) Similar to the *full two-way interaction QRs* (F2WI-QRs) in Figures 5.4 and 5.5, the two-way interaction IVs of both these regressions also have negative significant adjusted effects on the $log(Y1)$.
2) Unexpectedly, $log(X2)$ has different positive coefficients in the two regressions, but the other two IVs have the same coefficient in the six decimal points. Note that even though each of the variables, $log(X1)$ and $(log(X1) + 1)$, has the same correlation with $log(X2)$, but $log(X2)$ has different correlations with the interactions $log(X1)*log(X2)$ and $(log(X1) + 1)*log(X2)$, as presented in Figure 5.8.
3) Furthermore, note that both regressions present the same values of 11 other statistics, such as the values of $R$-squared up to the *Quasi-LR* statistic. Why? Because the second regression is obtained by translating the origin coordinate system $log(X1)$, $log(X2)$, and $log(Y1)$ from $(0,0,0)$ to $(-1,0,0)$.

| Correlation<br>t-statistic<br>Probability | LOG(X1) | LOG(X1)+1 | LOG(X1)*LOG(X2) | (LOG(X1)+1)*LOG(X2) |
|---|---|---|---|---|
| LOG(X2) | 0.368678 | 0.368678 | 0.569015 | 0.888643 |
| | 6.846659 | 6.846659 | 11.94502 | 33.45050 |
| | 0.0000 | 0.0000 | 0.0000 | 0.0000 |

**Figure 5.8**   Bivariate correlations between $log(X2)$, and $log(X1)$, $(log(X1) + 1)$.

Dependent Variable: LOG(Y1)
Method: Quantile Regression (Median)
Date: 05/22/18   Time: 14:56
Sample: 1 300
Included observations: 300
Huber Sandwich Standard Errors & Covariance
Sparsity method: Kernel (Epanechnikov) using residuals
Bandwidth method: Hall-Sheather, bw=0.14513
Estimation successfully identifies unique optimal solution

| Variable | Coefficient | Std. Error | t-Statistic | Prob. |
|---|---|---|---|---|
| C | 0.039294 | 0.343657 | 0.114340 | 0.9090 |
| LOG(X1)+1 | 0.943225 | 0.270515 | 3.486782 | 0.0006 |
| LOG(X2)+1 | 0.679688 | 0.192960 | 3.522435 | 0.0005 |
| (LOG(X1)+1)*(LOG(X2)+1) | -0.326779 | 0.147160 | -2.220560 | 0.0271 |

| | | | |
|---|---|---|---|
| Pseudo R-squared | 0.188384 | Mean dependent var | 1.735214 |
| Adjusted R-squared | 0.180158 | S.D. dependent var | 0.171367 |
| S.E. of regression | 0.138499 | Objective | 16.02560 |
| Quantile dependent var | 1.735189 | Restr. objective | 19.74530 |
| Sparsity | 0.338369 | Quasi-LR statistic | 87.94421 |
| Prob(Quasi-LR stat) | 0.000000 | | |

Dependent Variable: LOG((Y1-3)/(10-Y1))
Method: Quantile Regression (Median)
Date: 05/21/18   Time: 18:48
Sample: 1 300
Included observations: 300
Huber Sandwich Standard Errors & Covariance
Sparsity method: Kernel (Epanechnikov) using residuals
Bandwidth method: Hall-Sheather, bw=0.14513
Estimation successfully identifies unique optimal solution

| Variable | Coefficient | Std. Error | t-Statistic | Prob. |
|---|---|---|---|---|
| C | -5.123986 | 1.028359 | -4.982684 | 0.0000 |
| LOG(X1+1) | 4.465192 | 1.200995 | 3.717912 | 0.0002 |
| LOG(X2) | 3.224425 | 1.138063 | 2.833256 | 0.0049 |
| LOG(X1+1)*LOG(X2) | -2.651304 | 1.297816 | -2.042896 | 0.0419 |

| | | | |
|---|---|---|---|
| Pseudo R-squared | 0.181091 | Mean dependent var | -0.483878 |
| Adjusted R-squared | 0.172792 | S.D. dependent var | 0.641598 |
| S.E. of regression | 0.514934 | Objective | 58.99075 |
| Quantile dependent var | -0.483489 | Restr. objective | 72.03580 |
| Sparsity | 1.236859 | Quasi-LR statistic | 84.37539 |
| Prob(Quasi-LR stat) | 0.000000 | | |

**Figure 5.9**   Estimates of the QRs (5.12) and (5.13).

4) As additional comparisons, Figure 5.9 presents statistical results of two alternative QRs, with the following ESs:

$$\log(y1) \ c \ (\log(x1)+1) \ (\log(x2)+1) \ (\log(x1)+1)*(\log(x2)+1) \tag{5.12}$$

$$\log(y1) \ c \log(x1+1) \ \log(x2) \ \log(x1+1)*\log(x2) \tag{5.13}$$

Based on these results, the following findings and notes are presented:

4.1 Note that the QRs (5.9), (5.11), and (5.12) also have exactly the same statistical results. Because the (5.12) also is obtained by translating the origin of the coordinates system $\log(X1)$, $\log(X2)$, and $\log(Y1)$ from (0,0,0) to (−1,−1,0).

4.2 The QR (5.13) is quite a different translog regression. The three QRs (5.9), (5.11), and (5.12) can be considered translog interaction regressions based on $(X1,X2,Y1)$, but the QR (5.13) is a translog interaction regression based on $(X1+1,X2,Y1)$. So it is a different median regression of $\log(Y1)$. It is an acceptable interaction model, since the interaction $\log(X1+1)*\log(X2)$ in QR (5.13) has a negative significant adjusted effect on $\log(Y1)$, based on the $t$-statistic of $t_0 = -2.043$ with a $p$-value $= 0.0419/2 = 0.020\,95$. In addition, each of the main IVs, $\log(X1+1)$ and $\log(X2)$, has a positive significant adjusted effect on $\log(Y1)$, at the 1% level.

## 5.3   An Analysis Based on Mlogit.wf1

For a comparison of statistical results, this section presents alternative results from similar QRs, based on a special data file in EViews 8, Mlogit.wf1. I consider the data to be special data because the scores of most or all of the variables are not real observed scores. Some of the variables are generated using a special program, as shown in Figure 5.10, and the scores of Z1, Z2, and Z3 might be three random samples taken from a standardized normal distribution, with their graphs presented in Figure 5.11. And the variable LW is generated as a linear function of Z1, Z2, and Z3, using the following equations:

*Modified* 1 1000//$LW = 5 + 0.8*Z1 + 0.1*Z2 - Z3 + u2$

*Modified* 1 1000//@recode ($lfp = 1, lw, na$)

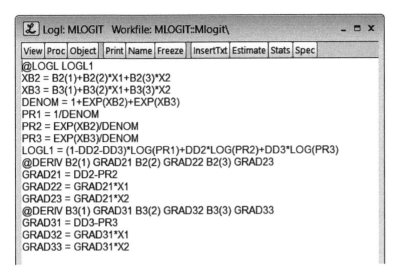

```
Logl: MLOGIT   Workfile: MLOGIT::Mlogit\                    _ ◻ X
View Proc Object  Print Name Freeze  InsertTxt Estimate Stats Spec
@LOGL LOGL1
XB2 = B2(1)+B2(2)*X1+B2(3)*X2
XB3 = B3(1)+B3(2)*X1+B3(3)*X2
DENOM = 1+EXP(XB2)+EXP(XB3)
PR1 = 1/DENOM
PR2 = EXP(XB2)/DENOM
PR3 = EXP(XB3)/DENOM
LOGL1 = (1-DD2-DD3)*LOG(PR1)+DD2*LOG(PR2)+DD3*LOG(PR3)
@DERIV B2(1) GRAD21 B2(2) GRAD22 B2(3) GRAD23
GRAD21 = DD2-PR2
GRAD22 = GRAD21*X1
GRAD23 = GRAD21*X2
@DERIV B3(1) GRAD31 B3(2) GRAD32 B3(3) GRAD33
GRAD31 = DD3-PR3
GRAD32 = GRAD31*X1
GRAD33 = GRAD31*X2
```

**Figure 5.10** A special program for developing several variables in Mlogit.wf1.

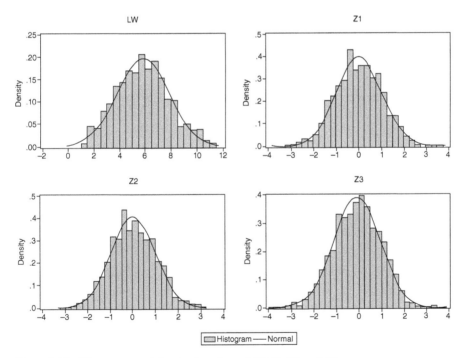

**Figure 5.11** Histograms and density functions of *LW, Z1, Z2,* and *Z3.*

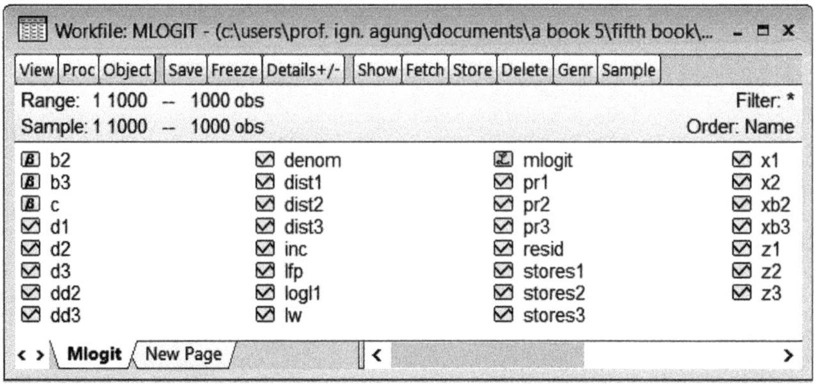

**Figure 5.12** A special Mlogit.wf1 in an EViews 8 data file.

Figure 5.12 shows the data file with 21 numerical variables. There are more than 1330 combinations of three variables that can be selected to develop alternative two-way interaction QRs, with various median regression functions, compared with the median regression functions obtained from Data_Faad.wf1. However, only special or specific statistical results are presented in the following examples, which are selected from various alternative models.

### 5.3.1 Alternative QRs of *LW*

This section presents special alternative QRs of the positive variable *LW* on various IVs. Since *Z1, Z2,* and *Z3* are Z-score variables, for the two-way interaction QR, 2WI-QR, I generated positive variables $Z1B = Z1 + 5$, $Z2B = Z2 + 5$, and $Z3B = Z3 + 5$. I did this to create two types of interaction models, using the interaction $Z1*Z2$ and $Z1B*Z2B$, to study their differences.

**Example 5.7** *Additive and Interaction Models of LW on (Z1,Z2) and (Z1B,Z2B)*
Table 5.1 presents the summary of the results of two pairs of additive and two-way interaction QRs of a positive variable *LW* on *(Z1,Z2)* and *(Z1B,Z2B)*, respectively. Based on this summary, the following findings and notes are presented:

1) Both additive QRs present the same slope coefficients and other statistics, such as the pseudo *R*-squared, adjusted *R*-squared, and Quasi-LR statistics, except their intercepts *C*. Hence, both regressions have the same quality level.
2) Based on the pairs of two-way interaction QRs, further findings and notes are as follows:
   2.1 Unexpectedly, only the coefficients of their interaction IVs have the same estimates, and the other statistics, such as the pseudo *R*-squared, adjusted *R*-squared, and Quasi-LR statistics. In addition, the interaction IVs of both QRs have significant positive adjusted effects on *LW*, based on the *t*-statistic of $t_0 = 2.0135$ with a *p*-value $= 0.0444/2$.
   2.2 Note that the positive coefficient of the interaction indicates that the effect of one of the variables is increasing with the increasing scores of the other variable.

**Table 5.1** Summary of the results of two pairs of QRs of *LW* on *(Z1,Z2)* and *(Z1B,Z2B)*.

| | Model based on (LW,Z1,Z2) | | | | Model based on (LW,Z1B,Z2B) | | | |
|---|---|---|---|---|---|---|---|---|
| **Add. model** | **C** | **Z1** | **Z2** | | **C** | **Z1B** | **Z2B** | |
| Coef. | 5.6731 | 0.7022 | −0.1773 | | 3.0487 | 0.7022 | −0.1773 | |
| *t*-stat | 60.7543 | 8.0539 | −1.7523 | | 4.3993 | 8.0539 | −1.7523 | |
| Prob. | 0.0000 | 0.0000 | 0.0801 | | 0.0000 | 0.0000 | 0.0801 | |
| Ps. $R^2$ | | 0.0641 | | | | 0.0641 | | |
| Adj. $R^2$ | | 0.0617 | | | | 0.0617 | | |
| Q-LR stat | | 65.4831 | | | | 65.4831 | | |
| Prob. | | 0.0000 | | | | 0.0000 | | |
| **2WI-QR** | **C** | **Z1** | **Z2** | **Z1*Z2** | **C** | **Z1B** | **Z2B** | **Z1B*Z2B** |
| Coef. | 5.7397 | 0.6219 | −0.2128 | 0.1831 | 8.2716 | −0.2935 | −1.1283 | 0.1831 |
| *t*-stat | 63.9313 | 8.0668 | −2.2275 | 2.0135 | 3.3897 | −0.6283 | −2.3797 | 2.0135 |
| Prob. | 0.0000 | 0.0000 | 0.0262 | 0.0444 | 0.0007 | 0.5300 | 0.0176 | 0.0444 |
| Ps. $R^2$ | | 0.0657 | | | | 0.0657 | | |
| Adj. $R^2$ | | 0.0622 | | | | 0.0622 | | |
| Q-LR stat | | 69.0375 | | | | 69.0375 | | |
| Prob. | | 0.0000 | | | | 0.0000 | | |

2.3 Referring to the pseudo *R*-squared and adjusted *R*-squared as the measures for the good of fit (GoF) models, both QRs have the same level of GoF.

2.4 Since the scores of *Z1*Z2* are considered to be misleading, in the mathematical sense, I would say that the interaction model based on *(Y1,Z1B,Z2B)* is a recommended model.

### Example 5.8 *Alternative Interaction QRs of LW on (Z1,Z3) and (Z1B,Z3B)*

As a comparison with previous statistical results, Table 5.2 presents only the summary of the results of a pairs of 2WI-QRs, of a positive variable *LW* on *(Z1,Z3)* and *(Z1B,Z3B)*, respectively. This is because the pairs of their additive models have the same results, as in the previous example. Based on this summary, the following findings and notes are presented:

1) The interaction IVs in both QRs have the same positive insignificant adjusted effects on *LW*, based on the *t*-statistic of $t_0 = 0.7081$ with a *p*-value $= 0.4791/2 = 0.239\,55$. So, the effect of one of the main variables will increase with the increasing scores of the other main variables, even though the interactions have insignificant effects.

Why the interactions *Z1*Z3* and *Z1B*Z3B* have the same coefficient can be explained by the following relationship, for $a = 0.0614$:

$$a\,(Z1B*Z3B) = a\,(Z1 + 5)*(Z3 + 5)$$

$$= a\,(Z1*Z3) + 5a*Z1 + 5a*Z3 + 25a$$

**Table 5.2** Summary of the results of 2WI-QRs of *LW* on *(Z1,Z3)* and *(Z1B,Z3B)*.

| 2W-IM | Model based on (LW,Z1,Z3) | | | | Model based on (LW,Z1B,Z3B) | | | |
|---|---|---|---|---|---|---|---|---|
| | C | Z1 | Z3 | Z1*Z3 | C | Z1B | Z3B | Z1B*Z3B |
| Coef. | 5.5339 | 0.7695 | −0.8393 | **0.0624** | 7.4429 | 0.4575 | −1.1513 | **0.0624** |
| *t*-stat. | 61.856 | 8.7713 | −9.9824 | **0.7081** | 3.4779 | 1.1026 | −2.5073 | **0.7081** |
| Prob. | 0.0000 | 0.0000 | 0.0000 | **0.4791** | 0.0005 | 0.2705 | 0.0124 | **0.4791** |
| Pseudo *R*-sq | **0.1470** | | | | **0.1470** | | | |
| Adj. *R*-sq | **0.1438** | | | | **0.1438** | | | |
| Sparsity | **4.7875** | | | | **4.7875** | | | |
| Q-LR stat | **159.66** | | | | **159.66** | | | |
| Prob. | **0.0000** | | | | **0.0000** | | | |

**Table 5.3** Summary of the results of three reduced models of the 2WI_QR of *LW* on *(Z1B,Z3B)*.

| Statistic | RM-1 | | | RM-2 | | | RM-3 | | |
|---|---|---|---|---|---|---|---|---|---|
| | C | Z1B | Z1B*Z3B | C | Z3B | Z1B*Z3B | C | Z1B | Z3B |
| Coef. | 2.227 | 1.461 | −0.157 | 9.581 | −1.607 | 0.158 | 5.930 | 0.744 | −0.817 |
| *t*-stat. | 5.178 | 13.477 | −10.846 | 23.906 | −12.73 | 9.412 | 10.527 | 9.056 | −9.861 |
| Prob. | 0.000 | 0.000 | 0.000 | 0.000 | 0.000 | 0.000 | 0.000 | 0.000 | 0.000 |
| Pseudo *R*-sq | 0.1416 | | | 0.1462 | | | 0.1468 | | |
| Adj. *R*-sq | 0.1394 | | | 0.1440 | | | 0.1447 | | |
| Sparsity | 4.7660 | | | 4.7212 | | | 4.8082 | | |
| QLR stat | 154.47 | | | 160.97 | | | 158.75 | | |
| Prob. | 0.0000 | | | 0.0000 | | | 0.0000 | | |

2) In order to have a better interaction model, in the statistical sense, that is the model having a significant interaction, specifically for the model of *LW* on *(Z1B,Z3B)*, Table 5.3 presents the results of two interaction reduced model, namely the QRs RM-1 and RM-2. Based on these QRs, the following findings and notes are presented:

2.1 The interaction IVs of both RM-1 and RM-2 have significant effects on *LW*. So, both QRs are acceptable regressions, in the statistical sense, to represent that the effect of one of the variables on *LW* depends the other variable. RM-2 has a better GoF, since it has greater values of pseudo *R*-squared and adjusted *R*-squared.

2.2 However, because the interactions have different signs in the two regressions, in the theoretical and substantial sense, only one of the two QRs should be the true regression in the corresponding population. Hence, you should use your best possible judgment, corresponding to the true or real measured variables used in your analysis.

3) To obtain an additional statistical result to use for making a selection, we can consider the estimate of another reduced QR, $\widehat{LW} = 5.951\,576 - 0.005\,993 * Z1B * Z3B$, where $Z1B * Z3B$ has an insignificant effect based on the $t$-statistic of $t_0 = -0.497\,575$ with *Prob.* = 0.6189.

4) For comparison, the table also presents the additive reduced model, RM-3, which has greater $R$-squared and adjusted $R$-squared than RM-2. Even though they have very small differences, RM-3 has a better GoF, in the statistical sense. However, you should not select between one of the two-way interaction QRs and the additive QR based on their statistical results because they present two distinct theoretical and statistical concepts, with their own specific objectives, as follows:

   4.1 The two-way interaction QR presents or tests the effect of one of the main IVs depends on the other main IV.

   4.2 On the other hand, the additive model presents or tests the effect of one of the main IVs, adjusted for the other main IV, or by taking into account its correlation with the other main variable, on the DV.

   However, we cannot say one of these two alternative models is the best model because there are even many other models, such as polynomial, nonlinear models, and models using the transformed DV. Additionally, we never know the true population model.

**Example 5.9** *Additive and Interaction Models of LW on (X1,Z1)*

As another type of statistical results, Table 5.4 presents the summary of the results of a F2WI-QR of *LW* on a positive variable *X1*, a *Z*-score variable *Z1*, and four of its reduced models, namely RM-1 to RM-4. Based on this summary, the findings and notes are as follows:

1) The set of IVs of each of the interaction models has significant joint effects on *LW*, based on the *QLR*-statistic with $p$-values = 0.0000. Hence, each of them is an acceptable model for representing that the effect of *X1* (or *Z1*) depends on *Z1* (or *X1*).

2) Since the interaction $X1*Z1$ in the F2WI-QR has an insignificant negative adjusted effect on *LW* with a very large $p$-value = 0.8992/2 = 0.4596, it is important to conduct the following testing hypothesis, based on the model:

$$LW = C(1) + C(2) * X1 + C(3) * Z1 + C(4) * X1 * Z1 + \varepsilon \tag{5.14}$$

   2.1 The null hypothesis $H_0$: $C(3) = C(4) = 0$ is rejected based on the $F$-statistic and Chi-squared statistic, as shown in Figure 5.13. Hence, we can conclude that the effect of *Z1* on *LW* is significantly dependent on a function of *X1*, namely $(C(3) + C(4)*X'1)$.

   2.2 On the other hand, based on the results in Figure 5.13, the null hypothesis $H_0$: $C(2) = C(4) = 0$ is accepted based on the $F$-statistic and Chi-squared statistic. Hence, we can conclude that the effect of *X1* on *LW* is insignificantly dependent on a function of *Z1*, namely $(C(2) + C(4)*Z1)$.

3) There is a similar problem with the reduced model RM-2. The interaction $X1*Z1$ has an insignificant negative effect on *LW*, but the null hypothesis $H_0$: $C(2) = C(3) = 0$ is rejected based on the $QLR$-statistic of $QLR_0 = 60.6856$ with $df = 2$ and $p$-value = 0.0000. Hence, the effect of *Z1* on *LW* is significantly dependent on $(C(2) + C(3)*X1)$, but the

**Table 5.4** Statistical results summary of the F2WI-QR of *LW* and four of its reduced models.

| F2WI_QR | C | X1 | Z1 | X1*Z1 | Ps. $R^2$ | Adj. $R^2$ | QLR stat | Prob. |
|---------|---|----|----|-------|-----------|------------|----------|-------|
| Coef. | 5.6581 | −0.0757 | 0.7189 | −0.0423 | 0.0610 | 0.0574 | 61.1672 | 0.0000 |
| *t*-stat | 27.8024 | −0.2239 | 3.3198 | −0.1267 | | | | |
| Prob. | 0.0000 | 0.8229 | 0.0009 | 0.8992 | | | | |
| **RM-1** | C | X1 | | X1*Z1 | | | | |
| Coef. | 5.8840 | −0.3839 | | 0.9783 | 0.0484 | 0.0460 | 48.9299 | 0.0000 |
| *t*-stat | 30.6333 | −1.1983 | | 7.4825 | | | | |
| Prob. | 0.0000 | 0.2311 | | 0.0000 | | | | |
| **RM-2** | C | | Z1 | X1*Z1 | | | | |
| Coef. | 5.6300 | | 0.7307 | −0.0280 | 0.0609 | 0.0585 | 60.6856 | 0.0000 |
| *t*-stat | 58.4108 | | 3.5209 | −0.0903 | | | | |
| Prob. | 0.0000 | | 0.0005 | 0.9281 | | | | |
| **RM-3** | C | | | X1*Z1 | | | | |
| Coef. | 5.6505 | | | 0.9119 | 0.0478 | 0.0466 | 47.8200 | 0.0000 |
| *t*-stat | 59.3888 | | | 7.2667 | | | | |
| Prob. | 0.0000 | | | 0.0000 | | | | |
| **RM-4** | C | X1 | Z1 | | | | | |
| Coef. | 5.6704 | −0.1016 | 0.6989 | | 0.0609 | 0.0585 | 61.1171 | 0.0000 |
| *t*-stat | 29.2243 | −0.3250 | 8.0158 | | | | | |
| Prob. | 0.0000 | 0.7452 | 0.0000 | | | | | |

Wald Test:
Equation: EQ01

| Test Statistic | Value | df | Probability |
|----------------|-------|-----|-------------|
| F-statistic | 32.67092 | (2, 786) | 0.0000 |
| Chi-square | 65.34185 | 2 | 0.0000 |

Null Hypothesis: C(3)=C(4)=0
Null Hypothesis Summary:

| Normalized Restriction (= 0) | Value | Std. Err. |
|------------------------------|-------|-----------|
| C(3) | 0.718865 | 0.216540 |
| C(4) | -0.042291 | 0.333786 |

Restrictions are linear in coefficients.

(a)

Wald Test:
Equation: EQ01

| Test Statistic | Value | df | Probability |
|----------------|-------|-----|-------------|
| F-statistic | 0.052482 | (2, 786) | 0.9489 |
| Chi-square | 0.104965 | 2 | 0.9489 |

Null Hypothesis: C(2)=C(4)=0
Null Hypothesis Summary:

| Normalized Restriction (= 0) | Value | Std. Err. |
|------------------------------|-------|-----------|
| C(2) | -0.075660 | 0.337962 |
| C(4) | -0.042291 | 0.333786 |

Restrictions are linear in coefficients.

(b)

**Figure 5.13** Two statistical Wald tests based on the F2WI-QR of *LW*.

effect of *X1* on *LW* is insignificantly dependent on *C(3)\*Z1,* based on the *t*-statistic of $t_0 = -0.0903$ with *p*-value = 0.9821.
4) For a comparison of the statistical results of the five QRs in Figure 5.4, do the models of *LW* on *(X1,Z1B)* as exercises.

**Example 5.10** *Additive and Interaction Translog QRs of LW on (Z2B,Z3B)*
For a prelimanary statistical analysis, Figure 5.14 shows the descriptive statistics of four positive variables *log(LW)*, *log(Z2B)*, *log(Z3B)*, and *log(Z2B)\*log(Z3B)* and their matrix correlations. Note that *log(LW)* has negative significant correlation with each of the other variables, at the 1% level of significance.

As special and unexpected statistical results, Table 5.5 presents the outputs summary of a *full-two-way-interaction translog QR(Median)* (F2WI-TLQR(Median)) of *LW* and four of its reduced models, *RM-1, RM-2, RM-3,* and *RM-4*. Note that a two-way interaction model is proposed with the main objective for testing the hypothesis that the effect of one of the main variables on the DV depends on the other main variable, which is valid for any regression having either two categorical or numerical IVs. Based on this summary, the following findings and notes are presented:

1) Unexpectedly, each IV of the F2WI-TLQR(Median) of *log(LW)* has a very large *p*-value because it has a significant negative correlation with *log(LW)*, as shown in Figure 5.14. Is this QR a good fit interaction model? The answer is yes, because of the following reasons:
  1.1 The three IVs have significant joint effects on *log(LW)*, based on the *QLR*-statistic of $QLR_0 = 78.110$ with *df* = 3 and *p*-value = 0.000.
  1.2 At the 5% level of significance, the two IVs, *log(Z2B)* and *log(Z2B\*log(Z3B)*, have significant joint effects, based on the output, shown in Figure 5.15a, of the redundant variables test (RVT) *QLR L*-statistic of $QLR\text{-}L_0 = 7.055\,269$ with *df* = 2 and *p*-value = 0.0294. So, we can conclude that the effect of *log(Z2B)* on *log(LW)* is significantly dependent on *(C(2) + C(4)\*log(Z3B))*, in the form of the following regression function:

$$\widehat{\log{(LW)}} = (2.358 - 0.153*\log{(Z3B)}) + (0.100 - 0.218*\log{(Z3B)})*\log{(Z2B)}$$

$$(5.15)$$

| | LOG(LW) | LOG(Z2B) | LOG(Z3B) | LOG(Z2B)* LOG(Z3B) |
|---|---|---|---|---|
| Mean | 1.690185 | 1.592404 | 1.571889 | 2.500803 |
| Median | 1.760101 | 1.609781 | 1.596925 | 2.506260 |
| Maximum | 2.422404 | 2.084915 | 2.120489 | 3.782158 |
| Minimum | 0.039201 | 0.701363 | 0.298883 | 0.504494 |
| Std. Dev. | 0.400969 | 0.205322 | 0.225384 | 0.476889 |
| Skewness | -0.905103 | -0.468018 | -0.893259 | -0.196435 |
| Kurtosis | 3.968305 | 3.230821 | 4.798583 | 3.126463 |
| | | | | |
| Jarque-Bera | 138.7260 | 38.72668 | 267.7727 | 7.097483 |
| Probability | 0.000000 | 0.000000 | 0.000000 | 0.028761 |
| | | | | |
| Sum | 1335.246 | 1592.404 | 1571.889 | 2500.803 |
| Sum Sq. Dev. | 126.8522 | 42.11510 | 50.74694 | 227.1956 |
| | | | | |
| Observations | 790 | 1000 | 1000 | 1000 |

| Correlation<br>t-Statistic | | | |
|---|---|---|---|
| Probability | LOG(LW) | LOG(Z2B) | LOG(Z3B) |
| LOG(LW) | 1.000000 | | |
| | ----- | | |
| | ----- | | |
| LOG(Z2B) | -0.120540 | 1.000000 | |
| | -3.408559 | ----- | |
| | 0.0007 | ----- | |
| LOG(Z3B) | -0.347867 | 0.018127 | 1.000000 |
| | -10.41562 | 0.508938 | ----- |
| | 0.0000 | 0.6109 | ----- |
| LOG(Z2B)*LOG(Z3B) | -0.343682 | 0.635232 | 0.779365 |
| | -10.27340 | 23.08860 | 34.91674 |
| | 0.0000 | 0.0000 | 0.0000 |

**Figure 5.14** Descriptive statistics and the correlation matrix of the four variables.

**Table 5.5** Statistical results summary of the F2WI-TLQR of *LW* and its reduced models.

| F2WI-TLQR | C | LOG(Z2B) | LOG(Z3B) | LOG(Z2B)*LOG(Z3B) | Ps. R^2 | Adj. R^2 | Quasi-LR | Prob. |
|---|---|---|---|---|---|---|---|---|
| Coef. | 2.358 | 0.100 | −0.153 | −0.218 | 0.06933 | 0.06578 | 78.110 | 0.000 |
| t-stat. | 2.498 | 0.177 | −0.240 | −0.566 | | | | |
| Prob. | 0.013 | 0.860 | 0.810 | 0.572 | | | | |
| **RM-1** | C | LOG(Z2B) | | LOG(Z2B)*LOG(Z3B) | Ps. R^2 | Adj. R^2 | Quasi-LR | Prob. |
| Coef. | 2.122 | 0.237 | | −0.306 | 0.06922 | 0.06686 | 78.045 | 0.000 |
| t-stat. | 15.949 | 2.503 | | −8.879 | | | | |
| Prob. | 0.000 | 0.013 | | 0.000 | | | | |
| **RM-2** | C | | LOG(Z3B) | LOG(Z2B)*LOG(Z3B) | Ps. R^2 | Adj. R^2 | Quasi-LR | Prob. |
| Coef. | 2.536 | | −0.265 | −0.155 | 0.06928 | 0.06691 | 77.959 | 0.000 |
| t-stat. | 29.421 | | −2.475 | −2.750 | | | | |
| Prob. | 0.000 | | 0.014 | 0.006 | | | | |
| **RM-3** | C | | | LOG(Z2B)*LOG(Z3B) | Ps. R^2 | Adj. R^2 | Quasi-LR | Prob. |
| Coef. | 2.396 | | | −0.264 | 0.06313 | 0.06194 | 69.907 | 0.000 |
| t-stat. | 31.782 | | | −8.283 | | | | |
| Prob. | 0.000 | | | 0.000 | | | | |
| **RM-4** | C | LOG(Z2B) | LOG(Z3B) | | Ps. R^2 | Adj. R^2 | Quasi-LR | Prob. |
| Coef. | 2.912 | −0.219 | −0.531 | | 0.06903 | 0.06667 | 77.748 | 0.000 |
| t-stat. | 18.005 | −2.624 | −9.142 | | | | | |
| Prob. | 0.000 | 0.009 | 0.000 | | | | | |

Note that this equation shows that the effect of *log(Z2B)* decreases with increasing scores of *log(Z3B)*, corresponding to the possible values of $(0.100 - 0.218*log(Z3B)) \leq 0.100 - 0.218*Min(log(Z3B)) = - 0.2582$. This shows the effects of *log(Z2B)* are negative for all scores of *log(Z3B)*.

1.3 At the 1% level, the two IVs, *log(Z3B)* and *log(Z2B*log(Z3B)*, have significant joint effects, based on the output in Figure 5.15b, of the RVT *QLR L*-statistic of $QLR-L_0 = 71.81\,905$ with $df = 2$ and *p*-value = 0.0000. Hence, we can conclude that the effect of *log(Z3B)* on *log(LW)* is significantly dependent on $(C(3) + C(4)*log(Z3B))$, in the form of the following regression function:

$$\widehat{log\,(LW)} = \left(2.358 - 0.100* \log{(Z2B)}\right) + \left(-0.153 - 0.218* \log{(Z2B)}\right) * \log{(Z3B)}$$

(5.16)

Compared with equation (5.15), this equation shows that the effect of *log(Z3B)* has significant negative effects for all scores of *log(Z2B)*, and it will decrease with increasing scores of *log(Z2B)*, corresponding to the possible values of $(- 0.153 - 0.218*log(Z2B)) \leq (-0.153 - 0.218*Min(logZ2B)) = -0.218\,18$, for all observed scores of *log(Z2B)*.

1.4 Hence, I consider the F2WI-TLQR to be a good fit interaction QR(Median), even though each of its IVs has a large *p*-value.

(a)
Redundant variables test
Equation: EQ06_TL
Redundant variables: LOG(Z2B) LOG(Z2B)*LOG(Z3B)
Specification: LOG(LW) C LOG(Z2B) LOG(Z3B) LOG(Z2B)*LOG(Z3B)
Null hypothesis: LOG(Z2B) LOG(Z2B)*LOG(Z3B) are jointly insignificant

|  | Value | df | Probability |
|---|---|---|---|
| QLR L-statistic | 7.055269 | 2 | 0.0294 |
| QLR lambda-statistic | 7.031638 | 2 | 0.0297 |

(b)
Redundant variables test
Equation: EQ06_TL
Redundant variables: LOG(Z3B) LOG(Z2B)*LOG(Z3B)
Specification: LOG(LW) C LOG(Z2B) LOG(Z3B) LOG(Z2B)*LOG(Z3B)
Null hypothesis: LOG(Z3B) LOG(Z2B)*LOG(Z3B) are jointly insignificant

|  | Value | df | Probability |
|---|---|---|---|
| QLR L-statistic | 71.81905 | 2 | 0.0000 |
| QLR lambda-statistic | 69.46618 | 2 | 0.0000 |

**Figure 5.15**   (a) The RVT of *log(Z2B) log(Z2B)*log(Z3b)*. (b) The RVT of *log(Z3B) log(Z2B)*log(Z3b)*.

2) However, as a comparative study, the four reduced models are presented. Based on their outputs, the findings and notes presented are as follows:

2.1 Referring to the values of the pseudo $R$-squared of the four 2WI-TLQR(Median), the F2WI_TLQR(Median) has the largest pseudo $R$-squared. So, in the statistical sense, the full model is the best QR among the four 2WI_TLQR(Median)s, even though they have very small differences.

2.2 However, based on the values of the adjusted $R$-squared, RM-2 is the best QR among the four reduced 2WI_TLQR(Median)s, since it has the largest adjusted $R$-squared.

2.3 RM-4 is an additive model, which does not belong to the group of 2WI-QRs. It has another main objective, which is to study and test the effect of an IV, adjusted for the other IVs. Based on its output, at the 1% level, we can conclude *log(Z2B)* has a significant effect on *log(LW)*, adjusted for *log(Z3B)*, based on the $t$-statistic of $t_0 = -2.624$ with $df = (790 - 3)$ and $p$-value $= 0.009$. In short, *log(Z2B)* has a significant adjusted effect on *log(LW)*.

## 5.3.2 Alternative QRs of *INC*

After doing QR analyses based on various alternative models, I can classify two groups of models. The first group is the models of *INC* with one or two Z-scores as predictors. The

QR functions obtained are poor regressions, since they have negative adjusted *R*-squared, as stated in the *EViews 8 Users Guide II*. The second group is the models using other selected numerical predictors.

### 5.3.2.1 Using *Z*-Scores Variables as Predictors

### Example 5.11 *Full Two-way Interaction QRs of INC with Selected Predictors*
Table 5.6 presents the statistical results summary of positive variable *INC* with four sets of two selected predictors with their interaction, with at least one being a Z-score variable. Based on these results, the following findings and notes are presented:

1) The four QRs have negative adjusted *R*-squared. So, the regressions are considered as poor regressions, in the statistical sense. In the *EViews 8 Users Guide II*, on page 13, it defines the adjusted *R*-squared as follows:

$$\overline{R}^2 = 1 - \left(1 - R^2\right) \times \frac{T-1}{T-k} \tag{5.17}$$

where *T* is the number of observations, and *k* is the number of independent variables of the model. The adjusted *R*-squared, $\overline{R}^2$, which is never larger than $R^2$, can decrease if you add regressors, and for poorly fitting models, it may be negative.

Since the four QRs in Table 5.6 have very small pseudo *R*-squareds, then each have a negative adjusted *R*-squared. For the F2WI-QR1, for example, by using $R^2$ = pseudo

**Table 5.6** Statistical results summary of four F2WI-QRs of *INC* with selected predictors.

| F2WI_QR1 | C | Z1B | Z2B | Z1B*Z2B | Ps. R^2 | Adj. R^2 | Quasi-LR | Prob. |
|---|---|---|---|---|---|---|---|---|
| Coef. | 5.870 | −0.190 | −0.168 | 0.030 | 0.000 | **−0.003** | 0.103 | 0.991 |
| *t*-stat | 1.737 | −0.286 | −0.269 | 0.241 | | | | |
| Prob. | 0.083 | 0.775 | 0.788 | 0.810 | | | | |
| **F2WI_QR2** | C | LOG(Z2B) | LOG(Z3B) | interaction | Ps. R^2 | Adj. R^2 | Quasi-LR | Prob. |
| Coef. | 2.723 | −0.659 | −0.719 | 0.412 | 0.000 | **−0.003** | 0.682 | 0.877 |
| *t*-stat | 1.655 | −0.627 | −0.698 | 0.626 | | | | |
| Prob. | 0.098 | 0.531 | 0.485 | 0.531 | | | | |
| **F2WI_QR3** | C | X1 | Z3 | X1*Z3 | Ps. R^2 | Adj. R^2 | Quasi-LR | Prob. |
| Coef. | 5.189 | −0.720 | −0.084 | 0.072 | 0.002 | **−0.001** | 2.526 | 0.471 |
| *t*-stat | 17.894 | −1.364 | −0.282 | 0.150 | | | | |
| Prob. | 0.000 | 0.173 | 0.778 | 0.881 | | | | |
| **F2WI_QR4** | C | DIST1 | Z2 | DIST1*Z2 | Ps. R^2 | Adj. R^2 | Quasi-LR | Prob. |
| Coef. | 5.039 | −0.036 | 0.189 | −0.044 | 0.001 | **−0.002** | 0.691 | 0.875 |
| *t*-stat | 14.400 | −0.640 | 0.617 | −0.874 | | | | |
| Prob. | 0.000 | 0.522 | 0.538 | 0.382 | | | | |

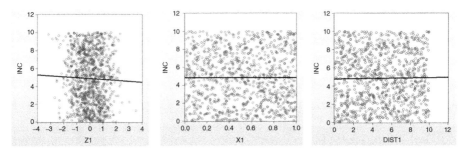

**Figure 5.16** Scatter graphs of *INC* on each variables *Z1*, *X1*, and *DIST*, with their regression lines.

$R$-squared $= 0.000$, we obtain the following negative $\overline{R}^2 <> -0.003$, as presented in the output. I really can't explain the cause of their difference.

$$\overline{R}^2 = 1 - (1 - 0.000) \times \frac{1000 - 1}{1000 - 3} = -0.002\,01$$

1) The four regressions have very small pseudo $R$-squareds, because the spreads of the scatter graphs of *INC* by each *Z1*, *X1*, and *DIST1*, as presented in Figure 5.16, with their regression lines.

Hence, it is acceptable that the QRs should have very small pseudo $R$-squareds and negative adjusted $R$-squareds. Then I would say that each of the QRs is an acceptable regression, in the statistical sense, even though it has a negative adjusted $R$-squared.

2) On the other hand, it is unexpected that each of the IVs of each QR has insignificant adjusted effects on *INC*, and all IVs also have insignificant joint effects. In addition, the reduced models of each of the QRs also have negative adjusted $R$-squareds. Do these as exercises and compare them to the following examples.

3) Since each of the QRs has a different set of IVs, each should be compared to its possible reduced models, including its additive reduced model. Do these as exercises.

### 5.3.2.2 Alternative QRs of *INC* on Other Sets of Numerical Predictors

**Example 5.12**  *Two-way Interaction QRs of INC on X1 and X2*
Table 5.7 presents the statistical results summary of the following F2WI-QR of *INC* with two of its interaction reduced models:

$$INC\ C\ X1\ X2\ X1^*X2 \tag{5.18}$$

Based on these results, the following findings and notes are presented:

1) Unexpectedly, the IVs of the three models have insignificant joint effects on *INC*, based on the *Quasi-LR* statistic, and the interaction *X1\*X2* also has an insignificant positive adjusted effect on *INC* in each of the three models. The causes are the spreads of the scatter graphs of *(X1,INC)* and *(X2,INC)*, as presented in Figure 5.17.

2) Similar to the models in the previous example, the three models also have very small pseudo $R$-squareds, but only the reduced model RM-2 has a negative adjusted $R$-squared.

**Table 5.7** Statistical results summary of the model (5.18) and two of its reduced models.

| Variable | Full model (5.18) | | | RM-1 | | | RM-2 | | |
|---|---|---|---|---|---|---|---|---|---|
| | Coef. | t-stat | Prob. | Coef. | t-stat | Prob. | Coef. | t-stat | Prob. |
| C | 5.7532 | 9.6069 | 0.0000 | 5.2547 | 18.1728 | 0.0000 | 5.0904 | 16.6316 | 0.0000 |
| X1 | −1.7380 | −1.5657 | 0.1177 | −1.0479 | −1.4893 | 0.1367 | | | |
| X2 | −1.2595 | −1.0876 | 0.2770 | | | | −0.1340 | −0.1825 | 0.8552 |
| X1*X2 | 2.2073 | 1.0132 | 0.3112 | 0.4092 | 0.3839 | 0.7012 | −0.7136 | −0.7415 | 0.4586 |
| Pseudo $R^2$ | 0.0039 | | | 0.0023 | | | 0.0014 | | |
| Adj. $R^2$ | 0.0009 | | | 0.0003 | | | −0.0006 | | |
| Quasi-LR | 4.1984 | | | 2.4639 | | | 1.4789 | | |
| Prob | 0.2408 | | | 0.2917 | | | 0.4774 | | |

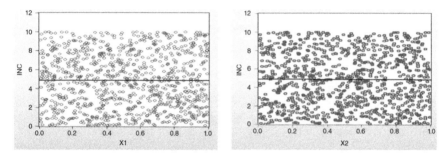

**Figure 5.17** Scatter graphs of *(X1,INC)* and *(X2,INC)* with regression lines.

3) In the statistical sense, the regression functions are acceptable QR(Median)s to represent that the effect of *X1 (or X2)* on *INC* is insignificantly dependent on *X2 (or X1)*, on the median of *INC*.

However, the full model should be considered the best, because it has the largest pseudo *R*-squared and adjusted *R*-squared.

**Example 5.13** *Two-way Interaction QRs of INC on LW and DENOM*
Table 5.8 presents the statistical results summary of the following F2WI-QR(Median) of *INC* with two of its interaction reduced models, RM-1 and RM-2:

$$INC\ C\ LW\ DENOM\ LW*DENOM \tag{5.19}$$

Based on these results, the following findings and notes are presented:

1) The IVs of all models have insignificant joint effects on *INC*, based on the *Quasi-LR* statistic. For the interaction *LW*DENOM* in the three models, these findings and notes are presented:

1.1 The interaction IV of the full model has an insignificant negative effect on *INC*, based on the *t*-statistic of $t_0 = -0.2914$ with p-value $= 0.7709/2$. Hence, we can conclude that the effect of *LW* will decrease insignificantly with the increasing scores of *DENOM*.

**Table 5.8** Statistical results summary of the model (5.19) and two of its reduced models.

| Variable | Full model (5.19) | | | RM-1 | | | RM-2 | | |
|---|---|---|---|---|---|---|---|---|---|
| | Coef. | t-stat | Prob. | Coef. | t-stat | Prob. | Coef. | t-stat | Prob. |
| C | 5.08904 | 1.25378 | 0.2103 | 5.56898 | 3.66772 | 0.0003 | 5.5251 | 11.781 | 0.0000 |
| LW | 0.09098 | 0.13521 | 0.8925 | | | | | | |
| DENOM | 0.14679 | 0.11173 | 0.9111 | −0.01482 | −0.0268 | 0.9786 | | | |
| LW*DENOM | −0.0622 | −0.2914 | 0.7709 | −0.03212 | −1.151 | **0.2501** | −0.0324 | −1.3127 | **0.1897** |
| Pseudo $R^2$ | 0.00275 | | | 0.00269 | | | 0.00269 | | |
| Adj. $R^2$ | −0.0011 | | | 0.00016 | | | 0.00143 | | |
| Quasi-LR Stat | 2.26763 | | | 2.21505 | | | 2.21316 | | |
| Prob | 0.51875 | | | 0.33038 | | | **0.13684** | | |

1.2 Each of the IVs of the full model has a large $p$-value, so, in general, we should explore an acceptable interaction reduced model with a significant interaction IV. For this reason, the results of two interaction reduced models, RM-1 and RM-2, are presented in Table 5.8.

1.3 At the 15% level of significance (used in Lepin 1973), the interaction IV of RM-1 has a significant negative effect on *INC*, based on the $t$-statistic of $t_0 = -1.1501$ with $p$-value $= 0.2501/2 = 0.12505 < 0.15$. The interaction IV of RM-2 has a significant negative effect on *INC*, at the 10% level, based on the $t$-statistic of $t_0 = -1.1327$ with $p$-value $= 0.1897/2 = 0.094855 < 0.10$. Thus, we can conclude that both reduced models show that the *LW* has a negative effect on *INC*, because *DENOM* is a positive variable.

1.4 It is important to note that the interaction *LW*DENOM* in the RM-2 has a probability of 0.1897 based on the $t$-statistic of $t_0 = -1.1327$, but it has a probability of 0.13684 based on the *Quasi-LR* statistic of $QLR_0 = 2.21316$. The relationship between the $t$-statistic and the *QLR*-statistic has not been found, but we know the probability of the $F$-statistic with $df = (1,n)$ is equal to the probability of the $t$-statistic with $df = n$, and it is approximately equal to the Chi-square with $df = 1$, as presented in Figure 5.18.

---

Wald test:
Equation: untitled

| Test statistic | Value | df | Probability |
|---|---|---|---|
| $t$-statistic | −1.312745 | 788 | 0.1897 |
| F-statistic | 1.723300 | (1, 788) | 0.1897 |
| Chi-square | 1.723300 | 1 | 0.1893 |

Null hypothesis: C(2) = 0

**Figure 5.18** The statistical results of the Wald test.

**Table 5.9** Statistical results summary of the QR(Median) (5.20) and two of its reduced models.

| Variable | Full model | | | RM-1 | | | RM-2 | | |
|---|---|---|---|---|---|---|---|---|---|
| | Coef. | t-stat | Prob. | Coef. | t-stat | Prob. | Coef. | t-stat | Prob. |
| C | 0.3882 | 95.751 | 0.0000 | 0.3230 | 111.101 | 0.0000 | 0.3893 | 189.299 | 0.0000 |
| DIST1 | 0.0002 | 0.3822 | 0.7024 | 0.0100 | 16.1925 | 0.0000 | | | |
| X1 | −0.1260 | −20.178 | 0.0000 | | | | −0.1274 | −32.055 | 0.0000 |
| DIST1*X1 | −0.0005 | −0.5243 | 0.6002 | −0.0194 | −31.923 | 0.0000 | −0.0002 | −0.4239 | 0.6717 |
| Pseudo R-sq | 0.5839 | | | 0.4263 | | | 0.5838 | | |
| Adj. R-sq | 0.5827 | | | 0.4251 | | | 0.5830 | | |
| Quasi-LR stat | 1419.8 | | | 848.35 | | | 1422.69 | | |
| Prob(Q-LR stat) | 0.0000 | | | 0.0000 | | | 0.0000 | | |

### 5.3.2.3 Alternative QRs Based on Other Sets of Numerical Variables

**Example 5.14** *Two-Way Interaction QR of PR1 on DIST1 and X1*
Through trial and error, I found the statistical results summary in Table 5.9, which is a good illustration of a two-way interaction QR and two of its reduced models, with the following ES of the full model:

$$PR1\ C\ DIST1\ X1\ DIST1*X1 \tag{5.20}$$

Based on this summary, the following findings and notes are presented:

1) The IVs of each of the models have significant joint effects on *PR1*, based on the *Quasi-LR* statistic with *p*-value = 0.0000, but for each IV of each model, we have the following unexpected findings and notes.
2) Since *PR1* and *DIST1*X1* have a significant negative correlation, as presented in Figure 5.19, it is unexpected that *DIST1*X1* has an insignificant negative adjusted effect in the full model and in RM-2. These cases show the unpredicted impacts of the multicollinearity or the bivariate correlations between the IVs.
3) Moreover, the *DIST1* has a significant positive adjusted effect in RM-1, but it has an insignificant negative correlation with *PR1*, based on the *t*-statistic of $t_0 = -0.242\,252$ with a *p*-value = 0.8086/2 = 0.4043, as presented in Figure 5.19.
4) Even though each of the IVs of RM-1 has a significant adjusted effect on *PR1*, it has the smallest adjusted R-squared. So, in the statistical sense, it is the worst GoF model among the three models.
5) On the other hand, for the full model, *X1* and *DIST1*X1* have significant partial joint effects on *PR1*, based on the Wald test *F*-statistic of $F_0 = 915.5892$ with $df = (2996)$ and *p*-value = 0.0000, and the *Chi-square* statistic of 1831.178 with $df = 2$ and *p*-value 0.0000. So, we have a good reason to keep *X1* in the full model. Hence, for the full model, we have the following QR function:

$$\widehat{PR1} = 0.3882 + 0.0002 \times DIST1 + (-0.1260 - 0.0005 \times DIST1) \times X1$$

| Correlation<br>t-statistic<br>Probability | DIST1 | X1 | DIST1*X1 | PR1 |
|---|---|---|---|---|
| PR1 | −0.007668 | −0.911896 | −0.602330 | 1.000000 |
| | −0.242252 | −70.19105 | −23.83759 | — |
| | 0.8086 | 0.0000 | 0.0000 | — |

**Figure 5.19** Bivariate correlations between *PR1* and each of *DIST1*, *X1*, and *DIST1*X1*.

which shows that the adjusted effect of *X1* on *PR1* is significantly dependent on $+(-0.1260 - 0.0005 \times DIST1)$, and it is always negative, because *DIST1* is a positive variable.

6) Additionally, RM-2 shows that the effect *X1* on *PR1* is significantly dependent on *DIST1*, in the form of $(-0.1274 - 0.0002 \times DIST1)$, which is negative for all scores of *DIST1*. Compared with the function of the full QR having various intercepts and slopes of *X1* as the functions of *DIST*, the RM-2 has the following restricted function, which has a single intercept for all possible values of *DIST1* and *X1*:

$$\widehat{PR1} = 0.3893 + (-0.1274 - 0.0002 \times DIST1) \times X1$$

**Example 5.15 Two-way Interaction QR(Median) of PR3 on XB2 and XB3**
Table 5.10 presents the statistical results summary of a F2WI-QR of *PR3* on *(XB2,XB3)*, with the following ES, and four of its reduced models.

$$PR3\ C\ XB2\ XB3\ XB2*XB3 \tag{5.21}$$

Based on this summary, the following findings and notes are presented:

1) Because the IVs of the *F2WI-QR, RM-2*, and *RM-4* have significant joint effects based on the *Quasi-LR* statistic, these three models, among the five, are acceptable models, in the statistical sense, to represent the median regressions of *INC*, based on the data set.

2) Note that the three models represent two types of the relationships between *XB2* and *XB3*, as presented in Figure 5.1, for *X1* and *X2*. In the first one, *XB2* and *XB3* don't have a causal relationship, as shown in Figure 5.1a. In this case, then, the additive model RM-4 is an acceptable model, in both the theoretical and statistical senses. In the second type, *XB2* and *XB3* have causal relationship, as shown in Figure 5.1b or c, so we have *F2WI-QR* and *RM-2* as two alternative acceptable interaction models to represent that the effect of either *XB2* or *XB3* on *PR3* depends on the other. However, they have different characteristics, such as follows:

2.1 The effect of *XB2*XB3* on *PR3* in the F2WI-QR is its effect adjusted for both *XB2* and *XB3*, which it is insignificant with a *p*-value = 0.2242. However, its effect in RM-4 is the effect adjusted for *XB2* only, which it is significant with a *p*-value 0.0011.

2.2 Even though the interaction IV of the F2WI-QR has an insignificant adjusted effect on *PR3*, at the 10% level, the joint effects of *XB3* and *XB2*XB3* on *PR3* have

**Table 5.10** Statistical results summary of the model (5.21) and its reduced models.

| F2WI-QR | C | XB2 | XB3 | XB2*XB3 | Ps. $R^2$ | Adj. $R^2$ | Sparsity | Quasi-LR | Prob. |
|---|---|---|---|---|---|---|---|---|---|
| Coef. | 3.9009 | −0.1835 | 0.2077 | 0.2515 | 0.0347 | 0.0318 | 0.2834 | 54.941 | 0.0000 |
| t-stat | 818.97 | −6.9703 | 3.1111 | 1.2161 | | | | | |
| Prob. | 0.0000 | 0.0000 | 0.0019 | **0.2242** | | | | | |
| **RM-1** | C | | XB3 | XB2*XB3 | | | | | |
| Coef. | 3.9148 | | 0.0453 | −0.0721 | 0.0009 | −0.0011 | 0.3749 | 1.0278 | **0.5982** |
| t-stat | 613.48 | | 0.9328 | −0.3506 | | | | | |
| Prob. | 0.0000 | | 0.3511 | **0.7259** | | | | | |
| **RM-2** | C | XB2 | | XB2*XB3 | | | | | |
| Coef. | 3.9066 | −0.1943 | | 0.6495 | 0.0255 | 0.0235 | 0.2813 | 40.646 | 0.0000 |
| t-stat | 851.22 | −7.0482 | | 3.2789 | | | | | |
| Prob. | 0.0000 | 0.0000 | | 0.0011 | | | | | |
| **RM-3** | C | | | XB2*XB3 | | | | | |
| Coef. | 3.9156 | | | −0.0660 | 0.0001 | −0.0009 | 0.3802 | 0.1419 | **0.7064** |
| t-stat | 608.90 | | | −0.3386 | | | | | |
| Prob. | 0.0000 | | | **0.7350** | | | | | |
| **RM-4** | C | XB2 | XB3 | | | | | | |
| Coef. | 3.9055 | −0.1942 | 0.3004 | | 0.0333 | 0.0313 | 0.2533 | 58.99 | 0.0000 |
| t-stat | 943.60 | −7.7004 | 5.4638 | | | | | | |
| Prob. | 0.0000 | 0.0000 | 0.0000 | | | | | | |

a significant effect, based on the Wald test $F$-statistic of $F_0 = 10.5758$ with $df = (2, 996)$ and $p$-value $= 0.0000$, and the *Chi-square* statistic of 21.1517 with $df = 2$ and $p$-value 0.0000. Then the effect of *XB3* on *PR3* is significantly dependent on the function $(C(3) + C(4)*XB2) = (0.2077 + 0.2515*XB2)$. And the joint effects of *XB2* and *XB2*XB3* on *PR3* have a significant effect, based on the Wald test $F$-statistic of $F_0 = 25.3415$ with $df = (2,996)$ and $p$-value $= 0.0000$, and the *Chi-square* statistic of 50.6830 with $df = 2$ and $p$-value 0.0000. Then the effect of *XB2* on *PR3* is significantly dependent on the function $(C(2) + C(4)*XB3)$. So, we have the following median regression function, which is a set of heterogeneous regression lines of $\widehat{PR3}$ on *XB2,* with various intercepts and various slopes for all values of *XB3*:

$$\widehat{PR3} = 3.9009 + 0.2077XB3 + (−0.1835 + 0.2515XB3) \times XB2$$

2.3 Compared with the previous regression function, the RM-2 has the following median regression function, which is a very or too specific set of regression lines of $\widehat{PR3}$ on *XB2,* with a single intercept of 3.9066 and various slopes for all values of *XB3*:

$$\widehat{PR3} = 3.9066 + (−0.1943 + 0.6495XB3) \times XB2$$

This regression, in fact, is an acceptable regression, in the statistical sense; however, it is never found or happens in practice or reality. For this reason, the F2WI-QR is acceptable, in both theoretical and statistical senses. In addition, it also has larger adjusted $R$-squared than RM-2.

3) However, the additive model RM-4 has the largest adjusted $R$-squared. So, the RM-4 is the best fit QR, in the statistical sense, but it shows the effect of *XB2* on *PR3* is significantly adjusted for *X3B*.

## 5.4   Polynomial Two-Way Interaction QRs

As an extension of the polynomial QRs presented in previous chapter, the following example presents selected polynomial two-way interaction QRs, using Data_Faad.wf1.

**Example 5.16**   *Third-degree Polynomial Interaction QR*
Figure 5.20 presents the output of a third-degree polynomial two-way interaction QR(Median) with the following ES:

$$Y1 \; C \; X1*X2 \; X1\char`\^2*X2 \; X1\char`\^3*X2 \; X1 \; X2 \tag{5.22}$$

Based on this output, the following findings and notes are presented:

1. At the 10% level of significance, each of the interaction IVs has either a positive or negative adjusted effect on *Y1*, with a $p$-value $=$ Prob/2 $< 0.10$. For instance, $X1\char`\^3*X2$ has a positive significant effect, based on the $t$-statistic of $t_0 = 1.5423$ with $df = (300 - 6)$ and $p$-value $= 0.1421/2 = 0.071\,55$.
2. Even though the main IV, *X2,* has a large $p$-value $= 0.4278$, the Full-QR(Median) is an acceptable model, because each of its interactions has either a positive or negative significant effect on *Y1*. In addition, at the 10% level, the output of the RVT shows that the three interactions have joint significant effects, based on the $QLR$ $L$-statistic of $QLR$-$L_0 = 7.771$ 690 with $df = 3$ and $p$-value $= 0.0510$.
3. As an additional illustration, Figure 5.21 presents the outputs of the Full-QR(0.25), which show that each interaction IV has either a positive or negative significant effect, at the 1% level, but for the Full-QR(0.75), each interaction IV has either a positive or negative insignificant effect, at the 10% level. Nonetheless, the three QRs presented in Figures 5.20 and 5.21 are acceptable models, because it is required they have the same set of IVs, for presenting the same statistical results of the quantile process, using each of the three QR(0.25), Q(0.50), and QR(0,75), specifically for the quantile process: QPE, QSET, and SQT, by using quantiles $= 4$, and also for quantiles $= 9$ for the nine values of tau from 0.1 to 0.9. We can easily obtain the simplest statistical results by inserting only one parameters C(2) to C(7) as the user specifified coefficient. Do as exercises.
4. As additional illustration, with the output of the QR(0.50) (5.22) on-screen, select the *Forecast* button, inserting the forecast name Y1F_FQR50, and click *OK* to obtain the output in Figure 5.22. Based on this output, the following findings and notes are presented:
   4.1 The forecast value variable of *Y1,* based on the Full_QR (5.22) in Figure 5.22, in fact is the same as its fitted value variable. The advantage of using the forecast

Dependent variable: Y1
Method: quantile regression (Median)
Included observations: 300
Huber sandwich standard errors and covariance
Sparsity method: Kernel (Epanechnikov) using residuals
Bandwidth method: Hall-Sheather, bw = 0.14513
Estimation successfully identifies unique optimal solution

| Variable | Coefficient | Std. error | t-statistic | Prob. |
|---|---|---|---|---|
| C | 1.568686 | 1.275909 | 1.229466 | 0.2199 |
| X1*X2 | 5.021074 | 3.651773 | 1.374969 | 0.1702 |
| X1^2*X2 | −3.609385 | 2.343856 | −1.539935 | 0.1247 |
| X1^3*X2 | 0.792930 | 0.514115 | 1.542320 | 0.1241 |
| X1 | 2.052514 | 0.917753 | 2.236457 | 0.0261 |
| X2 | −1.577804 | 1.987040 | −0.794047 | 0.4278 |

| | | | |
|---|---|---|---|
| Pseudo R-squared | 0.190803 | Mean dependent var | 5.751100 |
| Adjusted R-squared | 0.177041 | S.D. dependent var | 0.945077 |
| S.E. of regression | 0.768049 | Objective | 89.80065 |
| Quantile dependent var | 5.670000 | Restr. objective | 110.9750 |
| Sparsity | 1.925752 | Quasi-LR statistic | 87.96293 |
| Prob(quasi-LR stat) | 0.000000 | | |

**Figure 5.20** The output of the Full-QR(Median) in (5.22).

| Variable | Full-QR(0.25) | | | Full-QR(0.75) | | |
|---|---|---|---|---|---|---|
| | Coef. | t-stat. | Prob. | Coef. | t-stat. | Prob. |
| C | 0.5360 | 0.5196 | 0.6037 | 0.2372 | 0.1357 | 0.8922 |
| X1*X2 | 7.2964 | 2.3633 | 0.0188 | 2.3282 | 0.6866 | 0.4929 |
| X1^2*X2 | −5.6107 | −2.6256 | 0.0091 | −2.5435 | −1.0802 | 0.2810 |
| X1^3*X2 | 1.3040 | 2.7290 | 0.0067 | 0.6413 | 1.1820 | 0.2382 |
| X1 | 2.6353 | 3.2310 | 0.0014 | 3.4891 | 3.0261 | 0.0027 |
| X2 | −2.3651 | −1.5743 | 0.1165 | 0.4212 | 0.2436 | 0.8077 |
| Pse. R^2 | 0.2291 | | | 0.1978 | | |
| Adj. R^2 | 0.2159 | | | 0.1841 | | |
| S.E. of reg. | 0.9050 | | | 0.9357 | | |
| Quant DV | 5.3300 | | | 6.3300 | | |
| Sparsity | 2.3152 | | . | 2.2046 | | |
| QLR-stat. | 99.392 | | | 83.591 | | |
| Prob | 0.0000 | | | 0.0000 | | |

**Figure 5.21** The outputs of the Full-QR(0.25) and Full-QR(0.75).

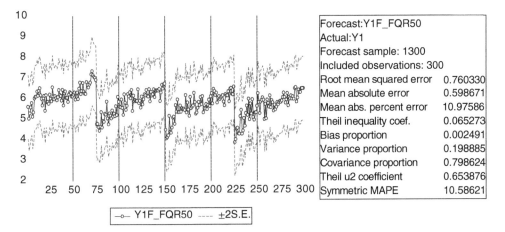

**Figure 5.22** Forecast evaluation of the Full-QR(0.50) in (5.22).

value variable is that we can use various useful statistics. One, in particular, is the TIC (*Theil inequality coefficient*) statistic, which is a relative GoF measure of the forecast variable, where TIC = 0 indicates the perfect forecast and TIC = 1 is the worst – refer to Theil (1966) and Bliemel (1973). Other statistics include RMSE (root-mean-square error), MAE (mean absolute error), and MAPE (mean absolute percentage error).

4.2 Note that the graph of *Y1F_FQR50* is divided into four parts. This is because the data are sorted by the two dichotomous factors *A* and *B* in the data set.

5. As a more advanced illustration, Figure 5.23 presents the graphs of *sorted Y1* and its forecast variables, *Y1F_FQR25, Y1F_FQR50*, and *Y1F_FQR75*. Based on these graphs, the findings and notes are as follows:

5.1 The sort of *Y1* presents a set of 19 monotone increase observed scores, but the Full_QR(0.50) has a set of 113 predicted values of *Y1*. These findings indicate that the same scores of *Y1* can have different predicted values, which are highly dependent on the scores of both predictors, *X1* and *X2*.

5.2 The forecast variable *Y1F_FQR50* tends to have larger values than the lower observed scores of *Y1,* and it tends to have smaller values than the upper observed scores of *Y1*. I would consider these patterns of the graphs deviations are a specific or special character of any QR(0.50).

## Example 5.17  *Two Alternative Reduced QRs of the QR (5.22)*

Because each two-way interaction IV of the Full-QR(0.5) in Figure 5.20 has an insignificant effect at the 10% level, then I present two alterantive reduced models by deleting either *X2* or *X1,* with their outputs summary presnted in Figure 5.24. Based on these outputs, the following findings and notes are presented:

1. The Reduced-QR1 is obtained by deleting *X2* from the Full-QR(Median) because *X2* has greater probability than *X1*. The Reduced-QR2 is obtained by deleting *X1*. Note that each IV of both Reduced-QRs has a significant effect, at the 1% or 5% level.

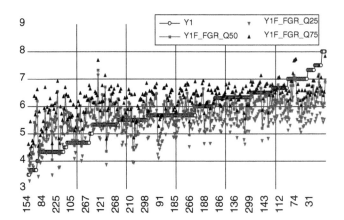

**Figure 5.23** The graphs of sorted *Y1*, and *Y1F_FQR(tau)*, for tau = 0.25, 0.50, and 0.75.

| Dependent variable: Y1 | | | | | | | | |
|---|---|---|---|---|---|---|---|---|
| Method: quantile regression (Median) | | | | | | | | |
| Included observations: 300 | | | | | | | | |
| Huber sandwich standard errors and covariance | | | | | | | | |
| Sparsity method: Kernel (Epanechnikov) using residuals | | | | | | | | |
| Bandwidth method: Hall-Sheather, bw = 0.14513 | | | | | | | | |
| Estimation successfully identifies unique optimal solution | | | | | | | | |
| Variable | Reduced-QR1 | | | | Reduced-QR2 | | | |
| | Coef. | Std. err. | *t*-stat. | Prob. | Coef. | Std. err. | *t*-stat. | Prob. |
| C | 0.702 | 0.868 | 0.809 | 0.419 | 4.354 | 0.269 | 16.175 | 0.000 |
| X1*X2 | 2.220 | 0.711 | 3.123 | 0.002 | 8.269 | 2.873 | 2.878 | 0.004 |
| X1^2*X2 | −1.890 | 0.814 | −2.321 | 0.021 | −5.117 | 1.997 | −2.562 | 0.011 |
| X1^3*X2 | 0.425 | 0.221 | 1.926 | 0.055 | 1.083 | 0.451 | 2.402 | 0.017 |
| X1 | 2.620 | 0.678 | 3.863 | 0.000 | | | | |
| X2 | | | | | −3.924 | 1.364 | −2.877 | 0.004 |
| Pse. *R*^2 | 0.189 | | | | 0.175 | | | |
| Adj. *R*^2 | 0.178 | | | | 0.164 | | | |
| S.E. of reg. | 0.769 | | | | 0.781 | | | |
| Quant. DV | 5.670 | | | | 5.670 | | | |
| Sparsity | 1.947 | | | | 1.923 | | | |
| QLR-stat | 86.24 | | | | 80.86 | | | |
| Prob | 0.000 | | | | 0.000 | | | |

**Figure 5.24** Statistical results summary of two Reduced-QRs of the QR (5.22).

Covariance analysis: ordinary
Date: 08/28/19   Time: 11:29
Sample: 1 300
Included observations: 300

| Correlation<br>t-statistic<br>Probability | Y1 | Y1F_FULL_QR | Y1F_RED_QR1 | Y1F_RED_QR2 |
|---|---|---|---|---|
| Y1 | 1.000000 | | | |
| | — | | | |
| | — | | | |
| Y1F_FULL_QR | 0.595204 | 1.000000 | | |
| | 12.78637 | — | | |
| | 0.0000 | — | | |
| Y1F_RED_QR1 | 0.593224 | 0.996206 | 1.000000 | |
| | 12.72068 | 197.6068 | — | |
| | 0.0000 | 0.0000 | — | |
| Y1F_RED_QR2 | 0.575600 | 0.977912 | 0.960753 | 1.000000 |
| | 12.15116 | 80.76552 | 59.78685 | — |
| | 0.0000 | 0.0000 | 0.0000 | — |

**Figure 5.25**   The bivariate correlation matrix between *Y1* and its three forecast variables.

2. In addition, note that the Reduced-QR1 has a slightly larger adjusted R-squared than the Full-QR(Median) and Reduced-QR2. However, the Full-QR(median) has a slightly larger pseudo R-squared than the Reduced-QR1. Then, we could have different opinions on the best fit QR among the three QR(Median)s. A note in the EViews 8 User Guide II recommends using the values of the adjusted $R$-squared, because the values of pseudo $R$-squared will always increase with the increasing number of the IVs. Hence, we can consider the Reduced-QR1 to be the best fit QR(Median), among the three QR(Median)s, and the Reduced-QR2 to be the worst.
3. As an additional illustration, Figure 5.25 presents the bivariate correlations matrix between the four variable: one *Y1* and its 3 forecast variables, which shows that each pair has a significant correlation with a $p$-value $= 0.0000$.
4. For the graphical illustrations, Figure 5.26a presents the graphs of three variables, *Y1*, *Y1F_Full_QR*, *Y1F_Reduced_QR1*, and Figure 5.2b presents the graphs of another three variables, *Y1*, *Y1F_Full-QR*, *Y1F_Reduced_QR2*, sort by *Y1*, which clearly show the forecast variables have greater values than lower scores of *Y1*, and they have smaller values than upper scores of *Y1*.

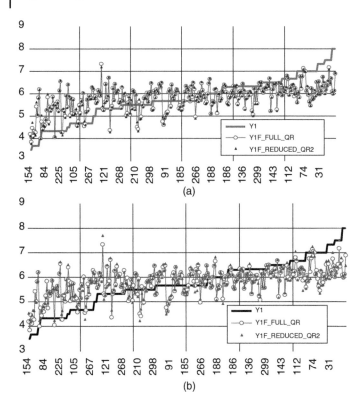

**Figure 5.26** (a) The graphs of *Y1, Y1F_Full_QR,* and *Y1F_Reduced_QR1,* sorted by *Y1.* (b) The graphs of *Y1, Y1F_Full_QR,* and *Y1F_Reduced_QR2,* sorted by *Y1.*

**Example 5.18** *An Application of the Quantile Slope Equality Test*

With the output of the Full-QR(Median) on-screen, select *View/Quantile Process/Slope Equality Test,* insert "0.1 0.9," as the *User-specified quantiles,* and then click *OK* to obtain the output in Figure 5.27, which shows the slope differences between two pairs of quantiles, (0.1, 0.5) and (0.5, 0.9), for the five IVs. Based on the Wald test (Chi-Square statistic) of $\chi_0^2 = 15.37314$ with $df = 10$ and $p$-value = 0.1190, we can conclude that the 10 pairs of slopes are jointly insignificant, at the 10% level. The IVs $X1^2*X2$ and $X1^3*X2$ have a significant difference between the pair of quantiles (0.1, 0.5), at the 10% and 5% levels of significance.

On the other hand, for testing a special pair of quantiles, such as (0.1, 0.9), the Full-QR(0.1) should be applied. With the output of the Full-QR(0.1) on-screen, select *View/Quantile Process/Slope Equality Test,* insert "0.1 0.9" as the *User-specified quantiles,* and then click *OK,* to obtain the results in Figure 5.28, which show that the five pairs of the slopes have insignificant differences, based on the Wald test (Chi-Square statistic) of $\chi_0^2 =$ of 6.624 759 with $df = 5$ and $p$-value = 0.2501, and two pairs of the slopes of $X1^2*X2$

Quantile slope equality test
Equation: UNTITLED
Specification: Y1 C X1*X2 X1^2*X2 X1^3*X2 X1 X2
Estimated equation quantile tau = 0.5
User-specified test quantiles: 0.1, 0.9
Test statistic compares all coefficients

| Test summary | Chi-Sq. statistic | Chi-Sq. d.f. | Prob. |
|---|---|---|---|
| Wald test | 15.37314 | 10 | 0.1190 |

Restriction detail: $b(tau\_h) - b(tau\_k) = 0$

| Quantiles | Variable | Restr. value | Std. error | Prob. |
|---|---|---|---|---|
| 0.1, 0.5 | X1*X2 | −6.591935 | 4.296201 | 0.1249 |
| | X1^2*X2 | 5.363458 | 2.802882 | 0.0557 |
| | X1^3*X2 | −1.310132 | 0.623473 | 0.0356 |
| | X1 | −1.104292 | 1.368963 | 0.4199 |
| | X2 | 2.166277 | 2.285864 | 0.3433 |
| 0.5, 0.9 | X1*X2 | −1.609831 | 4.758196 | 0.7351 |
| | X1^2*X2 | 1.359256 | 3.193280 | 0.6704 |
| | X1^3*X2 | −0.331493 | 0.734329 | 0.6517 |
| | X1 | −0.285156 | 1.283113 | 0.8241 |
| | X2 | 0.522151 | 2.451262 | 0.8313 |

**Figure 5.27** The output of a QSET of the Full-QR(Median).

and $X1^3*X2$, have significant differences, at the 10% level. Hence, we have a complete quantile slope equality test (QSETs) between three pairs of the quantiles 0.1, 0.5, and 0.9
Other pairs of quantiles also can easily be done. Do some as exercises.

## 5.5 Double Polynomial QRs

### 5.5.1 Additive Double Polynomial QRs

As an extension of third-degree polynomial QR of $Y1$ on $X1$ presented in the previous chapter, the following example presents an additive double third-degree polynomial QR

Quantile slope equality test
Equation: EQ38
Specification: Y1 C X1*X2 X1^2*X2 X1^3*X2 X1 X2
Estimated equation quantile tau = 0.1
User-specified test quantiles: 0.1, 0.9
Test statistic compares all coefficients

| Test summary | Chi-Sq. statistic | Chi-Sq. d.f. | Prob. |
|---|---|---|---|
| Wald test | 6.624759 | 5 | 0.2501 |

Restriction detail: $b(tau\_h) - b(tau\_k) = 0$

| Quantiles | Variable | Restr. value | Std. error | Prob. |
|---|---|---|---|---|
| 0.1, 0.9 | X1*X2 | −8.201766 | 5.359139 | 0.1259 |
| | X1^2*X2 | 6.722714 | 3.641803 | 0.0649 |
| | X1^3*X2 | −1.641625 | 0.842833 | 0.0514 |
| | X1 | −1.389448 | 1.714067 | 0.4176 |
| | X2 | 2.688427 | 2.707277 | 0.3207 |

**Figure 5.28** The output of a specific QSET of the Full-QR(0.1).

of *Y1* on *X1* and *X2*, with the following ES:

$$Y1 \ C \ X1 \ X1^2 \ X1^3 \ X2 \ X2^2 \ X2^3 \tag{5.23}$$

**Example 5.19** *Application of a QR(Median) in (5.23)*
Figure 5.29 presents the outputs summary of the double polynomial Full-QR($\tau$) in (5.23) for three $\tau = 0.1, 0.5$, and 0.9. Based on this output, the findings and notes presented are as follows:

1. Specific for the Full-QR(0.5), we have the following findings and notes:
   1.1 The IVs of the QR(Median) have significant joint effects, based on the *Quasi-LR* statistic of $QLR_0 = 91.41609$ with $df = 6$ and p-value $= 0.000000$. Each of the IVs, *X1, X1^2, X1^3*, and *X2*, has either a positive or negative significant effect, at the 5% or 10% level of significance.
   1.2 Based on the output of the RVT in Figure 5.30, at the 5% level, the three IVs *X2, X2^2*, and *X2^3* have significant joint effects on *Y1*, based on the *QLR L*-statistic of $QLRL_0 = 9.342027$ with $df = 3$ and p-value $= 0.0000$. So, we have a reason to use the three IVs in the Full-QR(0.5). In addition, Agung (2009a, 2011a, 2014) recommends keeping IVs with a probability $< 0.30$, since it has either positive or negative adjusted effect, at the 15% level with a p-value $=$ Prob/2 $< 0.15$. Hence, the double third-degree polynomial QR(Median) is an acceptable QR.
   1.3 As additional analysis, Figure 5.31 presents the output of residual analysis, namely the residuals histogram, with its normal density, and statistics, of the Full-QR(0.5).

| Variable | Full-QR(0.1) | | | Full-QR(0.5) | | | Full-QR(0.9) | | |
|---|---|---|---|---|---|---|---|---|---|
| | Coef. | t-stat | Prob. | Coef. | t-stat | Prob. | Coef. | t-stat | Prob. |
| C | 5.582 | 1.333 | 0.184 | −6.989 | −1.925 | 0.055 | −5.208 | −0.771 | 0.441 |
| X1 | −4.109 | −0.326 | 0.745 | 13.813 | 1.876 | 0.062 | 24.410 | 2.738 | 0.007 |
| X1^2 | 5.241 | 0.556 | 0.579 | −8.540 | −1.647 | 0.101 | −16.476 | −2.610 | 0.010 |
| X1^3 | −1.498 | −0.659 | 0.511 | 1.873 | 1.583 | 0.115 | 3.750 | 2.607 | 0.010 |
| X2 | −2.329 | −0.384 | 0.701 | 4.843 | 1.450 | 0.148 | −2.176 | −0.353 | 0.724 |
| X2^2 | 1.108 | 0.419 | 0.676 | −1.591 | −1.224 | 0.222 | 1.362 | 0.585 | 0.559 |
| X2^3 | −0.146 | −0.400 | 0.689 | 0.189 | 1.156 | 0.249 | −0.206 | −0.726 | 0.469 |
| Pse. R^2 | 0.189 | | | 0.194 | | | 0.186 | | |
| Adj. R^2 | 0.173 | | | 0.178 | | | 0.169 | | |
| S.E. of reg | 1.289 | | | 0.770 | | | 1.272 | | |
| Mean DV | 5.751 | | | 5.751 | | | 5.751 | | |
| Quant DV | 4.330 | | | 5.670 | | | 7.000 | | |
| Sparsity | 3.960 | | | 1.889 | | | 4.162 | | |
| QLR-stat | 52.657 | | | 91.416 | | | 46.937 | | |
| Prob | 0.000 | | | 0.000 | | | 0.000 | | |

**Figure 5.29** The outputs summary of three Full-QR($\tau$) in (5.23).

Redundant variables test
Null hypothesis: X2 X2^2 X2^3
Equation: EQ41
Specification: Y1 C X1 X1^2 X1^3 X2 X2^2 X2^3
QUANT

| | Value | df | Probability |
|---|---|---|---|
| QLR L-statistic | 9.342027 | 3 | 0.0251 |
| QLR lambda-statistic | 9.153739 | 3 | 0.0273 |

**Figure 5.30** A redundant variable test based on the QR(Median) in (5.23).

And Figure 5.32 presents its forecast evaluation, with a TIC = 0.065 285, where TIC = 0 indicates a perfect forecast and TIC = 1 indicates the worst; and the graph of its forecast value variables is sorted by *Y1F_EQ41*.

2. Even though both the Full-QR(0.1) and Full-QR(0.9) in Figure 5.29 have IVs with a large probabilities, we should not reduced the models, in order to conduct the QSET between the pair of quantiles "0.1 0.9" related to the QSET based on the Full –QR(0.5), similar to the QSET presented in the 5.18. Do this as an exercise.

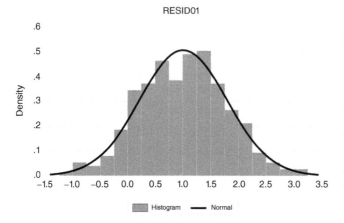

Figure 5.31   Residuals histogram, normal density, and statistics.

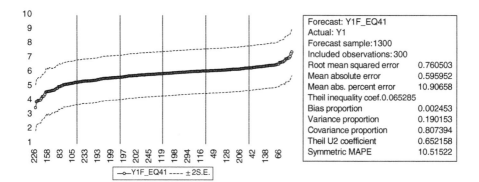

**Figure 5.32**   Forecast evaluation of the Full-QR(Median) in (5.23), sorted by *Y1F_EQ41*.

### 5.5.2   Interaction Double Polynomial QRs

As an extension of the additive double polynomial QR in (5.23), I present an interaction double third-degree polynomial QR in (5.23), with the following ES, with the outputs summary presented in Figure 5.33:

$$Y1\ C\ X1\ X2\ \ X1\char94 2*X2\ X1\char94 3*X2\ \ X1*X2\char94 2\ X1*X2\char94 3 \tag{5.24}$$

Based on this summary, the following findings and notes are presented:

1. Even though *X1* and *X2* in the Full-QR(0.5) have very large *p*-values, the QR is acceptable, since each of their interactions IVs has a probability. $< 0.30$ (refer to the note in Example 5.19, point 1.2), and at the 10% level, all interaction IVs have significant joint effects, based on the *QLR L*-statistics, as presented by the RVT in Figure 5.34.
2. Reduced-QR1 has a slight larger adjusted *R*-squared than the Full-QR. So, it is a better QR. And the Reduced-QR2 has a slightly larger adjusted *R*-squared than the Reduced QR1. So, the Reduced-QR2 is the best fit QR. Refer to the note in User Guide II, as presented in Example 5.17, point 2, which recommend of using the adjusted *R*-squared.

| Variable | Full QR(Median) | | | Reduced-QR1 | | | Reduced-QR2 | | |
|---|---|---|---|---|---|---|---|---|---|
| | Coef. | t-stat. | Prob. | Coef. | t-stat. | Prob. | Coef. | t-stat. | Prob. |
| C | 1.609 | 1.239 | 0.216 | 1.241 | 1.401 | 0.162 | 1.003 | 1.311 | 0.191 |
| X1 | −2.218 | −0.905 | **0.366** | −1.866 | - 0.793 | **0.429** | | | |
| X2 | −0.957 | −0.458 | **0.647** | | | | | | |
| X1*X2 | 8.490 | 1.850 | 0.065 | 6.643 | 2.575 | 0.011 | 4.851 | 3.484 | 0.001 |
| X1^2*X2 | −2.570 | −1.068 | 0.286 | −1.506 | −1.846 | 0.066 | −1.704 | −2.209 | 0.028 |
| X1^3*X2 | 0.554 | 1.060 | 0.290 | 0.334 | 1.510 | 0.132 | 0.382 | 1.802 | 0.073 |
| X1*X2^2 | −1.803 | −1.870 | 0.062 | −1.770 | −1.813 | 0.071 | −1.047 | −2.752 | 0.006 |
| X1*X2^3 | 0.213 | 1.784 | 0.076 | 0.210 | 1.730 | 0.085 | 0.126 | 2.280 | 0.023 |
| Pse. R-sq | 0.195 | | | 0.193 | | | 0.192 | | |
| Adj. R-sq | 0.175 | | | 0.177 | | | **0.179** | | |
| S.E. of reg | 0.775 | | | 0.774 | | | 0.771 | | |
| Quant DV | 5.670 | | | 5.670 | | | 5.670 | | |
| Sparsity | 1.887 | | | 1.906 | | | 1.903 | | |
| QLR stat | 91.57 | | | 90.06 | | | 89.759 | | |
| Prob | 0.000 | | | 0.000 | | | 0.000 | | |

**Figure 5.33** The output summary of the Full-QR(0.5) in (5.24) and its two reduced models.

Redundant variables test
Null hypothesis: X1*X2 X1^2*X2 X1^3*X2 X1*X2^2 X1*X2^3
Equation: EQ43
Specification: Y1 C X1 X2 X1*X2 X1^2*X2 X1^3*X2 X1*X2^2 X1*X2^3

| | Value | df | Probability |
|---|---|---|---|
| QLR L-statistic | 9.737566 | 5 | 0.0830 |
| QLR lambda-statistic | 9.614541 | 5 | 0.0869 |

**Figure 5.34** An RVT based on the Full-QR(0.5) in (5.24).

3. Because the Reduced-QR2 is the best fit QR among the three QRs, I won't provide additional analysis based on the Full-QR, bur rather use the Reduced-QR2.
4. In the previous example, I provided additional analysis based on the Full-QR(Median). In this one, the additional analysis will based on the Reduced-QR2(0.1), with the output in Figure 5.35. Based on this output, the following findings and notes are presented:
    4.1 Its IVs have significant joint effects, based on the $QLR$-statistic of $QLR_0 = 57.64990$ with $df = 5$ and $p$-value $= 0.0000$. Even though each of its IVs has a large or very large probability, it is acceptable as a group of Reduced-QR2($\tau$), for $\tau \, \varepsilon \, (0, 1)$.

Dependent variable: Y1
Method: quantile regression (tau = 0.1)
Date: 03/04/19   Time: 12:52
Sample: 1 300
Included observations: 300
Huber sandwich standard errors and covariance
Sparsity method: Kernel (Epanechnikov) using residuals
Bandwidth method: Hall-Sheather, bw = 0.051685
Estimation successfully identifies unique optimal solution

| Variable | Coefficient | Std. error | t-statistic | Prob. |
|---|---|---|---|---|
| C | 2.850284 | 0.853283 | 3.340373 | 0.0009 |
| X1*X2 | 0.232941 | 1.638066 | 0.142205 | 0.8870 |
| X1^2*X2 | 1.031298 | 0.869782 | 1.185697 | 0.2367 |
| X1^3*X2 | −0.356887 | 0.256347 | −1.392206 | 0.1649 |
| X1*X2^2 | −0.208440 | 0.545430 | −0.382158 | 0.7026 |
| X1*X2^3 | 0.019399 | 0.083824 | 0.231428 | 0.8171 |

| | | | |
|---|---|---|---|
| Pseudo R-squared | 0.195306 | Mean dependent var | 5.751100 |
| Adjusted R-squared | 0.181621 | S.D. dependent var | 0.945077 |
| S.E. of regression | 1.251104 | Objective | 39.89107 |
| Quantile dependent var | 4.330000 | Restr. objective | 49.57300 |
| Sparsity | 3.732078 | Quasi-LR statistic | 57.64990 |
| Prob(quasi-LR stat) | 0.000000 | | |

**Figure 5.35** The output of the Reduced-QR2(0.1).

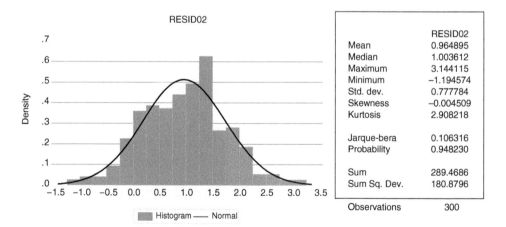

**Figure 5.36** The residuals histogram, its theoretical density, and statistics of the Reduced-QR2(0.1).

Quantile process estimates
Equation: EQ45
Specification: Y1 C X1*X2 X1^2*X2 X1^3*X2 X1*X2^2 X1*X2^3
Estimated equation quantile tau = 0.1
Number of process quantiles: 5
Display all coefficients

|  | Quantile | Coefficient | Std. error | *t*-statistic | Prob. |
|---|---|---|---|---|---|
| C | 0.100 | 2.850284 | 0.853283 | 3.340373 | 0.0009 |
|  | 0.200 | 1.740392 | 0.848094 | 2.052123 | 0.0410 |
|  | 0.400 | 0.599782 | 0.768826 | 0.780128 | 0.4359 |
|  | 0.600 | 0.262271 | 0.806420 | 0.325229 | 0.7452 |
|  | 0.800 | −0.012605 | 1.340973 | −0.009400 | 0.9925 |
| X1*X2 | 0.100 | 0.232941 | 1.638066 | 0.142205 | 0.8870 |
|  | 0.200 | 2.364559 | 1.598245 | 1.479472 | 0.1401 |
|  | 0.400 | 5.674806 | 1.462021 | 3.881482 | 0.0001 |
|  | 0.600 | 6.227517 | 1.476797 | 4.216909 | 0.0000 |
|  | 0.800 | 8.021426 | 2.043761 | 3.924835 | 0.0001 |
| X1^2*X2 | 0.100 | 1.031298 | 0.869782 | 1.185697 | 0.2367 |
|  | 0.200 | 0.185527 | 0.955754 | 0.194116 | 0.8462 |
|  | 0.400 | −2.267307 | 0.808986 | −2.802653 | 0.0054 |
|  | 0.600 | −2.206272 | 0.806601 | −2.735270 | 0.0066 |
|  | 0.800 | −3.233032 | 0.977617 | −3.307055 | 0.0011 |
| X1^3*X2 | 0.100 | −0.356887 | 0.256347 | −1.392206 | 0.1649 |
|  | 0.200 | −0.160145 | 0.277727 | −0.576629 | 0.5646 |
|  | 0.400 | 0.561623 | 0.220683 | 2.544932 | 0.0114 |
|  | 0.600 | 0.489458 | 0.221988 | 2.204887 | 0.0282 |
|  | 0.800 | 0.767636 | 0.242919 | 3.160057 | 0.0017 |
| X1*X2^2 | 0.100 | −0.208440 | 0.545430 | −0.382158 | 0.7026 |
|  | 0.200 | −0.719569 | 0.503506 | −1.429117 | 0.1540 |
|  | 0.400 | −1.276747 | 0.403895 | −3.161090 | 0.0017 |
|  | 0.600 | −1.391464 | 0.418715 | −3.323174 | 0.0010 |
|  | 0.800 | −1.852103 | 0.526365 | −3.518666 | 0.0005 |
| X1*X2^3 | 0.100 | 0.019399 | 0.083824 | 0.231428 | 0.8171 |
|  | 0.200 | 0.078631 | 0.080776 | 0.973449 | 0.3311 |
|  | 0.400 | 0.160917 | 0.059743 | 2.693502 | 0.0075 |
|  | 0.600 | 0.166257 | 0.061382 | 2.708555 | 0.0072 |
|  | 0.800 | 0.235466 | 0.075018 | 3.138805 | 0.0019 |

**Figure 5.37** A Quantile Process Estimates based on the Reduced-QR2(0.1).

Quantile slope equality test
Equation: EQ45
Specification: Y1 C X1*X2 X1^2*X2 X1^3*X2 X1*X2^2 X1*X2^3
Estimated equation quantile tau = 0.1
Number of test quantiles: 5
Test statistic compares all coefficients

| Test summary | Chi-Sq. statistic | Chi-Sq. d.f. | Prob. |
|---|---|---|---|
| Wald test | 54.48869 | 20 | 0.0000 |

Restriction detail: $b(tau\_h) - b(tau\_k) = 0$

| Quantiles | Variable | Restr. value | Std. error | Prob. |
|---|---|---|---|---|
| 0.1, 0.2 | X1*X2 | −2.131618 | 1.396902 | 0.1270 |
| | X1^2*X2 | 0.845771 | 0.773226 | 0.2740 |
| | X1^3*X2 | −0.196742 | 0.224958 | 0.3818 |
| | X1*X2^2 | 0.511129 | 0.448417 | 0.2543 |
| | X1*X2^3 | −0.059232 | 0.069523 | 0.3942 |
| 0.2, 0.4 | X1*X2 | −3.310247 | 1.372286 | 0.0159 |
| | X1^2*X2 | 2.452834 | 0.819988 | 0.0028 |
| | X1^3*X2 | −0.721768 | 0.237636 | 0.0024 |
| | X1*X2^2 | 0.557178 | 0.426649 | 0.1916 |
| | X1*X2^3 | −0.082285 | 0.067988 | 0.2262 |
| 0.4, 0.6 | X1*X2 | −0.552711 | 1.213422 | 0.6488 |
| | X1^2*X2 | −0.061035 | 0.664258 | 0.9268 |
| | X1^3*X2 | 0.072164 | 0.182002 | 0.6917 |
| | X1*X2^2 | 0.114717 | 0.340945 | 0.7365 |
| | X1*X2^3 | −0.005340 | 0.050177 | 0.9153 |
| 0.6, 0.8 | X1*X2 | −1.793909 | 1.787582 | 0.3156 |
| | X1^2*X2 | 1.026759 | 0.868096 | 0.2369 |
| | X1^3*X2 | −0.278178 | 0.221283 | 0.2087 |
| | X1*X2^2 | 0.460639 | 0.471551 | 0.3286 |
| | X1*X2^3 | −0.069210 | 0.067647 | 0.3063 |

**Figure 5.38** A quantile slope equality test based on the Reduced-QR2.

Ramsey RESET test
Equation: EQ45
Specification: Y1 C X1*X2 X1^2*X2 X1^3*X2 X1*X2^2 X1*X2^3
Omitted variables: squares of fitted values

|  | Value | df | Probability |
|---|---|---|---|
| QLR L-statistic | 0.519071 | 1 | 0.4712 |
| QLR lambda-statistic | 0.518488 | 1 | 0.4715 |

| L-test summary: | Value |
|---|---|
| Restricted objective | 39.89107 |
| Unrestricted objective | 39.80161 |
| Scale | 0.344713 |

| Lambda-test summary: | Value |
|---|---|
| Restricted log Obj. | 3.686153 |
| Unrestricted log Obj. | 3.683907 |
| Scale | 0.008661 |

Unrestricted test equation:
Dependent variable: Y1
Method: quantile regression (tau = 0.1)
Date: 03/07/19   Time: 10:33
Sample: 1 300
Included observations: 300
Huber sandwich standard errors and covariance
Sparsity method: Kernel (Epanechnikov) using residuals
Bandwidth method: Hall-Sheather, bw = 0.051685
Estimation successfully identifies unique optimal solution

| Variable | Coefficient | Std. error | t-statistic | Prob. |
|---|---|---|---|---|
| C | 0.655548 | 7.504118 | 0.087358 | 0.9304 |
| X1*X2 | 1.262362 | 4.508288 | 0.280009 | 0.7797 |
| X1^2*X2 | −1.647893 | 9.389775 | −0.175499 | 0.8608 |
| X1^3*X2 | 0.508844 | 3.003261 | 0.169431 | 0.8656 |
| X1*X2^2 | −0.262104 | 0.527022 | −0.497331 | 0.6193 |
| X1*X2^3 | 0.049062 | 0.114046 | 0.430191 | 0.6674 |
| **FITTED^2** | **0.233559** | **0.743558** | **0.314109** | **0.7537** |

**Figure 5.39** The Ramsey RESET test based on the Reduced-QR2(0.1).

| | | | |
|---|---|---|---|
| Pseudo *R*-squared | 0.197111 | Mean dependent var | 5.751100 |
| Adjusted *R*-squared | 0.180670 | S.D. dependent var | 0.945077 |
| S.E. of regression | 1.272283 | Objective | 39.80161 |
| Quantile dependent var | 4.330000 | Restr. objective | 49.57300 |
| Sparsity | 3.830144 | Quasi-LR statistic | 56.69291 |
| Prob(quasi-LR stat) | 0.000000 | | |

**Figure 5.39** (*continued*)

4.2 Figure 5.36 presents the output of its residual histogram, namely RESID02, its theoretical density, and statistics, which shows the normal distribution is very close to the residual histogram, and the Jarque-Bera test shows the normal distribution of RESID02 is accepted with very large p-value=0.948. Hence, we can say the data supports the residuals have normal distribution.

4.3 Figure 5.37 presents the output of quantile process estimates, with the 5 as the number of process quantiles, which shows a special list of five quantiles: 0.1, 0.2, 0.4, 0.6, and 0.8. Note that the quantiles have differential distances.

4.4 Corresponding to the output in Figure 5.37, Figure 5.38 shows the output of the QSET, which also presents a special pairs of quantiles.

4.5 Finally, Figure 5.39 presents the output of a Ramsey RESET test. RESET stands for *Regression Equation Specification Error Test*, and it was proposed by Ramsey (1969) to test the following statistical hypothesis:

$$H_0 : \varepsilon \sim N\left(0, \sigma^2 I\right) \text{ vs } H_1 : \varepsilon \sim N\left(\mu, \sigma^2 I\right) ; \mu \neq 0$$

Based on the *QLR L*-statistic of $QLRL_0 = 0.519\,071$ with $df = 1$, and p-value = 0.4712, the null hypothesis is accepted.

Note that the Ramsey RESET test is, in fact, an *Omitted Variable Test* of the variable "squares of fitted values," namely *FITTED^2*, of the Reduced-QR2(0.1), as mentioned in the output. Hence, as such, we can conclude that the *FITTED^2* does not need to be inserted as an additional IV of the Reduced-QR(0.1) in order to have a better fit model, based on the *t*-statistic of $t_0 = 0.314\,109$ with $df = 293 = (300 - 7)$ and a p-value = 0.7537.

# 6

# Quantile Regressions with Multiple Numerical Predictors

## 6.1 Introduction

As the extension of the quantile regressions (QRs) with two numerical predictors presented in previous chapter, this chapter presents illustrative examples of QRs with multiple numerical predictors, based on selected data sets. Only a few selected interaction QRs having more than three multiple numerical predictors are presented, because there will be many possible two-way interaction predictors, and three-way interaction predictors. For instance, having five predictors, there will be 10 possible three-way interaction, and 10 possible two-way interaction predictors. Hence, a researcher should develop the best possible path diagram (up-and-down or causal relationships) for the set of selected variables, which is acceptable and it is assumed to be true in the theoretical sense, such as presented in Figure 6.1, as the models based on four numerical variables (*X1,X2,X3,Y1*), and their extension for five numerical variables, as presented in Figures 6.35 and 6.36. Then based on each path diagram an equation specification of the QR, either additive or interaction QR, can easily be developed. Finally, selected additive, two-way interaction (2WI), and three-way interaction (3WI) QRs having more than four numerical IVs are presented.

## 6.2 Alternative Path Diagrams Based on *(X1,X2,X3,Y1)*

Figure 6.1 presents four alternative path diagrams, which show the up-and-down or causal relationships between a predicted *Y1* and three upper or cause variables or predictors *X1, X2*, and *X3*. Each of the path diagrams has specific characteristics or assumptions.

### 6.2.1 A QR Based on the Path Diagram in Figure 6.1a

This path diagram shows that each of the upper or cause variables *X1, X2,* and *X3* has a true direct effect on *Y1,* but the causal relationships between *X1, X2,* and *X3* are not defined or presented for the following possible reasons:

  (i) They really don't have causal relationships.
 (ii) It is assumed they don't have causal relationships, even though they might have a type of up-and-down or causal relationship.

*Quantile Regression: Applications on Experimental and Cross Section Data Using EViews,* First Edition.
I Gusti Ngurah Agung.
© 2021 John Wiley & Sons Ltd. Published 2021 by John Wiley & Sons Ltd.

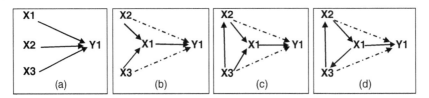

**Figure 6.1** Alternative path diagrams based on *(X1,X2,X3,Y1)*.

(iii) The researcher does not want to count on their causal relationships, even though they might have a type of up-and-down or causal relationship.

Based on this path diagram, we have an additive model, as the simplest model of *Y1* on *(X1,X2,X3)*, with one of the the following equation specifications (ESs):

$$Y1\ C\ X1\ X2\ X3 \tag{6.1a}$$

or

$$Y1 = C(1) + C(2) * X1 + C(3) * X2 + C(4) * X3 \tag{6.1b}$$

### 6.2.2 A QR Based on the Path Diagram in Figure 6.1b

This path diagram shows *X1* has a true direct effect on *Y1;* both *X2* and *X3* have a true direct effect on *X1;* and *X2* and *X3* doesn't have a causal relationship. In addition, both *X2* and *X3* have an indirect effect on *Y1* through *X1*. For this reason, the arrows are presented with dashed lines from *X2* and *X3* to *Y1*. In fact, those arrows can be omitted.

Based on this path diagram, we have a *partial two-way interaction QR* (P2WI-QR) of *Y1* on *(X1,X2,X3)*, with one of the following ESs for testing the hypothesis that the effect of *X1* on *Y1* depends on *X2* and *X3*:

$$Y1\ C\ X1 * X2\ X1 * X3\ X1\ X2\ X3 \tag{6.2a}$$

or

$$Y1 = C(1) + C(2) * X1 * X2 + C(3) * X1 * X3 + C(4) * X1 + C(5) * X2$$
$$+ C(6) * X3 \tag{6.2b}$$

where interactions *X1*X2* and *X1*X3* represent the indirect effects of each *X2* and *X3* on *Y1*, through *X1*. With five independent variables (IVs), at least one of the interaction IVs has a large or very large *p*-value, such as greater than 0.30, because in general the IVs are highly or significantly correlated. For this reason, an acceptable reduced interaction model should be explored, by deleting the main IVs using either the trial-and-error method or the manual *stepwise/multistage selection method* (Agung 2009a, 2011a, 2014, 2019). See the examples of the *two-way interaction QRs* (2WI-QRs) in the previous chapter and in the following examples.

### 6.2.3 QR Based on the Path Diagram in Figure 6.1c

As an extension of the path diagram in Figure 6.1b, the path diagram in Figure 6.1c presents special causal relationships between *X1, X2,* and *X3* called the *triangular causal effects*, where *X3* has a direct effect on *X2*, and *X2* and *X3* have true direct effects on *X1*. By having these direct effects, *X3* also has an indirect effect on *Y1* through two variables, *X2* and *X1*. This indirect effect should be represented as a three-way interaction IV of the statistical model. Hence, based on this path diagram, we have two possible interaction QRs of *Y1* on *(X1,X2,X3)*, discussed in the following two sections. For testing the hypothesis that the effect of *X1* on *Y1* depends on *X2* and *X3*, they are the same as the model (6.2). In fact, the model (6.2) is a reduced model of the following models.

### 6.2.3.1 A Full Two-Way Interaction QR

As an extension of the model (6.2), based on the path diagram in Figure 6.1c, we have a *full two-way interaction QR* (F2WI-QR), with one of the following ESs:

$$Y1 \, C \, X1 * X2 \; X1 * X3 \; X2 * X3 \; X1 \; X2 \; X3 \tag{6.3a}$$

or

$$Y1 = C\,(1) + C\,(2) * X1 * X2 + C\,(3) * X1 * X3 + C\,(4) * X2 * X3$$
$$+ C\,(5) * X1 + C\,(6) * X2 + C\,(7) * X3 \tag{6.3b}$$

As with the previous simpler model, we should apply the trial-and-error method or the manual stepwise/multistage selection method to explore a reduced two-way interaction model. An unexpected reduced model could be obtained because of the multicollinearity impact or bivariate correlations between the IVs. We may obtain insignificant interactions in the full and reduced model, as illustrated in previous example.

To obtain the best possible fit 2WI-QR, the following two stages of regression analysis should be done:

(i) Obtain the best fit model, namely QR-1, based on the following ES, so that each of its IVs has a *Prob.* <0.20 and their joint effects is significant. The reason for selecting the *Prob.* <0.20 is each of the IVs has either a positive or negative effect on *Y1*, at the 10% level of significance, with a *p*-value = *Prob.*/2 < 0.10.

$$Y1 \; C \; X1 * X2 \; X1 * X3 \; X2 * X3$$

(ii) Select the main variables *X1, X2,* and *X3* as additional IVs of QR-1, using the trial-and-error method.

### 6.2.3.2 A Full Three-Way Interaction QR

To represent the indirect effect of *X3* on *Y1* through *X2* and *X1*, we should consider a *full three-way interaction QR* (F3WI-QR), with one of the following ESs:

$$Y1 \, C \, X1 * X2 * X3 \; X1 * X2 \; X1 * X3 \; X2 * X3 \; X1 \; X2 \; X3 \tag{6.4a}$$

or

$$Y1 = C\,(1) + C\,(2) * X1 * X2 * X3 + C\,(3) * X1 * X2 + C\,(4) * X1 * X3$$
$$+ C\,(5) * X2 * X3 + C\,(6) * X1 + C\,(7) * X2 + C\,(8) * X3 \tag{6.4b}$$

Since the F3WI-QR is developed based on the same path diagram as the F2WI-QR, then whenever the interaction *X1\*X2\*X3* has an insignificant effect on *Y1* in the full and reduced models, we should conduct a reanalysis using the 2WI-QR.

To obtain the best possible fit three-way interaction QR (3WI-QR), the following three stages of regression analysis should be done:

(i) Find out whether the QR-1, *Y1 C X1\*X2\*X3,* has a significant IV. If it does, we continue to the second stage. Otherwise, we won't be able to obtain a 3WI-QR.

(ii) Using the trial-and-error method, select the two-way interactions, *X1\*X2, X1\*X3,* and *X2,\*X3,* as the additional IVs for QR-1. Then we have the QR-2 as an acceptable 3WI-QR.

(iii) Finally, using the trial-and-error method, select the main variables, *X1, X2,* and *X3,* as the additional IVs for QR-2.

### 6.2.4    QR Based on the Path Diagram in Figure 6.1d

The path diagram in Figure 6.1d presents a *circular up-and-down or causal* relationship, *X3 → X2 → X1 → X3,* as a true relationship between the three variables, in the theoretical sense. The arrows with dashed lines show that the corresponding pairs of variables also have up-and-down or causal relationships. Even without those dashed arrows, it should be well known or accepted that the corresponding pairs of variables at least have a direct up-and-down relationship. For instance, *X3* is an upper variable of both *X1* and *Y1.*

Now, note that the overall path relationships, indicated by arrows with solid and dashed lines, between the four variables in Figure 5.1c, d are, in fact, the same in the theoretical and substantial senses, even when the dashed arrows are omitted. Hence, based on the path diagram in Figure 6.1d, we also have two possible interaction QRs, as presented in (6.3) and (6.4).

## 6.3    Applications of QRs Based on Data_Faad.wf1

The data used is experimental data having four numerical variables, *X1, X2, Y1,* and *Y2,* with the design that *X1* and *X2* have direct effects on both *Y1* and *Y2,* and also *Y1* and *Y2* have reciprocal causal effects. The following two examples present the statistical results of two alternative QRs.

**Example 6.1    *Application of a Partial Two-way Interaction QR***
By not taking into account the up-and-down relationship between *X1* and *X2,* I propose a P2WI-QR with one of the following ESs. Refer to the path diagram in Figure 6.1b and the model (6.2).

$$Y2 \ C \ Y1 * X1 \ Y1 * X2 \ Y1 \ X1 \ X2 \tag{6.5a}$$

or

$$Y2 = C\,(1) + C\,(2) * Y1 * X1 + C\,(3) * Y1 * X2 + C\,(4) * Y1 + C\,(5) * X1$$
$$+ C\,(6) * X2 \tag{6.5b}$$

Table 6.1 shows the statistical results summary of the QR(Median) in (6.5a), with three of its reduced models, namely RM-1, RM-2, and RM-3. Based on these statistical results, the following findings and notes are presented:

**Table 6.1** Statistical result summary of the model (6.5) and three of its reduced models.

| P2WI-QR | C | Y1*X1 | Y1*X2 | Y1 | X1 | X2 | Ps. $R^2$ | Adj. $R^2$ | Q-LR | Prob |
|---------|---|-------|-------|----|----|----|-----------|------------|------|------|
| Coef. | −0.7255 | −0.2600 | 0.0729 | 0.8095 | 2.1488 | −0.0829 | 0.4042 | 0.3940 | 251.42 | 0.0000 |
| t-stat | −0.7512 | −1.8357 | 0.8788 | 4.2152 | 2.4394 | −0.1737 | | | | |
| Prob. | 0.4531 | 0.0674 | **0.3802** | 0.0000 | 0.0153 | **0.8622** | | | | |
| **RM-1** | C | Y1*X1 | Y1*X2 | Y1 | X1 | | Ps. $R^2$ | Adj. $R^2$ | Q-LR | Prob |
| Coef. | −0.8537 | −0.2526 | 0.0594 | 0.8319 | 2.1022 | | 0.4041 | 0.3960 | 251.04 | 0.0000 |
| t-stat | −0.9509 | −2.3277 | 3.3632 | 4.7099 | 3.1844 | | | | | |
| Prob. | 0.3424 | 0.0206 | 0.0009 | 0.0000 | 0.0016 | | | | | |
| **RM-2** | C | Y1*X1 | Y1*X2 | Y1 | | X2 | Ps. $R^2$ | Adj. $R^2$ | Q-LR | Prob |
| Coef. | 0.3773 | 0.0818 | −0.0277 | 0.6690 | | 0.4750 | 0.3987 | 0.3905 | 243.11 | 0.0000 |
| t-stat | 0.4064 | 2.7727 | −0.4010 | 3.5182 | | 1.2224 | | | | |
| Prob. | 0.6848 | 0.0059 | **0.6887** | 0.0005 | | **0.2225** | | | | |
| **RM-3** | C | Y1*X1 | Y1*X2 | | X1 | X2 | Ps. $R^2$ | Adj. $R^2$ | Q-LR | Prob |
| Coef. | 3.6602 | 0.0561 | 0.2089 | | 0.3998 | −0.8290 | 0.3880 | 0.3797 | 223.83 | 0.0000 |
| t-stat | 10.2969 | 0.3445 | 2.5091 | | 0.3903 | −1.6738 | | | | |
| Prob. | 0.0000 | **0.7307** | 0.0126 | | **0.6966** | **0.0952** | | | | |

1) The result of the P2WI-QR shows its IVs have significant joint effects, based on the *Quasi-LR* statistic of $QLR_0 = 251.42$ with a *p*-value = 0.0000. So, this is an acceptable model, in the statistical sense.

2) However, since the interaction *Y1*X1* is insignificant at the 5% level, and the interaction *Y1*X2* has a large *p*-value = 0.3802 > 0.30, the model is not a good model for presenting a P2W_IM. So, a reduced model should be explored, by deleting one of the main IVs, to obtain the best possible interaction QR. The first choice is the main variable *X2,* which should be deleted from the model because it has the greatest *p*-value.

3) We obtain the RM-1, in which each of the interactions *Y1*X1* and *Y1*X2* has significant adjusted effects, based on the *t*-statistics of −2.3277 and 3.3632, respectively. As comparisons, RM-2 and RM-3 are presented, which we obtain by deleting *X1* and *Y1*, respectively. We find RM-1 is the best QR for representing the data because it has the greatest adjusted *R*-squared among the four models, with the following regression function, which shows that the effect of *Y1* on the median of *Y2* depends on a function of *X1* and *X2*. It is significant because each *Y1*X1, Y1*X2,* and *Y1* in $(-0.2526X1 + 0.05935X2 + 0.8319) \times Y1$ has a significant adjusted effect, based on the t-statistic, as presented in Table 6.1, for RM-1.

$$\widehat{Y2} = (-0.8537 + 2.1021856X1) + (-0.2526X1 + 0.05935X2 + 0.8319) \times Y1$$

Additionally, the three variables, *Y1*X1, Y1*X2,* and *Y1,* have a joint significant effect, based on the Wald test *F*-statistic of $F_0 = 72.3476$ with $df = (3295)$ and a *p*-value = 0.0000, and the *Chi-square* statistic of 217.0427 with $df = 3$ and a *p*-value = 0.0000.

**Table 6.2** Statistical result summary of the manual multistage selection method, based on the F2WI-QR(Median) in (6.6).

| Full-QR | C | Y1*X1 | Y1*X2 | X1*X2 | Y1 | X1 | X2 |
|---|---|---|---|---|---|---|---|
| Coef. | −0.5389 | −0.2642 | 0.0954 | −0.0464 | 0.7498 | 2.3152 | −0.1606 |
| t-stat | −0.5647 | −1.8232 | 0.9329 | −0.1541 | 2.5341 | 2.1220 | −0.3319 |
| Prob. | 0.5727 | 0.0693 | 0.3516 | 0.8776 | 0.0118 | 0.0347 | 0.7402 |

| Ps. $R^2$ = 0.4042 | | Adj. $R^2$ = 0.3920 | | Quasi-LR = 251.62 | | Prob. = 0.0000 | |

| QR-1 | C | Y1*X1 | Y1*X2 | X1*X2 | | | |
|---|---|---|---|---|---|---|---|
| Coef. | 2.9118 | 0.2363 | 0.1385 | −0.3418 | | | |
| t-stat | 16.6043 | 6.2446 | 7.569 | −3.4291 | | | |
| Prob. | 0.00000 | 0.00000 | 0.00000 | 0.0007 | | | |

| Ps. $R^2$ = 0.3901 | | Adj. $R^2$ = 0.3839 | | Quasi-LR = 229.24 | | Prob. = 0.0000 | |

| QR-2 | C | Y1*X1 | Y1*X2 | X1*X2 | Y1 | | |
|---|---|---|---|---|---|---|---|
| Coef. | 0.2702 | −0.0636 | −0.0433 | 0.3494 | 0.9275 | | |
| t-stat | 0.3176 | −0.6144 | −0.7291 | 1.6119 | 3.0203 | | |
| Prob. | 0.751 | **0.5394** | **0.4665** | 0.1081 | 0.0027 | | |

| Ps. $R^2$ = 0.3982 | | Adj. $R^2$ = 0.3900 | | Quasi-LR = 245.21 | | Prob. = 0.0000 | |

| QR-3 | C | Y1*X1 | Y1*X2 | X1*X2 | | X1 | |
|---|---|---|---|---|---|---|---|
| Coef. | 1.4438 | 0.0504 | 0.2252 | −0.6637 | | 2.052 | |
| t-stat | 2.0183 | 0.4478 | 4.8195 | −3.3596 | | 1.8447 | |
| Prob. | 0.0445 | **0.6546** | 0 | 0.0009 | | 0.0661 | |

| Ps. $R^2$ = 0.3964 | | Adj. $R^2$ = 0.3882 | | Quasi-LR = 236.78 | | Prob. = 0.0000 | |

| QR-4 | C | Y1*X1 | Y1*X2 | X1*X2 | | | X2 |
|---|---|---|---|---|---|---|---|
| Coef. | 2.6534 | 0.2709 | 0.1271 | −0.4145 | | | 0.164 |
| t-stat | 3.5691 | 2.7305 | 2.4058 | −1.7106 | | | 0.2751 |
| Prob. | 0.0004 | 0.0067 | 0.0168 | 0.0882 | | | **0.7834** |

| Ps. $R^2$ = 0.3902 | | Adj. $R^2$ = 0.3819 | | Quasi-LR = 231.30 | | Prob. = 0.0000 | |

**Example 6.2** *An Application of a Full Two-Way Interaction QR*

By taking into account the up-and-down relationship between *X1* and *X2*, we have a F2WI-QR with the following equation, as an extension of the P2WI-QR (6.5a):

$$Y2 \ C \ Y1 * X1 \ Y1 * X2 \ X1 * X2 \ Y1 \ X1 \ X2 \tag{6.6}$$

Table 6.2 presents the statistical results summary of the Full QR(Median) and four of its possible Reduced-QRs, namely QR-1, QR-2, QR-3, and QR-4, which are obtained by using the manual multistage selection method, starting with a QR-1 of *Y2* on *C, Y1\*X1, Y1\*X2,* and *X1\*X2* only. This is the first stage of regression analysis, which shows each IV has significant effect at the 1% level. At the second stage of the regression analysis, the main variables *Y1, X1,* and *X2* are selected, using the trial-and error method, to be inserted as additional IVs of QR-1. We find that only one of the three main variables can be inserted as an additional IV for QR-1 to obtain the best fit model. For this reason, QRs 2, 3, and 4 are presented in Table 6.2, for comparison. Based on these results, the following findings and notes are presented:

1) It is common for the Full-QR to have the largest adjusted *R*-squared, with many insignificant adjusted effects. For this reason, the Reduced-QRs are presented.
2) The Full-QR and the four Reduced-QRs are acceptable QRs to represent the effects of *Y1\*X1, Y1\*X2,* and *X1\*X2* on *Y2,* because all IVs of each QR have significant joint effects on *Y2,* based on the *QLR* statistic with *p*-value $= 0.0000$.
3) Among the four Reduced-QRs, the reduced QR-2 has the largest adjusted *R*-squared $= 0.3920$, so it can be considered as the best fit reduced QR(Median). However, it worse than the Full QR, because it has a slightly smaller adjusted *R*-squared $= 0.3900$.
4) The QR-4 is the worst fit reduced QR because it has the smallest adjusted *R*-squared $= 0.3819$, even though each two-interaction has a significant adjusted effect, at the 1%, 5%, and 10% levels, respectively.
5) Hence, for additional analysis, we have a choice whether we want to use the Full-QR or QR-2, which the best reduced model, to study more characteristics of the IVs, mainly their two-way interaction effects. In this case, I would use the reduced QR-2 because its two-way interaction IV is insignificant, at the 10% level. For this, the following findings and notes are presented:

   7.1 The two-way interaction *Y1\*X1, Y1\*X2,* and *X1\*X2* are jointly significant, based on the Wald test *(Chi-square)* $\chi_0^2 = 24.664\,91$ with *df* $= 3$ and *p*-value $= 0.0000$.
   
   7.2 The IVs, *Y1\*X1, Y1\*X2,* and *Y1,* are jointly significant, based on the Wald Test *(Chi-square)* $\chi_0^2 = 128.3974$ with *df* $= 3$ and *p*-value $= 0.0000$. So, We can conclude the effect of *Y1* on *Y2* is significantly dependent on the linear combination of *X1* and *X2*, with the QR(Median) function as follows:

$$\widehat{Y2} = 0.2702 + 0.3494 \times X1 \times X2 + (0.9275 - 0.0636 \times X1 - 0.0433 \times X2) \times Y1$$

   Note that this function shows the effect of *Y1* on the median of *Y2* depends on $(0.9275 - 0.0636 \times X1 - 0.0433 \times X2)$, and it decreases with increasing values of *X1* or *X2* because *X1* and *X2* are positive variables and their coefficients are negative. In addition, we may consider the following QR function, which shows that the effect of *X2* on *Y2* depends on a function of *Y1* and *X1*. And we find that its effect on *Y2* is significantly dependent on *Y1* and *X1*, based on the Wald test *(Chi-square)* of $\chi_0^2 = 64.9355$ with *df* $= 3$ and a *p*-value $= 0.0000$.

$$\widehat{Y2} = (2.6534 + 0.2708Y1 \times X1) + (0.1271Y1 - 0.4145X1 + 0.1640) \times X2$$

   7.3 Based on the Full-QR, similar findings can easily be obtained. Do this as an exercise.

**Example 6.3** *Application of a Three-Way Interaction QR*

Referring to the path diagram in Figure 5.1c, using $Y1 = Y2$ and $(X1,X2,X3) = (Y1,X1,X2)$, this example presents the statistical summary of an acceptable 3WI-QR of $Y2$ on $(Y1,X1,X2)$, as an extension of an F2WI-QR in (6.6). The F3WI-QR has the following ES:

$$Y2 \ C \ Y1 * X1 * X2 \ Y1 * X1 \ Y1 * X2 \ X1 * X2 \ Y1 \ X1 \ X2 \tag{6.7}$$

Table 6.3 present the output of the Full-QR(Median) and five of its reduced QRs, QR-1 to QR-5. Based on these outputs, the finding and notes are as follows:

1) Each independent variable of the Full-QR have $p$-value $> 0.28$. For this reason, we have to use the manual multistage selection method or the trial-and-error method to develop the best possible 3WI-QR(Median), as follows:
2) At the first stage, we have the output of QR-1, $Y2 = C(1) + C(2)*Y1*X1*X2$, with $Y1*X1*X2$ significant based on the $t$-statistic of $t_0 = 13.258$ with $df = (300-2)$ and $p$-value $= 0.000$.
3) At the second stage, QR-2 is obtained by inserting the two-way interaction $Y1*X1, Y1*X2$, & $X1*X2$ as additional IVs for QR-1. The output of QR-2 shows $Y1*X1*X2$ has a negative significant adjusted effect, at 1% level, based on the $t$-statistic of $t_0 = -2.2579$ with $df = (300-4)$ and $p$-value $= 0.0117/2 = 0.005\,85$. Hence, this QR-2 is an acceptable 3WI-QR(Median), even though $X1*X2$ has large $p$-value $= 0.760$.
4) At the third stage, we insert two of the main variables, $Y1, X1,$ and $X2$, as the IVs of QR-2. That obtains the outputs QR-3, QR-4, and QR-5, which show QR-3 and QR-4 are acceptable 3WI-QR(Median), since $Y1*X1*X2$ has significant effects, at the 5% and 1% levels, respectively.
5) Using the adjusted $R$-square as a measure of GoF (Good of Fit), QR-3 is the best 3WI-QR(Median), since it has the largest adjusted $R$-square among the Full-QR and the five reduced QRs. Based on the output of QR-3, we have the following regression function and find the effect of $Y1$ on $Y2$ is significantly dependent on a function of $X1 \times X2, X1,$ and $X2$, based on the Wald test $F$-statistic of $F_0 = 42.0685$ with $df = (4, 293)$ and a $p$-value $= 0.0000$, and the *Chi-square* statistic of $168.2739$ with $df = 4$ and a $p$-value $= 0.0000$:

$$\widehat{Y2} = (1.9759 + 0.9150 X1 \times X2 - 1.3132 X2) +$$
$$(-0.1495 X1 \times X2 + 0.1098 X1 + 0.2732 X2 + 0.3438) \times Y1$$

6) For an additional reduced QR, we can delete $Y1$ from GR-3 and $X1$ from QR-4 to obtain the same regression, namely QR-6, with its output presented in Figure 6.2. Based on this output, the findings and notes are as follows:
   6.1 Each interaction IV of the QR-6 has significant effect, at the 1% or 5% level, and $X2$ has a negative significant effect, at the 1% level, based on the $t$-statistic of $t_0 = -2.4931$ with $df = (300-6)$ and $p$-value $= 0.0132/2 = 0.006\,06$.
   6.2 In addition, QR-6 has the largest adjusted $R$-squared of all the QRs in Table 6.3. Hence, QR-6 is the best 3WI-QR(Median) of $Y2$.
   6.3 Since each of the interactions $Y1*X1*X2, Y1*X1,$ and $Y1*X2$ has a significant effect on $Y2$, we can conclude directly that the three interactions have joint significant effects on $Y2$ without doing the Wald test. Furthermore, we also can conclude that

**Table 6.3** Statistical results summary based on the F3WI-QR in (6.7).

| Full-QR | C | Y1*X1*X2 | Y1*X1 | Y1*X2 | X1*X2 | Y1 | X1 | X2 |
|---|---|---|---|---|---|---|---|---|
| Coef. | 1.192 | −0.105 | −0.054 | 0.243 | 0.649 | 0.471 | 0.970 | −1.115 |
| *t*-stat | 0.432 | −0.663 | −0.129 | 1.076 | 0.703 | 0.823 | 0.401 | −0.957 |
| Prob. | 0.666 | 0.508 | 0.898 | 0.283 | 0.483 | 0.412 | 0.689 | 0.340 |

Ps. $R^2 = 0.4046$    Adj. $R^2 = 0.3903$    Quasi-LR = 254.66    Prob. = 0.0000

| QR-1 | C | Y1*X1*X2 | | | | | | |
|---|---|---|---|---|---|---|---|---|
| Coef. | 4.0762 | 0.0763 | | | | | | |
| *t*-stat | 28.782 | 13.258 | | | | | | |
| Prob. | 0.000 | 0.000 | | | | | | |

Ps. $R^2 = 0.3068$    Adj. $R^2 = 0.3045$    Quasi-LR = 162.37    Prob. = 0.0000

| QR-2 | C | Y1*X1*X2 | Y1*X1 | Y1*X2 | X1*X2 | | | |
|---|---|---|---|---|---|---|---|---|
| Coef. | 1.9761 | −0.0869 | 0.347 | 0.195 | −0.049 | | | |
| *t*-stat | 5.8606 | −2.5379 | 6.593 | 7.451 | −0.306 | | | |
| Prob. | 0 | 0.0117 | 0.000 | 0.000 | **0.760** | | | |

Ps. $R^2 = 03982$    Adj. $R^2 = 0.3901$    Quasi-LR = 242.92    Prob. = 0.0000

| QR-3 | C | Y1*X1*X2 | Y1*X1 | Y1*X2 | X1*X2 | Y1 | | X2 |
|---|---|---|---|---|---|---|---|---|
| Coef. | 1.9759 | −0.1495 | 0.110 | 0.273 | 0.915 | 0.344 | | −1.313 |
| *t*-stat | 1.5353 | −2.206 | 0.815 | 1.659 | 2.958 | 0.865 | | −1.815 |
| Prob. | 0.1258 | 0.0282 | **0.416** | 0.098 | 0.003 | **0.388** | | 0.071 |

Ps. $R^2 = 0.4044$    **Adj. $R^2 = 0.3921$**    Quasi-LR = 257.05    Prob. = 0.0000

| QR-4 | C | Y1*X1*X2 | Y1*X1 | Y1*X2 | X1*X2 | | X1 | X2 |
|---|---|---|---|---|---|---|---|---|
| Coef. | 3.4882 | −0.2162 | 0.361 | 0.369 | 1.182 | | −1.147 | −1.683 |
| *t*-stat | 3.6175 | −2.8437 | 1.636 | 4.508 | 1.888 | | −0.660 | −2.611 |
| Prob. | 0.0004 | 0.0048 | **0.103** | 0.000 | 0.060 | | **0.510** | 0.010 |

Ps. $R^2 = 0.4037$    Adj. $R^2 = 0.3914$    Quasi-LR = 255.02    Prob. = 0.0000

| QR-5 | C | Y1*X1*X2 | Y1*X1 | Y1*X2 | X1*X2 | Y1 | X1 | |
|---|---|---|---|---|---|---|---|---|
| Coef. | −0.8481 | 0.0069 | −0.271 | 0.061 | −0.048 | 0.827 | 2.224 | |
| *t*-stat | −0.7243 | 0.1081 | −0.997 | 0.760 | −0.117 | 2.400 | 1.382 | |
| Prob. | 0.4694 | **0.914** | 0.320 | **0.448** | **0.907** | 0.017 | **0.168** | |

Ps. $R^2 = 0.4041$    Adj. $R^2 = 0.3919$    Quasi-LR = 250.64    Prob. = 0.0000

Dependent Variable: Y2
Method: Quantile Regression (Median)
Date: 06/13/18   Time: 14:20
Sample: 1 300
Included observations: 300
Huber Sandwich Standard Errors & Covariance
Sparsity method: Kernel (Epanechnikov) using residuals
Bandwidth method: Hall-Sheather, bw=0.14513
Estimation successfully identifies unique optimal solution

| Variable | Coefficient | Std. Error | t-Statistic | Prob. |
|---|---|---|---|---|
| C | 2.937376 | 0.497339 | 5.906182 | 0.0000 |
| Y1*X1*X2 | -0.169590 | 0.046530 | -3.644713 | 0.0003 |
| Y1*X1 | 0.218383 | 0.065420 | 3.338151 | 0.0010 |
| Y1*X2 | 0.368672 | 0.077202 | 4.775434 | 0.0000 |
| X1*X2 | 0.781691 | 0.325171 | 2.403939 | 0.0168 |
| X2 | -1.462905 | 0.586772 | -2.493140 | 0.0132 |

| | | | |
|---|---|---|---|
| Pseudo R-squared | 0.403356 | Mean dependent var | 5.742600 |
| Adjusted R-squared | 0.393209 | S.D. dependent var | 1.062605 |
| S.E. of regression | 0.683165 | Objective | 78.30953 |
| Quantile dependent var | 5.750000 | Restr. objective | 131.2500 |
| Sparsity | 1.672389 | Quasi-LR statistic | 253.2448 |
| Prob(Quasi-LR stat) | 0.000000 | | |

**Figure 6.2** Statistical results based on QR-6 as a reduced model of QR-3 and QR-4.

the effect of *Y1* on *Y2* is significantly dependent on a function of $X1 \times X2$, *X1*, and *X2*.

7) For additional reduced QRs, we can delete *X1* or *Y1* from GR-5 to obtain two reduced models, namely QR-7 and QR-8, with the outputs presented in Figure 6.3. Based on these results, the findings and notes are as follows:

7.1 At the 10% level, the interaction *Y1*X1*X2* of QR-7 has a significant negative adjusted effect on *Y2*, based on the *t*-statistic of $t_0 = -1.586$ with a *p*-value $= 0.1193/2 = 0.05965 < 0.10$, and each of its two-way interaction IVs has a significant positive adjusted effect. Even though its main IV, *Y1*, has an insignificant adjusted effect, with a *p*-value $= 0.3687$, this model is an acceptable 3WI-QR, in the statistical sense, since its IVs have significant joint effects, based on the *Quasi-LR* statistic of 243.6258 with a *p*-value 0.0000. However, compared with QR-6, this is a worse model.

7.2 At the 1% level, the three-way interaction of QR-8 has a significant negative adjusted effect, and each of its two-way interactions has a significant positive adjusted effect. Even though its main IV, *X1*, has an insignificant adjusted effect, with a very large *p*-value $= 0.9736$, this model also is an acceptable 3WI-QR, in the statistical sense, since its IVs have significant joint effects, based on the *Quasi-LR* statistic of 243.1756 with a *p*-value 0.0000. Compared with the QR-7, this model is a worse QR, because it has a smaller adjusted *R*-squared.

8) Hence, based on the eight QRs, we can conclude that the QR-6 is the best QR because it has the greatest adjusted *R*-squared.

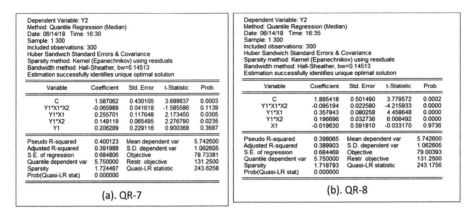

Figure 6.3   Statistical results of two reduced models of QR-5 in Table 6.2. (a) QR-7. (b) QR-8.

## 6.4   Applications of QRs Based on Data in Mlogit.wf1

Referring to the characteristics of the data in Mlogit.wf1 presented in the previous chapter, we can develop a lot of QRs, as extensions of the previous models, based on four numerical variables. In each of the following examples, unexpected statistical results of a special set of QRs are presented, with special notes and comments.

**Example 6.4   *A Special Set of QRs of LW on (Z1B,Z2B,Z3B)***
Table 6.4 presents six alternative QRs, as the results of the manual multistage selection and the trial-and-error methods, based on the F3WI-QR of *LW* on *(Z1B, Z2B,Z3B)* with the following ES. (Refer to the path diagram in Figure 6.1c, with *Y1 = LW*, and *(X1, X2, X3) = (Z1B, Z2B, Z3B)*.)

$$LW \; C \; Z1B * Z2B * Z3B \; Z1B * Z2B \; Z1B * Z3B \; Z2B * Z3B \; Z1B \; Z2B \; Z3B \qquad (6.8)$$

Based on these results, the following findings and notes are presented:

1) The first stage of analysis indicates that the QR-1 of *LW* has only the three-way interaction as an independent variable. In this case, at the 10% level of significance,

| | | | |
|---|---|---|---|
| Wald test: | | | |
| Equation: EQ16 | | | |

| Test Statistic | Value | df | Probability |
|---|---|---|---|
| *t*-statistic | −1.881130 | 788 | 0.0603 |
| *F*-statistic | 3.538651 | (1, 788) | 0.0603 |
| *Chi-square* | 3.538651 | 1 | 0.0600 |

Figure 6.4   The results of the Wald test for *Z1B*Z2B*Z3B* in the QR-1.

**Table 6.4** Statistical results summary of the manual multistage selection method, based on the F3WI-QR (6.8).

| QR-1 | C | Z1B*Z2B*Z3B | | | | | | | |
|------|------|------|------|------|------|------|------|------|------|
| Coef. | 6.2619 | −0.0039 | | | | | | | |
| t-stat. | 22.7820 | −1.8811 | | | | | | | |
| Prob. | 0.0000 | 0.0603 | | | | | | | |

|  | Ps. $R^2 = 0.0027$ | Adj. $R^2 = 0.0015$ | Quasi-LR = 2.6197 | Prob. = 0.1055 |
|---|---|---|---|---|

| QR-2 | C | Z1B*Z2B*Z3B | Z1B*Z2B | Z1B*Z3B | Z2B*Z3B | | | | |
|------|------|------|------|------|------|------|------|------|------|
| Coef. | 7.8918 | 0.0258 | 0.0519 | −0.0475 | −0.2267 | | | | |
| t-stat. | 5.9050 | 1.2182 | 1.0092 | −0.8330 | −4.1695 | | | | |
| Prob. | 0.0000 | **0.2235** | **0.3132** | **0.4051** | 0.0000 | | | | |

|  | Ps. $R^2 = 0.1453$ | Adj. $R^2 = 0.1409$ | Quasi-LR = 649.7887 | Prob. = 0.0000 |
|---|---|---|---|---|

| QR-3 | C | Z1B*Z2B*Z3B | Z1B*Z2B | Z1B*Z3B | Z2B*Z3B | Z1B | Z2B | Z3B |
|------|------|------|------|------|------|------|------|------|
| Coef. | −2.9179 | 0.0807 | −0.3957 | −0.3996 | −0.3985 | 2.7088 | 1.7674 | 1.1721 |
| t-stat. | −0.3245 | 1.2020 | −1.2993 | −1.0898 | −1.1354 | 1.5763 | 1.1037 | 0.6113 |
| Prob. | 0.7457 | **0.2297** | 0.1942 | **0.2761** | **0.2566** | 0.1154 | 0.2700 | 0.5412 |

|  | Ps. $R^2 = 0.1537$ | Adj. $R^2 = 0.1461$ | Quasi-LR = 172.17 | Prob. = 0.0000 |
|---|---|---|---|---|

| RM-1 | C | Z1B*Z2B*Z3B | Z1B*Z2B | Z1B*Z3B | Z2B*Z3B | Z1B | Z2B | |
|------|------|------|------|------|------|------|------|------|
| Coef. | 1.9426 | 0.0450 | −0.2551 | −0.1935 | −0.1954 | 1.8637 | 0.9557 | |
| t-stat. | 0.8800 | 2.1144 | −2.2468 | −2.1440 | −2.5653 | 2.9991 | 2.1279 | |
| Prob. | 0.3791 | 0.0348 | 0.0249 | 0.0323 | 0.0105 | 0.0028 | 0.0337 | |

|  | Ps. $R^2 = 0.1533$ | Adj. $R^2 = 0.1468$ | Quasi-LR = 172.17 | Prob. = 0.0000 |
|---|---|---|---|---|

| RM-2 | C | Z1B*Z2B*Z3B | Z1B*Z2B | Z1B*Z3B | Z2B*Z3B | Z1B | | |
|------|------|------|------|------|------|------|------|------|
| Coef. | 5.6673 | 0.0448 | −0.1314 | −0.2387 | −0.1440 | 1.4357 | | |
| t-stat. | 3.6490 | 2.1333 | −1.5712 | −2.6204 | −2.3473 | 2.6335 | | |
| Prob. | 0.0003 | 0.0332 | 0.1165 | 0.0090 | 0.0192 | 0.0086 | | |

|  | Ps. $R^2 = 0.1508$ | Adj. $R^2 = 0.1453$ | Quasi-LR = 170.00 | Prob. = 0.0000 |
|---|---|---|---|---|

| RM-3 | C | Z1B*Z2B*Z3B | Z1B*Z2B | Z1B*Z3B | Z2B*Z3B | | Z2B | |
|------|------|------|------|------|------|------|------|------|
| Coef. | 7.7457 | 0.0279 | 0.0344 | −0.0410 | −0.2435 | | 0.1175 | |
| t-stat. | 4.3371 | 1.3525 | 0.4631 | −0.5777 | −3.0677 | | 0.2200 | |
| Prob. | 0.0000 | 0.1766 | 0.6434 | 0.5636 | 0.0022 | | 0.8259 | |

|  | Ps. $R^2 = 0.1454$ | Adj. $R^2 = 0.1399$ | Quasi-LR = 162.40 | Prob. = 0.0000 |
|---|---|---|---|---|

Wald test:
Equation: EQ16_ST1

| Test statistic | Value | df | Probability |
|---|---|---|---|
| F-statistic | 32.34424 | (3, 785) | 0.0000 |
| Chi-square | 97.03271 | 3 | 0.0000 |

Null hypothesis: C(2) = C(3) = C(4) = 0

**Figure 6.5** The results of the Wald test for testing the joint effects of the interactions Z1B*Z2B*Z3B, Z1B*Z2B, and Z1B*Z3B.

the interaction has a significant effect, based on the $t$-statistic of $t_0 = -1.8811$ with a $p$-value $= 0.0603$; however, it is unexpected that it has an insignificant effect based on the $Quasi$-$LR$ statistic of $QLR_0 = 2.61$ with a $p$-value $= 0.1055$. Why? I have not found an explanation!

As a comparison, Figure 6.4 presents the Wald test for the interaction, using the $t$-statistic, $F$-statistic, and $Chi$-$square$ statistic, which have very close probabilities. It is defined or proved that $[t_0(df = n)]^2 = F_0(df = (1,n))$.

2) Whatever the results of QR-1, the second stage of analysis should be done by inserting the two-way interactions $Z1B*Z2B, Z1B*Z3B$ and $Z2B*Z3B$, as additional IVs for $QR$-$1$. Unexpectedly, each of the interaction IVs of QR-2 has an insignificant adjusted effect on $LW$, but their joint effects have a significant effect, based on the $Quasi$-$LR$ statistic of $QLR_0 = 649.78$ with a $p$-value $= 0.0000$. As a comparison, the Wald test also presents the IVs as having significant joint effects, based on the $F$-statistic of $F_0 = 65.5720$ with $df = (4,785)$ and a $p$-value $= 0.0000$, and the $Chi$-$square$ statistic of 278.2878 with $df = 4$ and a $p$-value $= 0.0000$. In addition, at the 15% level (used in Lapin 1973), we find that the three-way IV has a significant positive adjusted effect, based on the $t$-statistic of $t_0 = 1.2182$ with a $p$-value $= 0.2235/2 = 0.111\,75 < 0.15$. Hence, for these reasons, QR-2 is an acceptable model to present the true 3WI-QR.

As a supported statistical result on the acceptability of the QR-2, Figure 6.5 presents the results of a Wald test, which show that the three interactions $Z1B*Z2B*Z3B, Z1B*Z2B$, and $Z1B*Z3B$ have joint significant effects on $LW$, based on the Wald test $F$-statistic of $F_0 = 32.3442$ with $df = (3,785)$ and a $p$-value $= 0.0000$, and the $Chi$-$square$ statistic of 97.0327 with $df = 3$ and a $p$-value $= 0.0000$. This conclusion also indicates that the direct effect of $Z1B$ on $LW$ is significantly dependent on the following linear function: $(C(2)*Z2B*Z3B + C(3)*Z2B + C(4)*Z3B)$.

3) At the third stage of regression analysis, the three main variables are inserted as additional IVs for QR-2 because QR-2 is an acceptable regression to represent a true 3WI-QR. Then we have the QR-3, which is a GoF model, because its IVs have significant joint effects, based on the $Quasi$-$LR$ statistic of $QLR_0 = 172.17$ with a $p$-value 0.0000, even

Redundant Variables Test
Equation: EQ21
Specification: LW  C  Z1B*Z2B*Z3B Z1B*Z2B Z1B*Z3B Z2B*Z3B Z1B
Z2B Z3B
K

| | Value | df | Probability |
|---|---|---|---|
| QLR L-statistic | 86.77050 | 4 | 0.0000 |
| QLR Lambda-statistic | 83.02594 | 4 | 0.0000 |

L-test summary:

| | Value | df |
|---|---|---|
| Restricted Objective | 600.2451 | 786 |
| Unrestricted Objective | 549.9076 | 782 |
| Scale | 1.160246 | |

Lambda-test summary:

| | Value | df |
|---|---|---|
| Restricted Log Obj. | 6.397338 | 786 |
| Unrestricted Log Obj. | 6.309750 | 782 |
| Scale | 0.002110 | |

Restricted Test Equation:
Dependent Variable: LW
Method: Quantile Regression (Median)
Date: 06/15/18  Time: 13:38
Sample: 1 998
Included observations: 790
Huber Sandwich Standard Errors & Covariance
Sparsity method: Kernel (Epanechnikov) using residuals
Bandwidth method: Hall-Sheather, bw=0.1051
Estimation successfully identifies unique optimal solution

| Variable | Coefficient | Std. Error | t-Statistic | Prob. |
|---|---|---|---|---|
| C | 10.59132 | 2.676118 | 3.957719 | 0.0001 |
| Z2B*Z3B | 0.031307 | 0.105176 | 0.297658 | 0.7660 |
| Z2B | -0.288395 | 0.510698 | -0.564708 | 0.5724 |
| Z3B | -0.864373 | 0.554255 | -1.559520 | 0.1193 |

| | | | |
|---|---|---|---|
| Pseudo R-squared | 0.076246 | Mean dependent var | 5.822810 |
| Adjusted R-squared | 0.072720 | S.D. dependent var | 2.040345 |
| S.E. of regression | 1.894753 | Objective | 600.2451 |
| Quantile dependent var | 5.807151 | Restr. objective | 649.7887 |
| Sparsity | 4.938574 | Quasi-LR statistic | 80.25558 |
| Prob(Quasi-LR stat) | 0.000000 | | |

**Figure 6.6** The results of the redundant variables test for testing the joint effects of the IVs: *Z1B*Z2B*Z3B, Z1B*Z2B, Z1B*Z3B,* and *Z1B* of QR-3.

---

Dependent variable: LW

Method: quantile regression (Median)

Date: 06/16/18   Time: 06:49

Sample (adjusted): 1 998

Included observations: 790 after adjustments

Huber Sandwich standard errors and covariance

Sparsity method: Kernel (Epanechnikov) using residuals

Bandwidth method: Hall-Sheather, bw = 0.1051

Estimation successfully identifies unique optimal solution

| Variable | Coefficient | Std. error | t-statistic | Prob. |
|---|---|---|---|---|
| C | 6.891371 | 1.951107 | 3.532031 | 0.0004 |
| Z1B*Z2B*Z3B | −0.007461 | 0.003698 | −2.017272 | 0.0440 |
| Z1B*Z3B | 0.073627 | 0.078146 | 0.942169 | 0.3464 |
| Z1B | 0.533434 | 0.371057 | 1.437604 | 0.1509 |
| Z3B | −0.981796 | 0.412453 | −2.380382 | 0.0175 |

| | | | |
|---|---|---|---|
| Pseudo R-squared | 0.151429 | Mean dependent var | 5.822810 |
| Adjusted R-squared | 0.147106 | S.D. dependent var | 2.040345 |
| S.E. of regression | 1.747389 | Objective | 551.3915 |
| Quantile dependent var | 5.807151 | Restr. objective | 649.7887 |
| Sparsity | 4.674642 | Quasi-LR statistic | 168.3931 |
| Prob(quasi-LR stat) | 0.000000 | | |

**Figure 6.7** Statistical results of RM-4 as an alternative reduced model of QR-3.

though each of its IVs has an insignificant adjusted effect, at the 10% level. In addition, we find that the effect of *Z1B* on *LW* is significantly dependent on a function of *Z2B\*Z3B*, *Z2B,* and *Z3B,* using the *redundant variables test* (RVT), as an alternative to the Wald test, with the statistical results presented in Figure 6.6. Based on these results, the following findings and notes are presented:

3.1 The steps of the analysis are, with the output of QR-3 on-screen, select *View/Coefficient Diagnostics/Redundant Variables – Likelihood Ratio.* Then we can type or insert the IVs *"Z1B\*Z2B\*Z3B Z1B\*Z2B Z1B\*Z3B Z1B,"* and click OK. The results display on-screen.

3.2 The statistical hypothesis can be written as follows:

$$H_0; Resticted\ Equation : \text{``}LW\ C\ Z2B * Z3B\ Z2B\ Z3B\text{''}vs\ H_1 : QR - 3 \qquad (6.9)$$

and the results show the null hypothesis is rejected based on the *QLR L*-statistic and *QLR Lambda*-statistic of 86.77 and 83.03, respectively, with *df* = 4 and *p*-value = 0.0000.

---

Dependent variable: LW
Method: quantile regression (Median)
Date: 06/16/18   Time: 07:57
Sample (adjusted): 1 998
Included observations: 790 after adjustments
Huber Sandwich standard errors and covariance
Sparsity method: Kernel (Epanechnikov) using residuals
Bandwidth method: Hall-Sheather, bw = 0.1051
Estimation successfully identifies unique optimal solution

| Variable | Coefficient | Std. error | *t*-Statistic | Prob. |
| --- | --- | --- | --- | --- |
| C | 5.053878 | 0.726528 | 6.956210 | 0.0000 |
| Z1B*Z2B*Z3B | −0.007429 | 0.003659 | −2.030527 | 0.0426 |
| Z1B | 0.884741 | 0.115795 | 7.640575 | 0.0000 |
| Z3B | −0.592733 | 0.128012 | −4.630290 | 0.0000 |

| | | | |
| --- | --- | --- | --- |
| Pseudo *R*-squared | 0.150762 | Mean dependent var | 5.822810 |
| Adjusted *R*-squared | 0.147521 | S.D. dependent var | 2.040345 |
| S.E. of regression | 1.745640 | Objective | 551.8253 |
| Quantile dependent var | 5.807151 | Restr. objective | 649.7887 |
| Sparsity | 4.717462 | Quasi-LR statistic | 166.1289 |
| Prob(quasi-LR stat) | 0.000000 | | |

**Figure 6.8**   Statistical results of QR-5 as a reduced model of QR-4.

4) Since QR-3 is the F3WI-QR in (6.8), we then explore various reduced models, using the trial-and-error method. The best three for representing the 3WI-QRs were RM-1, RM-2, and RM-3. They are presented in Table 6.4, with the following characteristics:

4.1 The reduced model RM-1 is the best 3WI-QR, which is a really unexpected statistical result. We obtain it by deleting $Z3B$ from QR-3, since it has the greatest $p$-value. Note that each of the IVs of RM-1 has a significant adjusted effect, at the 1% or 5% level. And RM-1 is the best QR for representing the true three-way interaction, as it is presented by the path diagram in Figure 6.1c.

4.2 The reduced model RM-2 is obtained by deleting two of the three main IVs in QR-3, namely $Z2B$ and $Z3B$, which shows that the three-way interaction and two of the two-way interactions have significant adjusted effects, at the 1% or 5% level. In fact, at the 10% level, the interaction $Z1B*Z2B$ has a significant negative adjusted effect, based on the $t$-statistic of $t_0 = -1.5712$ with a $p$-value $= 0.1165/2 = 0.05825 < 0.10$.

4.3 We obtain the reduced model RM-3 by deleting two of the three main IVs in QR-3, namely $Z1B$ and $Z3B$, which showed that the three-way interaction has a significant positive adjusted effect, based on the $t$-statistic of $t_0 = 1.3525$ with a $p$-value $= 0.1766/2 = 0.0883 < 0.10$.

---

Dependent variable: LW
Method: quantile regression (Median)
Date: 06/16/18   Time: 08:49
Sample (adjusted): 1 998
Included observations: 790 after adjustments
Huber Sandwich standard errors and covariance
Sparsity method: Kernel (Epanechnikov) using residuals
Bandwidth method: Hall-Sheather, bw = 0.1051
Estimation successfully identifies unique optimal solution

| Variable | Coefficient | Std. error | t-Statistic | Prob. |
|---|---|---|---|---|
| C | 6.872331 | 0.751735 | 9.141963 | 0.0000 |
| Z1B | 0.719467 | 0.079571 | 9.041806 | 0.0000 |
| Z2B | −0.197270 | 0.094702 | −2.083048 | 0.0376 |
| Z3B | −0.781146 | 0.081113 | −9.630346 | 0.0000 |

| | | | | |
|---|---|---|---|---|
| Pseudo R-squared | 0.151603 | Mean dependent var | | 5.822810 |
| Adjusted R-squared | 0.148365 | S.D. dependent var | | 2.040345 |
| S.E. of regression | 1.745244 | Objective | | 551.2784 |
| Quantile dependent var | 5.807151 | Restr. objective | | 649.7887 |
| Sparsity | 4.686238 | Quasi-LR statistic | | 168.1694 |
| Prob(quasi-LR stat) | 0.000000 | | | |

**Figure 6.9** Statistical results of an additive QR of LW on *(Z1B,Z2B,Z3B)*.

5) After doing various additional trial-and-error selections of the IVs, we unexpectedly obtained the statistical result of a reduced model RM-4, as presented in Figure 6.7. Based on this result, the findings and notes are as follows:

   5.1 The three-way interaction IV of RM-4 is significant at the 5% level. In fact, it has a significant negative effect based on the $t$-statistic of $t_0 = -2.0173$ with a $p$-value $= 0.0440/2 = 0.0220 < 0.05$.

   5.2 Compared with the previous three reduced models, RM-4 is the simplest QR, but it has the greatest adjusted $R$-squared $= 0.1471$, while RM-1 has 0.1468. Even though they have a very small difference, RM-4 is a better fit QR than RM-1, in the statistical sense, to represent the true 3WI-QR.

   5.3 Since the interaction $Z1B*Z3B$ in RM-4 is insignificant, we try to reduce the model and find that RM-5 has a greater adjusted $R$-squared than RM-4, as shown in Figure 6.8. So, RM-5 is a better fit 3WI-QR than RM-4, in the statistical sense.

6) After doing exercises with a lot of possible 3WI-QRs, I consider RM-5 to be the simplest and the best 3WI-QR among all possible 3WI-QRs of *LW* on *(Z1B,Z2B,Z3B)*, specific for the data in Mlogit.wf1.

7) Aside from the 3WI-QRs, there are additive and alternative 2WI-QRs of *LW* on *(Z1B,Z2B,Z3B)* that can be done as exercises. We find that the additive QR has a greater adjusted $R$-squared than RM-5, and each of its IVs has a significant adjusted effect, as shown in Figure 6.9. Hence, the additive QR has a better fit model than all 3WI-QRs, but it does not represent the effect of *Z1B* on *LW* to be significantly dependent on *Z2B* and *Z3B*. The additive model presents the effect of an IV on the dependent variable (DV), adjusted for the other IVs. Such as the significant effect of *Z1B* on *LW* is an effect, adjusted for *Z2B* and *Z3B,* or it is an adjusted effect, in short.

## 6.5   QRs of *PR1* on *(DIST1,X1,X2)*

As an extension of the model (5.20), this example presents the application of the manual stepwise selection method based on a F3WI-QR of *PR1* on *(DIST1,X1,X2)* with the following ES:

$$PR1 \ C \ DIST1 * X1 * X2 \ DIST1 * X1 \ DIST1 * X2 \ X1 * X2 \ DIST1 \ X1 \ X2 \qquad (6.10)$$

Table 6.5 presents the statistical results of the manual three-stage selection method of the F3WI-QR (6.10), with three of its three-way interaction reduced models. Based on this summary, the following findings and notes are presented:

1) At the first and the second stages of analysis, QR-1 and QR-2 have significant IVs, with a $p$-value $= 0.0000$ each.

2) At the third stage, we get the estimate of the F3WI-QR in (6.10) by inserting the three main variables as additional IVs for QR-2. Because three of its interaction IVs have an insignificant effect, three of its reduced models, RM-1, RM-2, and RM-3, are presented. They have only two of the main variables, and their characteristics are as follows:

   2.1 Each of the IVs of RM-1 and RM-2 has a significant effect on *PR1* with a $p$-value $= 0.0000$, and the three-way interaction of RM-3 has a significant effect,

**Table 6.5** Statistical results summary of the manual three-stage selection method, the full model (6.10), and three of its three-way interaction reduced models.

| QR-1 | C | DIST1*X1*X2 | | | | | |
| --- | --- | --- | --- | --- | --- | --- | --- |
| Coef. | 0.3509 | −0.0174 | | | | | |
| *t*-stat. | 176.244 | −20.048 | | | | | |
| Prob. | 0.0000 | 0.0000 | | | | | |

| | Ps. $R^2$ = 0.2635 | Adj. $R^2$ = 2627 | Quasi-LR = 425.96 | Prob. = 0.0000 |
| --- | --- | --- | --- | --- |

| QR-2 | C | DIST1*X1*X2 | DIST1*X1 | DIST1*X2 | X1*X2 | | |
| --- | --- | --- | --- | --- | --- | --- | --- |
| Coef. | 0.3828 | 0.0177 | −0.0110 | −0.0012 | −0.1949 | | |
| *t*-stat. | 237.65 | 15.494 | −28.669 | −3.4653 | −32.424 | | |
| Prob. | 0.0000 | 0.0000 | 0.0000 | 0.0006 | 0.0000 | | |

| | Ps. $R^2$ | Adj. $R^2$ | Quasi-LR = 425.96 | Prob. |
| --- | --- | --- | --- | --- |

| F3WI-QR | C | DIST1*X1*X2 | DIST1*X1 | DIST1*X2 | X1*X2 | DIST1 | X1 | X2 |
| --- | --- | --- | --- | --- | --- | --- | --- | --- |
| Coef. | 0.4254 | 0.0000 | 0.0001 | 0.0000 | 0.0241 | −0.0001 | −0.1413 | −0.0704 |
| *t*-stat. | 2080.99 | −0.5512 | 1.0434 | 0.7373 | 51.334 | −1.4089 | −461.05 | −200.68 |
| Prob. | 0.0000 | **0.5817** | **0.2970** | **0.4611** | 0.0000 | 0.1592 | 0.0000 | 0.0000 |

| | Ps. $R^2$ = 0.9912 | Adj. $R^2$ = 0.9912 | Quasi-LR = 141 103.7 | Prob.0.0000 |
| --- | --- | --- | --- | --- |

| RM-1 | C | DIST1*X1*X2 | DIST1*X1 | DIST1*X2 | X1*X2 | DIST1 | X1 | |
| --- | --- | --- | --- | --- | --- | --- | --- | --- |
| Coef. | 0.3896 | 0.0142 | −0.0072 | −0.0101 | −0.0756 | 0.0051 | −0.0906 | |
| *t*-stat. | 242.742 | 31.4137 | −18.1807 | −44.4085 | −33.1051 | 20.038 | −35.625 | |
| Prob. | 0.0000 | 0.0000 | 0.0000 | 0.0000 | 0.0000 | 0.0000 | 0.0000 | |

| | Ps. $R^2$ = 0.9131 | Adj. $R^2$ = 0.9126 | Quasi-LR = 30 842.44 | Prob. = 0.0000 |
| --- | --- | --- | --- | --- |

| RM-2 | C | DIST1*X1*X2 | DIST1*X1 | DIST1*X2 | X1*X2 | DIST1 | | X2 |
| --- | --- | --- | --- | --- | --- | --- | --- | --- |
| Coef. | 0.3553 | 0.0279 | −0.0198 | −0.0139 | −0.1739 | 0.0099 | | 0.0281 |
| *t*-stat. | 92.809 | 32.2356 | −59.600 | −15.596 | −33.229 | 17.113 | | 4.6272 |
| Prob. | 0.0000 | 0.0000 | 0.0000 | 0.0000 | 0.0000 | 0.0000 | | 0.0000 |

| | Ps. $R^2$ = 0.8297 | Adj. $R^2$ = 0.8286 | Quasi-LR = 14 391.15 | Prob. = 0.0000 |
| --- | --- | --- | --- | --- |

| RM-3 | C | DIST1*X1*X2 | DIST1*X1 | DIST1*X2 | X1*X2 | | X1 | X2 |
| --- | --- | --- | --- | --- | --- | --- | --- | --- |
| Coef. | 0.4251 | 0.0001 | 0.0000 | 0.0000 | 0.0235 | | −0.1409 | −0.0699 |
| *t*-stat. | 3186.2 | 1.8494 | −0.7336 | −1.6186 | 70.353 | | −650.75 | −285.85 |
| Prob. | 0.0000 | 0.0647 | **0.4634** | **0.1059** | 0.0000 | | 0.0000 | 0.0000 |

| | Ps. $R^2$ = 0.9912 | Adj. $R^2$ = 0.9912 | Quasi-LR = 138 623 | Prob. = 0.0000 |
| --- | --- | --- | --- | --- |

| Redundant Variables Test | | | | Restricted Test Equation: | | | | |
|---|---|---|---|---|---|---|---|---|
| Equation: EQ23_ST4 | | | | Dependent Variable: PR1 | | | | |
| Specification: PR1 C DIST1*X1*X2 DIST1*X1 DIST1*X2 X1*X2 X1 X2 | | | | Method: Quantile Regression (Median) | | | | |
| K | | | | Date: 06/16/18 Time: 19.39 | | | | |
| | | | | Sample: 1 1000 | | | | |
| | | | | Included observations: 1000 | | | | |
| | Value | df | Probability | Huber Sandwich Standard Errors & Covariance | | | | |
| QLR L-statistic | 3.889180 | 3 | 0.2737 | Sparsity method: Kernel (Epanechnikov) using residuals | | | | |
| QLR Lambda-statistic | 3.883039 | 3 | 0.2744 | Bandwidth method: Hall-Sheather, bw=0.097156 | | | | |
| | | | | Estimation successfully identifies unique optimal solution | | | | |

L-test summary:

| | Value | df | | Variable | Coefficient | Std. Error | t-Statistic | Prob. |
|---|---|---|---|---|---|---|---|---|
| Restricted Objective | 0.152745 | 996 | | C | 0.425101 | 0.000141 | 3015.899 | 0.0000 |
| Unrestricted Objective | 0.152263 | 993 | | X1*X2 | 0.023920 | 0.000288 | 83.02079 | 0.0000 |
| Scale | 0.000248 | | | X1 | -0.140937 | 0.000203 | -692.5786 | 0.0000 |
| | | | | X2 | -0.070204 | 0.000213 | -328.9988 | 0.0000 |

Lambda-test summary:

| | Value | df | | | | | | |
|---|---|---|---|---|---|---|---|---|
| Restricted Log Obj. | -1.878985 | 996 | | Pseudo R-squared | 0.991184 | Mean dependent var | | 0.324000 |
| Unrestricted Log Obj. | -1.882145 | 993 | | Adjusted R-squared | 0.991158 | S.D. dependent var | | 0.040815 |
| Scale | 0.001627 | | | S.E. of regression | 0.000426 | Objective | | 0.152745 |
| | | | | Quantile dependent var | 0.323612 | Restr. objective | | 17.32660 |
| | | | | Sparsity | 0.001026 | Quasi-LR statistic | | 133909.5 |
| | | | | Prob(Quasi-LR stat) | 0.000000 | | | |

**Figure 6.10** A redundant variables test for testing the joint effects of $DIST1*X1*X2$, $DIST1*X1$, and $DIST1*X2$.

at the 10%, with a $p$-value $= 0.0647$ So, in the statistical sense, the three reduced models are acceptable models for representing the true 3WI-QRs.

2.2 Unexpectedly, RM-3 has the greatest adjusted $R$-squared of 0.9912, compared with RM-1 and RM-2, which have adjusted $R$-squareds of 0.9126 and 0.8286. Hence, RM-3 is the best fit 3WI-QR.

2.3 Note that RM-3 can be considered as the best fit QR because of its adjusted $R$-squared of $0.9912 \sim 1.0$. However, since two interaction IVs of RM-3 have large $p$-values of 0.4634 and 0.1059, we test the joint effects of the three interactions, $DIST1*X1*X2$, $DIST1*X1$, and $DIST1*X2$, for the effect of $DIST1$ on $PR1$, which is significantly dependent on $X1$ and $X2$, with the statistical hypothesis as follows:

$$H_0 : Restricted\ Equation : \text{“}PR1\ C\ X1\ X2\text{”} vs\ H_1 : RM-3 \qquad (6.11)$$

The RVT in Figure 6.10 shows that the null hypothesis is accepted, based on the $QLR$ $L$-statistic of 3.889 and $QLR$ $Lambda$-statistic of 3.883, with $df = 3$ and $p$-values of 0.2737 and 0.2744. Hence, the effect of $DIST1$ on $PR1$ is insignificantly dependent on $X*X2$, $X1$, and $X2$.

Then, so far, RM-1 should be considered the best QR among the three reduced models because it has a greater adjusted $R$-squared than RM-2. See an unexpected QR in the following example.

## Example 6.5 *Alternative Reduced Models of RM-3 in Table 6.5*

By using the trial-and-error method, we find various alternative three-way interaction reduced models of the RM-3, and Table 6.6 presents the five that I consider the best for a discussion. The characteristics of each of the QRs are as follows:

1) The interaction $DIST1*X1$ in RM-3.1 has a very large $p$-value $> 0.90$, but the joint effects of $DIST1*X1*X2$, $DIST1*X1$, and $DIST1*X2$ have a significant adjusted effect, based on the $QLR$ $L$-statistic of 864.8779 with $df = 3$ and a $p$-value $= 0.0000$. So, it is an acceptable QR for representing that the effect of $DIST1$ on $LW$ is significantly dependent on $X1*X2$, $X1$, and $X2$.

2) Similarly, the joint effects of $DIST1*X1*X2$, $DIST1*X1$, and $DIST1*X2$ in RM-3.2 have a significant adjusted effect, based on the $QLR$ $L$-statistic of 909.1596 with $df = 3$ and a $p$-value $= 0.0000$. So, it is also an acceptable QR for representing that the effect of $DIST1$ on $LW$ is significantly dependent on $X1*X2$, $X1$, and $X2$.

**Table 6.6** Statistical results summary of alternative reduced models of RM-3 in Table 6.5.

| RM-3.1 | C | DIST1*X1*X2 | DIST1*X1 | DIST1*X2 | X1*X2 | | X1 | |
|---|---|---|---|---|---|---|---|---|
| Coef. | 0.4098 | 0.0088 | 0.0000 | −0.0063 | −0.0757 | | −0.1190 | |
| t-stat. | 276.05 | 16.3581 | 0.0760 | −20.3082 | −23.136 | | −46.314 | |
| Prob. | 0.0000 | 0.0000 | **0.9394** | 0.0000 | 0.0000 | | 0.0000 | |

Ps. $R^2 = 0.8540$    Adj. $R^2 = 0.8532$    Quasi-LR $= 12\,785.9$    Prob. $= 0.0000$

| RM-3.2 | C | DIST1*X1*X2 | DIST1*X1 | DIST1*X2 | X1*X2 | | | X2 |
|---|---|---|---|---|---|---|---|---|
| Coef. | 0.3907 | 0.0168 | −0.0126 | 0.0004 | −0.1706 | | | −0.023 |
| t-stat. | 125.45 | 15.9447 | −21.522 | 1.2873 | −29.811 | | | −4.657 |
| Prob. | 0.0000 | 0.0000 | 0.0000 | **0.1983** | 0.0000 | | | 0.0000 |

Ps. $R^2 = 0.7138$    Adj. $R^2 = 0.7124$    Quasi-LR $= 5056.765$    Prob. $= 0.0000$

| RM-3.3 | C | DIST1*X1*X2 | DIST1*X1 | | | DIST1 | X1 | X2 |
|---|---|---|---|---|---|---|---|---|
| Coef. | 0.4215 | 0.0021 | −0.0013 | | | 0.0001 | −0.1274 | −0.064 |
| t-stat. | 1824.43 | 26.3027 | −13.4891 | | | 2.3319 | −221.69 | −165.3 |
| Prob. | 0.0000 | 0.0000 | 0.0000 | | | 0.0199 | 0.0000 | 0.0000 |

Ps. $R^2 = 0.9700$    Adj. $R^2 = 0.9698$    Quasi-LR $= 95\,006.6$    Prob. $= 0.0000$

| RM-3.4 | C | DIST1*X1*X2 | | DIST1*X2 | | DIST1 | X1 | X2 |
|---|---|---|---|---|---|---|---|---|
| Coef. | 0.4224 | 0.0021 | | −0.0011 | | 0.0000 | −0.1351 | −0.0583 |
| t-stat. | 1630.4 | 24.2395 | | −11.2483 | | −0.1489 | −331.71 | −98.67 |
| Prob. | 0.0000 | 0.0000 | | 0.0000 | | **0.8817** | 0.0000 | 0.0000 |

Ps. $R^2 = 0.9692$    Adj. $R^2 = 0.9690$    Quasi-LR $= 58\,626.51$    Prob. $= 0.0000$

| RM-3.5 | C | DIST1*X1*X2 | | DIST1*X2 | | | X1 | X2 |
|---|---|---|---|---|---|---|---|---|
| Coef. | 0.4224 | 0.0021 | | −0.0011 | | | −0.1351 | −0.0582 |
| t-stat. | 2163.5 | 24.0131 | | −16.6870 | | | −332.96 | −128.82 |
| Prob. | 0.0000 | 0.0000 | | 0.0000 | | | 0.0000 | 0.0000 |

Ps. $R^2 = 0.9692$    Adj. $R^2 = 0.9690$    Quasi-LR $= 58\,295.37$    Prob. $= 0.0000$

Note that RM-3.2 is worse than RM-3.1 because it has a smaller adjusted $R$-squared.

3) Very unexpectedly,we find RM-3.3, RM-3.4, and RM-3.5 to have the same adjusted $R$-squared of 0.97 to two decimal points, which is greater than RM-1 in Table 6.5. So, these reduced models are a better fit than RM-1.

4) Since RM-3.4 has an IV, *DIST1*, with a very large $p$-value $= 0.8817$, deleting *DIST1* from RM-3.4 obtains RM-3.5 with the same adjusted $R$-squared as RM-3.4 to four decimal

points. Since RM-3.3 has a greater adjusted *R*-squared than RM-3.5, even with a very small difference, RM-3.3 is the best QR among the 11 models in Tables 6.5 and 6.6, However, I prefer the RM-3.5 since it is the simplest model.

## 6.6 Advanced Statistical Analysis

With the output shown in Figure 6.11a on-screen, selecting *View/Quantile Process…* give us the options in Figure 6.11b, which show three statistical analysis methods, *Process Coefficients, Slope Equality Test*, and *Symmetric Quantile Test,* which will be discussed in the following subsections.

### 6.6.1 Applications of the Quantiles Process

#### 6.6.1.1 An Application of the Process Coefficients

**Example 6.6** *The Output of Process Coefficients*

After selecting *View/Quantile Process/Process Coefficients…*, we have the options in Figure 6.12a shown on the screen. Then, by inserting the option *Quantiles: 4 and clicking OK,* we obtain the output in Figure 6.12b. Based on this output, the findings and notes are as follows:

1) The output presents three QRs for Quantiles = 0.25, 0.50, and 0.75, with the tests for each their parameters. But there is no test of the joint effects of the IVs of each QR.
2) For testing the joint effects, we have to run each QR separately. Do this as exercises.
3) Note that the interaction $Z1B^*Z2B^*Z3B$ in the third regression for the Quantile = 0.75 has an insignificant adjusted effect with a very large *p*-value = 0.9506.

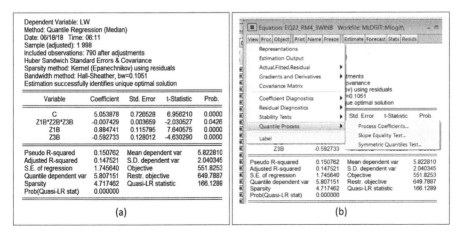

**Figure 6.11** The output of RM-5 in Figure 6.8 and options for advanced statistical analysis.

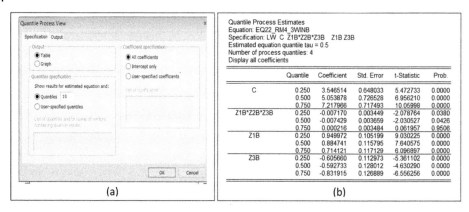

(a)                                                    (b)

**Figure 6.12**   The options of the quantile process View, and an output for quantiles = 4.

Quantile Slope Equality Test
Equation: EQ22_RM4_3WINB
Specification: LW C Z1B*Z2B*Z3B   Z1B Z3B
Estimated equation quantile tau = 0.5
Number of test quantiles: 4
Test statistic compares all coefficients

| Test Summary | Chi-Sq. Statistic | Chi-Sq. d.f. | Prob. |
|---|---|---|---|
| Wald Test | 6.920374 | 6 | 0.3283 |

Restriction Detail: b(tau_h) - b(tau_k) = 0

| Quantiles | Variable | Restr. Value | Std. Error | Prob. |
|---|---|---|---|---|
| 0.25, 0.5 | Z1B*Z2B*Z3B | 0.000259 | 0.003279 | 0.9371 |
| | Z1B | 0.065231 | 0.102338 | 0.5239 |
| | Z3B | -0.012926 | 0.112164 | 0.9083 |
| 0.5, 0.75 | Z1B*Z2B*Z3B | -0.007645 | 0.003293 | 0.0202 |
| | Z1B | 0.170620 | 0.107815 | 0.1135 |
| | Z3B | 0.239182 | 0.117246 | 0.0414 |

**Figure 6.13**   The output of the quantile slope equality test.

#### 6.6.1.2   An Application of the Quantile Slope Equality Test

**Example 6.7**   *The Output of a Quantile Slope Equality Test*
By selecting *View/Quantile Process/Quantile Slope Equality Test…*, we obtain the output in Figure 6.13. It shows that the six pairs of slopes of the three QRs' IVs have insignificant effects, based on the Wald test *Chi-square* statistic of 6.920 374, with $df = 6$ and $p$-value = 0.3263. However, only two pairs of the slopes have significant differences, at the 10% level, for the pair of quantiles (0.5, 0.75).

Symmetric Quantiles Test
Equation: EQ22_RM4_3WINB
Specification: LW C Z1B*Z2B*Z3B Z1B Z3B
Estimated equation quantile tau = 0.5
Number of test quantiles: 4
Test statistic compares all coefficients

| Test Summary | Chi-Sq. Statistic | Chi-Sq. d.f. | Prob. |
|---|---|---|---|
| Wald Test | 3.806332 | 4 | 0.4328 |

Restriction Detail: b(tau) + b(1-tau) - 2*b(.5) = 0

| Quantiles | Variable | Restr. Value | Std. Error | Prob. |
|---|---|---|---|---|
| 0.25, 0.75 | C | 0.656724 | 1.040938 | 0.5281 |
| | Z1B*Z2B*Z3B | 0.007904 | 0.005205 | 0.1289 |
| | Z1B | -0.105388 | 0.166054 | 0.5256 |
| | Z3B | -0.252108 | 0.182501 | 0.1672 |

**Figure 6.14** The output of the quantile slope equality test.

### 6.6.1.3 An Application of the Symmetric Quantiles Test

**Example 6.8** *The Output of a Symmetric Quantiles Test*
By selecting *View/Quantile Process/Symmetric Quantiles Test...,* we obtain the output in Figure 6.14 for testing the statistical hypothesis

$$H_0 : B_i(0.25) + B_i(0.75) - 2 * B_i(0.5) = 0 \text{ vs } H1 : Otherwise \qquad (6.12)$$

for all the variables, $i = 1, 2, 3$, and 4, which represent the four parameters of the model, and also for each variable.

The output shows that the null hypothesis for all four variables is accepted based on the Wald test *Chi-square* statistic of 3.806 332, with $df = 4$ and *p*-value $= 0.4326$. And the null hypothesis for each variable also is accepted, at the 10% level. Hence, we can conclude that the data supports that the quantiles are symmetrical.

### 6.6.2 An Application of the Ramsey RESET Test

**Example 6.9** *Output of a Ramsey RESET Test*
RESET stand for *regression specification test*. It was proposed by Ramsey (1969) with the assumption that the disturbance vector $\varepsilon$ of the following regression has the multivariate normal distribution, $N(0, \eth^2 I)$ (refer to the User Guide EViews 8. pp. 88–89):

$$Y = X\beta + \varepsilon \qquad (6.13)$$

And the hypothesis tested is as follows:

$$H_0 : \varepsilon \sim N\left(0, \eth^2 I\right) \text{ vs } H_1 :: \varepsilon \sim N\left(\mu, \eth^2 I\right), \mu \neq 0 \qquad (6.14)$$

| Ramsey RESET Test | | | |
|---|---|---|---|
| Equation: EQ22_RM4_3WINB | | | |
| Specification: LW C Z1B*Z2B*Z3B Z1B Z3B | | | |
| Omitted Variables: Squares of fitted values | | | |
| | Value | df | Probability |
| QLR L-statistic | 0.000103 | 1 | 0.9919 |
| QLR Lambda-statistic | 0.000103 | 1 | 0.9919 |

| L-test summary: | | |
|---|---|---|
| | Value | df |
| Restricted Objective | 551.8253 | 786 |
| Unrestricted Objective | 551.8252 | 785 |
| Scale | 1.179073 | |

| Lambda-test summary: | | |
|---|---|---|
| | Value | df |
| Restricted Log Obj. | 6.313231 | 786 |
| Unrestricted Log Obj. | 6.313231 | 785 |
| Scale | 0.002137 | |

Unrestricted Test Equation:
Dependent Variable: LW
Method: Quantile Regression (Median)
Date: 06/18/18   Time: 06:04
Sample: 1 998
Included observations: 790
Huber Sandwich Standard Errors & Covariance
Sparsity method: Kernel (Epanechnikov) using residuals
Bandwidth method: Hall-Sheather, bw=0.1051
Estimation successfully identifies unique optimal solution

| Variable | Coefficient | Std. Error | t-Statistic | Prob. |
|---|---|---|---|---|
| C | 5.020042 | 1.572227 | 3.192949 | 0.0015 |
| Z1B*Z2B*Z3B | -0.007307 | 0.006121 | -1.193872 | 0.2329 |
| Z1B | 0.869338 | 0.582163 | 1.493289 | 0.1358 |
| Z3B | -0.582357 | 0.398066 | -1.462964 | 0.1439 |
| FITTED^2 | 0.001411 | 0.054224 | 0.026021 | 0.9792 |

| | | | |
|---|---|---|---|
| Pseudo R-squared | 0.150762 | Mean dependent var | 5.822810 |
| Adjusted R-squared | 0.146435 | S.D. dependent var | 2.040345 |
| S.E. of regression | 1.746755 | Objective | 551.8252 |
| Quantile dependent var | 5.807151 | Restr. objective | 649.7887 |
| Sparsity | 4.716293 | Quasi-LR statistic | 166.1702 |
| Prob(Quasi-LR stat) | 0.000000 | | |

**Figure 6.15** Ramsey RESET test for the QR in Figure 6.10.

However, corresponding to the results in Figure 6.15, we see that the test, in fact, is an *omitted-variables test* (OVT): Squared of fitted values, *Fitted^2*, based on the following augmented model:

$$LW = C(1) + C(2) * Z1B * Z2b * Z3B + C(3) * Z1B + C(4) * Z3B$$
$$+ C(5) * Fitted^2 + \varepsilon \qquad (6.15)$$

So, the hypothesis could be written as

$$H_0 : C(5) = 0 \; vs \; H_1 :: C(5) \neq 0 \qquad (6.16a)$$

or

$$H_0 : LW = C(1) + C(2) * Z1B * Z2b * Z3B + C(3) * Z1B + C(4) * Z3B + \varepsilon, vs$$
$$H_1 : LW = C(1) + C(2) * Z1B * Z2b * Z3B + C(3) * Z1B + C(4) * Z3B$$
$$+ C(5) * Fitted^2 + \varepsilon \qquad (6.16b)$$

Then the null hypothesis $H_0$:$C(5) = 0$ or $H_0$: $\varepsilon \sim N(0,\eth^2 I)$ is accepted based on the *QLR L*-statistic and *QLR Lambda*-statistic of 0.000 103 with $df = 1$ and a $p$-value $= 0.9919$. Hence, the multivariate normal distribution of the error terms has a significant mean of zero. In addition, note that the output of the unrestricted test equation in Figure 6.15 also shows that the null hypothesis $C(5) = 0$ is accepted based on the $t$-statistic of $t_0 = 0.026\,021$ with a $p$-value $= 0.8782$, which looks like an RVT or OVT of the variable *Fitted²*.

As a comparison, Figure 6.16 presents the results of the RVT of the full model in hypothesis (6.16b), which shows different values of the *QLR L*-statistic and *QLR Lambda*-statistic compare to the Ramsey RESET test (RRT). On the other hand, the OVT based on the reduced model in hypothesis (6.16b) can't be obtained because the program presents the error message shown in Figure 6.17. To overcome this error message, we have to develop a complete Data_LW.wf, which we see in Section 6.8.

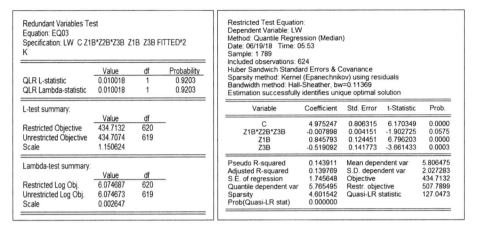

**Figure 6.16** The statistical results of an RVT of FITTED^2 of the full model.

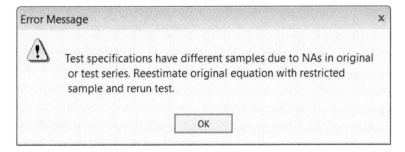

**Figure 6.17** An error message of the OVT.

Corresponding to this RRT, I have special notes as follows:

1) The test is not for testing the assumptions of the error terms of a regression. Agung (2009a, 2011a) mention that the basic assumptions should be accepted as a matter of course as they can't be tested.

2) The error terms of both models in the hypothesis (6.16b) are assumed to have multivariate normal distributions as presented in the hypothesis (6.14). And those assumptions also should be accepted as a matter of course because they are needed for the validity of the RRT.

3) We don't have to worry about the normality distributions of the observed scores of the variables used in the model. Hence, we don't need to test the normal distribution of each variable because it isn't required for the regression analysis.

4) Even the least squares estimation method of the model parameters doesn't need the normality assumption. The basic assumptions are needed for testing the model parameters, but the likelihood estimation method is invalid without the normal distribution. Refer to the misinterpretation of selected theoretical statistical concepts in Agung (2011a).

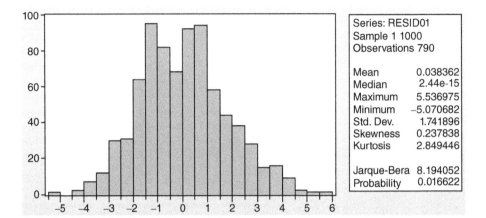

**Figure 6.18** Histogram and JB-normality test of the residual..

### 6.6.3 Residual Diagnostics

With the output of an QR on-screen, selecting *View/Residual Diagnostics* presents three options: *Correlogram-Q-statistic, Correlogram Squared Residual,* and *Histogram-Normality Test*. However, the first two are important only for the time series data. For this reason, the following example presents the third option.

**Example 6.10** *Histogram – Normality Test*
Figure 6.18 presents the histogram and the Jarque-Bera normality test for the residual of the QR in Figure 6.11, which shows that the data supports the normal distribution of the residual, at the 5% level of significance. Note that this test isn't for testing the basic assumptions of the error terms.

To conduct a more advanced or alternative residual analysis, as presented in Neter and Wasserman (1974), the variable residual should be generated first, as follows:

1. With the output on-screen, select *Proc/Make Residual Series…*. The dialog in Figure 6.19 appears.
2. Clicking OK inserts a new variable, *RESID03*, in the data file. You can rename the variable.

Then alternative residual analyses can be done for evaluating whether the model should be modified or additional IVs should be inserted, in order to obtain a better model, in the statistical sense. Please refer to the examples presented in Neter and Wasserman (1974).

## 6.7 Forecasting

In general, forecasting should be done based on time series data, as presented in (Agung 2019; Hankel and Reitsch 1992; Wilson and Keating 1994), specifically for the dynamic forecast, including the forecast beyond the sample period. However, the static option can be done based on all cross-section univariate regressions, including the binary choice models.

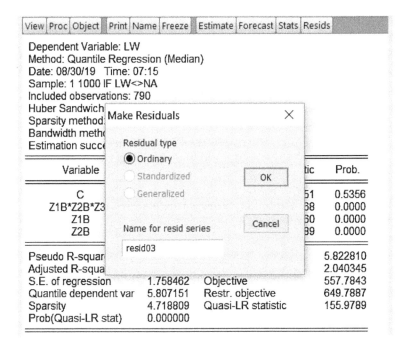

**Figure 6.19** The dialog for making residual series..

The following examples present the study on the differential between the fitted and forecast variables for QR. After forecasting based on all samples, and selected forecast sample from among all sample, I propose the basic forecasting and advanced forecasting in the following subsections because I found unexpected statistical results.

### 6.7.1 Basic Forecasting

#### Example 6.11 *Forecast for all Sample Points*

As an illustration, with the output in Figure 6.11a on-screen, select the *Forecast* button to obtain the options shown in Figure 6.20a. Then, using *lwf_all* as the forecast name with the forecast sample: 1 1000, and clicking *OK,* we obtain the output in Figure 6.20b. Based on this output, the following findings and notes are presented:

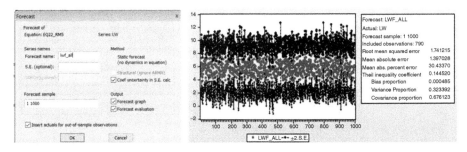

**Figure 6.20** The forecast options and their output.

1. Note the options that have been checked as the default options, mainly the Forecast evaluation, as follows (from the EViews 8 user guide):

   1) *Root Mean Squared Error (RMSE)* and *Mean Absolute Error (MAE)* are relative measures to compare forecasts for the same series across different models; the smaller the error, the better the forecasting ability of that model according to that criterion.

   2) *Mean Absolute Percent Error (MAPE)* and *Theil Inequality Coefficient (TIC)* are scale invariant. The Theil Inequality Coefficient (TIC) always lies between zero and one, where zero indicates a perfect forecast, and one is the worst forecast..

   3) In addition, the mean squared forecast error can be decomposed as *Bias Proportion (BP), Variance Proportion (VP),* and *Covariance Proportion (CP),* which have a total of one. The *BP* indicates how far the mean of the forecast is from the mean of the actual scores, the *VP* indicates how far the variance of the forecast is from the variance of the actual scores, and the *CP* measures the remaining unsystematic forecasting error.

   Agung (2019) has proposed a set of subjective classifications for the GoF of a forecast model based on its TIC. However, based on the QR, I propose the following specific classifications:

   i. TIC < 0.001 is a perfect forecast.
   ii. 0.001 ≤ TIC < 0.01 is an excellent forecast.
   iii. 0.01 ≤ TIC < 0.05 is a very good forecast.
   iv. 0.05 ≤ TIC < 0.10 is a good forecast.
   v. 0.10 ≤ TIC < 0.50 is an acceptable forecast.
   vi. TIC ≥ 0.50 is a miserable forecast.

2. For a clean presentation, Figure 6.21 presents a part of the graphs of *LW, LWF_ALL,* and *FITTED*, which shows the following characteristics:

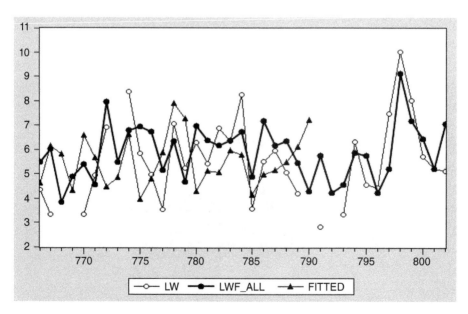

**Figure 6.21**   A part of the graphs of three variables *LW, LWF_ALL* and *FITTED*.

| Statistic | Common sample | | | Individual samples | | |
|---|---|---|---|---|---|---|
| | LW | LWF_ALL | FITTED | LW | LWF_ALL | FITTED |
| Mean | 5.8065 | 5.7839 | 5.8076 | 5.8228 | 5.6507 | 5.7844 |
| Median | 5.7799 | 5.8234 | 5.8217 | 5.8130 | 5.6609 | 5.8151 |
| Maximum | 11.2729 | 9.1495 | 9.1495 | 11.2729 | 9.1495 | 9.1495 |
| Minimum | 1.0400 | 2.4076 | 2.8358 | 1.0400 | 2.3089s | 2.4076 |
| Std. Dev. | 2.0273 | 1.0227 | 1.0439 | 2.0403 | 1.0652 | 1.0495 |
| Skewness | 0.1347 | −0.0331 | 0.1202 | 0.1554 | 0.1208 | 0.0792 |
| Kurtosis | 2.7155 | 2.8520 | 2.8676 | 2.6411 | 2.8988 | 2.9019 |
| Jarque-Bera | 3.9901 | 0.6837 | 1.9571 | 7.4199 | 2.8570 | 1.1433 |
| Probability | 0.1360 | 0.7104 | 0.3759 | 0.0245 | 0.2397 | 0.5646 |
| Sum | 3623.24 | 3609.14 | 3623.92 | 4600.02 | 5650.71 | 4569.71 |
| Sum Sq. Dev. | 2560.45 | 651.62 | 678.93 | 3284.61 | 1133.41 | 869.10 |
| Observations | 624 | 624 | 624 | 790 | 1000 | 790 |

**Figure 6.22** Descriptive statistics of the variables LW, LWF_ALL and FITTED.

2.1 Figure 6.22 presents an unexpected descriptive statistic of the three variables have only 624 observations, and based on the individual sample $LW$ and $FITTED$ have 790 observations, and 1000 observations for $LWF\_ALL$. *Why?* See the following explanations.

2.2 Note that there are no lines between two pairs of consecutive points of $LW$ within the first 790 observations, and the $FITTED$ graph has a complete points for the first 790 observations. There are 166 NAs for the $LW$ and $FITTED$ variables within the first 790 observations. To show this, I generated a dummy variable of the deviation of the two variables using the following equation, with its one-way tabulation presented in Figure 6.23, which clearly shows 166 missing observations (NAs):

$$D\_DEV = 1 * (LW - FITTED = NA) \qquad (6.17)$$

2.3 On the other hand, the forecast variable $LWF\_ALL$ has 1000 observations because the IVs of the regression function have 1000 complete observations. Hence, the 210 = (1000 − 790) missing observations (NAs) of $LW$ also have forecast values. Note the polygon graph of $LWF\_ALL$ is a complete graph with thick lines.

2.4 Note that the forecast values of $LWF\_ALL$, in fact, are not for the individuals in the sample, but the *predicted or abstract/unmeasured values* for 1000 individual out-of-sample observations having the same characteristics as the individuals in the sample, based on the IVs of the model. However, its graph is presented along with the graph of the variable $LW$. So the forecast values look like for the individuals

Tabulation of D_DEV
Date: 06/21/18   Time: 06:53
Sample: 1 1000 IF LW<>NA
Included observations: 790
Number of categories: 2

| Value | Count | Percent | Cumulative count | Cumulative percent |
|-------|-------|---------|------------------|--------------------|
| 0 | 624 | 78.99 | 624 | 78.99 |
| 1 | 166 | 21.01 | 790 | 100.00 |
| Total | 790 | 100.00 | 790 | 100.00 |

**Figure 6.23**   One-way tabulation of D_DEV.

in the sample. See the note *"Insert actuals for out-of-sample observations"* at the bottom of Figure 6.20a.

**Example 6.12**   *Forecast for a Forecast Sample: 1 25*
As an additional illustration, by inserting the *Forecast name*: LWF_25 and *Forecast sample*: 1 25, we obtain the output in Figure 6.24, which clearly shows 25 forecast values/points. Similar to the outputs in the previous Example 6.11, Figure 6.25 presents the graphs of *LW* and *LWF_25,* and Figure 6.26 presents the descriptive statistics of *LW, LWF_25,* and *FIT-TED,* for the common and individual samples. Based on these outputs, the following special findings and notes are presented:

1. The output in Figure 6.24 presents the *Forecast sample*: 1 25 with the *Included sample* = 20. It indicates that *LW* has five NAs within the first 25 observations, corresponding

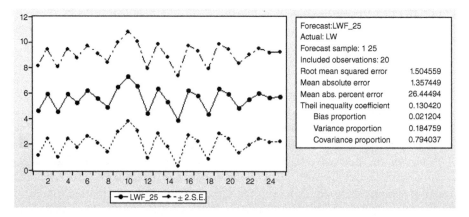

**Figure 6.24**   The output for the forecast sample 1 25.

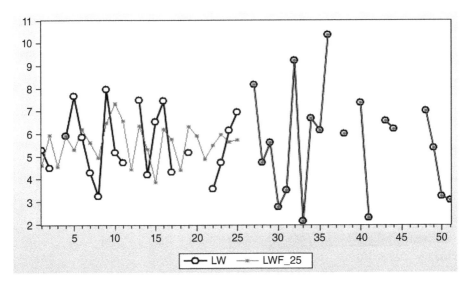

**Figure 6.25** A parth of the graphs of the variables *LW,* and *LWF_25.*

to its incomplete (broken) polygon graph in Figure 6.25. The forecast has RMSE = 1.505 and TIC = 0.1304, which are smaller than the *LWF_ALL* results.

2. The graphs in Figure 6.25 have the following characteristics:

2.1 Within the first 25 observations, there are two graphs of the variables *LW* and *LWF_25,* where the complete polygon graph is the forecast variable *LWF_25.* Afterward, the graphs of *LW* and *LWF_25* are presented as a single graph. Why? Unexpectedly, as there are no forecast values after the 25th observation, the graphs of *LW* and *LWF_25* are presented as a single graph. So that they are presented as broken polygon, representing the NAs of *LW*.

2.2 The descriptive statistics for *LW, LWF_25,* and *FITTED* in Figure 6.26 show 624 observations, based on the common sample, because the output shows the common sample without NAs, as shown previously in Figure 6.24. However, *LWF_25* has 795 observations, which is greater than the individual samples of *LW* and its fitted variable, because the forecast variable *LWF_25* has five values for the five NAs of *LW* within its first 25 observations. Note that the graph of *LW* has 20 points, and the graph of *LWF_25* presents points within the first 25 observations. So, I will present the following special section on advanced forecasting.

### 6.7.2 Advanced Forecasting

For doing this special forecasting, an additional variable, *ID,* should be generated using the following equation, with the scores from one to last (1000):

$$ID = @Trend + 1 \tag{6.18}$$

| Statistic | Common sample | | | Individual samples | | |
|---|---|---|---|---|---|---|
| | LW | LWF_25 | FITTED | LW | LWF_25 | FITTED |
| Mean | 5.8065 | 5.8135 | 5.8076 | 5.8228 | 5.8220 | 5.7844 |
| Median | 5.7799 | 5.7971 | 5.8217 | 5.8130 | 5.8072 | 5.8151 |
| Maximum | 11.2729 | 11.2729 | 9.1495 | 11.2729 | 11.2729 | 9.1495 |
| Minimum | 1.0400 | 1.0400 | 2.8358 | 1.0400 | 1.0400 | 2.4076 |
| Std. Dev. | 2.0273 | 2.0158 | 1.0439 | 2.0403 | 2.0270 | 1.0495 |
| Skewness | 0.1347 | 0.1289 | 0.1202 | 0.1554 | 0.1594 | 0.0792 |
| Kurtosis | 2.7155 | 2.7641 | 2.8676 | 2.6411 | 2.6863 | 2.9019 |
| Jarque-Bera | 3.9901 | 3.1744 | 1.9571 | 7.4199 | 6.6242 | 1.1433 |
| Probability | 0.1360 | 0.2045 | 0.3759 | 0.0245 | 0.0364 | 0.5646 |
| Sum | 3623.24 | 3627.62 | 3623.92 | 4600.02 | 4628.51 | 4569.71 |
| Sum Sq. Dev. | 2560.45 | 2531.64 | 678.93 | 3284.61 | 3262.31 | 869.10 |
| *Observations* | *624* | *624* | *624* | *790* | *795* | *790* |

**Figure 6.26** Statistic descriptive of the variables *LW, LWF_25*, and Fitted.

| Statistic | Common sample | | Individual samples | |
|---|---|---|---|---|
| | LW | LWF_25 | LW | LWF_25 |
| Mean | 5.5497 | 5.7688 | 5.5497 | 5.5794 |
| Median | 5.2204 | 5.8429 | 5.2204 | 5.7155 |
| Maximum | 7.9616 | 7.3222 | 7.9616 | 7.3222 |
| Minimum | 3.2439 | 3.8481 | 3.2439 | 3.8481 |
| Std. Dev. | 1.4222 | 0.7586 | 1.4222 | 0.8199 |
| Skewness | 0.2278 | −0.5474 | 0.2278 | −0.2164 |
| Kurtosis | 1.9074 | 3.8046 | 1.9074 | 2.6001 |
| Jarque-Bera | 1.1677 | 1.5381 | 1.1677 | 0.3617 |
| Probability | 0.5577 | 0.4634 | 0.5577 | 0.8345 |
| Sum | 110.99 | 115.38 | 110.99 | 139.49 |
| Sum Sq. Dev. | 38.4277 | 10.9350 | 38.4277 | 16.1326 |
| Observations | 20 | 20 | 20 | 25 |

**Figure 6.27** The descriptive statistics of *LW* and *LWF_25*, based on the sample {ID <26}.

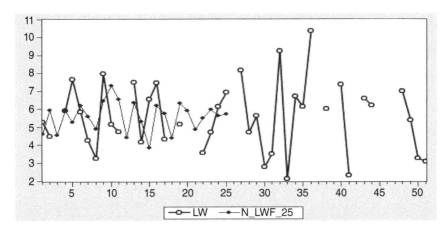

**Figure 6.28** Alternative graphs of the variables *LW* and *N_LWF_25*, as a better presentation of the graphs of *LW* and *LWF_25*.

### Example 6.13 *Modification of the Forecasting in Example 6.9*

Referring to the variable *LWF_25,* based on the forecast sample 1 25, Figure 6.27 presents the descriptive statistics of the variables *LW* and *LWF_25*, which show that *LW* and *LWF_25* have 20 observations, based on the common sample, and observations of 20 and 25, respectively, based on the individual samples. These numbers of observations are the same as the included observations and forecast sample 20 presented in the output of Figure 6.24. So, I would say that these outputs are more appropriate in representing the forecast values, with the forecast sample of 25.

In addition,we can develop alternative graphs of *LW* and *LWF_25* by using a new variable *N_LWF_25* = *LWF_25* for only *ID* < 26, as presented in Figure 6.28. Compared with the graphs in Figure 6.25, I would say that these graphs better represent the forecast values of *LWF_25* because they don't have the forecast values for *ID* > 25.

### Example 6.14 *An Application of the Forecast Sample 500 550*

As an additional illustration, Figure 6.29 presents the output of the forecasting of *LW* for the forecast sample 500 550, with *LWF_55* as its forecast name, and 42 included observations, which shows there are 8 (eight) NA observation scores.

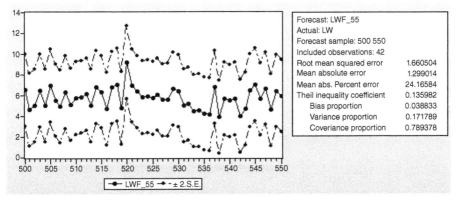

**Figure 6.29** The output for the forecast sample 500 550.

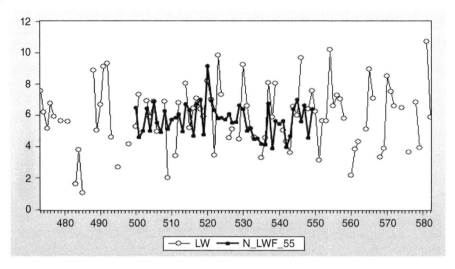

**Figure 6.30** An alternative graphs of the variables *LW* and *N_LWF_55*, as a better presentation of the graphs of *LW* and *N_LWF_55*.

Then, for the sample {ID >= 500 and ID < 550}, we can generate the new variable *N_LWF_55 = LWF_55* for developing the graphs of the variables *LW* and *LWF_55*, as presented in Figure 6.30.

## 6.8 Developing a Complete Data_LW.wf1

Data_LW.wf1 is a sub-data set of Mlogit.wf1, which can be developed using the following steps:

1. Select a sample for LW <> NA.
2. Show all variables which might will be used for alternative models of the variable *LW*.
3. Copy all variables to an Excel data file, Data_LW.xlsx.
4. Finally, Data_LW.xlsx can easily be transformed to a complete Data_LW.wf1 with 790 observations, as presented in Figure 6.31.

The following examples show that the RRT, OVT, and RVT present exactly the same values of the *QLR L*-statistic and *QLR Lambda*-statistic, based on the complete Data_LW.wf1.

**Example 6.15** *An Application of the Ramsey Test*
Having the output of the QR of *LW* on-screen, select *View/Stability Tests/Ramsey Test* and click *OK* to obtain the output that was presented in the Example 6.9. The hypothesis (6.16b), which also can be written as follows, is in fact a hypothesis for testing two nested models, which can be done using the OMT or RVT:

$$H_0 : Eq : LW\,C\,Z1B * Z2b * Z3B\,Z1B\,Z3B \tag{6.19}$$

$$H_1 : Eq : LW\,C\,Z1B * Z2B * Z3B\,Z1B\,Z3B\,Fitted\hat{\ }2$$

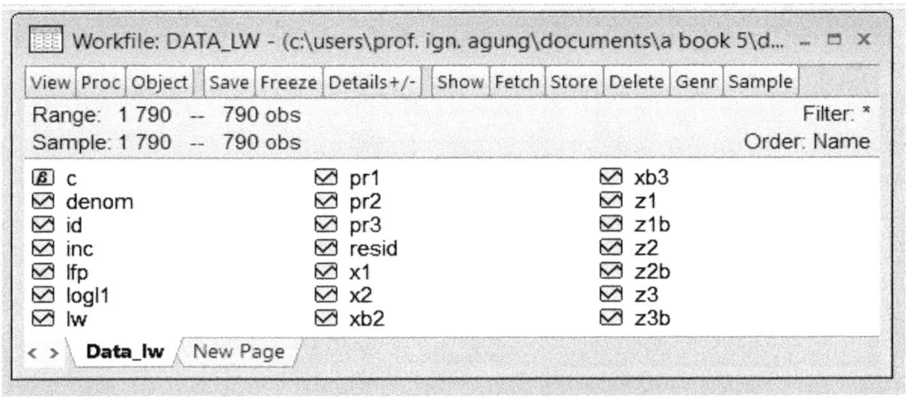

**Figure 6.31** A complete Data_LW.wf1 having most of the variables in Mlogit.wf1.

**Figure 6.32** The OMT of Fitted^2 based on the reduced model in $H_0$ (6.19).

However, in this case, we don't have yet the variable *FITTED* in Data_LW.wf1. For this reason, we have to generate the variable *FITTED* by using the output of the model in $H_0$. Then we perform the following tests.

### Example 6.16 *An Application of the Omitted Variables Test*

With the output of the equation "*LW C Z1B\*Z2B\*Z3B Z1B Z3B*" on-screen, we can select *View/Coefficient Diagnostics/Omitted Variables – Likelihood Ratio …*, insert *FITTED^2*, and click *OK* to obtain the output in Figure 6.32, which shows that the *QLR L*-statistic and *QLR Lambda*-statistics have the same values as the RRT in Example 6.9.

### Example 6.17 *Application of the Redundant Variables Test*

With the output of the equation "*LW C Z1B\*Z2b\*Z3B Z1B Z3B Fittted^2*" on-screen, select *View/Coefficient Diagnostics/Redundant Variables – Likelihood Ratio...*, insert *FITTED^2*,

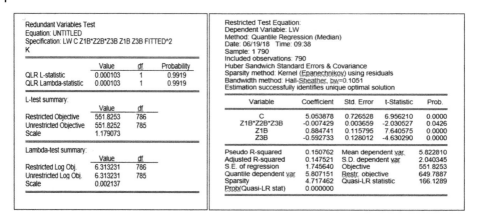

**Figure 6.33** The RVT of FITTED^2 based on the full model in (6.19).

and click *OK* to obtain the output in Figure 6.33, which shows that the *QLR L*-statistic and *QLR Lambda*-statistics have the same values as the RRT and OMT.

## 6.9 QRs with Four Numerical Predictors

As an extension of the models based on *(X1,X2,X3,Y1)*, we can have a lot of possible models of *Y1* on *(X1,X2,X3,X4)*. The F3WI-QR would have a total of 14 IVs, consisting of four three-way interactions, six two-way interactions, and four main variables. However, since those 14 variables are highly or significantly correlated, we generally would have an unexpected a Reduced-QR as an acceptable regression, in the statistical sense. In other words, researchers would never apply the F3WI-QR because most or all of its IVs would have large or very large probabilities, or stated differently they would have an insignificant adjusted effect with large *p*-values. For this reason, based on a specific set of five variables, I recommend generating a path diagram, which is accepted as the true up-and-down or causal relationship, in the theoretical sense. Then an ES of the model can be developed, based on the path diagram. Even so, it is likely that the statistical results would still present an unexpected regression function because of the uncontrollable multicollinearity or bivariate correlations between the IVs. These have been illustrated in previous examples and will be again in the following examples.

### 6.9.1 An Additive QR

The additive model with the following ES is the simplest model in a five-dimensional space. The graph of its regression function is called a *hyperplane*.

$$Y1 \ C \ X1 \ X2 \ X3 \ X4 \tag{6.20}$$

**Table 6.7** Statistical results summary of the full additive QR(Median) in (6.21) and three of its reduced models.

| Full-AQR | C | X1 | Z1 | Z2 | Z3 | Ps. $R^2$ | Adj. $R^2$ | Quasi-LR | Prob. |
|---|---|---|---|---|---|---|---|---|---|
| Coef. | 5.2132 | −0.8069 | −0.0096 | −0.0654 | −0.0729 | 0.0028 | **−0.0012** | 3.0008 | 0.5577 |
| t-stat. | 18.131 | −1.5542 | −0.0690 | −0.4343 | −0.4959 | | | | |
| Prob. | 0.0000 | 0.1205 | 0.9450 | 0.6642 | 0.6201 | | | | |
| **RM-1** | C | X1 | | Z2 | Z3 | Ps. $R^2$ | Adj. $R^2$ | Quasi-LR | Prob. |
| Coef. | 5.2288 | −0.8298 | | −0.0567 | −0.0680 | 0.0028 | **−0.0002** | 2.9942 | 0.3925 |
| t-stat. | 18.264 | −1.6009 | | −0.3784 | −0.4800 | | | | |
| Prob. | 0.0000 | 0.1097 | | 0.7052 | 0.6313 | | | | |
| **RM-2** | C | X1 | | | Z3 | Ps. $R^2$ | Adj. $R^2$ | Quasi-LR | Prob. |
| Coef. | 5.2216 | −0.7761 | | | −0.0628 | 0.0024 | 0.0004 | 2.5256 | 0.2829 |
| t-stat. | 18.069 | −1.4811 | | | −0.4396 | | | | |
| Prob. | 0.0000 | 0.1389 | | | 0.6603 | | | | |
| **RM-3** | C | X1 | | | | Ps. $R^2$ | Adj. $R^2$ | Quasi-LR | Prob. |
| Coef. | 5.2520 | −0.8010 | | | | 0.0023 | 0.0013 | 2.3841 | 0.1226 |
| t-stat. | 18.185 | −1.5356 | | | | | | | |
| Prob. | 0.0000 | 0.1250 | | | | | | | |

**Example 6.18   *Applications of an Additive Model***
Table 6.7 presents the statistical results summary of a full additive quantile-regression (Full-AQR), with the following ES, and three of its reduced models, namely RM-1, RM-2, and RM-3 for the data in Mlogit.wf1:

$$INC\ C\ X1\ Z1\ Z2\ Z3 \tag{6.21}$$

Based on these results, the following findings and notes are presented:

1. Unexpectedly, each of the IVs of the four models and the joint effects of the IVs have insignificant effects, at the 10% level of significance. Why? Refer to a similar case presented in Example 5.11.
   At the 10% level, only the variable *X1* has a significant negative (adjusted) effect on *INC* for the four models. For instance, based on the full additive model, *X1* has a significant negative adjusted effect, based on the *t*-statistic of $t_0 = −1.5542$ with a *p*-value $= 0.1205/2 = 0.06025 < 0.10$.
2. Both Full-AQR and RM-1 have negative adjusted *R*-squared. Hence, they are poor models. The negative values of the adjusted *R*-squared, and both RM-2 and RM-3, which have very small positive adjusted *R*-squareds, are highly dependent on the characteristics of the data, which should be accepted under the assumption the data are valid. (Refer to the scatter graphs in Figure 5.16 for an illustration.) So, they should be considered as acceptable QRs, in the statistical sense.

3. However, the four models are not comparable additive models, because they are different QRs with different sets of IVs. Compare this with the models RM-1, RM-2, and RM-3 in Table 5.7, which have the same set of variables. Hence, in this case, you should select the one to apply..

4. Note that RM-1 is obtained from the Full-AM by deleting the IV *Z1* having the greatest probability. Hence, RM-1 is not representative of the *INC* model on *X1, Z1, Z2,* and *Z3*, it is the QR of *INC* on *X1, Z2,* and *Z3*. Similarly RM-2 is the QR of *INC* on *X1* and *Z3*, and RM-3 is the QR of *INC* only on *X1*.

### Example 6.19   *Applications of an Alternative Additive Model*

As a comparison with the statistical results of the QR (6.21), Figure 6.34 presents the statistical results of an additive QR(Median), with the following ES, with its reduced model.

$$LW\ C\ X1\ Z1B\ Z2B\ Z3B \tag{6.22}$$

Based on these results, the following findings and notes are presented:

1. The IVs of both QRs have significant joint effects on *LW*, and each of the positive *Z*-variables, *Z1B, Z2B,* and *Z3B,* has an significant effect, at a 1% or 5% level of significance.

2. The QR (6.22) is a good fit additive model in five-dimensional space, even though *X1* has an insignificant adjusted effect, because of the other IVs has a significant effect, at a 1% or 5% level of significance. Its IVs have significant joint effects, based on the *Quasi-LR* statistic of 168.2235 with a *p*-value = 0.0000.

3. And the QR in Figure 6.34b is the best fit additive model in a four-dimensional space because each of all its IVs has a significant effect, at a 1% or 5% level, and its IVs have significant joint effects, based on the *Quasi-LR* statistic of 168.1694 with a *p*-value = 0.0000.

4. In addition, note that the QR in Figure 6.34a has an adjusted *R*-squared of 0.147 298, which is a little smaller than the 0.148 365 of the QR in Figure 6.33b. So, in the statistical sense, the variables *Z1, Z2,* and *Z3* are better predictors than the four variables *X1, Z1, Z2,* and *Z3* for the median of *LW*.

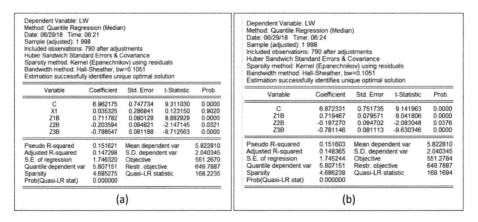

**Figure 6.34**   Statistical results of the additive QR(Median) in (6.22), and its reduced model..

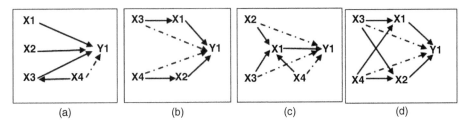

**Figure 6.35** Alternative path diagrams based on (*X1,X2,X3, X4,Y1*).

## 6.9.2 Alternative Two-Way Interaction QRs

Figure 6.35 presents only four alternative path diagrams, as an extension of the path diagrams in Figure 6.1, to present alternative 2WI-QRs. Note that the arrows with dotted lines represent the up-and-down relationships between the corresponding pairs of the variables, and the true causal effects are indicated by arrows with solid lines. For instance, in Figure 6.35a, *X4* has an indirect effect on *Y1* through *X3*, and then *X4* is an upper variable of *Y1*.

### 6.9.2.1 A Two-Way Interaction QR Based on Figure 6.35a
In Figure 6.35a, each of the variables, *X1, X2,* and *X3,* is defined to have a true direct effect on *Y1,* and *X4* has a true direct effect only on *X3*. The QR is the simplest 2WI_QR, with the following ES:

$$Y1\ C\ X3 * X4\ X1\ X2\ X3\ X4 \tag{6.23}$$

Note that two other equally simple QRs can be defined, whenever *X4* is defined to have a direct effect only on *X1* or *X2*.

### 6.9.2.2 A Two-Way Interaction QR Based on Figure 6.35b
In Figure 6.35b, both *X1* and *X2* are defined to have a true direct effect on *Y1, X3* has a true direct effect on *X1,* and *X4* has a true direct effect on *X2*. The QR is a 2WI-QR with two two-way interaction IVs and the following ES:

$$Y1\ C\ X1 * X3\ X2 * X4\ X1\ X2\ X3\ X4 \tag{6.24}$$

### 6.9.2.3 A Two-Way Interaction QR Based on Figure 6.35c
In Figure 6.35c, *X1* is defined to have a true direct effect on *Y1,* and X2, *X3,* and *X4* all have a true direct effect on *X1*. The QR is a 2WI-QR with three two-way interaction IVs and the following ES:

$$Y1\ C\ X1 * X2\ X1 * X3\ X1 * X4\ X1\ X2\ X3\ X4 \tag{6.25}$$

### 6.9.2.4 A Two-Way Interaction QR Based on Figure 6.35d
In Figure 6.35d, *X1* and *X2* are defined to have a true direct effect on *Y1,* and *X3* and *X4* have true direct effects on both *X1* and *X2*. The QR is a 2WI-QR with four two-way interaction IVs and the following ES:

$$Y1\ C\ X1 * X3\ X1 * X4\ X2 * X3\ X2 * X4\ X1\ X2\ X3\ X4 \tag{6.26}$$

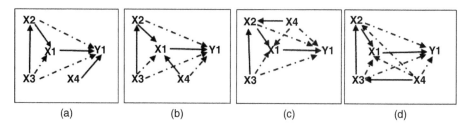

(a)                    (b)                    (c)                    (d)

**Figure 6.36** Alternative advanced path diagrams based on $(X1,X2,X3, X4,Y1)$.

### 6.9.3 Alternative Three-Way Interaction QRs

Figure 6.36 provides an extension of the path diagram in Figure 6.35 to represent alternative 3WI-QRs. It presents only four advanced path diagrams to represent four alternative 3WI-QRs, having possible three-way interaction IVs.

Note that in these path diagrams, the exogenous variable $X4$ is defined to have a true direct effect on only one of the variables, $X1, X2, X3$, or $Y1$. These path diagrams could be extended to the path diagrams where $X4$ has true direct effects on two or more variables.

In addition, the causal relationships between $X1, X2,$ and $X3$ can be modified or extended to the triangular and circular causal effects shown in Figure 6.1c, d.

Because we would have so many possible path diagrams, however, we should have expert judgement for selecting the best possible path diagram, which is assumed to be true in the theoretical sense.

#### 6.9.3.1 Alternative Models Based on Figure 6.36a

As an extension of the model (6.1a), we can have a 2WI-QR and a 3WI-QR, based on the path diagram in Figure 6.36a, as follows:

(i) The 2WI-QR has the following ES:

$$Y1 \, C \, X1 * X2 \, X1 * X3 \, X2 * X3 \, X1 \, X2 \, X3 \, X4 \tag{6.27}$$

where the two-way interactions represent the consecutive up-and-down or causal effects: $X2 \rightarrow X1 \rightarrow Y1, X3 \rightarrow X1 \rightarrow Y1,$ and $X3 \rightarrow X2 \rightarrow Y1$, respectively.

(ii) The three-way interaction is an extension of the 2W-QR in (6.27), with the following ES:

$$Y1 \, C \, X1 * X2 * X3 \, X1 * X2 \, X1 * X3 \, X2 * X3 \, X1 \, X2 \, X3 \, X4 \tag{6.28}$$

where the three-way interactions represent the true consecutive causal effects $X3 \rightarrow X2 \rightarrow X1 \rightarrow Y1$, as it was defined.

#### 6.9.3.2 Alternative Models Based on Figure 6.36b

Similar to the two models based on the path diagram in Figure 6.36a, with this path diagram, we have the following 2WI-QR and 3WI-QR:

(i) The 2WI-QR has the following ES:

$$Y1 \, C \, X1 * X2 \, X1 * X3 \, X1 * X4 \, X2 * X3 \, X1 \, X2 \, X3 \, X4 \tag{6.29}$$

where the two-way interactions represent the consecutive up-and-down or causal effects: $X2 \rightarrow X1 \rightarrow Y1, X3 \rightarrow X1 \rightarrow Y1, X4 \rightarrow X1 \rightarrow Y1,$ and $X3 \rightarrow X2 \rightarrow Y1,$ respectively.

(ii) The 3WI-QR is an extension of the 2WI-QR in (6.29) and has the following ES:

$$Y1\ C\ X1 * X2 * X3\ X1 * X2\ X1 * X3\ X1 * X4\ X2 * X3\ X1\ X2\ X3\ X4 \qquad (6.30)$$

where the three-way interactions represent the true consecutive causal effects $X3 \rightarrow X2 \rightarrow X1 \rightarrow Y1,$ as it was defined.

### 6.9.3.3 Alternative Models Based on Figure 6.36c

As a modification of the 2WI-R in (6.29), with this path diagram, we have the following 2WI-QR and 3WI-QR:

(i) The 2WI-QR has the following ES:

$$Y1\ C\ X1 * X2\ X1 * X3\ X1 * X4\ X2 * X3\ X2 * X4\ X1\ X2\ X3\ X4 \qquad (6.31)$$

where the two-way interactions represent the consecutive up-and-down or causal effects: $X2 \rightarrow X1 \rightarrow Y1, X3 \rightarrow X1 \rightarrow Y1, X4 \rightarrow X1 \rightarrow Y1, X3 \rightarrow X2 \rightarrow Y1,$ and $X4 \rightarrow X2 \rightarrow Y1,$ respectively.

(ii) The 3WI-QR is an extension tension of the 2WI-QR in (6.31), with the following ES:

$$Y1\ C\ X1 * X2 * X3\ X1 * X2 * X4\ X1 * X2\ X1 * X3\ X1 * X4$$
$$X2 * X3\ X2 * X4\ X1\ X2\ X3\ X4 \qquad (6.32)$$

where the three-way interactions represent the true consecutive causal effects $X3 \rightarrow X2 \rightarrow X1 \rightarrow Y1,$ as it was defined, and $X4 \rightarrow X2 \rightarrow X1 \rightarrow Y1,$ because it was defined that $X4$ has a direct effect on $X2$.

### 6.9.3.4 Alternative Models Based on Figure 6.36d

As a modification of the 2WI-QR in (6.31), with this path diagram, we have the following 2WI-QR and a 3WI-QR:

(i) The 2WI-QR has the following ES:

$$Y1\ C\ X1 * X2\ X1 * X3\ X1 * X4\ X2 * X3\ X2 * X4\ X3 * X4\ X1\ X2\ X3\ X4 \qquad (6.33)$$

where the two-way interactions represent the consecutive up-and-down or causal effects: $X2 \rightarrow X1 \rightarrow Y1, X3 \rightarrow X1 \rightarrow Y1, X4 \rightarrow X1 \rightarrow Y1, X3 \rightarrow X2 \rightarrow Y1, X4 \rightarrow X2 \rightarrow Y1,$ and $X4 \rightarrow X3 \rightarrow Y1,$ respectively. Note that this model has all possible two-way interactions based on the four variables $X1, X2, X3,$ and $X4,$ so that this model is called the F2WI-QR.

(ii) The 3WI-QR is an extension of the 2WI-QR in (6.33), with the following ES:

$$Y1\ C\ X1 * X2 * X3\ X1 * X2 * X4\ X1 * X3 * X4\ X1 * X2\ X1 * X3\ X1 * X4$$
$$X2 * X3\ X2 * X4\ X3 * X4\ X1\ X2\ X3\ X4 \qquad (6.34)$$

where the three-way interactions represent the true consecutive causal effects: $X3 \rightarrow X2 \rightarrow X1 \rightarrow Y1, X4 \rightarrow X2 \rightarrow X1 \rightarrow Y1,$ and $X4 \rightarrow X2 \rightarrow X1 \rightarrow Y1,$ respectively. Note that this model has only three of the four possible three-way interactions, based on $X1, X2, X3,$ and $X4.$

**Table 6.8** Statistical results summary of the manual multistage and trial-and error selection methods, based on the model (6.33).

| RM-1 | C | X1*Z1B | X1*Z2B | X1*Z3B | | | | |
|---|---|---|---|---|---|---|---|---|
| Coef. | 5.8378 | 1.0509 | −0.2041 | −0.934 | | | | |
| t-stat. | 33.371 | 9.0169 | −1.6243 | −7.3125 | | | | |
| Prob. | 0 | 0 | 0.1047 | 0 | | | | |
| | PS. $R^2$ = 0.1065 | | Adj. $R^2$ = 0.1031 | | Quasi-LR =112.96 | | Prob. = 0.0000 | |

| RM-2 | C | X1*Z1B | X1*Z2B | X1*Z3B | Z1B*Z2B | Z1B*Z3B | |
|---|---|---|---|---|---|---|---|
| Coef. | 6.2336 | 1.1483 | −0.8224 | −0.3547 | 0.0679 | −0.0919 | |
| t-stat. | 17.212 | 8.5237 | −3.6962 | −1.5726 | 2.6177 | −3.5433 | |
| Prob. | 0 | 0 | 0.0002 | 0.1162 | 0.009 | 0.0004 | |
| | PS. $R^2$ = 0.1177 | | Adj. $R^2$ = 0.112 | | Quasi-LR = 128.96 | | Prob. = 0.0000 |

| RM-3 | C | X1*Z1B | X1*Z2B | X1*Z3B | Z1B*Z2B | Z1B*Z3B | Z2B*Z3B |
|---|---|---|---|---|---|---|---|
| Coef. | 6.4112 | −0.0539 | −0.0204 | 0.0762 | 0.1242 | 0.0146 | −0.1719 |
| t-stat. | 16.44 | −0.2065 | −0.0696 | 0.3051 | 4.2329 | 0.4268 | −5.392 |
| Prob. | 0 | *0.8364* | *0.9445* | *0.7604* | 0 | *0.6696* | 0 |
| | PS. $R^2$ = 0.1443 | | Adj. $R^2$ = 0.1378 | | Quasi-LR = 159.18 | | Prob. = 0.0000 |

| RM-4 | C | X1*Z1B | X1*Z2B | X1*Z3B | Z1B*Z2B | Z1B*Z3B | | X1 |
|---|---|---|---|---|---|---|---|---|
| Coef. | 5.7102 | 1.0105 | −0.9012 | −0.4583 | 0.0754 | −0.0804 | | 1.6551 |
| t-stat. | 7.6108 | 3.7491 | −2.9504 | −1.7786 | 2.3426 | −2.8123 | | 0.6099 |
| Prob. | 0 | 0.0002 | 0.0033 | 0.0757 | 0.0194 | 0.005 | | **0.5421** |
| | PS. $R^2$ = 0.1182 | | Adj. $R^2$ = 0.1114 | | Quasi-LR = 128.68 | | Prob. = 0.0000 | |

**Example 6.20**   *An Application of a F2WI-QR in (6.33)*

As an illustration, Table 6.8 presents a statistical summary based on the following F2WI-QR, using the manual multistage and the trial-and-error selection methods. Based on this summary, the following findings and notes are presented:

$$LW \ C \ X1 * Z1B \ X1 * Z2B \ X1 * Z3B$$

$$Z1B * Z2B \ Z1B * Z3B \ Z2B * Z3B \ X1 \ Z1B \ Z2B \ Z3B \tag{6.35}$$

In this case, the true consecutive causal effects are defined as $Z3B \rightarrow Z2B \rightarrow Z1B \rightarrow X1 \rightarrow LW$, similar to the path diagram in Figure 6.36c. Hence, the main objective of this model is to test the hypothesis on the effect of *X1* depending on *Z1B, Z2B,* and *Z3B*.

1. The RM-1 and RM-2 are the first two stages of the regression analyses and are acceptable 2WI-QRs for the following reasons. At the 10% level, *X1*Z2B* in RM-1 has a

significant negative adjusted effect on *LW*, based on the *t*-statistic of $t_0 = -1.6243$ with a *p*-value $= 0.1047/2 < 0.10$. *X1*X3B* in RM-2 has a significant negative adjusted effect on *LW*, based on the *t*-statistic of $t_0 = -1.5726$ with a *p*-value $= 0.1162/2 < 0.10$, and each of the other IVs has a significant effect at the 1% level.

2. At the third stage of analysis, we obtain RM-3, with four IVs having very large probabilities. This indicates that *Z2B*Z3B* does not need to be inserted as an additional IV for RM-2.

3. Hence, at the final stage, the main variables *X1, Z1B, Z2B,* and *Z3B* are selected for the additional IVs of RM-2, using the trial-and-error method. We find only *X1* can be inserted as an IV, even though it has a large *p*-value $= 0.5421$, because each of its two-way interaction IVs has either a positive or negative significant effect, at the 1% or 5% level. For example, *X1*Z3B* has a negative significant adjusted effect on *LW,* based on the *t*-statistic of $t_0 = -1.7786$ with a *p*-value $= 0.0757/2 = 0.037\,85 < 0.05$.

4. Compared with the models in Table 6.7, the four models in this table have the same set of IVs, namely *X1, Z1B, Z2B,* and *Z3B*. So, they are comparable models, in both theoretical and statistical senses, to represent the QRs of *LW* on the four numerical predictors. Unexpectedly, the RM-3 has the greatest adjusted *R*-squared of 0.1378, so it is the best fit QR, in the statistical sense, even though four of its IVs have very large probabilities. However, RM-3 isn't suited to represent that the effect of *X1* depends on *Z1B, Z2B,* and *Z3B*.

5. Referring to the additive QR of *LW* on *X1, Z1B, Z2B,* and *Z3B* in Figure 6.34a, because it has the adjusted *R*-squared of 0.147 298, we can conclude that the additive model is a better fit than the four 2WI-QRs, in the statistical sense. However, note that the objective of an additive model is to study or test the effect of an IV, adjusted for the other IVs. On the other hand, an interaction QR is for studying or testing the effect of an IV being dependent on a least one of the other IVs. So, the additive model should not be compared to the interaction QR.

6. As an additional analysis, Figure 6.37 presents two RVTs for testing the effect of *X1* being dependent on *Z1B, Z2B,* and *Z3B*, based on the models RM-3 and RM-4, respectively. Based on these results: I present the following findings and notes:

   6.1 Figure 6.37a shows the RVT for the four IVs of RM-4, *X1*Z1B, X1*Z2B, X1*Z3B,* and *X1*, which shows that the effect of *X1* is significantly dependent on a linear function of *Z1B, Z2B,* and *Z2B*, namely *(C(2)*Z1B + C(3)*Z2B + C(4)*Z3B + C(7))*, where C(2), C(3), C(4), and C(7) are the parameters of the RM-4. Whenever *X1* is deleted from

Redundant Variables Test
Equation: UNTITLED
Specification: LW C X1*Z1B X1*Z2B X1*Z3B Z1B*Z2B Z1B*Z3B X1
RVT for: X1*Z1B X1*Z2B X1*Z3B X1

| | Value | df | Probability |
|---|---|---|---|
| QLR L-statistic | 91.48901 | 4 | 0.0000 |
| QLR Lambda-statistic | 87.38843 | 4 | 0.0000 |

(a). RVT based on RM-4

Redundant Variables Test
Equation: UNTITLED
Specification: LW C X1*Z1B X1*Z2B X1*Z3B Z1B*Z2B Z1B*Z3B Z2B*Z3B
RVT for: X1*Z1B X1*Z2B X1*Z3B

| | Value | df | Probability |
|---|---|---|---|
| QLR L-statistic | 0.129915 | 3 | 0.9880 |
| QLR Lambda-statistic | 0.129906 | 3 | 0.9880 |

(b). RVT based on RM-3

**Figure 6.37** Redundant ariables Tests for testing the effect of *X1* depends on *Z1B, Z2B* & *Z3B*. (a). RVT based on RM-4. (b). RVT based on RM-3.

RM-4, we have the RM-2, which has a slight greater adjusted $R$-squared than RM-4. So, both RM-2 and RM-4 are acceptable models to present the effect of $X1$ depending on $Z1B$, $Z2B$, and $Z3B$.

6.2 On the other hand, Figure 6.37b presents the RVT for the three IVs of RM-3, $X1*Z1B$, $X1*Z2B$, and $X1*Z3B$, which shows that the effect of $X1$ is insignificantly dependent on a linear function of $Z1B$, $Z2B$, and $Z2B$, namely $(C(2)*Z1B + C(3)*Z2B + C(4)*Z3B)$. So, RM-3 isn't an appropriate model to show the main objective of the analysis.

**Example 6.21** *An Application of a Special Manual Multistage Selection Method*
As an extension of the 2WI-QR, Table 6.9 presents a statistical results summary based on the following F3WI-QR, using a special manual multistage and the trial-and-error selection methods, which is different than I have presented in previous examples. Based on this summary, the following findings and notes are presented:

$$LW\ C\ X1*Z1B*Z2B\ X1*Z2B*Z3B\ Z1B*Z2B*Z3B\ X1*Z1B\ X1*Z2B$$

$$X1*Z3B\ Z1B*Z2B\ Z1B*Z3B\ Z2B*Z3B\ X1\ Z1B\ Z2B\ Z3B\quad (6.36)$$

1) At the first stage, RM-1 has only the three-way interaction IVs. At the 10% level, $Z1B*Z2B*Z3B$ has a negative significant effect, based on the $t$-statistic of $t_0 = -1.5489$ with a $p$-value $= 0.1218/2 = 0.0609 < 0.10$. As usual, if each of the IVs has *Prob.* $<0.30$ at the first stage of analysis, the model does not need to be reduced, because, at the 15% level, it will have either a positive or negative significant effect with a $p$-value $= Prob./2 < 0.15$.

2) At the second stage, three two-way interactions, $X1*Z1B$, $X1*Z2B$, and $X1*Z3B$, are inserted as additional IVs for RM-1. Unexpectedly, $Z1B*Z2B*Z3B$ has a small *Prob.* $= 0.0303$, but $X1*Z1B*Z2B$ has a very large *Prob.* $= 0.5460$. These results reflect the impact of the multicollinearity or bivariate correlations between the IVs. In this case, the RM-2 isn't an acceptable QR to present a 3WI-QR.

Usually, we have to select one or two of the two-way interactions as additional IVs for RM-1 to keep the three-way interactions with small $p$-values.

3) However, here we apply a special manual multistage selection method, by deleting $X1*Z1B*Z2B$ from RM-2, because I am very confident that $X1*Z1B*Z2B$ can be represented by both two-way interactions, $X1*Z1B$ and $X1*Z2B$. Then, as expected, we obtain RM-3 with each IV having either positive or negative significant effects at the 5% level. For instance, the interaction $X1*21B*Z3B$, with the largest probability, has a significant positive adjusted effect, based on the $t$-statistic of $t_0 = 1.6660$ with a $p$-value $= 0.0961/2 < 0.05$.

4) At the fourth stage, we have to select $Z1B*Z2B$, $Z1B*Z3B$, and $Z2B*Z3B$, as additional IVs for RM-3. We obtain an acceptable RM-4 with each of the IVs having either a positive or negative adjusted effect at the 1%, 5%, or 10% level.

5) At the fifth stage, we select one of the main variables, $X1$, $Z1B$, $Z2B$, and $Z3B$, as an additional IV for RM-4. We find that only $X1$ can be used as additional IV, but we obtain RM-5 with one of the interactions, $X1*Z1B$, having a very large *Prob.* $= 0.7904$. Then we have two alternative models as follows:

**Table 6.9** Statistical results summary of a special manual multistage method and the trial-and error selection method, based on the model (6.28).

| RM-1 | C | X1*Z1B*Z2B | X1*Z2B*Z3B | Z1B*Z2B*Z3B | | | |
|---|---|---|---|---|---|---|---|
| Coef. | 6.0199 | 0.1891 | −0.1898 | −0.0035 | | | |
| t-stat. | 24.144 | 9.6496 | −9.6442 | −1.5489 | | | |
| Prob. | 0.0000 | 0.0000 | 0.0000 | 0.1218 | | | |
| Ps. $R^2$ | 0.101906 | Adj. $R^2$ | 0.098478 | Quasi-LR | 110.12 | Prob. | 0 |

| RM-2 | C | X1*Z1B*Z2B | X1*Z2B*Z3B | Z1B*Z2B*Z3B | X1*Z1B | X1*Z2B | X1*Z3B |
|---|---|---|---|---|---|---|---|
| Coef. | 6.8036 | −0.0787 | 0.2228 | −0.0090 | 1.7307 | −0.6943 | −1.8051 |
| t-stat. | 13.247 | −0.6040 | 1.4488 | −2.1694 | 2.4410 | −1.9055 | −2.3288 |
| Prob. | 0.0000 | **0.5460** | 0.1478 | 0.0303 | 0.0149 | 0.0571 | 0.0201 |
| Ps. $R^2$ | 0.114157 | Adj. $R^2$ | 0.107369 | Quasi-LR | 123.94 | Prob. | 0 |

| RM-3 | C | | X1*Z2B*Z3B | Z1B*Z2B*Z3B | X1*Z1B | X1*Z2B | X1*Z3B |
|---|---|---|---|---|---|---|---|
| Coef. | 6.8212 | | 0.1603 | −0.0092 | 1.3950 | −0.7553 | −1.4838 |
| t-stat. | 14.053 | | 1.6660 | −2.3553 | 6.4557 | −2.1608 | −4.0742 |
| Prob. | 0.0000 | | 0.0961 | 0.0188 | 0.0000 | 0.0310 | 0.0001 |
| Ps. $R^2$ | 0114157 | Adj. $R^2$ | 0.107369 | Quasi-LR | 123.94 | Prob. | 0 |

| RM-4 | C | | X1*Z2B*Z3B | Z1B*Z2B*Z3B | X1*Z1B | X1*Z2B | X1*Z3B | Z1B*Z2B | Z1B*Z3B |
|---|---|---|---|---|---|---|---|---|---|
| Coef. | 3.9999 | | 0.1436 | −0.0491 | 0.5327 | −0.6133 | −0.6971 | 0.2007 | 0.1142 |
| t-stat. | 5.2161 | | 1.4757 | −4.8311 | 2.0097 | −1.6748 | −1.6325 | 5.1495 | 2.3790 |
| Prob. | 0.0000 | | 0.1404 | 0.0000 | 0.0448 | 0.0944 | 0.1030 | 0.0000 | 0.0176 |
| Ps. $R^2$ | 0.137706 | Adj. $R^2$ | 0.129987 | Quasi-LR | 152.433 | Prob. | 0 | | |

**Table 6.9** (Continued)

| RM-1 | C | X1*Z1B*Z2B | X1*Z2B*Z3B | Z1B*Z2B*Z3B |
|---|---|---|---|---|

**RM-5**

| | C | X1*Z2B*Z3B | ZB-Z28-Z38 | X1*Z1B | X1*Z2B | X1*Z3B | Z1B*Z2B | Z1B*Z3B | Z1B*Z4B |
|---|---|---|---|---|---|---|---|---|---|
| Coef. | 2.2560 | 0.7817 | -0.0693 | 0.0810 | -3.7672 | -3.7886 | 0.2963 | 0.1846 | 17.7971 |
| t-stat. | 2.2681 | 5.3086 | -6.4166 | 0.2659 | -5.5104 | -4.7352 | 6.9290 | 3.5323 | 4.1030 |
| Prob. | 0.0236 | 0.0000 | 0.0000 | **0.7904** | 0.0000 | 0.0000 | 0.0000 | 0.0000 | 0.0000 |

Ps. $R^2$ 0.148926 | Adj. $R^2$ 0.140208 | Quasi-LR 172.4909 | Prob. 0

**RM 6**

| | C | X1*Z2B*Z3B | Z1B*Z2B*Z3B | X1*Z2B | X1*Z3B | Z1B*Z2B | Z1B*Z3B | X1 | Z3B |
|---|---|---|---|---|---|---|---|---|---|
| Coef. | 7.2548 | 0.4108 | -0.0334 | -1.8182 | -1.8309 | 0.1158 | 0.1981 | 8.0522 | -1.0800 |
| t-stat. | 3.5310 | 1.9538 | -1.8719 | -1.8477 | -1.6244 | 1.5053 | 4.6869 | 1.5039 | -2.4639 |
| Prob. | 0.0004 | 0.0511 | 0.0616 | 0.0650 | 0.1047 | 0.1327 | 0.0000 | 0.1330 | 0.0140 |

Ps. $R^2$ 0.153562 | Adj. $R^2$ 0.144892 | Quasi-LR 175.165 | Prob. 0

**RM-7**

| | C | X1*Z2B*Z3B | Z1B*Z2B*Z3B | X1*Z2B | X1*Z3B | Z1B*Z2B | Z1B*Z3B | X1 | Z2B |
|---|---|---|---|---|---|---|---|---|---|
| Coef. | 2.8100 | 0.7379 | -0.0673 | -3.4125 | -3.5744 | 0.2945 | 0.1810 | 16.4291 | -0.1330 |
| t-stat. | 1.1824 | 3.9032 | -3.7478 | -3.6779 | -3.4501 | 10.1735 | 1.9753 | 3.1701 | -0.2902 |
| Prob. | 0.2374 | 0.0001 | 0.0002 | 0.0003 | 0.0006 | 0.0000 | 0.0486 | 0.0016 | **0.7717** |

Ps. $R^2$ 5.82281 | Adj. $R^2$ 2.040345 | Quasi-LR 173.08 | Prob. 0

Redundant Variables Test
Equation: UNTITLED
Specification: LW C X1*Z2B*Z3B Z1B*Z2B*Z3B X1*Z2B X1*Z3B Z1B*Z2B Z1B*Z3B X1
Z2B
RVT for : X1*Z2B*Z3B Z1B*Z2B*Z3B X1*Z2B Z1B*Z2B Z2B

|  | Value | Df | Probability |
|---|---|---|---|
| QLR L-statistic | 77.77123 | 5 | 0.0000 |
| QLR Lambda-statistic | 74.86572 | 5 | 0.0000 |

**Figure 6.38** A Redundant Variables Test based on the RM-7.

5.1 If we do not add a variable for RM-4, it is the final QR, which is an acceptable 3WI-QR.
5.2 As a special manual multistage selection method, deleting the interaction $X1*Z1B$ results in two alternative acceptable 3WI-QRs, RM-6 and RM-7, with two of the main variables as their IVs.
5.3 Even though $Z2B$ in RM-7 has a very large $Prob. = 0.7717$, it can be accepted because its interactions have significant adjusted effects. In addition, the five IVs, $X1*Z2B*Z3B$, $Z1B*Z2B*Z3B$, $X1*Z2B$, $Z1B*Z2B$, and $Z2B$, have joint significant effects on $LW$, based on the $QTR$ L-statistic of 77.7713 and $QLR$ $Lambda$-statistic of 74.865 72, with $df = 5$ and $p$-values $= 0.0000$, as presented in Figure 6.38
6) With the final QRs RM-6 and RM-7, we have an additional analysis to develop the path diagram for each QR, as follows:
6.1 First, let's see the RM-6 with following ES:

$$LW\,C\,X1 * Z2B * Z3B\,Z1B * Z2B * Z3B$$

$$X1 * Z2B\,X1 * Z3B\,Z1B * Z2B\,Z1B * Z3B\,X1\,Z3B$$

The three-way interactions, $X1*Z2B*Z3B$ and $Z1B*Z2B*Z3B$, show the consecutive effects $Z3B \rightarrow Z2B \rightarrow X1 \rightarrow LW$ and $Z3B \rightarrow Z2B \rightarrow Z1B \rightarrow LW$, respectively, which can be presented using the arrows with solid lines, as shown in Figure 6.39a. Furthermore, for the four two-way interactions, we have to draw two additional arrows to represent the consecutive $Z3B \rightarrow X1 \rightarrow LW$ and $Z3b \rightarrow Z1B \rightarrow LW$, respectively, as shown in Figure 6.39b, for the Stage-2. Finally, at Stage-3, we should draw an arrow from $Z3B$ to $LW$ to represent the direct effect of the main variable $Z3B$ on $LW$.
6.2 For the RM-7, we have exactly the same path diagrams as RM-6 to represent the three-way and two-way interaction IVs. However, at Stage-3, we have to draw an arrow with a dotted line from $Z2B$ to $LW$ to indicate that the effect of $Z2B$ on $LW$ is insignificant, as shown in Figure 6.40b.

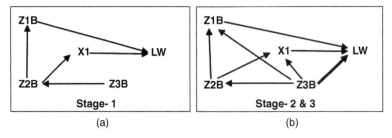

**Figure 6.39** Three stages in developing the path diagram of the RM-6.

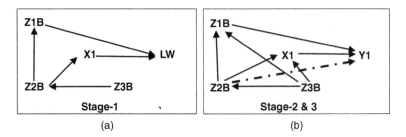

**Figure 6.40** Three stages in developing the path diagram of the RM-7.

## 6.10 QRs with Multiple Numerical Predictors

Suppose we want to develop QRs of $Y$ with 10 numerical predictors, $Xk$, $k = 1,2, ..., 10$. What are we planning to do? Do you want to develop a path diagram based on $Y$ and the 10 predictors? With 10 predictors, we would have 45 possible two way-interactions and 120 possible three-way interactions. So, we never develop a model with such a large number of interactions. Instead, I recommend developing either *a simple two-way interaction QR* (S2WI-QR), a *simple three-way interaction QR* (S3WI-QR), or *an additive model*. The steps in developing each of the models are presenting in the following three sections.

### 6.10.1 Developing an Additive QR

I consider an additive QR of $Y$ on a multivariate numerical to be the simplest multiple QR, and the data analysis can be done easily. For instance, the analysis of a QR of $Y$ on 10 numerical predictors, $Xk$, $k = 1, 2, ..., 10$, can easily be done using the following ES:

$$Y\ C\ X1\ X2\ X3\ X4\ X5\ X6\ X7\ X8\ X9\ X10 \tag{6.37}$$

However, in most cases, we obtain unexpected parameters estimates, where an important predictor, in the theoretical sense, has an insignificant adjusted effect with a large *p*-value, and the less important predictors have significant effects on $Y$. This is because of the *multicollinearity* or *bivariate correlations* between *X1* and *X10*, even though some of them might not be correlated in the theoretical sense. Note that all pairs of numerical variables, including the zero-one variables have quantitative values of correlations. For these reasons, I recommend the following manual multistage selection method of the data analysis.

1. Divide the 10 variables into at least three groups or subsets, where the first two subsets are classified as the first and the second most important predictors (upper, source, or cause variables) of *Y*. Without lossing generality, the 10 variables *X1 to X10* have the ranking from the most to the least important predictors for *Y*, in the theoretical sense. Then suppose we have three subsets, *X1 to X3, X4 to X7,* and *X8 to X10*.

2. We then can apply the manual three-stage selection method. Because we have used this method in several previous examples, I will not present it again.

3. However, in this section, I want to introduce or propose an alternative process as follows:

   3.1 The three stages of the stepwise least squares (STEPLS) regression can be applied to obtain an acceptable ordinary least squares (OLS) regression, ordinary regression, or mean regression (MR).

   3.2 Finally, the QR(Median) of *Y* can be obtained directly using the equation of the MR of *Y* to conduct the QR analysis.

## Example 6.22 *Additive Models Based on MLOGIT.wf1*

By doing experiments, using trial-and-error method, I have found outputs of several alternative additive MRs and QRs of *LW* on 10 numerical variables. One pair of their outputs is presented in Figure 6.41. I also obtain the error massage for some of the MRs.

1) The outputs are obtained using the following ES:

$$LW\ C\ PR1\ X1\ Z1\ Z2\ Z3\ XB2\ DIST1\ INC\ DIST2\ LOGL1 \tag{6.38}$$

2) Both MR and QR(Median) have the same set of IVs, *Z1, Z2,* and *Z3*, each of which has a significant effect on *LW* at the 1% and 5% levels. Note that *Z2* has *p*-values of 0.0340 and 0.0332 in MR and QR, respectively.

3) We find a large deviation between MR and QR on the probabilities (*p*-values) of *INC*, that is, 0.1845 and 0.5294, respectively. In addition, the MR has a greater adjusted *R*-squared than the QR, their values being 0.270 240 and 0.143 768, respectively.

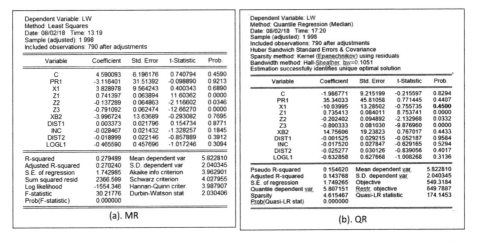

**Figure 6.41** Outputs of Mean and Quantile Regressions based on the ES (6.38). (a). MR. (b). QR.

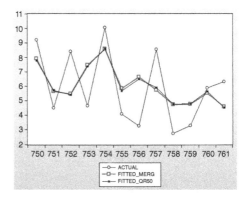

Covariance analysis: Ordinary
Date: 08/03/18 Time: 07:24
Sample (adjusted): 1790
Included observations: 790 after adjustments
Balanced sample (listwise missing value deletion)

| Correlation t-Statistic Probability | ACTUAL | FITTED_M... | FITTED_Q... |
|---|---|---|---|
| ACTUAL | 1.000000 ..... ..... | | |
| FITTED_MERG | 0.528668 17.48339 0.0000 | 1.000000 ..... ..... | |
| FITTED_QR50 | 0.523933 17.26721 0.0000 | 0.991045 208.3447 0.0000 | 1.000000 ..... ..... |

**Figure 6.42** Correlations between the three variables ACTUAL, FITTED_MR, and FITTED_QR, and a part of its graphs.

4) Figure 6.42 presents the correlations between the three variables *ACTUAL, FIT-TED_MR*, and *FITTED_QR50*, and a part of its graphs. Why use the variable *ACTUAL* instead of the original variable *LW*? The reason is that *LW* has $1000 - 790 = 210$ NAs, and the variable *ACTUAL* and the fitted values have 790 measured observations of the variable *LW*. So, the graph of the variable *ACTUAL* does not present any missing values. Do the graphs of the original variable *LW*, with its fitted value variables. You will see their differences. In addition, the graphs show that the *FITTED_MR* and *FITTED_QR* are very closed, with a very large correlation of 0.991 045. For this reason, their graphs are presented using different colors.

5) We can generate the variable *ACTUAL* as follows:

5.1 With the output of either MR or QR model on-screen, click *View/Actual, Fitted, Residual/Actual, Fitted, Residual Table* to display the output of Obs, Actual, Fitted, Residual, and Residual Plot on-screen.

5.2 We can copy the variable *ACTUAL* into the data file. Similarly for the FITTED variable, which can be rename.

6) As a comparison, Figure 6.43 presents a part of the *LW* data and its forecast variables, *LWF_MR*, and *LWF_QR50*. Based on these results, the following findings and notes are presented:

6.1 Their individual descriptive statistics present observations of 790, 1000, and 1000, for the variables *LW, LWF_MR,* and *LWF_QR50*, respectively.

6.2 Note *LW* has 210 =(1000-790) missing obsevations. And both forecast varaibles both *LWF_MR* and *LWF_QR50* have 1000 values, because the 10 IVs of the model have 1000 complete values.

6.3 Corresponding to the missing values of *LW*, its graph presents a broken polygon. In other words, there are no points for the missing values of *LW*.

**Example 6.23   *A Comparative Study***

As additional illustrations, Figure 6.44 presents a subset of the scores of the original variable *LW*, and its *ACTUAL* variable, *FITTED* variable, *FITTED_QR50*, and forecast variable, *LWF_QR50*, with part of their graphs. Based on these outputs, the following findings and notes are presented:

| Statistic | LW | LWF_MR | LWF_QR50 |
|---|---|---|---|
| Mean | 5.823 | 5.679 | 5.665 |
| Median | 5.813 | 5.667 | 5.657 |
| Maximum | 11.273 | 9.190 | 9.181 |
| Minimum | 1.040 | 1.673 | 1.753 |
| Std. Dev. | 2.040 | 1.104 | 1.109 |
| Skewness | 0.155 | 0.003 | 0.014 |
| Kurtosis | 2.641 | 2.898 | 2.862 |
| | | | |
| Jarque-Bera | 7.420 | 0.431 | 0.834 |
| Probability | 0.024 | 0.806 | 0.659 |
| Sum | 4600.0 | 5678.8 | 5664.6 |
| Sum Sq. Dev. | 3284.6 | 1217.3 | 1228.3 |
| | | | |
| Observations | 790 | 1000 | 1000 |

| Obs | LW | LWF_MR | LWF_QR50 |
|---|---|---|---|
| 10 | 5.154 | 7.488 | 7.338 |
| 11 | 4.729 | 6.518 | 6.610 |
| 12 | | 4.335 | 4.391 |
| 13 | 7.488 | 6.392 | 6.412 |
| 14 | 4.172 | 5.696 | 5.701 |
| 15 | 6.541 | 4.001 | 3.808 |
| 16 | 7.453 | 6.075 | 5.984 |
| 17 | 4.315 | 5.654 | 5.761 |
| 18 | | 4.815 | 4.459 |
| 19 | 5.152 | 6.477 | 6.438 |
| 20 | | 5.743 | 5.792 |
| 21 | | 4.733 | 4.798 |
| 22 | 3.554 | 5.886 | 5.639 |
| 23 | 4.709 | 5.919 | 5.879 |

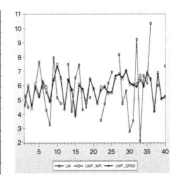

**Figure 6.43** The descriptive statistics of LW, LWF_MR & LWF_QR50, and a parts of their observed values, and a part of their graphs.

| obs. | LW | Actual | Fit_QR50 | LWF_QR50 |
|---|---|---|---|---|
| 785 | 3.561 | 3.268 | 4.414 | 4.768 |
| 786 | 5.502 | 3.682 | 4.954 | 7.301 |
| 787 | 5.959 | 4.378 | 4.808 | 6.455 |
| 788 | 5.055 | 5.748 | 5.509 | 6.452 |
| 789 | 4.204 | 6.636 | 6.331 | 5.130 |
| 790 | NA | 6.839 | 7.376 | 4.206 |
| 791 | 2.826 | NA | NA | 6.062 |
| 792 | NA | NA | NA | 4.134 |
| 793 | 3.334 | NA | NA | 4.619 |
| 794 | 6.306 | NA | NA | 5.988 |
| 795 | 4.560 | NA | NA | 5.623 |
| 796 | 4.443 | NA | NA | 4.306 |
| 797 | 7.493 | NA | NA | 5.674 |
| 798 | 10.032 | NA | NA | 9.181 |
| 799 | 8.014 | NA | NA | 7.107 |
| 800 | 5.710 | NA | NA | 6.960 |

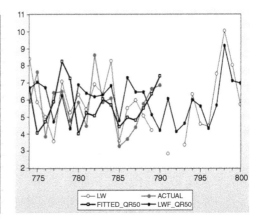

**Figure 6.44** A subset of scores of the original variable LW, and its ACTUAL and FITTED variable FITTED_QR50, and forecast variable LWF_QR50, with a part of their graphs.

1) The original variable *LW* has 1000 observations with 210 NAs, but its forecast variable *LWF_QR50* has 1000 complete observations.
2) The *ACTUAL* and *FITTED* variables have complete scores for the first 790 observations and 210 NAs afterward.
3) Their graphs include the graphs of four variables for the first 790 observations and only the two graphs of *LW* and *LWF_Q50* afterward.

## Example 6.24   *An Additional Comparison between MR and QR*

Figure 6.45 provides an additional comparison between the results of MR and QR, based on the reduced models in Figure 6.41. Based on these results, the following findings and notes are presented:

1) The output of the reduced MR is obtained using the STEPLS-Combinatorial and trial-and-error selection methods. The result is a good-fit MR having six IVs. Note that each of its IVs has either a positive of negative significant effect on *LW*, at the 1%, 5%,

**(a). Output of a Reduced MR**

Dependent Variable: LW
Method: Least Squares
Date: 09/01/18   Time: 16:24
Sample (adjusted): 1 790
Included observations: 790 after adjustments

| Variable | Coefficient | Std. Error | t-Statistic | Prob. |
|---|---|---|---|---|
| C | 4.460234 | 0.753040 | 5.922971 | 0.0000 |
| Z3 | -0.788924 | 0.062350 | -12.65322 | 0.0000 |
| Z1 | 0.737716 | 0.063696 | 11.58192 | 0.0000 |
| Z2 | -0.136713 | 0.064523 | -2.118818 | 0.0344 |
| INC | -0.028789 | 0.021395 | -1.345562 | 0.1788 |
| X1 | 2.801804 | 1.616446 | 1.733310 | 0.0834 |
| XB2 | -2.580765 | 1.608204 | -1.604750 | 0.1090 |

| | | | | |
|---|---|---|---|---|
| R-squared | 0.277655 | Mean dependent var | | 5.822810 |
| Adjusted R-squared | 0.272120 | S.D. dependent var | | 2.040345 |
| S.E. of regression | 1.740739 | Akaike info criterion | | 3.955318 |
| Sum squared resid | 2372.625 | Schwarz criterion | | 3.996716 |
| Log likelihood | -1555.351 | Hannan-Quinn criter. | | 3.971230 |
| F-statistic | 50.16154 | Durbin-Watson stat | | 0.564132 |
| Prob(F-statistic) | 0.000000 | | | |

**(b). Output of a Reduced QR**

Dependent Variable: LW
Method: Quantile Regression (Median)
Date: 09/01/18   Time: 17:46
Sample (adjusted): 1 790
Included observations: 790 after adjustments
Huber Sandwich Standard Errors & Covariance
Sparsity method: Kernel (Epanechnikov) using residuals
Bandwidth method: Hall-Sheather, bw=0.1051
Estimation successfully identifies unique optimal solution

| Variable | Coefficient | Std. Error | t-Statistic | Prob. |
|---|---|---|---|---|
| C | 5.201071 | 1.097718 | 4.738075 | 0.0000 |
| Z3 | -0.809430 | 0.082160 | -9.851829 | 0.0000 |
| Z1 | 0.719512 | 0.080950 | 8.888382 | 0.0000 |
| Z2 | -0.212711 | 0.096048 | -2.214619 | 0.0271 |
| INC | -0.006611 | 0.027862 | -0.237275 | 0.8125 |
| X1 | 0.882754 | 2.348303 | 0.375912 | 0.7071 |
| XB2 | -0.909125 | 2.403634 | -0.378229 | 0.7054 |

| | | | | |
|---|---|---|---|---|
| Pseudo R-squared | 0.151830 | Mean dependent var | | 5.822810 |
| Adjusted R-squared | 0.145331 | S.D. dependent var | | 2.040345 |
| S.E. of regression | 1.747169 | Objective | | 551.1310 |
| Quantile dependent var | 5.807151 | Restr. objective | | 649.7887 |
| Sparsity | 4.714981 | Quasi-LR statistic | | 167.3945 |
| Prob(Quasi-LR stat) | 0.000000 | | | |

**Figure 6.45** Outputs of Mean and Quantile Regressions based on the reduced MR and QR in ES (6.38). (a). Output of a Reduced MR. (b). Output of a Reduced QR.

or 10% level. For instance, at the 10% level, *INC* has a negative significant effect, based on the *t*-statistic of $t_0 = -1346$ with $df = (790 - 7)$ and a *p*-value $= 0.1788/2 < 0.10$.

2) Then we use the ES of the MR to obtain the output of the Reduced-QR, which shows that each of the IVs, *INC, X1,* and *XB2,* has a very large *p*-value. So, the Reduced-QR isn't a good-fit QR, in the statistical sense.

3) Based on these findings, we can't use the IVs of a good-fit MR directly for a good-fit or acceptable QR, in the statistical sense. Even so, if we have a large number of IVs, then we can use the IVs of a good-fit MR as a guide in developing a good-fit QR. By using the STEPLS regression with the multistage selection method, we can easily develop a good-fit MR.

## 6.10.2   Developing a Simple Two-Way Interaction QR

Referring to the alternative 2WI-QRs of Y on *(X1,X2,X3,X4)* presented previously, I recommend the following steps of analysis in developing a simple 2WI-QR:

1) As a preliminary step, a path diagram, such as the alternative path diagrams presented in Figure 6.35. The diagram is dvveloped, and assumed to be true, in the theoretical sense.

2) Then, at the first stage of analysis, we should use a 2WI-QR, based on the path diagram and having at least one of the six possible two-way interactions, namely *X1\*X2, X1\*X3, X1\*X4, X2\*X3, X3\*X4,* and *X2\*X4*.

3) As an example, assume the effect of *X1* on *Y1* depends on *X2, X3,* and *X4*. Then, at the first stage of analysis, the following ES should be applied:

$$Y1\ C\ X1 * X2\ X1 * X3\ X1 * X4 \tag{6.39}$$

4) This obtains an acceptable QR function, called RM-1, of the 2WI-QR of *Y* on the 10 variables. Note that each of the IVs of RM-1 should have the *Prob.* <0.30, since at the

15% level, which is applied in Lapin (1973), the IV would have a significant adjusted effect with a *p*-value = *Prob./2 < 0.15*.

5) At the second stage of analysis, we may have one or more of the other three two-way interactions, namely *X2\*X3, X2\*X4,* and *X3\*X4,* as additional IVs for RM-1, corresponding to the path diagram we developed. So, we might obtain an RM-2, as an upper model of RM-1.

6) Finally, the 10 main variables should be inserted as additional IVs of RM-2. In this case, I recommend making groups of the 10 main variables, with the first group is defined as the most important to be selected for additional variables of RM-2, even though it will be very subjective.

7) As an alternative method, we could select the 10 variables using the STEPLS-Combinatorial method, with RM-2 as the *Equation specification* and the 10 variables as the *List of search regressors*. In general, the MR obtained would have fewer than the 10 main variables as its IVs. So, we would have smaller number of main variables, which should be selected for the QR. Refer to the results presented in Figure 6.44a, which shows that only 6 of 10 main variables are available for developing the QR.

### 6.10.3  Developing a Simple Three-Way Interaction QR

As an extension of the 2WI-QRs presented in previous section, we can identify a 3WI-QR to apply, based on the path diagram which has been developed. For instance, if the ES (6.39) is a true relationship, then we can have at least one of the three-way interactions, *X1\*X2\*X3, X1\*X2\*X4,* and *X1\*X3\*X4,* to use in the first stage of the regression analysis. Hence, as an extension of the ES (6.39), I recommend applying the following ES, at the first stage of analysis:

$$Y \ C \ X1 * X2 * X3 \ X1 * X2 * X4 \ X1 * X3 * X4 \qquad (6.40)$$

An acceptable QR function obtained is called RM-1 of the 3WI-QR of *Y* on the 10 variables. Note that each of the IVs of the acceptable RM-1 should has the *Prob.* <0.30, since at the 15% level, which is applied in Lapin (1973), the IV would have a significant adjusted effect with a *p*-value = *Prob./2 < 0.15*.

Then, at the second stage, we have a three-way interaction, *X2\*X3\*X4,* whenever it is supported by the defined path diagram. Otherwise, we have six two-way interactions, which should be selected for the additional IVs of RM-1. Note that whenever *X1\*X2\*X3* is a significant IV of the RM-1, we might not need to insert all three two-way interactions, *X1\*X2, X1\*X3,* and *X2\*X3,* as additional IVs for RM-1. Hence, in general, a 3WI-QR will be a *nonhierarchical model*, because it will not have the complete two-way interactions of a three-way interaction. The acceptable 3WI-QR is called RM-2.

Finally, we have the 10 main variables, which should be selected using the same method as presented in previous section, for the 2WI-QR.

### Example 6.25  *Application of a 3WI-QR*

Based on the MR of *LW* on the 10 variables presented in Figure 6.41, and its reduced MR of *LW* on the six variables presented in Figure 6.44a, let's select *Z1, Z2, Z3,* and *INC* as the first four important predictors of *LW*. Furthermore, assume that the effect of *Z1* on *LW* depends

**Table 6.10** Statistical results summary of the reduced models RM-1, RM-2, and RM-3.

| | RM-1 | | | RM-2 | | | RM-3 | | |
|---|---|---|---|---|---|---|---|---|---|
| Variable | Coef. | t-stat. | Prob. | Coef. | t-stat. | Prob. | Coef. | t-stat. | Prob. |
| C | 5.8249 | 58.995 | 0.0000 | 5.8084 | 59.2440 | 0.0000 | 5.8300 | 62.5610 | 0.0000 |
| Z1*Z3*INC | −0.0286 | −1.4841 | 0.1382 | −0.0320 | −1.6336 | 0.1027 | −0.0239 | −1.3364 | 0.1818 |
| Z1*Z2*Z3 | −0.0589 | −0.4958 | 0.6202 | | | | | | |
| Z1*Z2*INC | 0.0074 | 0.3163 | 0.7519 | | | | | | |
| Z1*Z2 | | | | | | | 0.1792 | 2.2020 | 0.0280 |
| Z2*Z3 | | | | | | | −0.2914 | −3.7533 | 0.0002 |
| Z2*INC | | | | | | | −0.0614 | −3.6005 | 0.0003 |
| Pseudo R-sq | 0.0031 | | | 0.0020 | | | 0.0189 | | |
| Adj. R-sq | −0.0007 | | | 0.0008 | | | 0.0139 | | |
| S.E. of reg | 2.0385 | | | 2.0371 | | | 2.0182 | | |
| Quantile DV | 5.8072 | | | 5.8072 | | | 5.8072 | | |
| Sparsity | 5.4315 | | | 5.4472 | | | 5.1481 | | |
| Quasi-LR stat. | 2.9409 | | | 1.9248 | | | 19.034 | | |
| Prob. | 0.4008 | | | 0.1653 | | | 0.0008 | | |

on *Z2*, *Z3*, and *INC*. Referring to the ES (6.40), we then have the following ES at the first stage of analysis:

$$LW\ C\ Z1 * Z2 * Z3\ Z1 * Z2 * INC\ Z1 * Z3 * INC \tag{6.41}$$

Table 6.10 presents the statistical results summary of three reduced models, RM-1, RM-2, and RM-3, of the 3WI-QR(Median) of *LW* on 10 selected predictors in (6.38). Based on this summary, the following findings and notes are presented:

1) At the first stage of analysis, the output of (6.39), that is RM-1, shows that only one of the IVs has *p*-value $< 0.30$.
2) So, at the second stage, we find that RM-2 has only a single three-way interaction, *Z1*Z3*INC*, which, at the 10% level, has a negative significant effect, based on the *t*-statistic of $t_0 = -1.6636$ with a *p*-value $= 0.1027/2 < 0.10$. In fact, as a result of using the STEPLS-Combinatorial method, the good-fit MR also has the single IV, *Z1*Z3*INC*, having the *t*-statistic of $t_0 = -1.960\,347$ with a *Prob.* $= 0.0503$. So, in this case, the result based on the MR can be used directly for the QR(Median).
3) At the third stage of analysis, two alternative methods have exercises, as follows:
   3.1 First, if we try to insert the two-way interactions of the three-way interaction *Z1*Z3*INC*, namely *Z1*Z3, Z1*INC*, and *Z3*INC*, we find that none can be inserted as an additional IV for RM-2. If *Z3*INC* is inserted for RM-2, based on the STEPLS-Combinatorial method, the three-way interaction *Z1*Z3*INC* becomes insignificant with a *Prob.* $>0.3$ in the MR and a *Prob.* $>0.4$ in the QR(Median). So, in this case, the ES of the MR also can be used directly for QR(Median).

3.2 Then we try to use the other two-way interactions, with the output presented as the RM-3, which shows, at the 10% level, the three-way interaction $Z1*Z3*INC$ has a negative significant adjusted effect, based on the $t$-statistic of $t_0 = -1.3364$ with a $p$-value $= 0.1818/2 < 0.10$. As a comparison, if we use the STEPLS-Combinatorial method, the interaction $Z1*Z3*INC$ in the MR has the $t$-statistic of $t_0 = -1.770\,797$ with a *Prob.* $= 0.0755$.

4) Finally, we have the main variables as additional IVs for RM-3. In this case, we find none of the main variables can be used as an additional IV for RM-3. Then, RM-3 can be considered the best-fit 3WI-QR(Median), in the statistical sense.

5) As an exercise, do this for the two-way interaction MR and QR. Refer to the ES (6.38).

**Example 6.26** *Alternative 3WI-MR and 3WI-QR(Median)*
As each of the three-way interaction IVs of RM-1 in Table 6.11 has an insignificant effect, I think the main cause is that the ordering scores of the three interactions are miserable or inconsistent. This is because the interactions $Z1*Z2*INC$ and $Z1*Z3*INC$ can be very large for negative pairs *(Z1,Z2)* and *(Z1,Z3)*, and they even can be greater than all the interactions for the pairs of their positive scores since *INC* is a positive variable. The case is similar for the interaction $Z1*Z2*Z3$ if *Z1, Z2,* or *Z3* is positive. For these reasons, let's transform *Z1,*

**Table 6.11** Statistical results summary of STEPLS-Regression, Reduced STEPLS-Regression, and QREG – Quantile Regression (Included LAD).

| Variable | STEPLS-Reg | | | Reduced STEPLS-Reg | | | QREG | | |
| --- | --- | --- | --- | --- | --- | --- | --- | --- | --- |
| | Coef. | t-stat. | Prob.* | Coef. | t-stat. | Prob.* | Coef. | t-stat. | Prob.* |
| C | 5.607 | 9.430 | 0.000 | 5.656 | 10.965 | 0.000 | 5.738 | 7.707 | 0.000 |
| PZ1*PZ2*PZ3 | −0.001 | −0.167 | **0.867** | | | | | | |
| PZ1*PZ2*INC | −0.023 | −2.614 | 0.009 | −0.023 | −2.610 | 0.009 | −0.028 | −2.539 | 0.011 |
| PZ1*PZ3*INC | 0.016 | 1.741 | 0.082 | 0.015 | 1.787 | 0.074 | 0.024 | 2.296 | 0.022 |
| PZ2*PZ3 | −0.113 | −3.471 | 0.001 | −0.111 | −2.534 | 0.012 | −0.166 | −3.154 | 0.002 |
| PZ1*PZ2 | 0.126 | 4.287 | 0.000 | −0.117 | −6.619 | 0.000 | −0.108 | −4.716 | 0.000 |
| PZ1*INC | 0.065 | 2.305 | 0.021 | 0.122 | 6.100 | 0.000 | 0.101 | 4.862 | 0.000 |
| PZ3*INC | −0.112 | −2.528 | 0.012 | 0.065 | 2.307 | 0.021 | 0.067 | 1.524 | 0.128 |
| PZ2*INC | 0.079 | 1.913 | 0.056 | 0.080 | 1.951 | 0.051 | 0.116 | 2.502 | 0.013 |
| R-squared | 0.275 | | | 0.275 | | | Pseudo R-sq. | | 0.154 |
| Adj. R-sq. | 0.268 | | | 0.269 | | | Adj. R-sq. | | 0.147 |
| S.E. of reg | 1.746 | | | 1.745 | | | S.E. of reg. | | 1.753 |
| Sum sq. resid | 2381.2 | | | 2381.2 | | | Quantile DV | | 5.807 |
| Log likelihood | −1556.8 | | | −1556.8 | | | Sparsity | | 4.563 |
| F-statistic | 37.041 | | | 42.381 | | | Quasi-LR stat | | 175.62 |
| Prob(F-stat.) | 0.000 | | | 0.000 | | | Prob(QLR stat) | | 0.000 |

Dependent Variable: LW
Method: Least Squares
Date: 09/03/18   Time: 19:03
Sample (adjusted): 1 790
Included observations: 790 after adjustments

| Variable | Coefficient | Std. Error | t-Statistic | Prob. |
|---|---|---|---|---|
| C | 6.272107 | 0.223769 | 28.02932 | 0.0000 |
| PZ1*PZ2*PZ3 | -0.003642 | 0.001747 | -2.084459 | 0.0374 |
| PZ1*PZ2*INC | 0.007704 | 0.001772 | 4.346738 | 0.0000 |
| PZ1*PZ3*INC | -0.008136 | 0.001868 | -4.356398 | 0.0000 |

| | | | |
|---|---|---|---|
| R-squared | 0.033098 | Mean dependent var | 5.822810 |
| Adjusted R-squared | 0.029408 | S.D. dependent var | 2.040345 |
| S.E. of regression | 2.010120 | Akaike info criterion | 4.239317 |
| Sum squared resid | 3175.899 | Schwarz criterion | 4.262972 |
| Log likelihood | -1670.530 | Hannan-Quinn criter. | 4.248410 |
| F-statistic | 8.968555 | Durbin-Watson stat | 0.065138 |
| Prob(F-statistic) | 0.000008 | | |

*(a). 3WI-LSREG*

Dependent Variable: LW
Method: Quantile Regression (Median)
Date: 09/03/18   Time: 19:38
Sample (adjusted): 1 790
Included observations: 790 after adjustments
Huber Sandwich Standard Errors & Covariance
Sparsity method: Kernel (Epanechnikov) using residuals
Bandwidth method: Hall-Sheather, bw=0.1051
Estimation successfully identifies unique optimal solution

| Variable | Coefficient | Std. Error | t-Statistic | Prob. |
|---|---|---|---|---|
| C | 6.123845 | 0.280169 | 21.85771 | 0.0000 |
| PZ1*PZ2*PZ3 | -0.003489 | 0.002221 | -1.570964 | 0.1166 |
| PZ1*PZ2*INC | 0.008512 | 0.002074 | 4.104050 | 0.0000 |
| PZ1*PZ3*INC | -0.007927 | 0.002501 | -3.168903 | 0.0016 |

| | | | |
|---|---|---|---|
| Pseudo R-squared | 0.019194 | Mean dependent var | 5.822810 |
| Adjusted R-squared | 0.015450 | S.D. dependent var | 2.040345 |
| S.E. of regression | 2.012034 | Objective | 637.3168 |
| Quantile dependent var | 5.807151 | Restr. objective | 649.7887 |
| Sparsity | 5.291193 | Quasi-LR statistic | 18.85675 |
| Prob(Quasi-LR stat) | 0.000293 | | |

*(b). 3WI-QR*

**Figure 6.46**   The output of LS-Regression and QR(Median) based on the ES (6.42). (a). 3WI-LSREG. (b). 3WI-QR.

Z2, and Z3 to positive variables, namely *PZ1, PZ2,* and *PZ3,* which we can generate using the following equation:

$$PZk = Zk + 5, for\ k = 1, 2\&3$$

Then, instead of the ES (6.39), we have the following ES:

$$LW\ C\ PZ1 * PZ2 * PZ3\ PZ1 * PZ2 * INC\ PZ1 * PZ3 * INC \tag{6.42}$$

As the first stage of analysis, for comparison, Figure 6.46 presents the outputs of the MR and the QR(Median), which show much better results than the RM-1 in Table 6.10, in the statistical sense. In addition, these results also show that the equation of an MR can be used directly for the QR(Median), since each of its IVs has either a positive or negative significant effect, at the 1% or 10% level of significance. For instance, *PZ1*PZ2*PZ3* has a negative significant effect, based on the *t*-statistic of $t_0 = -1.571$ with a *p*-value $= 0.1166/2 < 0.10$. So these findings show that we can use the STEPLS regression to select IVs to use, at least as a guide to selecting appropriate IVs for the QR(Median).

At the second stage, the STEPLS regression is applied for selecting additional IVs of the MR obtained at the first stage. For this analysis, the ES of the MR in Figure 6.46, "*LW C Z1*Z2*Z3 Z1*Z2*INC Z1*Z3*INC*", is applied, and the six two-way interactions of *Z1, Z2, Z3,* and *INC* are applied as the list of search regressors. The output of the STEPLS regression is presented in the first block of Table 6.11, with one of the three-way interactions having a very large probability. Each of the other seven IVs has a significant effect, at the 1%, 5%, or 10% level.

After observing the output of several alternative LS-Regressions, including the LS-Regressions having all three-way interactions, we select the *Reduced STEPLS-Regression* having only two three-way interactions as its IVs, as shown in the second block of Table 6.11. Based on the results in this table, the following findings and notes are presented:.

1) If all three-way interactions should be used as the IVs, with the criterion that each should have a *Prob.* <0.30, then only one two-way interaction, namely *PZ2*PZ3*, can be inserted

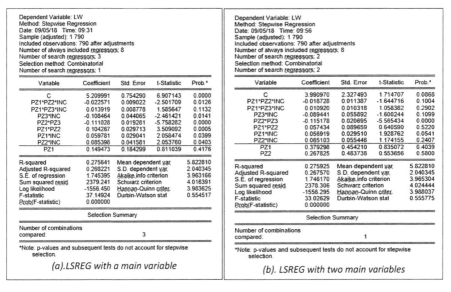

**Figure 6.47** The outputs of two acceptable STEPLS-Regressions. (a).LSREG with a main variable. (b). LSREG with two main variables.

as an additional IV using the STEPLS-Combinatorial method. If two of the two-way interactions were inserted, then the three-way interactions would have probabilities of 0.44, 0.51, and 0.74, respectively. So it is not an acceptable 3WI-MR.

2) So, at the second stage, we apply the STEPLS regression, with *LW C PZ1\*PZ2\*INC PZ1\*PZ3\*INC* as the ES, and the six two-way interactions as the list of search regressors. This obtains the *Reduced STEPLS-Reg* results in Table 6.11 as a good-fit model with five two-way interactions.

3) Finally, we obtain the acceptable 3WI-QR(Median) by using the ES of the reduced STEPLS regression, with one of the two-way interaction IVs, *PZ2\*INC*, having *Prob.* 0.128. But at the 10% level, it has a positive significant adjusted effect, based on the *t*-statistic of $t_0 = 1.524$ with a *p*-value $= 0.128/2 < 0.10$.

4) Hence, based on these findings, I recommend applying the STEPLS-Combinatorial and trial-and-error methods to select the IVs for the QR(Median), in cases in which a large number of IVs should be selected, because we never can predict the impacts of the multicollinearity of the IVs.

5) For the third stage, we do the following:

5.1 We divide the 10 main variables into two groups. The first group is the four variables *PZ1, PZ2, PZ3,* and *INC*, which are defined as the most important predictors. The others are the second group.

5.2 Using the STEPLS-Combinatorial method, we use the *Reduced STEPLS-Reg* in Table 6.11 as the ES and *PZ1, PZ2, PZ3,* and *INC* as the list of search regressors. Then inserting 1 (one) as the number of regressors to select results in the variable *INC* being selected, and the output on-screen shows that the three-way interaction *PZ1\*PZ3\*INC* has a large *p*-value $= 0.5482$. So, *INC* should not be used as an additional IV.

**Table 6.12** Statistical results summary of the final mean regression and QR(Median).

| Variable | Mean-Reg | | | QR(Median) | | |
|---|---|---|---|---|---|---|
| | Coef. | t-stat. | Prob. | Coef. | t-stat. | Prob. |
| C | 4.196 | 0.667 | 0.505 | −4.033 | −0.493 | 0.622 |
| PZ1*PZ2*INC | −0.023 | −2.552 | 0.011 | −0.022 | −1.953 | 0.051 |
| PZ1*PZ3*INC | 0.014 | 1.595 | 0.111 | 0.020 | 1.752 | 0.080 |
| PZ3*INC | −0.108 | −2.449 | 0.015 | −0.140 | −2.579 | 0.010 |
| PZ2*PZ3 | −0.114 | −5.890 | 0.000 | −0.109 | −4.567 | 0.000 |
| PZ1*PZ2 | 0.108 | 3.628 | 0.000 | 0.099 | 2.787 | 0.006 |
| PZ1*INC | 0.060 | 2.049 | 0.041 | 0.057 | 1.285 | 0.199 |
| PZ2*INC | 0.088 | 2.102 | 0.036 | 0.088 | 1.870 | 0.062 |
| PZ1 | 0.138 | 0.747 | 0.456 | 0.015 | 0.070 | **0.944** |
| DIST1 | 0.007 | 0.320 | 0.749 | −0.001 | −0.042 | **0.966** |
| DIST2 | −0.024 | −1.060 | 0.289 | −0.022 | −0.769 | **0.442** |
| LOGL1 | −0.390 | −0.849 | 0.396 | −0.756 | −1.254 | 0.210 |
| X1 | 3.526 | 0.368 | 0.713 | −12.992 | −1.074 | 0.283 |
| PR1 | −2.792 | −0.088 | 0.930 | 45.685 | 1.122 | 0.262 |
| XB2 | −3.610 | −0.264 | 0.792 | 18.992 | 1.092 | 0.275 |
| R-squared | 0.2810 | | | Pseudo R-sq. | | 0.1573 |
| Adj. R-sq. | 0.2680 | | | Adj. R-sq | | 0.1420 |
| S.E. of reg. | 1.7456 | | | S.E. of reg. | | 1.7571 |
| Sum sq. resid | 2361.5 | | | Quantile DV | | 5.8072 |
| Log likelihood | 1553.5 | | | Sparsity | | 4.4712 |
| F-statistic | 21.638 | | | Quasi-LS stat | | 182.84 |
| Prob(F-stat) | 0.0000 | | | Prob(QLR stat) | | 0.0000 |

5.3 For the this reason, rerun the same combinatorial selection method, by clicking the *Estimate* button and the *Options* tab, deleting *INC* from the list of regressors, and then clicking *OK*. The new output displays on-screen and shows that the two three-way interactions have negative and positive significant effects, respectively, as shown in Figure 6.47a. For instance, at the 10% level, *PZ1*PZ3*INC* has a positive significant effect with a p-value = 0.1132/2 < 0.10. So, this LSREG is an acceptable 3WI-LSREG, even though *PZ1* has a large p-value = 0.4718.

5.4 With the output still on-screen, click the *Estimate* button and the *Options* tab, enter 2(two) as the number of regressors, and click *OK*. The output displays on-screen and shows that *PZ11*PZ3*INC* has a *Prob.* = 0.3613 > 0.30. So the model should be modified as follows.

5.5 With the output still on-screen, click the *Estimate* button and the *Options* tab, delete *PZ3* from the list of regressors, and click *OK*. This results in the output in

Figure 6.47b, which shows *PZ1\*PZ2\*INC* and *PZ1\*PZ3\*INC* to have probabilities of 0.1004 and 0.2902 < 0.30, respectively. So the MR is an acceptable 3WI-LSREG since, at the 15% level, *PZ1\*PZ3\*INC* has a positive significant effect with a *p*-value = 0.2902/2 < 0.15, even though both *PZ1* and *PZ2* have large *p*-values.

6) With the two LSREGs in Figure 6.47, we in fact have a choice of which one we will use for the fourth stage of analysis, by inserting the other six main variables. However, in this case, we are doing an analysis based on both MRs in Figure 6.47, so we use the LSREG in Figure 6.47a. The final results summary is presented in Table 6.12. Based on this summary, the findings and notes presented as follows:

1. For the *Mean-Reg*, the six main variables in the second group can be used as additional IVs for the 3WI-MR, even though each of the main variables has large or very large *p*-values, because the two three-way interactions have negative and positive significant effects, at the 5% and 10% levels of significance, respectively, with a *p*-values of 0.011/2 = 0.0055 < 0.01, and 0.111/2 = 0.0555 < 0.10.

2. Similarly, for the QR(Median), obtained directly using the ES of the MR, the two three-way interaction IVs of the QR(Median), at the 5% level, have negative and positive significant effects, with *p*-values of 0.051/2 and 0.080 = 0.040 < 0.05, respectively. Even though three of the main IVs have large and very large *p*-values, this QR is a good-fit 3WI-QR. Based on this QR, we have additional findings as follows:

    2.1 At the 15% level, each of the main IVs, *LOGL1, X1, PR1,* and *XB2,* has either a negative or positive significant effect with a *p*-value = *Prob.*/2 < 0.15.

    2.2 All IVs of this QR, which are defined based on the 10 main variables, have significant joint effects, based on the *Quasi-LR* statistic of $QLR_0 = 182.84$ with a *p*-value = 0.0000.

    2.3 By using the RVT, we find that the five IVs, *PZ1\*PZ2\*INC, PZ1\*PZ3\*INC, PZ1\*PZ2, PZ1\*INC,* and *PZ1,* have significant joint effects on *LW,* based on the *QLR L*-statistic of 94.215 54 with *df* = 5 and *p*-value = 0.0000. Hence, we can conclude that the effect of *PZ1* on *LW* is significantly dependent on *PZ2, PZ3,* and *INC,* in the form of a function of *PZ2\*INC, PZ3\*INC, PZ2,* and *INC.*

    2.4 In addition, we also find that the eight IVs, "*PZ1\*PZ2\*INC, PZ1\*PZ3\*INC, PZ3\*INC, PZ2\*PZ3, PZ1\*PZ2, PZ1\*INC, PZ2\*INC,* and *PZ1,*" have significant joint effects on *LW,* adjusted for the six main variables, *DIST1, DIST2, LOGL1, X1, PR1,* and *XB2,* and based on the *QLR L*-statistic of 177.6477 with *df* = 8 and *p*-value = 0.0000.

# 7

# Quantile Regressions with the Ranks of Numerical Predictors

## 7.1 Introduction

The quantile-regression having numerical independent variables has been called the *semiparametric-quantile-regression (SPQR)*, because the quantiles of the DV are estimated or predicted using the numerical IV. This chapter presents quantile regression having the ranks of numerical variables as predictors or independent variables, which is called the *nonparametric-quantile-regression (NPQR)*. And special notes and comments are presented for SPQR and NPQR. In fact the NPQRs have been presented in the chapter 2, as the QRs (Quantile-Regressions) having categorical IVs. It is recognized that the equations of all Ordinary-Regression (OR) can directly be used to conducting the Quantile-Regression (QR) analysis. In addition, they also can easily be transformed to the *Nonparametric-Quantile-Regressions* (NPQRs), by replacing their numerical IVs with their ranks. So we can have a lot of possible NPQRs. However, this chapter present only selected alternative NPQRs, starting with the simplest model based on bivariate numerical variables *(X,Y)*.

## 7.2 NPQRs Based on a Single Rank Predictor

All QRs of $Y$ on a numerical variable $X$, based on selected data sets from the previous chapters, can easily be transformed to NPQRs using the ranks of $X$, which can be generated using the following equation:

$$Rank\_X = @Ranks\ (X, a)\ \ \ \ \ \ \ \ \ \ \ \ \ \ \ \ (7.1a)$$

Since *Ranks_X* might have non-integer scores we then transform it to the variable *R_X* or *RX,* with integer scores, using the following equation:

$$RX = @Round\ (Ranks\_X)\ \ \ \ \ \ \ \ \ \ \ \ \ \ \ \ (7.1b)$$

Then *RK* can be applied as a set of dummy IVs of alternative QRs or NPQRs, by using the function *@Expand(RX)* as the predictors. See an example below. However, in this section, I will present examples based only on the two variables *ACCEL* and *MILLI* in MCYCLE.wfl because they have special characteristics, as shown by the scatter graph of *ACCEL* on *MILLI* with their polynomial fit, kernel fit, and piece-wise linier or polynomial graphs, presented in previous chapter.

*Quantile Regression: Applications on Experimental and Cross Section Data Using EViews,* First Edition.
I Gusti Ngurah Agung.
© 2021 John Wiley & Sons Ltd. Published 2021 by John Wiley & Sons Ltd.

| Obs. | MILLI | RANK_MIL | R_MILLI | | Obs. | MILLI | RANK_MIL | R_MILLI |
|------|-------|----------|---------|---|------|-------|----------|---------|
| 10 | 8.2 | 10 | 10 | | 21 | 13.8 | 21 | 21 |
| 11 | 8.8 | 11.5 | 12 | | 22 | 14.6 | 24.5 | 25 |
| 12 | 8.8 | 11.5 | 12 | | 23 | 14.6 | 24.5 | 25 |
| 13 | 9.6 | 13 | 13 | | 24 | 14.6 | 24.5 | 25 |
| 14 | 10 | 14 | 14 | | 25 | 14.6 | 24.5 | 25 |
| 15 | 10.2 | 15 | 15 | | 26 | 14.6 | 24.5 | 25 |
| 20 | 13.6 | 20 | 20 | | 27 | 14.6 | 24.5 | 25 |

**Figure 7.1**   Selected scores of the two variables *Rank_Mil* and *R_Milli*.

We generate the ranks of *MILLI* using the following equations:

$$Rank\_Mil = @Ranks\,(Milli, a)) \qquad (7.2a)$$

$$R\_Milli = @Round(Rank\_Mil) \qquad (7.2b)$$

Figure 7.1 presents some selected scores of the variable *Rank_Mil*, which have non-integer scores. They are then transformed to *R_Milli* with integer scores. Note the values of *R_Milli*, corresponding to the same observed scores of the variable *MILLI*, for the observation numbers 11–12, and 22–27.

**Example 7.1**   *Scatter Graphs with Fitted Variables*
As a preliminary analysis, Figure 7.2 presents the scatter graphs of *ACCEL* and *R_Milli* with their three fitted variables, namely kernel fit, LOWESS linear fit, and linear sixth degree polynomial fit. Based on these graphs, the following findings and notes are presented:

1) The scatter graphs of *ACCEL* on *R_Milli* look similar to the ones we saw previously for *ACCEL* on *MILLI* in Figure 4.17.
2) Even though the scatter graphs of *ACCEL* on *R_Milli* are similar to the ones for *ACCEL* on *MILLI,* their kernel and LOWESS linear fits have slightly different fits. So, all models of *ACCEL* on *MILLI* presented in Chapter 4 can be applied for the NPQR.
3) *R_Milli* has 94 different ranks among the 133 observations.

### 7.2.1   Alternative Piecewise NPQRs of *ACCEL* on *R_Milli*

The following NPQR examples correspond to the QR examples presented in Chapter 4.

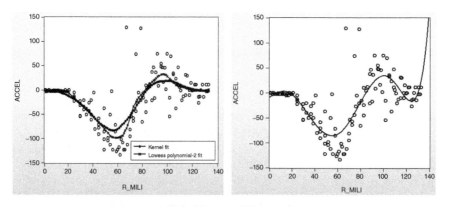

**Figure 7.2**   Scatter graphs of *ACCEL* on *R_Milli* with kernel fit, LOWESS linear fit, and the sixth-degree polynomial mean regression.

## Example 7.2 An NPQR Corresponding to QR (4.17)

Corresponding to the QR (4.17), Figure 7.3 presents the output of an NPQR, using the following equation specifications (ESs). In addition, this figure also presents the graphs of *ACCEL*, its fitted variable, *Fitted_73*, and *Fitted_417* as the fitted variable of the QR(Median) in (4.17):

$$ACCEL\ C@Expand\ (Int4b, @Dropfirst)\ R\_Milli*@Expand\ (Int4b) \tag{7.3a}$$

$$Int4b = 1 + 1*(milli >= 13) + 1*(milli >= 21) + 1*(milli >= 29) \tag{7.3b}$$

Based on these statistical results, the following findings and notes are presented:

1) This NPQR has an adjusted *R*-squared = 0.4627, which is greater than the QR (4.17)'s adjusted *R*-squared of 0.4381. So, this NPQR(Median) is a better fit model than its corresponding QR.
2) At the 5% level of significance, $R\_Milli*(Int4b = 4)$ has a significant negative effect, based on the *t*-statistic of $t_0 = -2.269$ with a *p*-value = 0.025/2, but $Milli*(Int4b = 4)$ has an insignificant negative effect, based on the *t*-statistic of $t_0 = -1.183$ with a *p*-value with a *Prob.* = 0.2391/2.
3) Additionally, Figure 7.4 presents the bivariate correlations between *ACCEL Fitted_417* and *Fitted_73*, which shows that *Fitted_417* and *Fitted_73* have a very large correlation of 0.987 935.

## Example 7.3 An NPQR Corresponding to the QR (4.25)

Corresponding to the QR (4.25), Figure 7.5 presents the output of an NPQR using the following ESs. In addition, this figure also presents the graphs of *ACCEL*, its fitted variable of the NPQR (7.4), namely *Fitted_74*, and the fitted variable *Fitted_425*, based on the QR (4.25).

$$ACCEL\ C@Expand\ (Int4d, @Dropfirst)\ R\_Milli*@Expand\ (Int4d)$$

$$R\_Milli\textasciicircum 2*@Expand\ (Int4d, @Dropfirst) \tag{7.4a}$$

$$Int4d = 1 + 1*(milli >= 15) + 1*(milli >= 26) + 1*(milli >= 33) \tag{7.4b}$$

Dependent variable:ACCEL
Method: quantile regression (median)
Date: 08/19/18  Time: 06:19
Sample: 1133
Included observations: 133
Huber sandwitch standard errors & covariance
Sparsity method: Kernel(Epanechnikov) using residuals
Bandwidth method: Hall-Sheather, bw=0.19033
Estimation successfully identifies unique optimal solution

| Variable | Coefficient | Std.error | t-Statistic | Prob. |
|---|---|---|---|---|
| C | -2.700000 | 6.660935 | -0.405349 | 0.6859 |
| INT4B=2 | 64.71250 | 11.95298 | 5.413924 | 0.0000 |
| INT4B=3 | -455.9545 | 59.07059 | -7718808 | 0.0000 |
| INT4B=4 | 1304056 | 54.01393 | 2.414295 | 0.0172 |
| R_MILLI*(INT4B=1) | -1.37E-16 | 0.613204 | -2.24E-16 | 1.0000 |
| R_MILLI*(INT4B=2) | -3.012500 | 0.268755 | -11.20911 | 0.0000 |
| R_MILLI*(INT4B=3) | 5.472727 | 0.768945 | 7.117191 | 0.0000 |
| R_MILLI*(INT4B=4) | -1.005556 | 0.443106 | -2.269332 | 0.0250 |

| | | | | |
|---|---|---|---|---|
| Pseudo R-squared | 0.491161 | Mean dependent var | -20.73083 | |
| Adjusted R-squared | 0.462666 | S.D. dependent var | 50.94502 | |
| S.E. of regression | 35.39193 | Objective | 1268.485 | |
| Quantile dependent var | -5.400000 | Restr. objective | 2492.900 | |
| Sparsity | 48.99799 | Quasi-LR ststistic | 199.9127 | |
| Prob(Quasi-LR stat) | 0.000000 | | | |

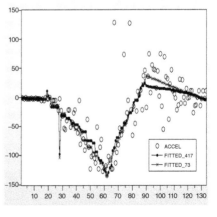

**Figure 7.3** Output of the NPQR (7.3a), and the graphs of *ACCEL*, *Fitted_417*, and *Fitted_73*.

Covariance Analysis: Ordinary
Date: 01/24/21  Time: 19:00
Sample: 1 133
Included observations: 133

Correlation
t-Statistic

| Probability | ACCEL | FITTED_417 |
|---|---|---|
| ACCEL | 1.000000 | |
| | ----- | |
| | ----- | |
| FITTED_417 | 0.745022 | 1.000000 |
| | 12.78358 | ----- |
| | 0.0000 | ----- |
| FITTED_73 | 0.750299 | 0.987935 |
| | 12.98983 | 73.01250 |
| | 0.0000 | 0.0000 |

**Figure 7.4**  Bivariate correlation of *ACCEL*, *Fitted_417* and *Fitted_73*

Dependent variable:ACCEL
Method: quantile regression (median)
Date: 08/19/18   Time: 06:35
Sample: 1 133
Included observations: 133
Huber sandwitch standard errors & covariance
Sparsity method: Kernel(Epanechnikov) using residuals
Bandwidth method: Hall-Sheather, bw=0.19033
Estimation successfully identifies unique optimal solution

| Variable | Coefficient | Std. error | t-Statistic | Prob. |
|---|---|---|---|---|
| C | -0.900000 | 5.230728 | -0.172060 | 0.8667 |
| INT4D=2 | 376.8311 | 96.88780 | 3.889355 | 0.0002 |
| INT4D=3 | -285.0031 | 2623.622 | -0.108630 | 0.9137 |
| INT4D=4 | 369.3907 | 585.7261 | 0.630654 | 0.5294 |
| R_MILLI*(INT4D=1) | -0.180000 | 0.347987 | -0.517261 | 0.6059 |
| R_MILLI*(INT4D=2) | -17.17383 | 3.725788 | -4.609449 | 0.0000 |
| R_MILLI*(INT4D=3) | 2.507500 | 59.50843 | 0.042137 | 0.9665 |
| R_MILLI*(INT4D=4) | -5.748626 | 9.910528 | -0.58052 | 0.5629 |
| R_MILLI^2*(INT4D=2) | 0.152629 | 0.033902 | 4.502051 | 0.0000 |
| R_MILLI^2*(INT4D=3) | 0.010625 | 0.335776 | 0.031643 | 0.9748 |
| R_MILLI^2*(INT4D=4) | 0.022253 | 0.041808 | 0.532264 | 0.5955 |

| | | | | |
|---|---|---|---|---|
| Pseudo R-squared | 0.474671 | Mean dependent var | -20.73083 | |
| Adjusted R-squared | 0.431611 | S.D. dependent var | 50.94502 | |
| S.E. of regression | 36.59118 | Objective | 1309.593 | |
| Quantile dependent var | -5.400000 | Restr. objective | 2492.900 | |
| Sparsity | 48.92605 | Quasi-LR ststistic | 193.4851 | |
| Prob(Quasi-LR stat) | 0.000000 | | | |

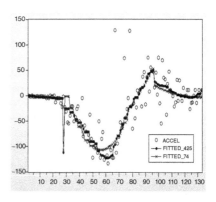

**Figure 7.5**  Output of the NPQR (7.4a), and the graphs of *ACCEL*, *Fitted_425*, and *Fitted_74*.

Based on these statistical results, the following findings and notes are presented:

1) Compared with the QR (4.25), which has an adjusted $R$-squared of 0.4772, this NPQR has a smaller adjusted $R$-squared of 0.4318. So, this NPQR is a worse fit model than its corresponding QR.

2) At the 10% level of significance, $R\_Milli\hat{}\,2*(Int4b = 3)$ and $R\_Milli\hat{}\,2*(Intab = 4)$ have insignificant effects with very large $p$-values of 0.975 and 0.5629, respectively. But $Milli\hat{}\,2*(Int4b = 3)$ and $Milli\hat{}\,2*(Int4b = 4)$ have smaller $p$-values of 0.502 and 0.182, respectively. So, at the 10% level, $Milli\hat{}\,2*(Int4b = 4)$ has a significant positive effect, based on the $t$-statistic of $t_0 = 1.342$ with a $p$-value with a $p$-value $= 0.182/2 < 0.10$.

3) As an additional output, Figure 7.6 presents the bivariate correlations between *ACCEL* *Fitted_425* and *Fitted_74,* which shows that *Fitted_425* and *Fitted_74* have a very large correlation of 0.983 043.

**Figure 7.6** Bivariate correlation of *ACCEL*, *Fitted_425* and *Fitted_74*

Covariance Analysis: Ordinary
Date: 08/23/18 Time: 07:59
Sample: 1 133
Included observations: 133

| Correlation<br>t-Statistic<br>Probability | ACCEL | FITTED_425 |
|---|---|---|
| ACCEL | 1.000000 | |
| | ---- | |
| | ---- | |
| FITTED_425 | 0.756758 | 1.000000 |
| | 13.25005 | ---- |
| | 0.0000 | ---- |
| FITTED_74 | 0.739085 | 0.983043 |
| | 12.55800 | 61.35839 |
| | 0.0000 | 0.0000 |

## Example 7.4 *An NPQR Corresponding to the QR (4.28)*

Corresponding to the QR (4.28), Figure 7.7 presents the output of an NPQR using the following ESs. This figure also presents the graphs of *ACCEL*, its fitted variable *Fitted_75*, and *Fitted_428* as the fitted variable of QR(Median) in (4.28):

$$ACCEL = (C(10) + C(11)*R\_MILLI)*(INT3 = 1)$$
$$+ (C(20) + C(21)*R\_MILLI + C(22)*R\_MILLI^2)*(INT3 = 2)$$
$$+ (C(30) + C(31)*R\_MILLI + C(32)*R\_MILLI^2 + C(33)*R\_MILLI^3$$
$$+ C(34)*R\_MILLI^4 + C(35)*R\_MILLI^5)*(INT3 = 3) \tag{7.5a}$$

$$INT3 = 1 + 1*(Milli >= 15) + 1*(Milli >= 27) \tag{7.5b}$$

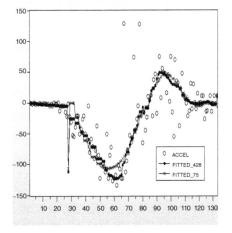

Dependent variable: ACCEL
Method: quantile regression (median)
Date: 08/19/18   Time: 07:24
Sample: 1 133
Included observations: 133
Huber sandwich standard errors & covariance
Sparsity method: Kernel (Epanechnikov) using residuals
Bandwidth method: Hall-Sheather, bw=0,19033
Estimation successfully identifies unique optimal solution
ACCEL=(C(10)+C(11)*R_MILLI)*(INT3=1)+(C(20)+C(21)*R_MILLI
+C(22)*R_MILLI^2)*(INT3=2)+(C(30)+C(31)*R_MILLI+C(32)
*R_MILLI^2+C(33)*R_MILLI^3+C(34)*R_MILLI^4+C(35)
*R_MILLI^5)*(INT3=3)

| | Coefficient | Std. error | t-Statistic | Prob. |
|---|---|---|---|---|
| C(10) | -0.900000 | 5.364239 | -0.167778 | 0.8670 |
| C(11) | -0.180000 | 0.355319 | -0.506588 | 0.6134 |
| C(20) | 388.5978 | 91.30048 | 4.256251 | 0.0000 |
| C(21) | -17.74221 | 3.453244 | -5.137837 | 0.0000 |
| C(22) | 0.158875 | 0.030603 | -5.191524 | 0.0000 |
| C(30) | 87068.94 | 99934.63 | 0.871259 | 0.3853 |
| C(31) | -4375.662 | 4800.007 | -0.911595 | 0.3638 |
| C(32) | 86.56614 | 91.57601 | 0.945293 | 0.3464 |
| C(33) | -0.843402 | 0.867550 | -0.972165 | 0.3329 |
| C(34) | 0.004051 | 0.004082 | 0.992548 | 0.3229 |
| C(35) | -7.69E-06 | -7.69E-06 | -1.007053 | 0.3159 |

| | | | |
|---|---|---|---|
| Pseudo R-squared | 0.470748 | Mean dependent var | -20.73083 |
| Adjusted R-squared | 0.427367 | S.D. dependent var | 50.94502 |
| S.E of regression | 36.95390 | Objective | 1319.372 |
| Quantile dependent var | -5.400000 | Restr. objective | 2492.900 |
| Sparsity | 48.53025 | | |

**Figure 7.7** Output of the NPQR (7.5a), and the graphs of *ACCEL*, *Fitted_428*, and *Fitted_75*.

Covariance Analysis: Ordinary
Date: 08/23/18 Time: 08:06
Sample: 1 133
Included observations: 133

| Correlation<br>t-Statistic<br>Probability | ACCEL | FITTED_428 |
|---|---|---|
| ACCEL | 1.000000 | |
| | ---- | |
| | ---- | |
| FITTED_428 | 0.756623 | 1.000000 |
| | 13.24451 | ---- |
| | 0.0000 | ---- |
| FITTED_75 | 0.736412 | 0.981956 |
| | 12.45856 | 59.43037 |
| | 0.0000 | 0.0000 |

**Figure 7.8** Bivariate correlation of *ACCEL*, *Fitted_428* and *Fitted_75*

Based on these statistical results, the following findings and notes are presented:

1) Compared with the QR (4.28), which has the adjusted $R$-squared of 0.4772, this NPQR has a smaller adjusted $R$-squared of 0.4274. So, this NPQR(Median) is a worse fit model than its corresponding QR.

2) At the 10% level of significance, each of the IVs of the fifth-degree polynomial NPQR in the interval INT3 = 3 has an insignificant adjusted effect. Similarly, each of the IVs of the fifth-degree polynomial QR (4.28) in the interval INT3 = 3 also has an insignificant adjusted effect.

3) As an additional output, Figure 7.8 presents the bivariate correlations between *ACCEL* *Fitted_428* and *Fitted_75*, which shows *Fitted_428* and *Fitted_75* have a very large correlation of 0.981 956. So, both QR and NPQR are acceptable median regressions, in the statistical sense.

**Example 7.5** *An Application of a Special NPQR of ACCEL*

Figure 7.9 presents a part of the output of a special NPQR of *ACCEL* on @Expand(R_Milli), using the following ES:

$$ACCEL\ C\ @Expand\ (R\_Milli, @Dropfirst) \tag{7.6}$$

Based on this output, the following findings and notes are presented:

1) In previous models in (7.5), the variable *R_Milli* is used as a numerical variable, but in this model, *R_Milli* is used as a categorical variable having 94 levels, with 133 observations.

2) Compared with all the median regressions of *ACCEL*, this NPQR has the greatest pseudo $R$-squared of 0.797 4065. So, its IVs, *@Expand(R_Milli,@Dropfirst)* as a set of (94-1) dummy variables, can explain 79.7% the total variance of *ACCEL*.

3) However, this NPQR has the smallest adjusted $R$-squared, because it has a large number of IVs, that is (94-1) dummies.

Dependent Variable: ACCEL
Method: Quantile Regression (Median)
Date: 08/20/18 Time: 09:21
Sample: 1 133
Included observations: 133
Huber Sandwich Standard Errors & Covariance
Sparsity method: Kernel (Epanechnikov) using residuals
Bandwidth method: Hall-Sheather, bw=0.19033
Estimation successful but solution may not be unique

| Variable | Coefficient | Std. Error | t-Statistic | Prob. |
|---|---|---|---|---|
| C | 0.000000 | 10.57863 | 0.000000 | 1.0000 |
| R_MILLI=2 | -1.300000 | 14.96044 | -0.086896 | 0.9312 |
| R_MILLI=3 | -2.700000 | 14.96044 | -0.180476 | 0.8577 |
| R_MILLI=4 | 0.000000 | 14.96044 | 0.000000 | 1.0000 |
| R_MILLI=5 | -2.700000 | 14.96044 | -0.180476 | 0.8577 |
| R_MILLI=6 | -2.700000 | 14.96044 | -0.180476 | 0.8577 |
| R_MILLI=7 | -2.700000 | 14.96044 | -0.180476 | 0.8577 |
| R_MILLI=8 | -1.300000 | 14.96044 | -0.086896 | 0.9312 |
| R_MILLI=9 | -2.700000 | 14.96044 | -0.180476 | 0.8577 |
| R_MILLI=10 | -2.700000 | 14.96044 | -0.180476 | 0.8577 |

| Variable | Coefficient | Std. Error | t-Statistic | Prob. |
|---|---|---|---|---|
| R_MILLI=120 | -1.300000 | 14.96044 | -0.086896 | 0.9312 |
| R_MILLI=121 | 0.000000 | 14.96044 | 0.000000 | 1.0000 |
| R_MILLI=122 | 10.70000 | 14.96044 | 0.715220 | 0.4787 |
| R_MILLI=123 | 10.70000 | 14.96044 | 0.715220 | 0.4787 |
| R_MILLI=125 | -14.70000 | 14.93771 | -0.984086 | 0.3311 |
| R_MILLI=126 | -13.30000 | 14.96044 | -0.889011 | 0.3794 |
| R_MILLI=127 | 0.000000 | 14.96044 | 0.000000 | 1.0000 |
| R_MILLI=128 | 10.70000 | 14.96044 | 0.715220 | 0.4787 |
| R_MILLI=129 | -14.70000 | 14.96044 | -0.982591 | 0.3319 |
| R_MILLI=131 | 10.70000 | 15.71809 | 0.680744 | 0.5001 |
| R_MILLI=132 | -2.700000 | 14.96044 | -0.180476 | 0.8577 |
| R_MILLI=133 | 10.70000 | 14.96044 | 0.715220 | 0.4787 |

| | | | |
|---|---|---|---|
| Pseudo R-squared | 0.797465 | Mean dependent var | -20.73083 |
| Adjusted R-squared | 0.314496 | S.D. dependent var | 50.94502 |
| S.E. of regression | 39.74419 | Objective | 504.9000 |
| Quantile dependent var | -5.400000 | Restr. objective | 2492.900 |
| Sparsity | 26.27967 | Quasi-LR statistic | 605.1826 |
| Prob(Quasi-LR stat) | 0.000000 | | |

**Figure 7.9** A part of the output of the NPQR in (7.6).

4) Based on the *Quasi-LR* statistic of $QLR_0 = 605.1828$ with a *Prob.* = 0.000 000, we can conclude that medians of *ACCEL* have significant differences between the 94 levels of *R_Milli*. In other words, *@Expand(R_Milli,@Dropfirst)* has a significant effect on *ACCEL*. There are 88 of 133 observations where the scores of *ACCEL* = *FITTED,* which can be identified by generating a dummy variable *DAF = 1\*(ACCEL = FITTED)* and then a one-way tabulation for *DAF* can show the number of *DAF* = 1.

5) The *t*-statistic of each of the IVs can be used to test the median difference of *ACCEL* with *R_Milli* = 1 as the reference level or group.

6) As an additional illustration, Figure 7.10 presents the graphs of *ACCEL* and *FITTED,* which is the fitted value variable of the NPQR in (7.6).

7) As more detailed information, Table 7.1 presents three selected groups of the observed scores of *ACCEL* and its fitted value variable, *FITTED*. Based on this table, the following findings and notes are presented:

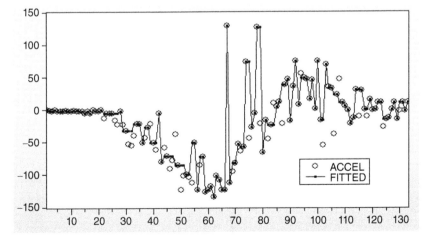

**Figure 7.10** The graphs of *ACCEL* and *FITTED*, which is the fitted value variable of NPQR (7.6).

**Table 7.1** Three selected groups of the *ACCEL* and *FITTED* scores.

| | First group | | | Second group | | | Third group | |
|---|---|---|---|---|---|---|---|---|
| **Obs** | **ACCEL** | **FITTED** | **Obs** | **ACCEL** | **FITTED** | **Obs** | **ACCEL** | **FITTED** |
| 1 | 0 | 0 | 22 | −13.3 | −5.4 | 67 | 128.5 | 128.5 |
| 2 | −1.3 | −1.3 | 23 | −5.4 | −5.4 | 68 | −112.5 | −112.5 |
| 3 | −2.7 | −2.7 | 24 | −5.4 | −5.4 | 69 | −95.1 | −81.8 |
| 4 | 0 | 0 | 25 | −5.3 | −5.4 | 70 | −81.8 | −81.8 |
| 5 | −2.7 | −2.7 | 26 | −16 | −5.4 | — | | |
| 6 | −2.7 | −2.7 | 27 | −22.8 | −5.4 | 78 | 127.1 | 127.1 |
| 7 | −2.7 | −2.7 | 28 | −2.7 | −2.7 | 79 | −21.5 | 127.1 |
| 8 | −1.3 | −1.3 | 29 | −22.8 | −32.1 | — | | |
| 9 | −2.7 | −2.7 | 30 | −32.1 | −32.1 | 82 | −45.6 | −24.2 |
| 10 | −2.7 | −2.7 | 31 | −53.5 | −32.1 | 83 | −24.2 | −24.2 |
| 11 | −1.3 | −1.3 | 32 | −54.9 | −32.1 | 84 | 9.5 | −24.2 |

7.1 The first group presents the observed scores *ACCEL* equal to its fitted value variable.

7.2 The second group presents an unexpected *FITTED* score of −5.4, corresponding to various six observed scores of *ACCEL* from −22.8 to −5.3. Similarly for the four *FITTED* scores −32.1.

7.3 And the third group presents an unbelievable *FITTED* score of 127.1, corresponding to the observed scores of 127.1 and −21.5. As a comparison, note that the two observed scores of *ACCEL* = 128.5 and −112.5 for observations number 67 and 68 are equal to the fitted values. Whatever the results, they should be accepted, in the statistical sense, because the results are obtained using the valid specific estimation method for the QR, and the results also are highly dependent on the data used.

### 7.2.2 Polynomial NPQRs of *ACCEL* on *R_Milli*

As the extension or modification of the polynomial QRs of *ACCEL* on *Milli* in the ES (4.29), this section presents the *k*-th degree polynomial NPQR models with the following general ES. The integer *k* should be obtained using the trial-and-error method.

$$ACCEL\ C\ R\_Milli\ R\_Milli^{\wedge}2 \ldots R\_Milli^{\wedge}k \tag{7.7}$$

**Example 7.6** *An Application of the Sixth-Degree Polynomial NPQR in (7.7)*
Table 7.2 presents the statistical results summary of a sixth-degree polynomial NPQR(tau) for tau = 0.1, 0.5, and 0.9. Based on this summary, the following findings and notes are presented:

1. Note that the joint effects of the IVs of each polynomial NPQR are significant based on the *Quasi-LR* statistic with a *p*-value = 0.0000.

**Table 7.2** Statistical results summary the sixth-degree polynomial NPQR(tau)s for tau = 0.1, 0.5, and 0.9.

| Variable | NPQR(tau = 0.1) | | | NPQR(tau = 0.5) | | | NPQR(tau = 0.9) | | |
|---|---|---|---|---|---|---|---|---|---|
| | Coef. | t-Stat. | Prob. | Coef. | t-Stat. | Prob. | Coef. | t-Stat. | Prob. |
| C | −9.514 | −1.017 | 0.311 | 2.510 | 0.246 | 0.806 | 0.159 | 0.008 | 0.993 |
| R_MILLI | 1.411 | 0.546 | 0.586 | −2.567 | −1.032 | 0.304 | −0.239 | −0.038 | 0.970 |
| R_MILLI^2 | −0.028 | −0.137 | 0.891 | 0.367 | 1.962 | 0.052 | 0.086 | 0.155 | 0.877 |
| R_MILLI^3 | −0.006 | −0.930 | 0.354 | −0.019 | −3.227 | 0.002 | −0.006 | −0.331 | 0.741 |
| R_MILLI^4 | 0.000 | 1.780 | 0.077 | 0.000 | 4.170 | 0.000 | 0.000 | 0.507 | 0.613 |
| R_MILLI^5 | 0.000 | −2.349 | 0.020 | 0.000 | −4.797 | 0.000 | 0.000 | −0.665 | 0.507 |
| R_MILLI^6 | 0.000 | 2.706 | 0.008 | 0.000 | 5.199 | 0.000 | 0.000 | 0.804 | 0.423 |
| Pseudo R-sq. | 0.587 | | | 0.443 | | | 0.325 | | |
| Adj. R-sq | 0.567 | | | 0.417 | | | 0.293 | | |
| S.E. of reg | 50.692 | | | 36.411 | | | 59.619 | | |
| Quantile DV | −101.90 | | | −5.400 | | | 36.200 | | |
| Sparsity | 85.473 | | | 53.580 | | | 117.90 | | |
| Quasi-LR stat | 196.70 | | | 164.95 | | | 71.190 | | |
| Prob | 0.0000 | | | 0.0000 | | | 0.0000 | | |

2. Specific for the NPQR(0.5), each of its IVs has a significant effect at the 1% or 5% levels except that R_Milli has an insignificant effect with a $p$-value = 0.304. And each R_Milli^k, for $k > 3$ of the NPQR(tau = 0.1), has a significant effect at the 1%, 5%, or 10% level. So, both NPQR(0.1) and NPQR(0.5) are acceptable, in the statistical sense, to present sixth-degree polynomial NPQRs.

3. On the other hand, each IV of the NPQR(0.9) has a large $p$-value, with the smallest $p$-value = 0.423 for Milli^6. So, this regression is not a good NPQR. Then, if we try to explore their reduced models, we get the statistical results summary presented in Table 7.3. Based on this summary, the following findings and notes are presented:

1) The variables $(R\_Milli/10)^k$, for $k = 5$ and 6, are used as IVs, in order to have sufficiently large coefficients, compared with the corresponding R_Milli^k in the full models.

2) Each of the IVs of the three reduced NPQRs has either a positive or negative significant effect on ACCEL at the 1% or 5% level. For instance, the IV R_Milli^2 of the NPQR(0.1) has a positive significant effect, based on the $t$-statistic of $t_0 = 1.7709$ with a $p$-value $= 0.0790/2 = 0.0395 < 0.05$. So, the three reduced NPQRs are better sixth-degree polynomial NPQRs, in the statistical sense, specifically for the NPQR(tau = 0.9), compared with their full models.

3) It is unexpected that the reduced NPQR(0.1) and NPQR(0.9) have adjusted $R$-squared of 0.5967 and 0.2962, respectively, which are greater than their full models in Table 7.2, with adjusted $R$-squared of 0.567 and 0.293, respectively.

**Table 7.3** Statistical results summary the reduced NPQRs in Table 7.2.

| Variable | Reduced NPQR(tau = 0.1) | | | Reduced NPQR(tau = 0.5) | | | Reduced NPQR(tau = 0.9) | | |
|---|---|---|---|---|---|---|---|---|---|
| | Coef. | t-Stat. | Prob. | Coef. | t-Stat. | Prob. | Coef. | t-Stat. | Prob. |
| C | −4.0050 | −0.9607 | 0.3385 | −5.9280 | −1.1089 | 0.2696 | 7.6514 | 1.4553 | 0.1480 |
| R_MILLI^2 | 0.0869 | 1.7709 | 0.0790 | 0.1511 | 2.8053 | 0.0058 | | | |
| R MILLI^3 | −0.0097 | −4.1824 | 0.0001 | −0.0116 | −4.3324 | 0.0000 | −0.0022 | −2.3810 | 0.0187 |
| R MILLI^4 | 0.0002 | 5.2696 | 0.0000 | 0.0002 | 5.0553 | 0.0000 | 0.0001 | 2.9878 | 0.0034 |
| (R MILLI/10)^5 | −0.1823 | −5.6626 | 0.0000 | −0.1986 | −5.3810 | 0.0000 | −0.0720 | −3.3999 | 0.0009 |
| (R MILLI/10)^6 | 0.0052 | 5.7440 | 0.0000 | 0.0056 | 5.5053 | 0.0000 | 0.0023 | 3.6429 | 0.0004 |
| Pseudo R-sq | 0.5860 | | | 0.4382 | | | 0.3175 | | |
| Adj. R-sq | 0.5697 | | | 0.4161 | | | 0.2962 | | |
| S.E. of reg | 49.0406 | | | 35.8456 | | | 63.7670 | | |
| Quantile DV | −101.900 | | | −5.4000 | | | 36.2000 | | |
| Sparsity | 83.9527 | | | 57.9536 | | | 120.833 | | |
| Quasi-LR stat | 199.961 | | | 150.806 | | | 67.8224 | | |
| Prob. | 0.0000 | | | 0.0000 | | | 0.0000 | | |

4) Figure 7.11 shows the graphs of *ACCEL* and three its fitted value variables, based on the full and reduced NPQRs. The findings and notes from these graphs are as follows:

4.1 The two sets of graphs have slightly different patterns, especially at the first and last few observations.

4.2 Figure 7.12 presents some of the score differences between the observed scores of *ACCEL* and between pairs of its fitted value variables.

4.3 The fitted value of the reduced model (RM) for tau = 0.90, namely *Fit_RDM_Q90* = −2.7, is exactly equal to the observed *ACCEL*'s score = −2.7, the observation number 28.

4.4 Similar to the QRs presented in the previous chapter 4, these NPQRs also present unexpected results. The fitted variables of the full model (FM) and the reduced model (RM) for *tau* = 0.10, namely *Fit_FM_Q10* and *Fit_RDM_Q10*, have very large negative values of −126.9613 and −126.8114, respectively. We see similar results for the fitted values of *Fit_FM_Q50* = −96.8202 and *Fit_RDM_Q50* = −92.9361 and for the fitted value of *Fit_FM_Q90* = −15.5598.

### 7.2.3 Special Notes and Comments

Referring to all SPQRs and NPQRs, I have the following special notes and comments:

1) Whatever the statistical results are, they are acceptable in the statistical sense, under the assumption the model is a good QR, in the theoretical sense.

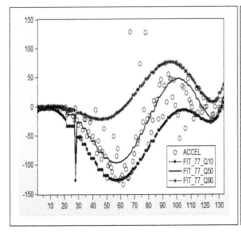 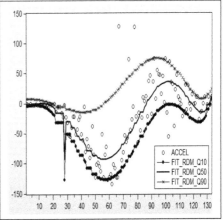

**Figure 7.11** Graphs of *ACCEL* and three of its fitted value variables based on the full and reduced NPQRs.

| Obs. | ACCEL | FIT_FM_Q10 | FIT_RDM_Q10 | FIT_FM_Q50 | FIT_RDM_Q50 | FIT_FM_Q90 | FIT_RDM_Q90 |
|---|---|---|---|---|---|---|---|
| 1 | 0 | −8.1371 | −3.9276 | 0.2917 | −5.7883 | 0.0000 | 7.6493 |
| 2 | −1.3 | −6.8506 | −3.7315 | −1.3 | −5.4128 | −0.0223 | 7.6348 |
| 3 | −2.7 | −5.6843 | −3.4672 | −2.3662 | −4.8628 | 0.0595 | 7.5912 |
| 4 | 0 | −4.6646 | −3.1805 | −2.9998 | −4.1944 | 0.2160 | 7.5272 |
| 5 | −2.7 | −3.8146 | −2.9126 | −3.2862 | −3.4581 | 0.4206 | 7.4170 |
| ... | | | | | | | |
| 2.7 | −22.8 | −34.4046 | −32.2369 | −13.3 | −15.8701 | −5.4 | −5.4 |
| **28** | **−2.7** | **−126.9613** | **−126.8314** | **−96.8202** | **−92.9361** | **−15.5598** | **−27** |
| 29 | −22.8 | −53.5 | −50.8666 | −27.2873 | −30.3312 | −10.7184 | −9.8460 |
| ... | | | | | | | |
| 130 | −2.7 | −9.2080 | −10.6067 | −2.7 | −2.7 | 22.38544 | 14.95315 |
| 131 | 10.7 | −9.2080 | −10.6067 | −2.7 | −2.7 | 22.38544 | 14.95315 |
| 132 | −2.7 | −2.7 | −2.7 | 6.0450 | 3.6197 | 28.81463 | 19.04728 |
| 133 | 10.7 | 5.1538 | 6.8503 | 16.9698 | 11.5778 | 36.72282 | 24.21908 |

**Figure 7.12** Some selected scores of *ACCEL* and three pairs of its fitted variables.

2) Referring to the ranks of *MILLI* from 1 to 94 for its 133 observed scores, most of the scores are single scores for predicting the quantiles of *ACCEL*. For example, in observation number 28, *ACCEL* has a single score of −2.7, as presented in the previous example. Only 26 of the ranks have from two to six multiple scores of the dependent variable (DV) *ACCEL*. Hence, (94 − 26) = 68 quantiles are estimates based on single observations. So the fitted values of *ACCEL* are estimated using 94 scores of R_Milli in *NPQR*, and they are estimated using 133 scores of *Milli* in SPQR.

3) So, in most cases, the fitted values of an SPQR(tau) of a numerical DV, say *Y*, for any values of 0 < tau <1, are estimated based on some or many single observed scores of *Y*. Then, in general, the fitted values are abstract predicted scores since they are not observed scores. For this reason, I propose another special NPQR on group of R_Milli, which I'll present in the following section.

## 7.3 NPQRs on Group of *R_Milli*

In order to be able to compute the statistical values of quantiles *tau* = 0.1 to 0.9, I used the following equation to generate a group variable based on *R_Milli* or the original variable *Milli*, *G_Milli*, with at least 10 observed scores within each level of *G_Milli*:

$$G\_Milli = 1 + 1^* (R\_milli > 10) + 1^* (R\_milli > 20) + 1^* (R\_milli > 30)$$
$$+ 1^* (R\_milli > 40) + 1^* (R\_milli > 50) + 1^* (R\_milli > 60)$$
$$+ 1^* (R\_milli > 70) + 1^* (R\_milli > 80) + 1^* (R\_milli > 90)$$
$$+ 1^* (R\_milli > 100) + 1^* (R\_milli > 110) + 1^* (R\_milli > 120) \qquad (7.8)$$

Note that since *G_Milli* is an ordinal categorical variable, it can be used either as dummy IVs or as a numerical IV of the NPQRs of *ACCEL*. See the following examples.

### 7.3.1 An Application of the *G_Milli* as a Categorical Variable

As opposed to the NPQR of *ACCEL* on the categorical variable *R_Milli* in (7.6), the following ES presents the NPQR of *ACCEL* on the categorical *G_Milli*:

$$ACCEL@Expand (G\_Milli) \qquad (7.9)$$

**Example 7.7** *Application of the NPQR (7.9)*
As an illustration, Table 7.4 presents the statistical summary of the NPQR (7.9), for the quantiles *tau* = 0.25, 0.50, and 0.75. Based on these results, the following findings and comments are presented:

1) The coefficients of each *G_Milli* = *k* are the quantiles or the descriptive statistics of *ACCEL* for *tau* = 0.25, 0.50, and 0.75, among the set of the observed scores of *ACCEL* within the *k*-th level of *G_Milli*. For instance, we have the quantiles of −2.7, and −1.3 for *tau* = 0.25, 0.50, and 0.75, respectively, within the level *G_Milli* = 1.
2) Figure 7.13 provides better information in the descriptive statistics of *ACCEL*, for the quantiles *tau* = 0.25, 0.50, and 0.75 within the level *G_Milli* = 1, as well as with a part of the scatter graphs of *ACCEL* on *G_Milli* with its fitted values. Based on the results in this figure, the following findings and notes are presented:
   2.1 Within the 10 observations in *G_Milli* = 1, there are only three different scores: 0, −1.3, and −2.7. So, the scatter graphs present three points, and all three fitted values are equal to the observed scores of *ACCEL*.
   2.2 In addition, Table 7.4 and the graph in Figure 7.13 show the three fitted values in *G_Milli* = 2 are equal to −2.7, and three different fitted values of −32.1, −16, and −5.4 in *G_Milli* = 3, with *ACCEL* has scores maximum = 0, and minimum = −54.9.
3) Furthermore, Figure 7.14 presents the scatter graphs of *ACCEL* and three of its fitted value variables, *Fitted_Q25*, *Fitted_Q50*, and *Fitted_Q75*, for the whole sample. Based on these graphs, the findings and notes are as follows:
   3.1 Unexpectedly, the graphs present double lines from the quantiles within the levels of *G_Milli* = 3 to 6. I can't find an explanation for this in the EViews User Guide.

**Table 7.4** Statistical results summary based on the NPQR (7.9), for three quantiles tau = 0.25, 0.50, and 0.75.

| Variable | QREG(tau = 0.25) | | | QREG(tau = 0.50) | | | QREG(tau = 0.75) | | |
|---|---|---|---|---|---|---|---|---|---|
| | Coef. | t-Stat. | Prob. | Coef. | t-Stat. | Prob. | Coef. | t-Stat. | Prob. |
| G_MILLI = 1 | −2.7 | −0.7238 | 0.4706 | −2.7 | −0.6585 | 0.5115 | −1.3 | −0.3665 | 0.7146 |
| G_MIILI = 2 | −2.7 | −0.7229 | 0.4711 | −2.7 | −0.6576 | 0.5121 | −2.7 | −0.7594 | 0.4491 |
| G_MILLI = 3 | −32.1 | −2.5558 | 0.0118 | −16 | −2.4126 | 0.0174 | −5.4 | −0.8508 | 0.3966 |
| G_MILLI = 4 | −50.8 | −6.1867 | 0.0000 | −40.2 | −4.2584 | 0.0000 | −21.5 | −2.5602 | 0.0117 |
| G_MILLI = 5 | −91.1 | −9.7088 | 0.0000 | −77.7 | −8.1528 | 0.0000 | −59 | −2.8082 | 0.0058 |
| G_MILLI = 6 | −123.1 | −12.7377 | 0.0000 | −99.1 | −7.1872 | 0.0000 | −72.3 | −3.1262 | 0.0022 |
| G_MILLI = 7 | −123.1 | −16.0893 | 0.0000 | −108.4 | −13.218 | 0.0000 | −95.1 | −8.4069 | 0.0000 |
| G_MILLI = 8 | −57.6 | −6.6160 | 0.0000 | −26.8 | −1.3842 | 0.1689 | −5.4 | −0.1999 | 0.8419 |
| G_MILLI = 9 | −21.5 | −2.2399 | 0.0269 | 4 | 0.2694 | 0.7880 | 12 | 0.9553 | 0.3414 |
| G_MILLI = 10 | 16 | 1.0084 | 0.3153 | 45.6 | 5.0413 | 0.0000 | 54.9 | 5.5077 | 0.0000 |
| G_MILLI = 11 | −16 | −0.4309 | 0.6673 | 10.7 | 0.6646 | 0.5076 | 34.8 | 3.1570 | 0.0020 |
| G_MILLI = 12 | −10.7 | −1.7282 | 0.0865 | −1.3 | −0.1738 | 0.8623 | 14.7 | 1.1593 | 0.2486 |
| G_MILLI = 13 | −13.3 | −1.9324 | 0.0557 | 0.00 | 0.0000 | 1.0000 | 10.7 | 1.9735 | 0.0507 |
| Pseudo R-sq | 0.608 | | | 0.485 | | | 0.3812 | | |
| Adj. R-sq | 0.569 | | | 0.433 | | | 0.3194 | | |
| S.E. of reg | 43 | | | 36.53 | | | 38.342 | | |
| Quantile OV | −50.8 | | | −5.4 | | | 0.000 | | |
| Mean DV | −20.73 | | | −20.73 | | | −20.73 | | |
| S.D. DV | 50.95 | | | 50.95 | | | 50.945 | | |
| Sparsity | 56.74 | | | 49.87 | | | 60.051 | | |

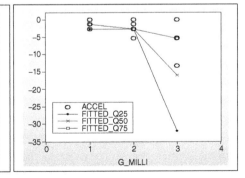

| Obs. | ACCEL | tau=0.25 | tau=0.50 | tau=0.75 |
|---|---|---|---|---|
| 1 | 0 | | | |
| 2 | −1.3 | | | |
| 3 | −2.7 | −2.7 | | |
| 4 | 0 | | | |
| 5 | −2.7 | | −2.7 | |
| 6 | −2.7 | | | |
| 7 | −2.7 | | | |
| 8 | −1.3 | | | 1.3 |
| 9 | −2.7 | | | |
| 10 | −2.7 | | | |

**Figure 7.13** Descriptive statistics of *ACCEL*, the quantiles tau = 0.25, 0.50, and 0.75 within the level *G_Milli* = 1, and a part of the scatter graphs of *ACCEL* and its fitted variables on *G_Milli*.

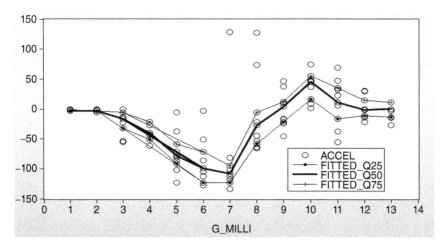

**Figure 7.14** The scatter graphs of *ACCEL* and three of its fitted value variables and of *ACCEL* on the ordinal categorical *G_Milli* in Table 7.4.

    3.2 Compared with the graphs in Figure 7.13, these graphs are worse in presenting the observed scores of *ACCEL* and its fitted values within the two levels *G_Milli* = 1 and 2.

4) We also can generate a group alternative, *GA_Milli* with 10 levels, directly, based on the variable *Milli*, using the following equation:

$$GA\_Milli = 1 + 1*(Milli > @Quantile\,(Milli, 0.1))$$

$$+\,1*(Milli > @Quantile\,(Milli, 0.2)) + 1*(Milli > @Quantile\,(Milli, 0.3))$$

$$+\,1*(Milli > @Quantile\,(Milli, 0.4)) + 1*(Milli > @Quantile\,(Milli, 0.5))$$

$$+\,1*(Milli > @Quantile\,(Milli, 0.6)) + 1*(Milli > @Quantile\,(Milli, 0.7))$$

$$+\,1*(Milli > @Quantile\,(Milli, 0.8)) + 1*(Milli > @Quantile\,(Milli, 0.9))$$

$$(7.10)$$

### 7.3.2   The *k*th-Degree Polynomial NPQRs of *ACCEL* on *G_Milli*

Referring to the *k*th-degree polynomial model of *ACCEL* on *Milli* in ES (4.9), this section presents an alternative *k*th-degree polynomial SPQR, with the following general ES, with *G_Milli* used as a numerical variable:

$$ACCEL\ C\ G\_Milli\ G\_Milli\,\hat{}\,2\ldots G\_Milli\,\hat{}\,k \tag{7.11}$$

**Example 7.8**   *An Application of the SPQR in (9.11) for* **k = 6**
As an illustration, Table 7.5 presents the statistical results summary based on the sixth-degree polynomial SPQR in (7.10), for three quantiles tau = 0.10, 0.50, and 0.90. Based on this summary, the following findings and notes are presented:

1) SPQR(tau = 0.5) is an acceptable sixth-degree polynomial model because each of its IVs has a significant adjusted effect at the 1% level of significance.

**Table 7.5** Statistical results summary based on the SPQR (7.11), for three quantiles tau = 0.10, 0.50, and 0.90.

| Variable | SPQR(tau = 0.1) | | | SPQR(tau = 0.5) | | | SPQR(tau = 0.9) | | |
|---|---|---|---|---|---|---|---|---|---|
| | Coef. | t-Stat. | Prob. | Coef. | t-Stat. | Prob. | Coef. | t-Stat. | Prob. |
| C | −8.192 | −0.160 | 0.873 | 81.236 | 2.080 | 0.040 | 0.671 | 0.006 | 0.995 |
| G_MILLI | −5.876 | −0.063 | 0.950 | −182.088 | −2.661 | 0.009 | −7.369 | −0.038 | 0.970 |
| G_MILLI^2 | 23.621 | 0.416 | 0.678 | 137.376 | 3.383 | 0.001 | 11.842 | 0.097 | 0.923 |
| G_MILLI^3 | −14.926 | −0.968 | 0.335 | −45.667 | −4.216 | 0.000 | −6.317 | −0.186 | 0.853 |
| G_MILLI^4 | 2.890 | 1.399 | 0.164 | 6.903 | 4.853 | 0.000 | 1.273 | 0.282 | 0.779 |
| G_MILLI^5 | −0.224 | −1.682 | 0.095 | −0.474 | −5.275 | 0.000 | −0.104 | −0.368 | 0.714 |
| G_MILLI^6 | 0.006 | 1.855 | 0.066 | 0.012 | 5.532 | 0.000 | 0.003 | 0.438 | 0.662 |
| Pseudo R-sq | 0.565 | | | 0.461 | | | 0.314 | | |
| Adj. R-sq | 0.544 | | | 0.435 | | | 0.281 | | |
| S.E. of reg | 51.067 | | | 37.525 | | | 63.843 | | |
| Quantile DV | −101.900 | | | −5.400 | | | 36.200 | | |
| Sparsity | 91.508 | | | 48.833 | | | 121.499 | | |
| Quasi-LR stat | 176.746 | | | 188.116 | | | 66.686 | | |
| Prob(Quasi-L) | 0.000 | | | 0.000 | | | 0.000 | | |

2) Specific for SPQR(tau = 0.1), because $G\_Milli^6$ has a significant effect on *ACCEL* at the 10% level, this is an acceptable sixth-degree polynomial SPQR, in the statistical sense, even though each of $G\_Milli^k$, for $k < 5$, has an insignificant effect. By deleting $G\_Milli$ from the model, each IV of the reduced model has significant effect at the 1% level of significance, as presented in Figure 7.15a. The fitted variables of the full SPQR and

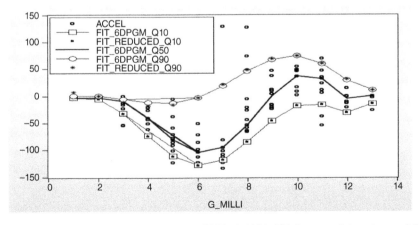

**Figure 7.15** The outputs of the reduced SPQRs in Table 7.5, for tau = 0.1 and tau = 0.9. (a) Reduced SPQR(tau = 0.1). (b) Reduced SPQR(tau = 0.9).

its reduced model have a correlation coefficient of 0.999 971. So, the full SPQR can be replaced by its reduced model, as a good-fit model. In addition, the reduced model has a greater adjusted *R*-squared than the full SPQR.

3) On the other hand, the SPQR(tau = 0.9) is not a good SPQR, because each of its IVs has large and very large *p*-values. However, by deleting *G_Milli* and *G_Milli^2* from the model, each IV of the reduced model also has a significant effect at the 1% level, as presented in Figure 7.15b. Since the fitted variables of the full SPQR and its reduced model have a very large correlation coefficient of 0.994 809, we can consider the reduced SPQR to be a good-fit sixth-degree polynomial SPQR.

4) As an additional illustration, Figure 7.16 presents the scatter graph of *ACCEL*, the fitted value variables of the SPQR in Table 7.5 (*Fit_6DPGM_Q10*, *Fit_6DPGM_Q50*, and *Fit_6DPGM_Q90*) and the fitted value variables of the reduced SPQR in Figure 7.15 (*Fit_Reduced_Q10*, and *Fit_Reduced_Q90*). These graphs clearly show most of the scores of the fitted variables of the reduced models are equal or very close to their corresponding full models.

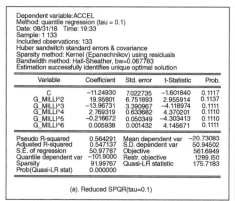

Dependent variable:ACCEL
Method: quantile regression (tau = 0.1)
Date: 08/31/18   Time: 19:33
Sample: 1 133
Included observations: 133
Huber sandwitch standard errors & covariance
Sparsity method: Kernel (Epanechnikov) using residuals
Bandwidth method: Hall-Sheather, bw=0.067783
Estimation successfully identifies unique optimal solution

| Variable | Coefficient | Std. error | t-Statistic | Prob. |
|---|---|---|---|---|
| C | −11.24930 | 7.022735 | −1.601840 | 0.1117 |
| G_MILL^2 | 19.95801 | 6.751893 | 2.955914 | 0.1137 |
| G_MILL^3 | −13.96731 | 3.390967 | −4.118974 | 0.1111 |
| G_MILL^4 | 2.769319 | 0.633682 | 4.370201 | 0.1110 |
| G_MILL^5 | −0.216672 | 0.050349 | −4.303413 | 0.1110 |
| G_MILL^6 | 0.005938 | 0.001432 | 4.145671 | 0.1111 |

| | | | |
|---|---|---|---|
| Pseudo R-squared | 0.564291 | Mean dependent var | −20.73083 |
| Adjusted R-squared | 0.547137 | S.D. dependent var | 50.94502 |
| S.E. of regression | 50.97787 | Objective | 561.6949 |
| Quantile dependent var | −101.9000 | Restr. objective | 1289.l50 |
| Sparsity | 91.99767 | Quasi-LR statistic | 175.7183 |
| Prob(Quasi-LR stat) | 0.000000 | | |

(a). Reduced SPQR(tau=0.1)

Dependent variable:ACCEL
Method: quantile regression (tau = 0.9)
Date: 08/31/18 Time: 19:36
Sample: 1 133
Included observations: 133
Huber sandwitch standard errors & covariance
Sparsity method: Kernel (Epanechnikov) using residuals
Bandwidth method: Hall-Sheather, bw=0.067783
Estimation successfully identifies unique optimal solution

| Variable | Coefficient | Std. error | t-Statistic | Prob. |
|---|---|---|---|---|
| C | 9.095920 | 5.402951 | 1.683509 | 0.0947 |
| G_MILL^3 | −2.285977 | 0.728503 | −3.137911 | 0.0021 |
| G_MILL^4 | 0.700503 | 0.217459 | 3.221308 | 0.10016 |
| G_MILL^5 | −0.067123 | 0.022229 | −3.1019628 | 0.0031 |
| G_MILL^6 | 0.002059 | 0.000749 | 2.750414 | 0.0068 |

| | | | |
|---|---|---|---|
| Pseudo R-squared | 0.300112 | Mean dependent var | −20.73083 |
| Adjusted R-squared | 0.278240 | S.D. dependent var | 50.94502 |
| S.E. of regression | 63.58362 | Objective | 812.9762 |
| Quantile dependent var | 36.20000 | Restr. objective | 1161.580 |
| Sparsity | 120.4540 | Quasi-LR statistic | 64.31292 |
| Prob(Quasi-LR stat) | 0.000000 | | |

(b). Reduced SPQR(tau=0.9)

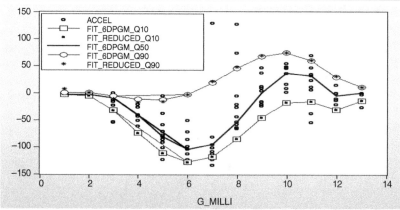

**Figure 7.16** The scatter graphs of *ACCEL*, fitted value variables of SPQRs in Table 7.5, and fitted value variables of the reduced SPQR in Figure 7.15 on *G_Milli*.

## 7.4 Multiple NPQRs Based on Data-Faad.wf1

### 7.4.1 An NPQR Based on a Triple Numerical Variable (*X1,X2,Y*)

As the modification of the two-way interaction QRs of *Y1* on *X1* and *X2* in Section 5.2.2, based on the triple numerical variables (*X1,X2,Y1*), this section presents a two-way interaction NPQR of *Y1* on the ranks of *X1* and *X2*, *RX1* and *RX2*, which are generated using the following equation, for $k = 1$ and 2:

$$RXk = @Round\,(@Ranks\,(Xk,a)) \tag{7.12}$$

Hence the two-way NPQR has the following ES:

$$Y1\ RX1{*}RX2\ RX1\ RX2\ C \tag{7.13}$$

### 7.4.2 NPQRs with Multi-Rank Predictors

Referring to the three-way and two-way interaction QRs of *Y1* on multiple numerical predictors *Xk*, for various integers *k*, presented in ESs (5.6) and (6.4), this section presents alternative NPQRs of *Y1* on multi-rank predictors *RXk*.

**Example 7.9** *An Application of a Full Two-Way Interaction NPQR*
Referring to the full two-way interaction quantile regression (F2WI-QR) of *Y2* on *Y1*, *X1*, and *X2*, based on Data_faad.wf1, in (6.6), this example presents the application of a full two-way interaction nonparametric quantile regression (F2WI-NPQR), with the following ES:

$$Y2\ C\ RY1/10{*}RX1/10\ \ RY1/10{*}RX2/10\ \ RX1/10{*}RX2/10$$
$$RY1/10\ \ RX1/10\ \ RX2/10 \tag{7.14}$$

Note that the scaled variables *RY1/10*, *RX1/10*, and *RX2/10* should be used to ensure the coefficients of the regression are sufficiently large. Table 7.6 presents the statistical results summary of five reduced 2WI-NPQR(Median)s in (7.14), namely RM-1 to RM-5, which are obtained using the manual multistage selection method (MMSM). Based on this summary, the following findings and notes are presented:

1) At the first stage, RM-1 is obtained as the simplest 2WI-NPQR(Median) having only two-way interaction IVs. Since each IV has a significant effect at the 1% level, RM-1 is an acceptable 2WI-QR(Median).
2) At the second stage, inserting each of the main variables, *RY1/10*, *RX1/10*, and *RX2/10*, as an additional IV for RM-1 results in RM-2, RM-3, and RM-4. In this case, we can consider RM-3 and RM-4 to be acceptable 2WI-QRs, because each of their IVs has either a positive or negative significant effect at 1%, 2%, or 10% level. For instance, at the 10% level, the interaction *RY1\*RX1/100* in RM-3 has a positive significant effect based on the *t*-statistic of $t_0 = 1.4870$ with $df4 = 95$ and a *p*-value $= 0.1381/2 = 0.0695$.
3) At the final stage, RM-5 is obtained by using both variables *RX1/10* and *RX2/10* as additional IVs for RM-1. Note that, at the 10% level, the interaction *RY1\*RX1/100* in RM-5 has a positive significant effect based on the *t*-statistic of $t_0 = 1.3710$ with $df = 294$ and a *p*-value $= 0.1714/2 = 0.0857$. So, RM-5 is the final 2WI-NPQR(Median).

**Table 7.6** Statistical result summary of the manual multistage selection method, based on the F2WI-NPQR(Median) in (7.14).

| RM-1 | C | RY1/10*RX1/10 | Y1/10*X2/10 | X1/10*X2/10 | | | |
|---|---|---|---|---|---|---|---|
| Coef. | 4.7549 | 0.0024 | 0.0024 | −0.0012 | | | |
| t-Stat | 55.100 | 7.4112 | 7.1645 | −3.2368 | | | |
| Prob. | 0.0000 | 0.0000 | 0.0000 | 0.0013 | | | |
| Ps. R^2 = 0.3558 | | Adj R^2 = 0.3493 | | | Quasi-LR = 203.72 | | Prob. = 0.0000 |
| **RM-2** | C | RY1/10*RX1/10 | Y1/10*X2/10 | X1/10*X2/10 | RY1/10 | | |
| Coef. | 4.1149 | −0.0003 | −0.0006 | 0.0013 | 0.1016 | | |
| t-Stat | 27.3484 | −0.6295 | −0.9858 | 2.2722 | 5.4007 | | |
| Prob. | 0.0000 | **0.5295** | **0.3250** | 0.0238 | 0.0000 | | |
| Ps. R^2 = 0.3957 | | Adj R^2 = 0.3875 | | | Quasi-LR = 243.01 | | Prob. = 0.0000 |
| **RM-3** | C | RY1/10*RX1/10 | Y1/10*X2/10 | X1/10*X2/10 | | RX1/10 | |
| Coef. | 4.4424 | 0.0009 | 0.0034 | −0.0024 | | 0.0515 | |
| t-Stat | 24.8922 | 1.4870 | 6.8016 | −4.1300 | | 2.5610 | |
| Prob. | 0.0000 | 0.1381 | 0.0000 | 0.0000 | | 0.0109 | |
| Ps. R^2 = 0.3719 | | Adj R^2 = 0.3634 | | | Quasi-LR = 214.49 | | Prob. = 0.0000 |
| **RM-4** | C | RY1/10*RX1/10 | Y1/10*X2/10 | X1/10*X2/10 | | | RX2/10 |
| Coef. | 4.7259 | 0.0025 | 0.0023 | −0.0013 | | | 0.0039 |
| t-Stat | 29.3808 | 4.9333 | 4.1712 | −2.1997 | | | 0.2257 |
| Prob. | 0.0000 | 0.0000 | 0.0000 | 0.0286 | | | **0.8216** |
| Ps. R^2 = 0.3559 | | Adj R^2 = 0.3472 | | | Quasi-LR = 203.75df | | Prob. = 0.0000 |
| **RM-5** | C | RY1/10*RX1/10 | Y1/10*X2/10 | X1/10*X2/10 | | RX1/10 | RX2/10 |
| Coef. | 4.4152 | 0.0009 | 0.0033 | −0.0025 | | 0.0512 | 0.0046 |
| t-Stat | 18.4515 | 1.3710 | 5.2639 | −3.2840 | | 2.5532 | 0.2638 |
| Prob. | 0.0000 | 0.1714 | 0.0000 | 0.0011 | | 0.0112 | **0.7921** |
| Ps. R^2 = 0.3721 | | Adj R^2 = 0.3614 | | | Quasi-LR = 215.13 | | Prob. = 0.0000 |

**Example 7.10** *An Application of Three-Way Interaction NPQRs*

As an extension of the F2WI-NPQR in (7.14), Table 7.7 presents the statistical results summary of four reduced models of the full three-way interaction nonparametric-quantile regression (F3WI-NPQR(0.5)), namely RM-1 to RM-4, with the following ES:

$$Y2\ C\ RY1*RX1*RX2/1000\ RY1*RX1/100\ RY1*RX2/100$$
$$RX1*RX2/100\ RY1/10\ RX1/10\ RX2/10 \tag{7.15}$$

Based on this summary, the findings and notes are as follows:

1) The three-way interaction IV of each of the four median regressions has either a positive or negative significant effect on $Y2$ at the 1% or 10% level of significance. Note that, at the

**Table 7.7** Statistical results summary of the reduced F3WI-NPQR(0.5) in (7.15).

| RM-1 | C | RY1*RX1*RX2/ 1000 | RY1*RX1/ 100 | RY1*RX2/ 100 | RX1*RX2/ 100 | | |
|---|---|---|---|---|---|---|---|
| Coef. | 4.465 25 | −0.000 16 | 0.003 59 | 0.003 49 | 0.000 59 | | |
| t-Stat | 33.055 22 | −3.365 94 | 5.589 49 | 7.871 58 | 0.841 86 | | |
| Prob. | 0.000 00 | 0.000 90 | 0.000 00 | 0.000 00 | **0.400 60** | | |
| Ps. $R^2 = 0.3748$ | | Adj $R^2 = 0.3663$ | | Quasi-LR $= 213.66$ | | Prob. $= 0.0000$ | |
| RM-2 | C | RY1*RX1*RX2/ 1000 | RY1*RX1/ 100 | RY1*RX2/ 100 | RX1*RX2/ 100 | RX1/ 10 | |
| Coef. | 4.374 078 | −6.47E-05 | 0.001 62 | 0.003 71 | −0.001 45 | 0.040 96 | |
| t-Stat | 25.4498 | −1.384 98 | 1.749 91 | 7.478 58 | −1.447 82 | 1.807 36 | |
| Prob. | 0.000 00 | 0.167 10 | 0.081 20 | 0.000 00 | 0.148 70 | 0.071 70 | |
| Ps. $R^2 = 0.3772$ | | Adj $R^2 = 0.3666$ | | Quasi-LR $= 220.23$ | | Prob. $= 0.0000$ | |
| RM-4 | C | RY1*RX1*RX2/ 1000 | RY1*RX1/ 100 | RY1*RX2/ 100 | RX1*RX2/ 100 | RX1/ 10 | RX2/ 10 |
| Coef. | 4.554 992 | −1.72E-04 | 0.002 94 | 0.004 33 | 0.001 44 | 0.003 60 | −0.022 81 |
| t-Stat | 18.450 98 | −2.848 37 | 3.124 15 | 5.719 38 | 0.918 02 | 0.148 16 | −1.072 53 |
| Prob. | 0.000 00 | 0.0047 | 0.002 00 | 0.000 00 | **0.359 40** | **0.882 30** | **0.284 40** |
| Ps. $R^2 = 0.3792$ | | Adj $R^2 = 0.3664$ | | Quasi-LR $= 221.73$ | | Prob. $= 0.0000$ | |

10% level, the three-way interaction IV of RM-2 has a negative significant effect, based on the $t$-statistic of $t_0 = -1.385$ with $df = 294$ and a $p$-value $= 0.167\,10/2 = 0.083\,55$. So, in the statistical sense, each of the median regressions is an acceptable three-way interaction median regression, even though some of the two-way interactions and the main variables have large $p$-values.

2) The results are obtained using the MMSM, as follows:

    2.1 The reduced model RM-1 is the result of the first two-stages of analysis. Even though *RX1*RX2/100* has a large $p$-value, it can be kept as an IV because the three-way interaction has a significant effect.

    2.2 At the third stage, inserting each of the main variables *RY1/10*, *RX1/10*, and *RX2/10* as an additional IV for RM-1 obtains RM-2 and RM-3 as acceptable three-way interaction quantile regressions (3WI-QR(0.5)s) because their three-way interaction IVs have a significant effect, as mentioned previously.

    2.3 At the fourth stage, RM-4 is obtained by using both variables *RX1/10* and *RX2/10* as additional IVs for RM-1. Even though three of its IVs have large $p$-values, they can be kept in the model because the three-way interaction has a significant effect at the 1% level.

3) Since RM-3 has the largest adjusted $R$-squared, we can consider it to be the best and final 3WI-NPQR(Median) among the four RMs. Based on the results in Figure 7.6, the following findings and notes can be presented:

    3.1 Its IVs have a joint significant effect, based on the *Quasi-LR* statistic of 221.78 with the $p$-value $= 0.0000$.

3.2 At the 10% level, *RX2/10* has a negative significant effect, based on the *t*-statistic of $t_0 = -1.365$ with $df = 294$ and a *p*-value $= 0.1735/2 = 0.086\,75$.

3.3 Note that *RY1\*RX1\*RX2/1000, RY1\*RX1/100,* and *RY1\*RX2/100* all have a significant effect at the 1% level. Then we can conclude directly that the effect of *RY1* on *Y2* is significantly dependent on the function *(a\*RX1\*RX2+b\*RX1+c\*RX2)*.

3.4 Furthermore, additional analysis and testing can be done easily based on RM-3, as has been presented in the previous examples. Do this as exercises.

## 7.5 Multiple NPQRs Based on MLogit.wf1

Various QRs have been presented in Chapter 6 based on Mlogit.wf1, even for the QRs of *PRI* on *(DIST1,X1,X2)*, as presented in Tables 6.4 and 6.5. This section presents only two additional examples using the rank variables.

### Example 7.11  *A 3WI-NPQR of PRI on* (RDIST1,RX1,RX2)

As a modification of the QR of *PRI* on *(DIST1,X1,X2)* in (6.10), we have the following ES for a 3WI-NPQR. Since the parameters' estimates are likely to be zeros, we should apply the scaled IVs, as we did in Example 7.9.

$$PR1 \ C \ RDIST1*RX1*RX2 \ RDIST1*RX1 \ RDIST1*RX2$$

$$RX1*RX2 \ RDIST1 \ RX1 \ RX2 \tag{7.16}$$

Table 7.8 presents the statistical results summary of a mean regression, NPQR(Median), and its Reduced-NPQR, based on ES (7.16) using the scaled IVs. Based on this summary, the findings and notes are as follows:

1) The statistical results in this example use a different estimation process than the estimation process or method presented previously for the QR analysis based on the ES (6.10), such as explained in the follows comments.
2) First of all, it is important to note the different scaled IVs are used to make sure the parameters' estimates do not result in zeros.
3) Because the model has a single three-way interaction IV, at the first stage of analysis, we obtain the mean regression by using the STEPLS-Combinatorial method, discussed in Chapter 6, with *"PRI C RDIST1\*RX1\*RX2"* as the *Equation specification*, and the other IVs in (7.16) as the *List of search regressors*. Then by inserting alternative numbers of regressors to select, we finally obtain the mean regression in Table 7.14 as the best-fit out of six possible models. Note that the three-way interaction has a *Prob.* $< 0.30$, so at the 15% level, it has a positive significant adjusted effect on *PRI,* based on the *t*-statistic of $t_0 = 1.078$ with a *p*-value $= 0.281/2 < 0.15$.

   Even though two of its IVs have large probabilities, this mean regression is a good-fit model because IVs' three-way interaction has a positive significant effect.
4) At the second stage, we use the ES of the MR directly to obtain the output of the NPQR(Median), which shows the three-way interaction has a *Prob.* $> 0.30$. Hence a reduced model should be explored.

**Table 7.8** Statistical results summary based on the mean regression, 3WI-NPQR (median), and its Reduced-QR.

| Variable | Mean-Regression | | | QREG | | | Reduced QREG | | |
|---|---|---|---|---|---|---|---|---|---|
| | Coef. | t-Stat. | Prob.[a] | Coef. | t-Stat. | Prob.[a] | Coef. | t-Stat. | Prob.[a] |
| C | 0.423 | 3817.4 | 0.000 | 0.423 | 2785.4 | 0.000 | 0.423 | 3104.4 | 0.000 |
| RDIST1*RX1*RX2/10^12 | 0.446 | 1.078 | 0.281 | 0.397 | 0.833 | 0.405 | 0.397 | 1.103 | 0.270 |
| RX1/10^2 | −0.014 | −631.1 | 0.000 | −0.014 | −596.3 | 0.000 | −0.014 | −901.5 | 0.000 |
| RX2/10^2 | −0.007 | −482.0 | 0.000 | −0.007 | −394.4 | 0.000 | −0.007 | −395.9 | 0.000 |
| RX1*RX2/10^4 | 0.000 | 70.790 | 0.000 | 0.000 | 75.060 | 0.000 | 0.000 | 84.935 | 0.000 |
| RDIST1*RX1/10^8 | −0.018 | −0.542 | **0.588** | 0.000 | −0.011 | **0.991** | | | |
| RDIST1/10^6 | −0.065 | −0.464 | **0.643** | −0.066 | −0.366 | **0.715** | −0.069 | −0.483 | **0.629** |
| R-squared | 1.000 | | | Pseudo R-sq | | 0.985 | 0.985 | | |
| Adj. R-sq. | 1.000 | | | Adj. R-sq | | 0.985 | 0.985 | | |
| S.E. of reg. | 0.001 | | | S.E. of reg | | 0.001 | 0.001 | | |
| Sum sq. resid | 0.000 | | | Quantile DV | | 0.324 | 0.324 | | |
| Log likelihood | 5925.2 | | | Sparsity | | 0.002 | 0.002 | | |
| F-statistic | 659 030.4 | | | Quasi-LR stat | | 81 021.4 | 81 000.1 | | |
| Prob(F-stat.) | 0.000 | | | Prob(Q-LR stat) | | 0.000 | 0.000 | | |

5) Finally, a reduced QR(Median) is obtained. Its three-way interaction IV has a *Prob.* < 0.30. So, this QR(Median) is a good-fit model, even though two of its other IVs have large $p$-values, similar to the MR. As an additional analysis, deleting $DIST1/10^6$ from the Reduced-QR obtains a better Reduced-QR, in the statistical sense, with its three-way interaction IV having the $t$-statistic of $t_0 = 1.286\,832$ with a *Prob.* = 0.1985. So, at the 10% level, the three-way interaction IV has a positive significant effect with a $p$-value $0.1985/2 = 0.099\,25 < 0.10$.

### Example 7.12 *A Modification of the 3WI-QR in Example 6.25*

Compared with the limitation of the three-way interaction ES (6.39), presented in Example 6.25, I consider the 3WI-NPQRs of *LW* on the ranks of *Z1, Z2, Z3*, and *INC*, (*RZ1, RZ2, RZ3*, and *RINC*) to be better QRs than those presented in Example 6.25. The ranks variables are generated using the equation $RV = @Round(@Ranks(V.a))$. So, instead of the ES (6.39), we will have the following ES:

$$LW \ C \ RZ1*RZ2*RZ3/10^9 \ RZ1*RZ2*RINC/10^9 \ RZ1*RZ3*RINC/10^9 \qquad (7.17)$$

where the scale $V/10^9$ of $V$ is used in order to have the parameters' estimates are not appear to be zeros. Figure 7.17 present the outputs of the 3WI-LSR and 3WI-NPQR based on the ES (7.17). Based on these outputs, the following findings and notes are presented:

1) Each IV of the least-square-regression has either a positive or negative significant effect on *LW* at the 1% or 5% level. For instance, at the 5% level, $RZ1*RZ3*RINC/10^9$

| Dependent Variable: LW |
|---|
| Method: Least Squares |
| Date: 09/02/18  Time: 21:12 |
| Sample (adjusted): 1 790 |
| Included observations: 790 after adjustments |

| Variable | Coefficient | Std. Error | t-Statistic | Prob. |
|---|---|---|---|---|
| C | 5.840042 | 0.104478 | 55.89714 | 0.0000 |
| RZ1*RZ2*RZ3/10^9 | -2.017447 | 0.619237 | -3.257954 | 0.0012 |
| RZ1*RZ2*RINC/10^9 | 2.876919 | 0.583315 | 4.932014 | 0.0000 |
| RZ1*RZ3*RINC/10^9 | -1.238035 | 0.648615 | -1.908737 | 0.0567 |

| | | | |
|---|---|---|---|
| R-squared | 0.037070 | Mean dependent var | 5.822810 |
| Adjusted R-squared | 0.033395 | S.D. dependent var | 2.040345 |
| S.E. of regression | 2.005987 | Akaike info criterion | 4.235200 |
| Sum squared resid | 3162.853 | Schwarz criterion | 4.258856 |
| Log likelihood | -1668.904 | Hannan-Quinn criter. | 4.244293 |
| F-statistic | 10.08623 | Durbin-Watson stat | 0.075908 |
| Prob(F-statistic) | 0.000002 | | |

(a). 3WI-LSR

| Dependent Variable: LW |
|---|
| Method: Quantile Regression (Median) |
| Date: 09/02/18  Time: 21:14 |
| Sample (adjusted): 1 790 |
| Included observations: 790 after adjustments |
| Huber Sandwich Standard Errors & Covariance |
| Sparsity method: Kernel (Epanechnikov) using residuals |
| Bandwidth method: Hall-Sheather, bw=0.1051 |
| Estimation successfully identifies unique optimal solution |

| Variable | Coefficient | Std. Error | t-Statistic | Prob. |
|---|---|---|---|---|
| C | 5.754918 | 0.135254 | 42.54889 | 0.0000 |
| RZ1*RZ2*RZ3/10^9 | -1.975192 | 0.921132 | -2.144310 | 0.0323 |
| RZ1*RZ2*RINC/10^9 | 2.990299 | 0.682210 | 4.383256 | 0.0000 |
| RZ1*RZ3*RINC/10^9 | -0.762566 | 0.956906 | -0.796908 | 0.4257 |

| | | | |
|---|---|---|---|
| Pseudo R-squared | 0.024081 | Mean dependent var | 5.822810 |
| Adjusted R-squared | 0.020356 | S.D. dependent var | 2.040345 |
| S.E. of regression | 2.007776 | Objective | 634.1410 |
| Quantile dependent var | 5.807151 | Restr. objective | 649.7887 |
| Sparsity | 5.233070 | Quasi-LR statistic | 23.92115 |
| Prob(Quasi-LR stat) | 0.000026 | | |

(b). 3WI-NPQR (Median)

**Figure 7.17** The outputs of 3WI-LSR and 3WI-NPQR based on the ES (7.17) (a) 3WI-LSR. (b) 3WI-NPQR (Median).

has a negative significant effect, based on the $t$-statistic of $t_0 = -1.909$ with a $p$-value $= 0.0567/2 < 0.05$.

2) However, the first two IVs of the NPQR have significant effects at the 5% and 1% levels, respectively, and the interaction $RZ1*RZ3*RINC/10^9$ has a large $p$-value $= 0.4527$. So, to continue the analysis by using the two-way interactions, we have to start with the model having two three-way interactions with the ES as follows:

$$LW\ C\ RZ1*RZ2*RZ3/10^9\ RZ1*RZ2*RINC/10^9 \qquad (7.18)$$

3) Figure 7.18a presents the output of an LS-regression, which is obtained by using the STEPLS-Combintorial regression, with the ES (6.18) as the *Equation specification* and the six possible two-way interactions of *RZ1, RZ2, RZ3,* and *RINC* as the *List of search regressors*. Since the three-way interaction has *Prob.* $< 30$, and at the 10% level, it has a positive significant effect, the LS-regression is a good-fit 3WI-LSR, even though one of its two-way interactions, $RZ1*RZ3/10^6$, has a large *Prob.* $= 0.3629$.

However, if $RZ1*RZ3/10^6$ is deleted from the model, the three-way interaction $RZ1*RZ2*RZ3/10^9$ unexpectedly has a greater *Prob.* $= 0.2802$. So, I prefer to use the LS-regression in Figure 7.18a.

4) Then the ES of the LS-regression is used directly to obtain the output of the 3WI-NPQR in Figure 7.18b, where the two three-way interactions have positive and negative significant effects with the $p$-values of $0.0966/2 < 0.05$, and $0.0052/2 < 0.01$, respectively. Similar to the LS-regression, this NPQR also has the two-way interaction $RZ1*RZ3/10^6$ with a *Prob.* $= 0.3609$. This result shows that the IVs of the NPQR are the IVs of the LS-regression, which can easily be obtained using the STEPLS-Combinatorial method. So, based on these findings, whenever we have a large number of IVs to be selected for a QR, I recommend first selecting the IVs using the STEPLS regression.

However, if $RZ1*RZ3/10^6$ is deleted from the NPQR, the three-way interaction $RZ1*RZ2*RZ3/10^9$ unexpectedly has a greater *Prob.* $= 0.2342$.

**(a). 3WI-LSR**

Dependent Variable: LW
Method: Least Squares
Date: 09/03/18  Time: 10:29
Sample (adjusted): 1 790
Included observations: 790 after adjustments

| Variable | Coefficient | Std. Error | t-Statistic | Prob. |
| --- | --- | --- | --- | --- |
| C | 5.738829 | 0.199214 | 28.80729 | 0.0000 |
| RZ1*RZ2*RZ3/10^9 | 2.278900 | 1.600628 | 1.423753 | 0.1549 |
| RZ1*RZ2*RINC/10^9 | -5.792980 | 1.551144 | -3.734649 | 0.0002 |
| RZ2*RZ3/10^6 | -2.815470 | 0.689070 | -4.085901 | 0.0000 |
| RZ1*RZ2/10^6 | 2.722158 | 0.674006 | 4.038776 | 0.0001 |
| RZ1*RINC/10^6 | 4.266291 | 0.707280 | 6.031965 | 0.0000 |
| RZ3*RINC/10^6 | -2.678839 | 0.533090 | -5.025119 | 0.0000 |
| RZ2*RINC/10^6 | 0.844076 | 0.637806 | 1.323406 | 0.1861 |
| RZ1*RZ3/10^6 | -0.673320 | 0.724357 | -0.929542 | 0.3529 |

| R-squared | 0.262395 | Mean dependent var | 5.822810 |
| --- | --- | --- | --- |
| Adjusted R-squared | 0.254839 | S.D. dependent var | 2.040345 |
| S.E. of regression | 1.761281 | Akaike info criterion | 3.981286 |
| Sum squared resid | 2422.747 | Schwarz criterion | 4.034512 |
| Log likelihood | -1563.608 | Hannan-Quinn criter. | 4.001745 |
| F-statistic | 34.72902 | Durbin-Watson stat | 0.528499 |
| Prob(F-statistic) | 0.000000 | | |

**(b). 3WI-NPQR**

Dependent Variable: LW
Method: Quantile Regression (Median)
Date: 09/03/18  Time: 10:22
Sample (adjusted): 1 790
Included observations: 790 after adjustments
Huber Sandwich Standard Errors & Covariance
Sparsity method: Kernel (Epanechnikov) using residuals
Bandwidth method: Hall-Sheather, bw=0.1051
Estimation successfully identifies unique optimal solution

| Variable | Coefficient | Std. Error | t-Statistic | Prob. |
| --- | --- | --- | --- | --- |
| C | 5.669370 | 0.245411 | 23.10149 | 0.0000 |
| RZ1*RZ2*RZ3/10^9 | 3.257684 | 1.958021 | 1.663763 | 0.0966 |
| RZ1*RZ2*RINC/10^9 | -5.691145 | 2.029747 | -2.803869 | 0.0052 |
| RZ2*RZ3/10^6 | -2.902731 | 0.797246 | -3.640950 | 0.0003 |
| RZ1*RZ2/10^6 | 2.098025 | 0.829588 | 2.528996 | 0.0116 |
| RZ1*RINC/10^6 | 4.601001 | 0.905692 | 5.080097 | 0.0000 |
| RZ3*RINC/10^6 | -3.010248 | 0.625404 | -4.813283 | 0.0000 |
| RZ2*RINC/10^6 | 1.247635 | 0.731791 | 1.704906 | 0.0886 |
| RZ1*RZ3/10^6 | -0.856203 | 0.936539 | -0.914220 | 0.3609 |

| Pseudo R-squared | 0.150235 | Mean dependent var | 5.822810 |
| --- | --- | --- | --- |
| Adjusted R-squared | 0.141531 | S.D. dependent var | 2.040345 |
| S.E. of regression | 1.765414 | Objective | 552.1677 |
| Quantile dependent var | 5.807151 | Restr. objective | 649.7887 |
| Sparsity | 4.522528 | Quasi-LR statistic | 172.6839 |
| Prob(Quasi-LR stat) | 0.000000 | | |

**Figure 7.18** The outputs of 3WI-LSR and 3WI-NPQR, the extensions of ES (7.18). (a) 3WI-LSR. (b) 3WI-NPQR.

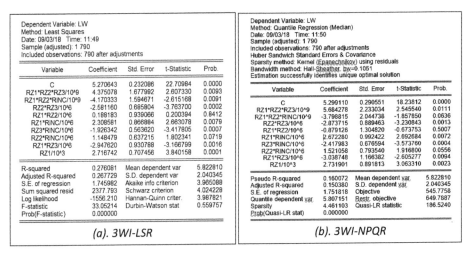

**(a). 3WI-LSR**

Dependent Variable: LW
Method: Least Squares
Date: 09/03/18  Time: 11:49
Sample (adjusted): 1 790
Included observations: 790 after adjustments

| Variable | Coefficient | Std. Error | t-Statistic | Prob. |
| --- | --- | --- | --- | --- |
| C | 5.270643 | 0.232086 | 22.70984 | 0.0000 |
| RZ1*RZ2*RZ3/10^9 | 4.375078 | 1.677992 | 2.607330 | 0.0093 |
| RZ1*RZ2*RINC/10^9 | -4.170333 | 1.594671 | -2.615168 | 0.0091 |
| RZ2*RZ3/10^6 | -2.581160 | 0.685804 | -3.763700 | 0.0002 |
| RZ1*RZ2/10^6 | 0.188183 | 0.939066 | 0.200394 | 0.8412 |
| RZ1*RINC/10^6 | 2.308581 | 0.866884 | 2.663078 | 0.0079 |
| RZ3*RINC/10^6 | -1.926342 | 0.563620 | -3.417805 | 0.0007 |
| RZ2*RINC/10^6 | 1.148479 | 0.637215 | 1.802341 | 0.0719 |
| RZ1*RZ3/10^6 | -2.947620 | 0.930788 | -3.166799 | 0.0016 |
| RZ1/10^3 | 2.716742 | 0.707456 | 3.840158 | 0.0001 |

| R-squared | 0.276081 | Mean dependent var | 5.822810 |
| --- | --- | --- | --- |
| Adjusted R-squared | 0.267729 | S.D. dependent var | 2.040345 |
| S.E. of regression | 1.745982 | Akaike info criterion | 3.965088 |
| Sum squared resid | 2377.793 | Schwarz criterion | 4.024228 |
| Log likelihood | -1556.210 | Hannan-Quinn criter. | 3.987821 |
| F-statistic | 33.05214 | Durbin-Watson stat | 0.559757 |
| Prob(F-statistic) | 0.000000 | | |

**(b). 3WI-NPQR**

Dependent Variable: LW
Method: Quantile Regression (Median)
Date: 09/03/18  Time: 11:50
Sample (adjusted): 1 790
Included observations: 790 after adjustments
Huber Sandwich Standard Errors & Covariance
Sparsity method: Kernel (Epanechnikov) using residuals
Bandwidth method: Hall-Sheather, bw=0.1051
Estimation successfully identifies unique optimal solution

| Variable | Coefficient | Std. Error | t-Statistic | Prob. |
| --- | --- | --- | --- | --- |
| C | 5.299110 | 0.290551 | 18.23812 | 0.0000 |
| RZ1*RZ2*RZ3/10^9 | 5.684278 | 2.233034 | 2.545540 | 0.0111 |
| RZ1*RZ2*RINC/10^9 | -3.798815 | 2.044738 | -1.857850 | 0.0636 |
| RZ2*RZ3/10^6 | -2.873715 | 0.889463 | -3.230843 | 0.0013 |
| RZ1*RZ2/10^6 | -0.879126 | 1.304820 | -0.673753 | 0.5007 |
| RZ1*RINC/10^6 | 2.672280 | 0.992422 | 2.692684 | 0.0072 |
| RZ3*RINC/10^6 | -2.417983 | 0.676594 | -3.573760 | 0.0004 |
| RZ2*RINC/10^6 | 1.521058 | 0.793540 | 1.916800 | 0.0556 |
| RZ1*RZ3/10^6 | -3.038748 | 1.166382 | -2.605277 | 0.0094 |
| RZ1/10^3 | 2.731901 | 0.891813 | 3.063310 | 0.0023 |

| Pseudo R-squared | 0.160072 | Mean dependent var | 5.822810 |
| --- | --- | --- | --- |
| Adjusted R-squared | 0.150380 | S.D. dependent var | 2.040345 |
| S.E. of regression | 1.751818 | Objective | 545.7758 |
| Quantile dependent var | 5.807151 | Restr. objective | 649.7887 |
| Sparsity | 4.461103 | Quasi-LR statistic | 186.5240 |
| Prob(Quasi-LR stat) | 0.000000 | | |

**Figure 7.19** The outputs of 3WI-LSR and 3WI-NPQR, the extensions of the regressions in Figure 7.18. (a) 3WI-LSR. (b) 3WI-NPQR.

5) At the final stage of analysis, we have six main variables as presented in Figure 6.44, or 10 main variables as presented in Figure 6.40, to be selected for the 3WI-LSR in Figure 7.18a. This is because the six main variables, *Z1, Z2, Z2,INC, X1,* and *XB2,* have been selected using the STELS regression from the 10 main variables.

Figure 7.19 presents the final outputs the 3WI-LSR and 3WI-NPQR as the extension of the regressions in Figure 7.18. Based on these outputs, the findings and notes are as follows:

5.1 The output of the 3WI-LSR is obtained using the STEPLS-Combinatorial method, with the 3WI_LSR in Figure 7.18a as the equation specification and the main variables, *RZ1/10^3, RZ2/10^3, RZ2/10^3, RINC/10^3, RX1/10^3,* and *RXB2/10^3,* as the search regressors. Note that only one out of the six main variables, namely *RZ1/10^3,* can be inserted as an additional variable so each of the three-way interaction IVs has a significant effect.

In addition, note that *RZ1\*RZ3/10^6* has a significant effect at the 1% level, but in the LS regression in Figure 7.18a has a large *p*-value = 0.3629. On the other hand, one of the other two-way interactions, *RZ1\*RZ2/10^6,* has a very large *p*-value = 0.8412. Even though, this LS regression is an acceptable 3WI-LSR, and the interaction *RZ1\*RZ2/10^6* does not have to be deleted from the model.

5.2 The output of the 3WI-NPQR is obtained by directly using the 3WI-LSR's ES. The output obtained is an acceptable 3WI-NPQR because only the two-way interaction *RZ1\*RZ2/10^6* has an insignificant effect with a *p*-value = 0.5007. If the interaction *RZ1\*RZ2/10^6* is deleted from the model, then each of the IVs of the reduced model has a significant effect at the 1% level, except the interaction *RZ2\*RINC/10^6* has a *Prob.* = 0.0434.

# 8

# Heterogeneous Quantile Regressions Based on Experimental Data

## 8.1 Introduction

It is well known that an experiment is conducted with the main objective of studying the treatment factors, either nominal or ordinal, on the response or objective variables by taking into account the effects of covariates. In general, the covariates also are the cause factors or variables of the response variables, so we can study the differential effects of covariates on the objective variables by the factors. We do this using *heterogeneous quantile regressions* (HQRs) of an objective variable by the factors considered.

The applications of quantile regressions (QRs) presented in previous chapters have been based on experimental data in Data-Faad.wf1 containing four numerical variables, two covariates $X1$ and $X2$, two response variables $Y1$ and $Y2$, and two dichotomous factors $A$ and $B$, as presented in Figure 1.1. In addition, to present various alternative QRs, four ordinal variables $G2$, $G4$, $H2$, and $H4$ were generated based on the covariates $X1$ and $X2$, respectively, as presented in Chapter 1. The following examples present illustrations based on selected HQRs.

## 8.2 HQRs of *Y1* on *X1* by a Cell-Factor

### 8.2.1 The Simplest HQR

The name *Cell-Factor* (CF) is used to represent one or more factors that can be generated based on either numerical or categorical variables. Hence, the simplest HQR of $Y1$ on $X1$ by *CF* is the *heterogeneous linear QR* (HLQR) with the following alternative general equation specifications (ESs):

$$Y1 @ Expand\,(CF)\ X1 * @ Expand\,(CF) \tag{8.1a}$$

$$Y1\ C\ @ Expand\,(CF, @ Droplast)\ X1 * @ Expand\,(CF) \tag{8.1b}$$

$$Y1\ C\ X1 @ Expand\,(CF, @ Droplast)\ X1 * @ Expand\,(CF, @ Droplast) \tag{8.1c}$$

I recommend applying the model without an intercept "C" in (8.1a) with the main objective of seeing and writing the QR functions for a set of quantiles, which are presented by the output of *quantile process estimates* (QPEs). The others are the models with intercepts, which should be used to apply the *quantile slope quality test* (QSET)—refer to Example 1.4.

*Quantile Regression: Applications on Experimental and Cross Section Data Using EViews*, First Edition.
I Gusti Ngurah Agung.
© 2021 John Wiley & Sons Ltd. Published 2021 by John Wiley & Sons Ltd.

In addition, the model (8.1c) is most advantageous for testing hypotheses using the *redundant variables test* (RVT). See the following example.

The models could be extended to HLQRs based on *(X1,log(Y1))* for the positive variable *Y1; (log(X1),log(Y1))* for the positive variables *X1* and *Y1*; the latent variables *(X1,Y1)*; and the models with the ranks of *X1* as independent variables (IVs).

**Example 8.1**   *Application of the Quantile Process of a HLQR (8.1a) by Factor A*
The analysis is done using the following ES:

$$Y1@Expand\,(A)\;\;X1*@Expand\,(A) \tag{8.2}$$

Having the output of the HLQR(0.5) in (8.2) on the screen, by selecting *View/Quantile Process/Process Coefficients* ..., there are three alternative options shown on the screen, those are two *Output*, Table of Graph, two *Quantile Specification*, number of Quantiles or User specified quantiles, and three Coefficient Specifications, all coefficients, Intercept only, User specified coefficients. So we can have many possible statistical results. However, in this example only several selected options will be applied, such as follows.

1) The full output of the results presented in Figure 8.1 is obtained by using the default options, that is by selecting *View/Quantile Process/Process Coefficients* and clicking *OK*. Based on these statistical results, the findings and notes are as follows:
    1.1 This figure presents only 9 of 18 median regressions, specific for the level of $A = 1$. Based on the full output, we can test the linear effect of *X1* on *Y1* in each of the 18 median regressions for each tau $= 0.1$ to 0.9 and each level of the factor *A*.
    1.2 However, we can't test the linear effects differences of *X1* on *Y1* between the quantiles. See the following illustration. .
2) By selecting only *Graph* under *Output* and clicking *OK*, we obtain the graphs in Figure 8.2, which show the coefficients of the 18 QR(tau)s, for tau $= 0.1$ to 0.9. The graphs show the results of the full output of Figure 8.1, with the graphs of their Median $+/- 2se$.
3) By entering 0.1 0.25 0.75 0.9 as the *User-specified quantiles*, C(3) C(4) as the *Coefficient specification*, and click *OK,* we obtain the output in Figure 8.3. Based on this output, the following findings and notes are presented:
    3.1 Note that the output presents the slopes of *X1* by the factor *A* for five QRs, including tau $= 0.5$, even though only four quantiles are specified.
    3.2 However, if the QPE runs based on the HLQR(0.1), the output presents exactly four quantile regressions. Do this as an exercise.
4) As an additional illustration, Figure 8.4 presents the scatter graph of *Y1* and its fitted value variables Mean-R and Median-R on *X1*. It shows that Mean-R and Median-R are very close, with a correlation of 0.992. And Figure 8.5 presents the scatter graph of *Y1* and its fitted value variables of the two HLQR(0.1) and HLQR(0.9) on *X1*, *Fit_82_Q10* and *Fit_82_90*, based on the ES (8.2). It shows two regression lines for each level of the factor *A*.
5) As another illustration, with the output of HLQR(0.1) of the ES (8.2) on-screen, we can select *View/Quantile Process/Symmetric Quantile Test*, insert the user-specified test quantiles 0.1 0.75, and insert the user-specified coefficients C(3) C(4) to obtain the output in Figure 8.6. Based on this output, the following findings and notes are presented:

Quantile process estimates

Equation: EQ03_81A

Specification: Y1 @EXPAND(A) X1*@EXPAND(A)

Estimated equation quantile tau = 0.5

Number of process quantiles: 10

Display user-specified coefficients: A = 1 X 1*(A = 1)

|  | Quantile | Coefficient | Std. error | t-Statistic | Prob. |
|---|---|---|---|---|---|
| A = 1 | 0.100 | 2.706731 | 0.401845 | 6.735757 | 0.0000 |
|  | 0.200 | 2.465652 | 0.387221 | 6.367560 | 0.0000 |
|  | 0.300 | 2.341579 | 0.363511 | 6.441558 | 0.0000 |
|  | 0.400 | 3.421429 | 0.484934 | 7.055453 | 0.0000 |
|  | 0.500 | 3.828571 | 0.415557 | 9.213099 | 0.0000 |
|  | 0.600 | 3.921875 | 0.367582 | 10.66940 | 0.0000 |
|  | 0.700 | 4.247143 | 0.367088 | 11.56981 | 0.0000 |
|  | 0.800 | 4.179388 | 0.413822 | 10.09949 | 0.0000 |
|  | 0.900 | 5.237500 | 0.794612 | 6.591268 | 0.0000 |
| X1*(A = 1) | 0.100 | 1.596154 | 0.326862 | 4.883266 | 0.0000 |
|  | 0.200 | 1.942029 | 0.283020 | 6.861803 | 0.0000 |
|  | 0.300 | 2.197368 | 0.229822 | 9.561196 | 0.0000 |
|  | 0.400 | 1.704082 | 0.299798 | 5.684090 | 0.0000 |
|  | 0.500 | 1.519481 | 0.263395 | 5.768825 | 0.0000 |
|  | 0.600 | 1.562500 | 0.236715 | 6.600760 | 0.0000 |
|  | 0.700 | 1.482143 | 0.238751 | 6.207897 | 0.0000 |
|  | 0.800 | 1.693878 | 0.277955 | 6.094077 | 0.0000 |
|  | 0.900 | 1.250000 | 0.472894 | 2.643298 | 0.0086 |

**Figure 8.1** A part of the quantile process estimates based on an HQR(0.5) in (8.2) specific for the level A = 1.

5.1 The symbols C(3) and C(4), respectively, are the parameters of the IVs: $X1*(A = 1)$ and $X1*(A = 2)$. This test is considered to be a specific symmetric quantiles test (SQT). That is because of entering 0.1 0.75 as the user-specified test quantiles, with the results for two pairs of quantiles (0.1, 0.9) and (0.25, 0.75), for which each pair has a total of one.

5.2 The statistical hypothesis, for each tau = 0.10 and tau = 0.75 and each C(3) and C(4), can be presented as follows, where $H_0$ indicates the pair of quantiles tau and (1-tau) are symmetric:

$$H_0 : b(tau) + b(1 - tau) - 2*b(0.5) = 0 \text{ vs } H_1 : Otherwise$$

5.3 At the 10% level of significance, the conclusions are
(i) The multiple/multivariate null hypothesis is accepted based on the Wald test (*Chi-square* statistic) of 7.3655 with $df = 4$ and p-value = 0.1178.

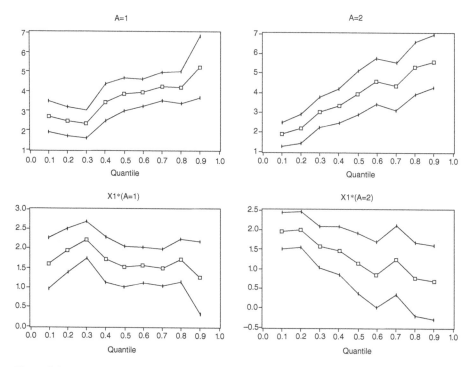

**Figure 8.2** The graphs of the coefficients of 18 quantile regressions in (8.2).

Quantile process estimates Equation: EQ03_81A
Specification: Y1 @EXPAND(A) X1*@EXPAND(A)
Estimated equation quantile tau = 0.5
User-specified process quantiles: 0.1 0.25 0.75 0.9
Display user-specified coefficients: X1*(A = 1) X1*(A = 2)

|  | Quantile | Coefficient | Std. Error | t-Statistic | Prob. |
|---|---|---|---|---|---|
| X1*(A = 1) | 0.100 | 1.596154 | 0.326862 | 4.883266 | 0.0000 |
|  | 0.250 | 2.231707 | 0.216276 | 10.31880 | 0.0000 |
|  | 0.500 | 1.519481 | 0.263395 | 5.768825 | 0.0000 |
|  | 0.750 | 1.477778 | 0.274311 | 5.387242 | 0.0000 |
|  | 0.900 | 1.250000 | 0.472894 | 2.643298 | 0.0086 |
| X1*(A = 2) | 0.100 | 1.960784 | 0.235196 | 8.336800 | 0.0000 |
|  | 0.250 | 1.686869 | 0.264490 | 6.377809 | 0.0000 |
|  | 0.500 | 1.136364 | 0.394606 | 2.879740 | 0.0043 |
|  | 0.750 | 1.123596 | 0.420461 | 2.672293 | 0.0080 |
|  | 0.900 | 0.666667 | 0.480430 | 1.387647 | 0.1663 |

**Figure 8.3** A specific quantile process estimate of an HQR(0.5) in (8.2).

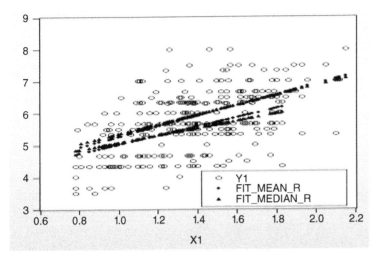

**Figure 8.4** The scatter graphs of *Y1* and its fitted value variable HLQR(0.5) on *X1*.

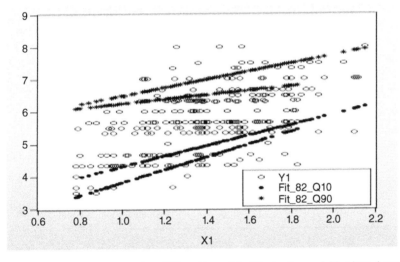

**Figure 8.5** The scatter graphs of *Y1* and two of its fitted value variables based on the HLQR(0,1) and HLQR(0.9) of *Y1* on *X1*.

(ii) Three univariate $H_0$ hypotheses are accepted, but one is rejected. That is, the univariate $H_0$ for tau = 0.25 and C(3) with *p*-value = 0.0752. So only the slopes of $X1^*(A = 1)$ for the pair of quantiles 0.25 and 0.75 are significantly asymmetric. Note that the output doesn't present the statistic for testing the univariate hypothesis; however, I am sure it uses the *Chi-square* or *QRL* statistic since it is a nonparametric test.

6) *Application of the Wald test.*

The Wald test can be applied for testing the linear effects of *X1* on *Y1* between the levels of the factor *A*, with the output presented in Figure 8.7, based on the HLQR(0.5). This shows that *X1* has an insignificant linear effect on *Y1,* based on the *Chi-square* statistic

Symmetric quantiles test
Equation: EQ03_81A
Specification: Y1 @EXPAND(A) X1*@EXPAND(A)
Estimated equation quantile tau = 0.1
User-specified test quantiles: 0.1 0.75
Test statistic compares user-specified coefficients: C(3) C(4)

| Test summary | Chi-Sq. statistic | Chi-Sq. d.f. | Prob. |
|---|---|---|---|
| Wald test | 7.365459 | 4 | 0.1178 |

Restriction detail: b(tau) + b(1-tau) − 2*b(0.5) = 0

| Quantiles | Variable | Restr. value | Std. Error | Prob. |
|---|---|---|---|---|
| 0.1, 0.9 | X1*(A = 1) | −0.192807 | 0.610997 | 0.7523 |
|  | X1*(A = 2) | 0.354724 | 0.750111 | 0.6363 |
| 0.25, 0.75 | X1*(A = 1) | 0.670524 | 0.376880 | 0.0752 |
|  | X1*(A = 2) | 0.537737 | 0.571702 | 0.3469 |

**Figure 8.6** A specific symmetric quantile test based on HLQR(0.1).

Wald test:
Equation: EQ03_81A

| Test statistic | Value | Df | Probability |
|---|---|---|---|
| t-statistic | 0.807518 | 296 | 0.4200 |
| F-statistic | 0.652085 | (1, 296) | 0.4200 |
| Chi-square | 0.652085 | 1 | 0.4194 |

Null Hypothesis: C(3) = C(4)
Null Hypothesis summary:

| Normalized restriction (= 0) | Value | Std. Err. |
|---|---|---|
| C(3) − C(4) | 0.383117 | 0.474438 |

Restrictions are linear in coefficients.

**Figure 8.7** The output of a Wald test based on HLQR(0.5) in (8.2).

of 0.6521 with $df = 1$ and $p$-value $= 0.4194$. So, in the statistical sense, the HLQR(0.5) can be presented as a homogeneous linear QR or *analysis of covariance* (ANCOVA)-QR(0.5), with the following alternative ESs:

$$Y1\ X1 @ Expand\ (A) \tag{8.3a}$$

$$Y1\ C\ X1 @ Expand\ (A, @ Droplast) \tag{8.3b}$$

$$Y1\ C\ X1\ (A = 1) \tag{8.3c}$$

7) Furthermore, based on the ANCOVA-QR(0.5) in (8.3a), we can test its intercepts difference using the Wald test. We find they have a significant difference, based on the $t$-statistic of $t_0 = 3.4491$ with $df = 297$ and $p$-value $= 0.0003$, or the *Chi-square* statistic of 13.3159 with $df = 1$ and $p$-value 0.0003. So, the ANCOVA-QR(0.5) presents significant parallel lines.

**Example 8.2**  *An Application of HLQR (8.1b) by the Factor* **G4**

Figure 8.8 presents the output of the HLQR (8.1b) using the following ES, where *G4* is an ordinal variable with four levels that was generated based on *X1*, as mentioned earlier:

$$Y1\ C @ Expand\ (G4, @ Droplast)\ X1 * @ Expand\ (G4) \tag{8.4}$$

Dependent variable: Y1
Method: Quantile Regression (Median)
Date: 10/18/18   Time: 20:15
Sample: 1 300
Included observations: 300
Huber Sandwich Standard Errors and Covariance
Sparsity method: Kernel (Epanechnikov) using residuals
Bandwidth method: Hall-Sheather, bw = 0.14513
Estimation successfully identifies unique optimal solution

| Variable | Coefficient | Std. Error | t-Statistic | Prob. |
|---|---|---|---|---|
| C | 3.787045 | 1.245504 | 3.040573 | 0.0026 |
| G4 = 1 | −3.307586 | 1.550589 | −2.133116 | 0.0337 |
| G4 = 2 | −1.264738 | 4.021620 | −0.314485 | 0.7534 |
| G4 = 3 | 1.882955 | 2.767698 | 0.680332 | 0.4968 |
| X1*(G4 = 1) | 4.513514 | 0.882214 | 5.116119 | 0.0000 |
| X1*(G4 = 2) | 2.538462 | 2.887928 | 0.878991 | 0.3801 |
| X1*(G4 = 3) | −6.11E−17 | 1.655192 | −3.69E−17 | 1.0000 |
| X1*(G4 = 4) | 1.522727 | 0.698567 | 2.179787 | 0.0301 |

| | | | |
|---|---|---|---|
| Pseudo R-squared | 0.155772 | Mean dependent var | 5.751100 |
| Adjusted R-squared | 0.135534 | S.D. dependent var | 0.945077 |
| S.E. of regression | 0.823465 | Objective | 93.68815 |
| Quantile dependent var | 5.670000 | Restr. objective | 110.9750 |
| Sparsity | 1.991149 | Quasi-LR statistic | 69.45477 |
| Prob(Quasi-LR stat) | 0.000000 | | |

**Figure 8.8**   The output of the HLQR(Median) in (8.4).

Quantile slope equality test
Equation: EQ04_G4
Specification: Y1 C @EXPAND(G4,@DROPLAST) X1*@EXPAND(G4)
Estimated equation quantile tau = 0.5
User-specified test quantiles: 0.1 0.9
Test statistic compares user-specified coefficients: C(5) C(6) C(7) C(8)

| Test summary | Chi-Sq. Statistic | Chi-Sq. d.f. | Prob. |
|---|---|---|---|
| Wald test | 21.66070 | 8 | 0.0056 |

Restriction Detail: b(tau_h) − b(tau_k) = 0

| Quantiles | Variable | Restr. value | Std. Error | Prob. |
|---|---|---|---|---|
| 0.1, 0.5 | X1*(G4 = 1) | −0.180180 | 1.329289 | 0.8922 |
|  | X1*(G4 = 2) | −0.109890 | 3.349380 | 0.9738 |
|  | X1*(G4 = 3) | 5.000000 | 1.659117 | 0.0026 |
|  | X1*(G4 = 4) | −1.522727 | 1.064048 | 0.1524 |
| 0.5, 0.9 | X1*(G4 = 1) | 1.881935 | 1.193894 | 0.1150 |
|  | X1*(G4 = 2) | 3.482906 | 2.875493 | 0.2258 |
|  | X1*(G4 = 3) | −6.058824 | 1.803196 | 0.0008 |
|  | X1*(G4 = 4) | −0.750000 | 0.775114 | 0.3332 |

**Figure 8.9** A QSET based on the median-regression in Figure 8.8.

All analyses and testing hypotheses presented based on the HLQR (8.1a) can be done based on this HLQR. So, only selected analysis and testing will be presented based on this model, and the QSET, which can't be tested based on HLQR (8.1a).

We obtain the output of the QSET presented in Figure 8.9 by selecting *View/Quantile Process/Slope Equality Test*, inserting the user-specified test quantiles 0.1 0.9, and using the user-specified coefficients C(5) C(6) C(7) C(8), which are slope parameters of *X1*(G4 = k)* for $k = 1, 2, 3$, and 4. Based on this output, the findings and notes are as follows:

1) Even though the user-specified test quantiles 0.1 0.9 are inserted, the output does not present the slope equality test for the pair of quantiles (0.1, 0.9). Instead, it presents two tests of the pairs of quantiles (0.1, 0.5) and (0.5, 0.9). So based on the test statistics, we have the following conclusions:

   1.1 At the 1% level, the multivariate $H_0$: b(tau_h) − b(tau_k) = 0 of the eight univariate hypotheses is rejected based on the *Chi-square* statistic of 21.6607 with df = 8 and *p*-value = 0.0056.

1.2 Only two of the eight univariate null hypotheses are rejected at the 1% level, for the IV, $X1*(G4 = 3)$, for both pairs of quantiles.

2) To obtain the testing for the pair of quantiles (0.1, 0.9), the QSET should be based on the HLQR(0.1) or the HLQR(0.9), with the output presented in Figure 8.10. Based on this output, the findings and notes are as follows:

2.1 At the 10% level, the multivariate $H_0$: b(tau_h) − b(tau_k) = 0 of the four univariate hypotheses is accepted based on the *Chi-square* statistic of 6.2294 with $df = 4$ and p-value = 0.1827.

2.2 Only one of the four univariate null hypotheses is rejected at the 10% level, namely the slopes of $X1*(G4 = 4)$ for the pair of quantiles (0.1, 0.9), with p-value = 0.0545. So, the linear effect of $X1*(G4 = 4)$ on $Y1$ has a significant difference between the pair of quantiles (0.1, 0.9).

3) As additional statistical results, Figure 8.11 presents the scatter graphs of $Y1$ and its fitted values variables of the HLQR(Median) in (8.4), which show a piecewise median regression because $G4$ is an ordinal variable generated based on the numerical variable $X1$. They have significant slope differences, based on the Wald test (*Chi-square* statistic) of 12.1315 with $df = 3$ and p-value = 0.0069.

Quantile slope equality test
Equation: EQ04_G4

Specification: Y1 C @EXPAND(G4,@DROPLAST) X1*@EXPAND(G4)
Estimated equation quantile tau = 0.1
User-specified test quantiles: 0.1 0.9
Test statistic compares user-specified coefficients: C(5) C(6) C(7) C(8)

| Test summary | Chi-Sq. statistic | Chi-Sq. d.f. | Prob. |
|---|---|---|---|
| Wald test | 6.229356 | 4 | 0.1827 |

Restriction detail: b(tau_h) − b(tau_k) = 0

| Quantiles | Variable | Restr. value | Std. Error | Prob. |
|---|---|---|---|---|
| 0.1, 0.9 | X1*(G4 = 1) | 1.701754 | 1.657141 | 0.3045 |
| | X1*(G4 = 2) | 3.373016 | 3.212178 | 0.2937 |
| | X1*(G4 = 3) | −1.058824 | 1.727150 | 0.5398 |
| | X1*(G4 = 4) | −2.272727 | 1.182124 | 0.0545 |

**Figure 8.10** A special QSET based on the HLQR(0.1) in (8.4).

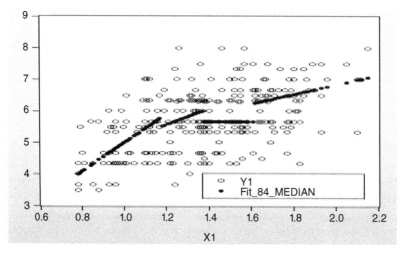

**Figure 8.11**  Scatter graphs of *Y1* and its fitted value variable on *X1*.

### Example 8.3   *An Application of the Special HLQR (8.1c) by the Factor G4*

As a modification of the ES (8.4), we have the following ES, which is important to illustrate a RVT of the linear effect of *X1* on *Y1* between the levels of *G4*:

$$Y1 \; C \; X1 \; @Expand\,(G4, @Droplast) \; X1*@Expand\,(G4, @Droplast) \qquad (8.5)$$

Figure 8.12 presents the output of the HLQR (8.5). Based on this output, the following findings and notes are presented:

1) The linear effects of *X1* on *Y1* between two levels of *G4*, with *G4* = 4 as the reference level, can be tested using the *t*-statistic in the output. For instance, at the 10% level, the linear effects of *X1* on *Y1* between two levels of *G4* = 1 and *G4* = 4 have an insignificant difference based on the *t*-statistic of $t_0 = 0.6803$ with $df = 292$ and *p*-value = 0.4938

2) All IVs have significant joint effects, based on the *Quasi LR* statistic of 69.4548, with *Prob.* = 0.0000.

3) Figure 8.13 presents the output of a RVT based on the median regression in Figure 8.12, which can't be done based on the QR (8.4). The output shows that, at the 5% level of significance, the effect of *X1* on *Y1* has significant differences between the four levels of *G4*, based on a *QLR L*-statistic of 10.5193 with $df = 3$ and *p*-value = 0.0147. So, we can conclude that the linear effect of *X1* on *Y1* is significantly dependent on *G4*.

4) Compared with the graphs of HLQR in Figures 8.4 and 8.5, the graphs in Figure 8.11 in fact is the graph of a special HLQR, called *piecewise linear quantile regression* (PWL-QR), because *G4* is generated based on *X1* as the numerical IV.

Dependent Variable: Y1
Method: Quantile Regression (Median)
Date: 10/17/18   Time: 13:54
Sample: 1 300
Included observations: 300
Huber Sandwich Standard Errors and Covariance
Sparsity method: Kernel (Epanechnikov) using residuals
Bandwidth method: Hall-Sheather, bw = 0.14513
Estimation successful but solution may not be unique

| Variable | Coefficient | Std. Error | t-Statistic | Prob. |
|---|---|---|---|---|
| C | 3.787045 | 1.245504 | 3.040573 | 0.0026 |
| X1 | 1.522727 | 0.698567 | 2.179787 | 0.0301 |
| G4 = 1 | −3.307586 | 1.550589 | −2.133116 | 0.0337 |
| G4 = 2 | −1.264738 | 4.021620 | −0.314485 | 0.7534 |
| G4 = 3 | 1.882955 | 2.767698 | 0.680332 | 0.4968 |
| X1*(G4 = 1) | 2.990786 | 1.125299 | 2.657770 | 0.0083 |
| X1*(G4 = 2) | 1.015734 | 2.971216 | 0.341858 | 0.7327 |
| X1*(G4 = 3) | −1.522727 | 1.796568 | −0.847576 | 0.3974 |

| | | | |
|---|---|---|---|
| Pseudo R-squared | 0.155772 | Mean dependent var | 5.751100 |
| Adjusted R-squared | 0.135534 | S.D. dependent var | 0.945077 |
| S.E. of regression | 0.823465 | Objective | 93.68815 |
| Quantile dependent var | 5.670000 | Restr. objective | 110.9750 |
| Sparsity | 1.991149 | Quasi-LR statistic | 69.45477 |
| Prob(Quasi-LR stat) | 0.000000 | | |

**Figure 8.12** Statistical results of the HQR(Median) in (8.5).

Redundant variables test
Equation: EQ02
Specification: Y1 C X1 @EXPAND(G4,@DROPLAST) X1
    *@EXPAND(G4,@DROPLAST)
K

| | Value | df | Probability |
|---|---|---|---|
| QLR L-statistic | 10.51393 | 3 | 0.0147 |
| QLR Lambda-statistic | 10.36978 | 3 | 0.0157 |

**Figure 8.13** The RVT of *X1*@Expand(G4,@Droplast)*.

**Example 8.4** *An Application of HLQR (8.1c) by (A,B)*

Figure 8.14 presents the output of the model in (8.1c) for *CF* generated by *(A,B)*, with the following ES:

$$Y2\ C\ X2\ @Expand\ (A, B, @Droplast)\ X2^*@Expand\ (A, B, @Droplast) \tag{8.6}$$

Based on this output, the following findings and notes are presented:

1) At the 1% level, *X2* has a positive significant effect on *Y2*, based on the *t*-statistic of $t_0 = 3.606\,419$ with $df = 300 - 8 = 292$ and *p*-value $0.0004/2 = 0.0002$, within the cell $(A = 2, B = 2)$

2) At the 10% level, the effect of *X2* on *Y2* in $(A = 1, B2)$ is significantly greater than in $(A = 2, B = 2)$, based on the *t*-statistic of $t_0 = 1.343\,315$ with $df = 292$ and *p*-value $=0.1802/2 = 0.0901 < 0.10$.

3) At the 5% level, the effect of *X2* on *Y2* in $(A = 2, B = 1)$ is significantly smaller than in $(A = 2, B = 2)$ based on the *t*-statistic of $t_0 = -2.061\,036$ with $df = 292$ and *p*-value $0.0402/2 = 0.0201 < 0.05$.

4) We also find that, at the 15% level, *X2* has significant different effects on *Y2* between the four cells of *(A,B)*, based on the *QLR L*-statistic of $QLR\ L_0 = 5.962\,127$ with $df = 3$ and *p*-value $= 0.1135$.

Dependent variable: Y2
Method: quantile regression (median)
Included observations: 300

| Variable | Coefficient | Std. Error | t-Statistic | Prob. |
|---|---|---|---|---|
| C | 3.613636 | 0.476686 | 7.580752 | 0.0000 |
| X2 | 0.681818 | 0.189057 | 3.606419 | 0.0004 |
| A = 1,B = 1 | 2.261364 | 1.044459 | 2.165106 | 0.0312 |
| A = 1,B = 2 | −0.649351 | 0.770767 | −0.842473 | 0.4002 |
| A = 2,B = 1 | 1.886364 | 0.816476 | 2.310372 | 0.0216 |
| X2*(X2*(A = 1),B = 1) | −0.306818 | 0.357220 | −0.858905 | 0.3911 |
| X2*(X2*(A = 1),B = 2) | 0.389610 | 0.290037 | 1.343315 | 0.1802 |
| X2*(X2*(A = 2),B = 1) | −0.681818 | 0.330813 | −2.061036 | 0.0402 |

| | | | |
|---|---|---|---|
| Pseudo R-squared | 0.255396 | Mean dependent var | 5.742600 |
| Adjusted R-squared | 0.237546 | S.D. dependent var | 1.062605 |
| S.E. of regression | 0.871058 | Objective | 97.72930 |
| Quantile dependent var | 5.750000 | Restr. objective | 131.2500 |
| Sparsity | 1.988296 | Quasi-LR statistic | 134.8721 |
| Prob(Quasi-LR stat) | 0.000000 | | |

**Figure 8.14** The output of the HLQR in (8.6).

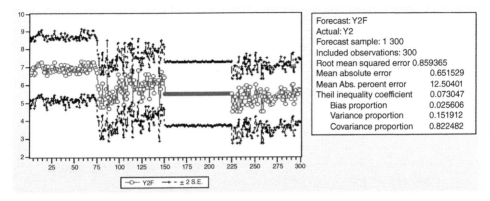

**Figure 8.15** The graph and statistics of the forecast variable of the HLQR(0.5) in (8.6).

5) Several additional analyses and tests can be done as exercises. However, some unexpected forecasting results are presented in Figure 8.15. The graph of the forecast variable *Y2F* in the cell *(A = 2, B = 1)* is a horizontal line, and we find that *Y2F* has a constant value = 5.5 within the cell *(A = 2, B = 1)*.

We also find its fitted variable have a constant value of 5.5, which is equal to the median of *Y2* in the cell *(A = 2,B = 1)*. This is a very unexpected result because the observed scores of *Y2* have Mean = 5.2812, Maximum. = 7.0, Minimum = 3.5, and Std. Dev. = 0.901 724.

6) As additional illustrations, Figure 8.16 presents the graphs of the sorted *Y2* and its fitted variable, *Fit_Y2*, specific for the two cells *(A = 2,B = 1)* and *(A = 2,B = 2)*. These graphs present special characteristics, as follows:

6.1 Since scores of *Fit_Y2* = 5.5 within the cell (A = 2,B = 1), the expected medians of *Y2* are greater than all observed scores of *Y2 < 5.5*, but they are smaller than all observed scores of Y2 > 5.5.

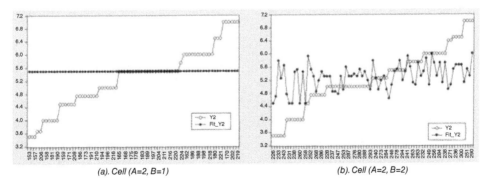

(a). Cell (A=2, B=1)    (b). Cell (A=2, B=2)

**Figure 8.16** The graphs of sorted *Y2* and its fitted variable, *Fit_Y2*, for the cells *(A = 2, B = 1)* and *(A = 2, B = 2)*.

6.2 Within the cell $(A = 2, B = 2)$, all scores of *Fit_Y2* or the expected medians of *Y2* have Maximum. $= 6.0$, Minimum $= 4.5$, and Std. Dev. $= 0.387\,140$. For all $Y2 < 5$, the scores of *Fit_Y2* are greater or equal to the scores of *Y2*, but most of the scores of *Fit_Y2* are smaller than the scores of $Y2 > 5.75$.

6.3 These findings show unbelievable statistical results for a QR, based on specific sample data, since the expected quantiles are conditional for the set of fixed scores of the IVs or predictors. Agung (2011a, p. 7) presents a definition of a sample as a set of multidimensional scores, values, or measurements of a finite number of variables, which happen to be selected or available for the researcher. Hence, whatever the conditional quantiles, they should be acceptable only in the statistical sense, and they can't be generalized.

**Example 8.5    *An Unexpected Statistical Result***

Figure 8.17 presents the output of a $2 \times 2 \times 2$ factorial HLQR(Median) using the following ES:

$$Y2\ C\ X2\ @expand\ (A, B, G2, @droplast)\ X2^*\ @expand\ (A, B, G2, @droplast) \qquad (8.7)$$

Based on this output, the findings and notes are as follows:

1) Unexpectedly, each IV has an insignificant effect at the 10% level, except $(A = 1, B = 1, G2 = 2)$, with large or very large *p*-values. However, the joint effects of all IVs are significant on *Y2*, based on the *Quasi-LR* statistic of $QLR_0 = 173.4170$ with $df = 15$ (# of IVs) and *p*-value $= 0.000\,000$.

2) In addition, $X2^*$ *@Expand(A,B,G2,@Droplast)* has a significant effect, based on the *QLR L*-statistic of $QLR\text{-}L_0 = 23.391\,34$ with $df = 7$ and *p*-value $= 0.0015$, as presented in Figure 8.18.

3) As an additional illustration, Figure 8.19 presents the graphs of sorted *Y2* and its fitted variable or expected Median(*Y2*), *Fit_Y2_ABG*. They show that most of the scores of *Fit_Y2_ABG* are greater than the observed scores of *Y2*, for small scores of *Y2*; but, for large scores of Y2, most of the scores of *Fit_Y2_ABG* are smaller than the observed scores of *Y2*.

## 8.2.2    A Piecewise Quadratic QR

Observing the graph Figure 8.11, I propose a specific *piecewise quadratic quantile regression* (PWQ-QR) of *Y1* on *X1* by *G2*, with the following alternative *explicit* ESs, where $G2 = 1 + 1^*(G4 > 2)$ is generated:

$$Y1 = (C\,(10) + C\,(11)\,{}^*X1 + C\,(12)\,{}^*X1^{\wedge}2)\,{}^*\,(G2 = 1)$$
$$+ (C\,(20) + C\,(21)\,{}^*X1 + C\,(22)\,{}^*X1^{\wedge}2)\,{}^*\,(G2 = 2) \qquad (8.8a)$$

Dependent Variable: Y2
Method: Quantile Regression (Median)
Sample: 1 300
Bandwidth method: Hall-Sheather, bw = 0.14513
Estimation successful but solution may not be unique

| Variable | Coefficient | Std. Error | t-Statistic | Prob. |
|---|---|---|---|---|
| C | 3.812500 | 1.138977 | 3.347302 | 0.0009 |
| X2 | 0.625000 | 0.459484 | 1.360221 | 0.1748 |
| A = 1,B = 1,G2 = 1 | −0.812500 | 1.740452 | −0.466833 | 0.6410 |
| A = 1,B = 1,G2 = 2 | 3.187500 | 1.381379 | 2.307477 | 0.0217 |
| A = 1,B = 2,G2 = 1 | −0.912500 | 1.279069 | −0.713409 | 0.4762 |
| A = 1,B = 2,G2 = 2 | 1.587500 | 1.770527 | 0.896626 | 0.3707 |
| A = 2,B = 1,G2 = 1 | −0.043269 | 1.644886 | −0.026305 | 0.9790 |
| A = 2,B = 1,G2 = 2 | 1.687500 | 1.275052 | 1.323476 | 0.1867 |
| A = 2,B = 2,G2 = 1 | −0.125000 | 1.250973 | −0.099922 | 0.9205 |
| X2*(X2*(X2*(A = 1),B = 1),G2 = 1) | 0.625000 | 0.610109 | 1.024407 | 0.3065 |
| X2*(X2*(X2*(A = 1),B = 1),G2 = 2) | −0.625000 | 0.542629 | −1.151799 | 0.2504 |
| X2*(X2*(X2*(A = 1),B = 2),G2 = 1) | 0.375000 | 0.507998 | 0.738192 | 0.4610 |
| X2*(X2*(X2*(A = 1),B = 2),G2 = 2) | −0.291667 | 0.647387 | −0.450529 | 0.6527 |
| X2*(X2*(X2*(A = 2),B = 1),G2 = 1) | −0.048077 | 0.725048 | −0.066309 | 0.9472 |
| X2*(X2*(X2*(A = 2),B = 1),G2 = 2) | −0.625000 | 0.517745 | −1.207159 | 0.2284 |
| X2*(X2*(X2*(A = 2),B = 2),G2 = 1) | −4.55E-17 | 0.502235 | −9.07E-17 | 1.0000 |

| | | | | |
|---|---|---|---|---|
| Pseudo R-squared | 0.310040 | Mean dependent var | | 5.742600 |
| Adjusted R-squared | 0.273598 | S.D. dependent var | | 1.062605 |
| S.E. of regression | 0.826325 | Objective | | 90.55728 |
| Quantile dependent var | 5.750000 | Restr. objective | | 131.2500 |
| Sparsity | 1.877220 | Quasi-LR statistic | | 173.4170 |
| Prob(Quasi-LR stat) | 0.000000 | | | |

Figure 8.17 The output of the HLQR(Median) in (8.7).

$$Y1 = (C(10) + C(11)*X1 + C(12)*X1^2)$$
$$+ (C(20) + C(21)*X1 + C(22)*X1^2)*(G2 = 2) \qquad (8.8b)$$

Note that the model in (8.8a), without an intercept, can't be applied to do the QSET. But the model in (8.8b), with an intercept C(10), can be applied to do the QSET, as presented in the Example 8.2.

Redundant variables test
Equation: EQ03_ABG
Specification: Y2 C X2 @EXPAND(A,B,G2,@DROPLAST) X2
    *@EXPAND(A,B,G2,@DROPLAST)
K

|  | Value | df | Probability |
|---|---|---|---|
| QLR L-statistic | 23.39134 | 7 | 0.0015 |
| QLR Lambda-statistic | 22.70985 | 7 | 0.0019 |

**Figure 8.18**   The output RVT of $X2^*@Expand(A,B,G2,@Droplast)$.

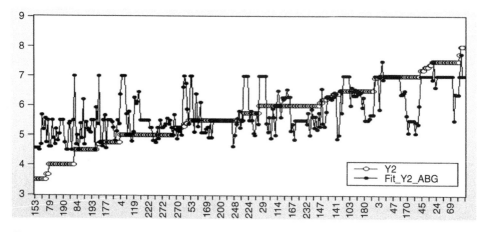

**Figure 8.19**   The graphs of sorted $Y2$ and the fitted variable of the HLQR(Median) in (8.7).

### Example 8.6   *An Application of the PWQ-QR(Median) in(8.8a)*

Table 8.1 presents the statistical results summary of the PWQ_QR(Median) in (8.8a), and its reduced model (RM).

Based on the output of the RM in Table 8.1 and the scatter graphs in Figure 8.20, the following findings and notes are presented:

1) Since the IVs $X1$ and $X^2$ of the QR(Median), within the level $G2 = 2$, have insignificant effects, its acceptable RM is presented in the table. Note that the RM has a slight greater adjusted $R$-squared, so it can be considered a better fit QR than the full model.
2) The graphs in Figure 8.20 show that the graph of the RM function within the level $G2 = 1$ is the left part of a parabolic curve having a maximum value because $X1\hat{\ }2$ has a negative coefficient. On the other hand, the graph of the RM function within the level $G2 = 2$ is the right part of a parabolic curve, with a minimum value $Y1 = 4.3480$ for $X1 = 0$, because $X1^2$ has a positive coefficient.

**Table 8.1** Statistical results summary of the Median-QR in (8.12a) and its reduced model.

| Parameter | Median-QR(8.5a) | | | Reduced model (RM) | | |
|---|---|---|---|---|---|---|
| | Coef. | t-Stat. | Prob. | Coef. | t-Stat. | Prob. |
| C(10) | −3.0122 | −0.9560 | 0.3398 | −3.0122 | −0.9566 | 0.3395 |
| C(11) | 12.1785 | 2.0835 | 0.0381 | 12.1785 | 2.0848 | 0.0379 |
| C(12) | −4.0877 | −1.5389 | 0.1249 | −4.0877 | −1.5397 | 0.1247 |
| C(20) | −0.7682 | −0.1462 | 0.8838 | 4.3480 | 13.1403 | 0.0000 |
| C(21) | 6.1066 | 0.9860 | 0.3250 | | | |
| C(22) | −1.1493 | −0.6367 | 0.5248 | 0.6288 | 5.3565 | 0.0000 |
| Pseudo R-sq | 0.1456 | | | 0.1432 | | |
| Adjusted R-sq | 0.1310 | | | 0.1316 | | |
| S.E. of reg | 0.8047 | | | 0.8029 | | |
| Quantile DV | 5.6700 | | | 5.6700 | | |
| Sparsity | 2.0795 | | | 2.0797 | | |

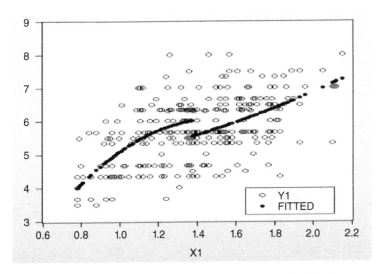

**Figure 8.20** The scatter graphs of *Y1* and its fitted RM in Table 8.1.

## Example 8.7   *An Alternative PWQ-QR*

Table 8.2 presents the statistical results summary of the outputs of a median regression and its reduced model, with the following ES, where $G2a = 1 + 1*(G4 > 3)$, as a modification of the PWQ_QR in (8.12a):

$$Y1 = (C(10) + C(11)*X1 + C(12)*X1{\char`\^}2)*(G2a = 1)$$
$$+ (C(20) + C(21)*X1 + C(22)*X1{\char`\^}2)*(G2a = 2) \tag{8.9}$$

**Table 8.2** Statistical results summary of the Median-QR in (8.9) and its RM.

| Parameter | Quadratic HQR | | | Reduced model | | |
|---|---|---|---|---|---|---|
| | Coef. | t-Stat. | Prob. | Coef. | t-Stat. | Prob. |
| C(10) | −3.1057 | −1.5818 | 0.1148 | −3.1057 | −1.5810 | 0.1149 |
| C(11) | 12.7757 | 3.9071 | 0.0001 | 12.7757 | 3.9042 | 0.0001 |
| C(12) | −4.5956 | −3.4478 | 0.0006 | −4.5956 | −3.4443 | 0.0007 |
| C(20) | 12.9446 | 0.8629 | 0.3889 | 5.2553 | 8.6546 | 0.0000 |
| C(21) | −4.1572 | −0.5088 | 0.6113 | | | |
| C(22) | 2.6112 | 0.5905 | 0.5553 | 0.3919 | 2.0766 | 0.0387 |
| Pseudo R-sq | 0.1451 | | | 0.1445 | | |
| Adjusted R-sq | 0.1306 | | | 0.1329 | | |
| S.E. of reg | 0.8214 | | | 0.8228 | | |
| Quantile DV | 5.6700 | | | 5.6700 | | |
| Sparsity | 2.0291 | | | 2.0268 | | |

Based on this summary, the following findings and notes are presented:

1) Comparing the scatter graphs of *Y1* with its fitted value variables in Figure 8.20 and Figure 8.21, which one do you judge as a better fit median regression? However, in the statistical sense, the Reduced-QR of (8.9) in Table 8.2 is a better fit QR(Median) because it has greater pseudo and adjusted *R*-squared than the Reduced-QR of (8.12a) in Table 8.1.

2) As an additional illustration, Figure 8.22 presents the graph of sorted *Y1* and its fitted variable of the Median-QR (8.10), *Fit_G2a*. Based on this graph, the following steps, findings and notes are presented:

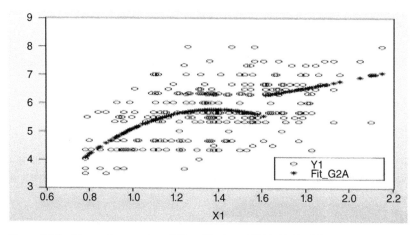

**Figure 8.21** The scatter graphs of *Y1* and its fitted RM in Table 8.2.

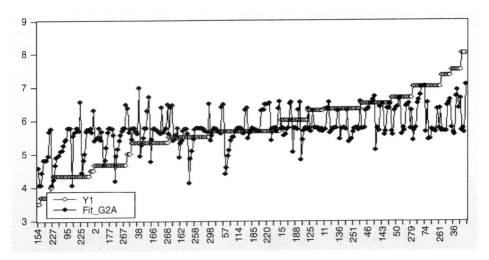

**Figure 8.22** A special graph of the sorted *Y1* and *Fit_G2a*.

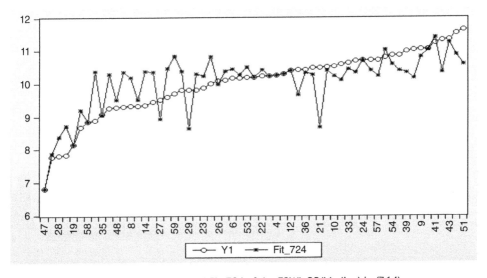

**Figure 8.23** The graph of sorted Y1, and Fit_724 of the F2WI_QR(Median) in (7.14).

2.1 With the data of *Y1* and *Fit_G2a* on-screen, select *View/Graph* and click *OK,* their lines graph appears on-screen. Then by clicking the caption *Freeze/OK*, a new graph appears on-screen.

2.2 With the new graph on-screen, right-click and select *Sort, OK, and Yes* to obtain the pure lines graph with sorted *Y1*.

2.3 Finally, the graph can be modified, so that we have the graph in Figure 8.16.

2.4 It is really unexpected that the scores of *Fit_G2a* tend to be greater than the observed scores of *Y1*, for small scores of *Y1*, say *Y1* < 6. We find that 137 (80.59%) of the 170 scores

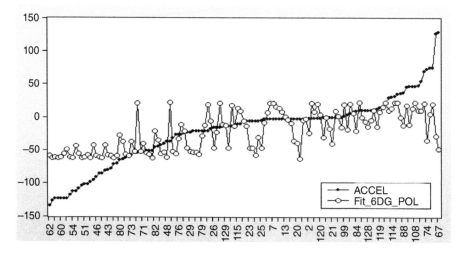

**Figure 8.24** The graph of sorted ACCEL, and Fit_6dg_Pol, based on the Mcycle.wf1.

of *Fit_G2a* are greater than the scores $Y1 < 6$. However, for $Y1 \geq 6$, only 12 (9.23%) of the 130 scores are greater than the scores of *Y1*.

2.5 The scores of *Fit_G2a* are the medians of *Y1*, conditional for the set of fixed observed scores of *X1* and *G2a*, called *conditional medians Y1; therefore*, they should not be compared to the set of fixed observed scores of *Y1* to evaluate whether the median regression is a good fit.

2.6 Under the assumption the model in (8.10) is an acceptable or true model, the *conditional median regression*, or the *conditional quantile regression* in general, is acceptable in the statistical sense because it is estimated using the well-defined and developed least absolute deviation (LAD) estimation method.

2.7 As comparisons, I have tried several alternative quantile regressions and found similar graph patterns. Two of them are presented in figures 8.23 and 8.24. Figure 8.24

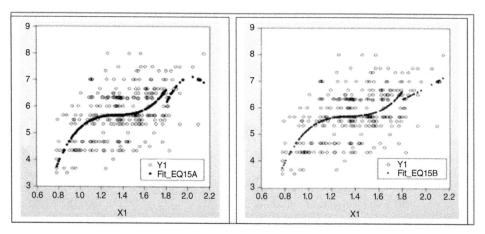

**Figure 8.25** The scatter graphs of Y1 and their fitted variables of the Median-QR (8.11a) and (8.11b).

shows the first graph, based on the F2WI-QR of *Y1* in (7.14), using the data in Data_Faad.wf1. And Figure 8.25 presents the graph of *ACCEL* and its fitted variable of the sixth-degree polynomial QR(Median) of *ACCEL* on *Milli,* based on Mcycle.wf1 (refer to Table 7.4). So, based on these findings, I would say that all QRs will present similar graph patterns.

### 8.2.3  A Piecewise Polynomial Quantile Regression

As an illustration, the following general ES shows a *two piecewise polynomial QR* (PWP-QR) of *Y1* on *X1* by a dichotomous factor *G,* having *m* and *n* degrees, respectively:

$$Y1 = (C(10) + C(11) *X1 + C(12) *X1\char`^2 + \ldots + C(1\,m) *X1\char`^m) * (G = 1)$$
$$+ (C(20) + C(21) *X1 + C(22) *X1\char`^2 + \ldots + C(2n) *X1\char`^n) * (G = 2) \qquad (8.10)$$

**Example 8.8    *An Application of a Specific PWP-QR in (8.10)***
As a modification of the PWQ-QR in Tables 8.1 and 8.2, Table 8.3 presents the statistical results summary of (a) PWP-QR(Median) of *Y1* on *X1* by a dichotomous *GA* and (b) its reduced model:

$$Y1 = (C(10) + C(11) *X1 + C(12) *X1\char`^2 + C(13) *X1\char`^3) * (GA = 1)$$
$$+ (C(20) + C(21) *X1 + C(22) *X1\char`^2) * (GA = 2) \qquad (8.11a)$$
$$Y1 = (C(10) + C(11) *X1 + C(12) *X1\char`^2 + C(13) *X1\char`^3) * (GA = 1)$$
$$+ (C)20) + C(22) *X1\char`^2)*)GA = 2) \qquad (8.11b)$$

Based on this summary, the findings and notes are as follows:

**Table 8.3**   Statistical results summary of the PWP-QR(Median) (8.11a) and (8.11b).

| Parameter | Median-QR (8.11a) | | | Median-QR (8.11b) | | |
|---|---|---|---|---|---|---|
| | Coef. | t-Stat. | Prob. | Coef. | t-Stat. | Prob. |
| C(10) | −18.4719 | −3.1107 | 0.0021 | −18.4719 | −3.1107 | 0.0021 |
| C(11) | 53.0161 | 3.6762 | 0.0003 | 53.0161 | 3.6761 | 0.0003 |
| C(12) | −38.9550 | −3.4339 | 0.0007 | −38.9550 | −3.4339 | 0.0007 |
| C(13) | 9.5799 | 3.3063 | 0.0011 | 9.5799 | 3.3063 | 0.0011 |
| C(20) | −66.8457 | −0.8578 | 0.3917 | 4.3942 | 3.2388 | 0.0013 |
| C(21) | 72.5093 | 0.9090 | 0.3641 | | | |
| C(22) | −17.7719 | −0.8738 | 0.3829 | 0.5909 | 1.6410 | 0.1019 |
| Pseudo R-sq | 0.1408 | | | 0.1370 | | |
| Adj. R-sq | 0.1232 | | | 0.1224 | | |
| S.E. of reg | 0.8163 | | | 0.8152 | | |
| Quantile DV | 5.6700 | | | 5.6700 | | |
| Sparsity | 2.0900 | | | 2.1128 | | |

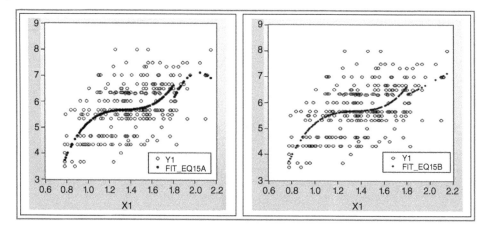

**Figure 8.26** The scatter graphs of Y1 and their fitted variables of the Median-QR (8.11a) and (8.11b).

1) The factor $GA = 1 + 1*(X1 > 1.8)$ is used as one of several alternative factors, which gives the best possible output of the QR(Median) in (8.11a). Since it has a smaller pseudo $R$-squared and adjusted $R$-squared than the PWP-QR in Tables 8.1 and 8.2, this QR is worse than the PWQ-QR, even though they have a very small difference. However, the fitted variable of this QR has a different growth pattern.

2) Based on the output of the QR(Median) in (8.11a), the following findings and notes are presented:

   2.1 In fact, the QR(Median) in (8.11a) is a reduced model of the third-degree polynomial QR(Median) by $GA$, but each IV of the QR in $GA = 2$ has a very large $p$-value.

   2.2 Since each of the QRs in $GA = 2$ has $p$-value >0.30, let's consider a reduced QR (8.11b), which at the 10% level, shows $X1\hat{\ }2$ has a positive significant effect, based on the $t$-statistic of $t_0 = 1.6410$ with $df = 294$ and $p$-value $=0.1019/2 = 0.0546$.

3) By comparing both scatter graphs in Figure 8.26, I would select the QR(Median) in (8.11a) as the best PWP-QR, specific for the third-degree polynomial QR, based on Data_Faad.wf1, which is different from the PWQ-QRs.

## 8.3 HLQR of Y1 on (X1,X2) by the Cell-Factor

Referring to the path diagram we saw earlier in Figure 5.1, which are representing the additive QR and interaction QR of an endogenous variable Y1 on the exogenous (X1,X2), this section presents their extension to an additive HLQR and an interaction HLQR, by the bivariate factors (A,B), which are presented in the following sections.

### 8.3.1 Additive HLQR of *Y1* on (*X1,X2*) by CF

The additive HLQR (AHLQR) of *Y1* on *X1* and *X2,* by bivariate factor *CF,* has the following alternative ESs:

$$Y1 \; C \; X1 \; X2 \; @Expand \, (CF, \; @Dropfirst) \; X1^* \; @ \; Expand \, (CF, \; @Dropfirst)$$
$$X2^*@ \; Expand \, (CF, @ \; Dropfirst) \tag{8.12a}$$
$$Y1 \; C @Expand \, (CF, @ \; Dropfirst) \; X1^*@ \; Expand \, (CF) \; X2^*@ \; Expand \, (CF) \tag{8.12b}$$

**Example 8.9** *An Application of AHLQR (8.12a)*
Figure 8.27 presents the output of the AHLQR(Median) in (8.12a), using *CF* = (*A,B*). Aside from the testing hypotheses which can be done directly using the statistics presented in the output, I present only the testing hypothesis as follows:

1) Specific for the median regression within the cell (*A* = 1, *B* = 1), at the 1% level, the null hypothesis C(2) = C(3) = 0 is rejected, based on the Wald test (*Chi-square* statistic) of 15. 016 35 with *df* = 9 and *p*-value = 0.0054. Hence, both variables, *X1* and *X2,* have significant joint effects on *Y1* within the cell (*A* = 1,*B* = 1).
2) At the 10% level, the output RVT of *X1*\*@Expand(A,B,@Dropfirst)* shows it has an insignificant effect on *Y1* based on the *QLR L*-statistic of 1.638 02 with *df* = 3 and *p*-value = 0.6530. Hence, we can conclude that *X1* has insignificant different joint effects on *Y1* between the four cells of (*A,B*).
3) At the 10% level, *X1*\*@Expand(A,B,@Dropfirst)& X2*\*@Expand(A,B,@Dropfirst)* has insignificant joint effects on *Y1* based on the *QLR L*-statistic of 4.972 326 with *df* = 6 and *p*-value = 0.5475, as presented in Figure 8.28. Hence, we can conclude that *X1* and *X2* have insignificant different joint effects on *Y1* between the four cells of (*A,B*). Based on this conclusion, in the statistical sense, the QR could be reduced to a homogeneous LQR by the factor (A,B), or a two-way ANCOVA LQR, with the following ES:

$$Y1 \; C \; X1 \; X2 \; @Expand \, (A, B, @Dropfirst) \tag{8.13}$$

4) As additional illustration, Figure 8.28 presents the graphs of sorted *Y1* and its fitted variables based on the AHLQR (8.12a). The graph clearly shows most of the fitted scores of the AHLQR are greater than the small observed scores of *Y1*, and most of the fitted scores are smaller than the large observed scores of *Y1*.
5) As an additional illustration, Figure 8.29 presents the graphs of sorted *Y1* and its fitted variables, *FV_Q25* and *FV_Q75*, of the AHLQR(tau) in (8.12a) for tau = 0.25 and 0.75, respectively.

Redundant Variables Test

Equation: EQ35_NOV28

Specification: Y1 C X1 X2 @EXPAND(A,B,@DROPFIRST) X1
   *@EXPAND(A,B,@DROPFIRST) X2*@EXPAND(A,B,@DROPFI RST)

RVT: X1*@Expand(A,B,@Dropfirst)

|  | Value | df | Probability |
|---|---|---|---|
| QLR L-statistic | 1.632802 | 3 | 0.6520 |
| QLR Lambda-statistic | 1.629253 | 3 | 0.6528 |

Redundant Variables Test

Equation: EQ35_NOV28

RVT: X1*@Expand(A,B,@Dropfirst) X2*@Expand,@Dropfirst

|  | Value | df | Probability |
|---|---|---|---|
| QLR L-statistic | 4.971326 | 6 | 0.5475 |
| QLR Lambda-statistic | 4.938624 | 6 | 0.5517 |

**Figure 8.27** Two outputs of RVT based on the QR in Figure 8.26.

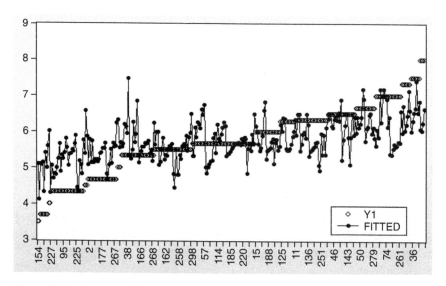

**Figure 8.28** The graph of the sorted Y1 and its fitted variable of the AHLQR(0.5) in Figure 8.26.

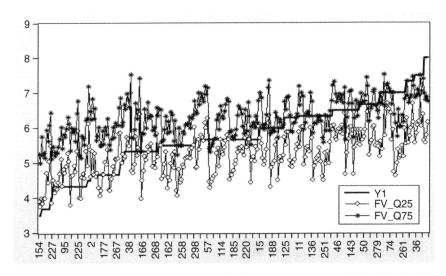

**Figure 8.29** The graph of the sorted Y1 and its fitted variable of the AHLQR(0.25 & 0.75).

**Example 8.10** *An Application of the QSET Based on the AHLQR in Figure 8.27*
As an illustration, Figure 8.30 presents two types of the QSETs based on the AHLQR (8.12a), which are obtained using the following steps:

1) With the output in Figure 8.26 on-screen, select *View/Quantile Process/Slope Equality Test,* insert 0.25 0.75 as the user-specified test quantiles, use C(2) C(3) as the user-specified coefficients, and click *OK*. This obtains the output in Figure 8.30a. Based on this output, the following findings and notes are presented:
   1.1 The output presents the test for two pairs of quantiles (0.25, 0.5) and (0.5, 0.75), for each of *X1* and *X2*. It does not present the test for the pair of quantiles (0.25, 0.75).
   1.2 For testing the pair of (0.25, 0.75), the QSET should be done using the AHLQR(0.25), with the output presented in Figure 8.30b.
   1.3 The output does not present the test statistic for each *X1* and *X2*, but includes only their restriction value, standard error, and probability. I am very sure the *Chi-square* test statistic is applied as the basic nonparametric test.
2) The output in Figure 8.30b is obtained using the steps as follows:
   2.1 To test the pair of quantiles (0.25, 0.75), first we should run a quantile regression(tau) for any tau ≤0.25. However, in this case, it is done using tau = 0.25.
   2.2 With the output of the QR(0.25) on-screen, select *View/Quantile Process/Slope Equality Test,* insert 0.25 0.75 as the user-specified test quantiles, use C(2) C(3) as the user-specified coefficients, and click *OK*. The output is in Figure 8.30b. Note the output presents the statement "*Estimated equation quantile tau* =0.25."

(a)
Quantile Slope Equality Test
Equation: EQ35_NOV28
Specification: Y1 C X1 X2 @EXPAND(A,B,@DROPFIRST) X1
   *@EXPAND(A,B,@DROPFIRST) X2*@EXPAND(A,B,@DROPFIRST)
Estimated equation quantile tau = 0.5
User-specified test quantiles: 0.25 0.75
Test statistic compares user-specified coefficients: C(2) C(3)

| Test Summary | Chi-Sq. Statistic | Chi-Sq. d.f. | Prob. |
|---|---|---|---|
| Wald Test | 5.013112 | 4 | 0.2860 |

Restriction Detail: b(tau_h) – b(tau_k) = 0

| Quantiles | Variable | Restr. Value | Std. Error | Prob. |
|---|---|---|---|---|
| 0.25, 0.5 | X1 | −0.037667 | 0.454970 | 0.9340 |
|  | X2 | 0.017651 | 0.205038 | 0.9314 |
| 0.5, 0.75 | X1 | 0.993956 | 0.452896 | 0.0282 |
|  | X2 | 0.080632 | 0.261960 | 0.7582 |

(b)
Quantile Slope Equality Test
Equation: EQ35_NOV28
Specification: Y1 C X1 X2 @EXPAND(A,B,@DROPFIRST) X1
*@EXPAND(A,B,@DROPFIRST) X2*@EXPAND(A,B,@DROPFIRS T)
Estimated equation quantile tau = 0.25
User-specified test quantiles: 0.25 0.75
Test statistic compares user-specified coefficients: C(2) C(3)

| Test Summary | Chi-Sq. Statistic | Chi-Sq. d.f. | Prob. |
|---|---|---|---|
| Wald Test | 2.557529 | 2 | 0.2784 |

Restriction Detail: b(tau_h) – b(tau_k) = 0

| Quantiles | Variable | Restr. Value | Std. Error | Prob. |
|---|---|---|---|---|
| 0.25, 0.75 | X1 | 0.956289 | 0.648720 | 0.1405 |
|  | X2 | 0.098283 | 0.331011 | 0.7665 |

**Figure 8.30** (a) The output of a QSET based on the AHLQR(0.5) in Figure 8.26. (b) The output a QSET based on the AHLQR(0.1) in Figure 8.26.

### 8.3.2 A Two-Way Interaction Heterogeneous-QR of Y1 on *(X1,X2)* by CF

As the extension of the previous AHLQR in (8.12a & b), I propose a two-way interaction heterogeneous QR (2WI-HQR) of *Y1* on *X1* and *X2,* by *CF*, with the following alternative ESs:

$$Y1\ C@Expand\,(CF,@Dropfirst)\ X1*X2*@Expand\,(CF)\ X1*@Expand\,(CF)$$
$$X2*@Expand\,(CF) \tag{8.14a}$$
$$Y1\ C\,X1*X2\,X1\,X2\,@Expand\,(CF,@Dropfirst)$$
$$X1*X2*@Expand\,(CF,@Dropfirst)\ X1*@Expand\,(CF)\ X2*@Expand\,(CF) \tag{8.14b}$$

**Example 8.11** *Application of TWI-HQR (8.14a)*
Table 8.4 presents the statistical results summary of the QR(0.5) in (8.14a), for *CF* = *A,B* and two of its RMs, RM-1 and RM-2. Based on this summary, the findings and notes are as follows:

1) It is important to note the coefficient of the interaction *X1*X2* within the cell (*A* = 2, *B* = 2). It has negative coefficient in the full model, but it has positive coefficient in both RM-1 and RM-2. So, we have the following differential characteristics.
2) The full model in *(A* = 2, *B* = 2) has the following equation:

$$Y1 = 1.049 + 2.352*X1 + 1.126*X2 - 0.437*X1*X2$$
$$Y1 = 1.0491 + 2.352*X1 + (1.126 - 0.437*X1)\,*X2$$

   This equation shows that the effect of *X2* on *Y1* depends on the function (1.126 − 0.437* *X1*), which indicates that the effect of *X2* decreases with increasing positive scores of *X1*. Note that its effects are positive for *X1* < 1.126/0.437 = 2.5767 and negative for *X1* > 2.5767. their is similar for the QRs within the other three cells, either for the full model or the reduced models, RM-1 and RM-2.
   However, for both RM-1 and RM-2 within the cell (*A* = 2, *B* = 2), the coefficients of *X1*X2* are positive, which indicates that the effects of *X2* increase with increasing positive scores of *X1*.
   In this case, we should use the theoretical or true causal effects between the variables within the set, to choose either the full or reduced models as the final QR. Referring to the negative coefficients of *X1*X2*, the first three cells of *(A,B)* or the three QR(Median)s, in this case, I would choose the full model as the final and acceptable QR(Median), even though *X1*X2* has an insignificant effect within the cell (*A* = 2, *B* = 2).
3) As an additional analysis, Figure 8.31 presents the results of a QSET between a special pair of quantiles (0.1, 0.9) for the interaction *X1*X2*, only by the factors *A* and *B*. The method is as follows:
   3.1 We should run the ES (8.14a) for tau = 0.1.

**Table 8.4** Statistical results summary of the QR(0.5) in (8.14a) and its reduced models.

| Variable | Median-QR (8.14a) | | | RM-1 | | | RM-2 | | |
|---|---|---|---|---|---|---|---|---|---|
| | Coef. | t-Stat | Prob. | Coef. | t-Stat | Prob. | Coef. | t-Stat | Prob. |
| C | −1.850 | −0.644 | 0.520 | −1.850 | −0.643 | 0.521 | −1.850 | −0.643 | 0.521 |
| A = 1,B = 2 | 1.154 | 0.324 | 0.746 | 1.154 | 0.324 | 0.746 | 1.154 | 0.323 | 0.747 |
| A = 2,B = 1 | −0.414 | −0.095 | 0.924 | −0.414 | −0.095 | 0.924 | −0.414 | −0.095 | 0.924 |
| A = 2,B = 2 | 2.899 | 0.880 | 0.380 | 5.263 | 1.777 | 0.077 | 5.532 | 1.891 | 0.060 |
| X1*X2*(A = 1),B = 1) | −0.988 | −1.444 | 0.150 | −0.988 | −1.444 | 0.150 | −0.988 | −1.444 | 0.150 |
| X1*X2*(A = 1),B = 2) | −0.908 | −1.341 | 0.181 | −0.908 | −1.339 | 0.182 | −0.908 | −1.335 | 0.183 |
| X1*X2*(A = 2),B = 1) | −1.946 | −1.821 | 0.070 | −1.946 | −1.824 | 0.069 | −1.946 | −1.832 | 0.068 |
| X1*X2*(A = 2),B = 2) | −0.437 | −0.757 | **0.450** | 0.345 | 1.774 | 0.077 | 0.481 | 2.572 | 0.011 |
| X1*(A = 1),B = 1) | 4.171 | 2.094 | 0.037 | 4.171 | 2.094 | 0.037 | 4.171 | 2.093 | 0.037 |
| X1*(A = 1),B = 2) | 4.155 | 2.597 | 0.010 | 4.155 | 2.594 | 0.010 | 4.155 | 2.585 | 0.010 |
| X1*(A = 2),B = 1) | 5.179 | 2.383 | 0.018 | 5.179 | 2.387 | 0.018 | 5.179 | 2.397 | 0.017 |
| X1*(A = 2),B = 2) | 2.352 | 1.747 | 0.082 | 0.708 | 0.821 | **0.412** | | | |
| X2*(A = 1),B = 1) | 2.104 | 2.101 | 0.037 | 2.104 | 2.100 | 0.037 | 2.104 | 2.098 | 0.037 |
| X2*(A = 1),B = 2) | 1.575 | 1.690 | 0.092 | 1.575 | 1.688 | 0.093 | 1.575 | 1.683 | 0.094 |
| X2*(A = 2),B = 1) | 2.987 | 1.814 | 0.071 | 2.987 | 1.817 | 0.070 | 2.987 | 1.827 | 0.069 |
| X2*(A = 2),B = 2) | 1.126 | 1.531 | 0.127 | | | | 0.088 | 0.232 | **0.816** |
| Pseudo R-sq | 0.220 | | | 0.215 | | | 0.213 | | |
| Adj. R-sq | 0.179 | | | 0.176 | | | 0.174 | | |
| S.E. of reg | 0.762 | | | 0.761 | | | 0.765 | | |
| Quantile DV | 5.670 | | | 5.670 | | | 5.670 | | |
| Sparsity | 1.860 | | | 1.899 | | | 1.875 | | |
| QLR-stat | 105.19 | | | 100.41 | | | 100.66 | | |
| Prob | 0.000 | | | 0.000 | | | 0.000 | | |

    3.2 With the output on-screen, select *View/Quantile Process/ Slope Equality Test*, enter the user-specified quantiles 0.1 0.9, and insert the user-specified coefficients C(5) C(6) C(7) C(8).

    3.3 Finally, click *OK* to obtain the output.

4) Based on the output in Figure 8.31, the following findings and notes are presented:

    4.1 Note the estimated equation quantile tau = 0.1, which indicates the output is obtained based on the QR(0.1). If the output of QR(Median) is used, the output of the QSET would present two tests for pairs of quantiles (0.1, 0.5) and (0.5, 0.9). Do this as an exercise.

    4.2 At the 10% level, only *X1*X2*(A = 1, B = 2)* has a significant difference between the quantiles 0.1 and 0.9.

Quantile slope equality test
Equation: EQ14
Specification: Y1 C @EXPAND(A,B,@DROPFIRST) X1*X2 *@EXPAND(A,B)
   X1*@EXPAND(A,B) X2*@EXPAND(A,B)
Estimated equation quantile tau = 0.1
User-specified test quantiles: 0.1 0.9
Test statistic compares user-specified coefficients: C(5) C(6) C(7) C(8)

| Test summary | Chi-Sq. statistic | Chi-Sq. d.f. | Prob. |
|---|---|---|---|
| Wald test | 4.133774 | 4 | 0.3882 |

Restriction detail: b(tau_h) − b(tau_k) = 0

| Quantiles | Variable | Restr. Value | Std. Error | Prob. |
|---|---|---|---|---|
| 0.1, 0.9 | X1*X2*(X1*X2*(A = 1),B = 1) | 0.136325 | 2.504085 | 0.9566 |
| | **X1*X2*(X1*X2*(A = 1),B = 2)** | **1.643898** | **0.908784** | **0.0705** |
| | X1*X2*(X1*X2*(A = 2),B = 1) | 0.914087 | 0.986954 | 0.3544 |
| | X1*X2*(X1*X2*(A = 2),B = 2) | 0.030215 | 1.003842 | 0.9760 |

**Figure 8.31**   The output of a quantile slope equality test for the interaction *X1*X2*.

### 8.3.3   An Application of Translog-Linear QR of *Y1* on *(X1,X2)* by CF

As a more advanced model, this section presents the translog linear heterogeneous QR (TLHQR) of *Y1* on *(X1,X2)* by *CF*, with the following ES:

$$log\,(Y1)\ C\ @expand\,(CF, @dropfirst)$$
$$log\,(X1)*@expand\,(CF)\ log\,(X2)*@expand\,(CF) \tag{8.15}$$

With this model, we could do all the possible analysis and testing, based on each QR(tau), presented in previous examples.

**Example 8.12**   *Applications of ES (8.15)*
Table 8.5 presents the statistical results summary of the QR (8.15), with $CF = (A,B)$ for tau = 0.1, 0.5, and 0.9. Based on this summary, the following findings and notes are presented.

1) Based on the TLHQR(0.5), at the 10% level, the linear effect of each of $log(X1)$ and $log(X2)$ has a positive significant effect on $log(Y1)$, within each cell $(A = i, B = j)$. For instance, in $(A = 1, B = 1)$, $log(x2)$ has a positive significant effect on $log(Y1)$ based on the *t*-statistic of $t_0 = 1.461$ with $df = 272$ and *p*-value = 0.145/2 = 0.0725.

**Table 8.5** Statistical results summary of the TLHQR (8.15) for tau = 0.1, 0.5, and 0.9.

| H Variable | TLHQR(0.1) Coef. | t-Stat. | Prob. | TLHQR(0.5) Coef. | t-Stat. | Prob. | TLHQR(0.9) Coef. | t-Stat. | Prob. |
|---|---|---|---|---|---|---|---|---|---|
| C | 1.418 | 9.352 | 0.000 | 1.668 | 10.330 | 0.000 | 1.850 | 23.189 | 0.000 |
| A = 1,B = 2 | −0.204 | −0.928 | 0.354 | −0.341 | −1.850 | 0.065 | −0.215 | −1.628 | 0.105 |
| A = 2,B = 1 | −0.207 | −1.230 | 0.220 | −0.292 | −1.475 | 0.141 | −0.257 | −1.482 | 0.139 |
| A = 2,B = 2 | −0.020 | −0.129 | 0.897 | −0.320 | −1.852 | 0.065 | −0.197 | −1.277 | 0.203 |
| LOG(X1)*(A = 1,B = 1) | 0.453 | 1.841 | 0.067 | 0.202 | 1.886 | 0.060 | 0.093 | 1.110 | 0.268 |
| LOG(X1)*(A = 1,B = 2) | 0.422 | 1.022 | 0.308 | 0.331 | 4.104 | 0.000 | 0.348 | 4.441 | 0.000 |
| LOG(X1)*(A = 2,B = 1) | 0.306 | 1.489 | 0.138 | 0.454 | 4.358 | 0.000 | 0.222 | 0.887 | **0.376** |
| LOG(X1)*(A = 2,B = 2) | 0.527 | 6.601 | 0.000 | 0.145 | 1.636 | 0.103 | 0.159 | 1.019 | **0.309** |
| LOG(X2)*(A = 1,B = 1) | 0.062 | 0.326 | **0.745** | 0.179 | 1.461 | 0.145 | 0.129 | 2.350 | 0.020 |
| LOG(X2)*(A = 1,B = 2) | 0.273 | 0.854 | **0.394** | 0.342 | 3.735 | 0.000 | 0.156 | 1.524 | 0.129 |
| LOG(X2)*(A = 2,B = 1) | 0.188 | 1.268 | 0.206 | 0.165 | 1.603 | 0.110 | 0.230 | 1.793 | 0.074 |
| LOG(X2)*(A = 2,B = 2) | −0.092 | −1.252 | 0.212 | 0.318 | 5.104 | 0.000 | 0.151 | 0.943 | **0.346** |
| Pseudo R-squared | 0.324 | | | 0.313 | | | 0.280 | | |
| Adjusted R-sq | 0.298 | | | 0.287 | | | 0.252 | | |
| S.E. of regression | 0.241 | | | 0.146 | | | 0.222 | | |
| Quantile DV | 1.504 | | | 1.749 | | | 1.974 | | |
| Sparsity | 0.785 | | | 0.319 | | | 0.527 | | |
| QLR-stat | 103.84 | | | 183.24 | | | 101.34 | | |
| Prob. | 0.000 | | | 0.000 | | | 0.000 | | |

Wald Test:
Equation: TLQR(0.5)

| Test Statistic | Value | df | Probability |
|---|---|---|---|
| F-statistic | 1.421729 | (6, 288) | 0.2060 |
| Chi-square | 8.530372 | 6 | 0.2018 |

Null Hypothesis: C(5)=C(6)=C(7)=C(8),
  C(9)=C(10)=C(11)=C(12)

**Figure 8.32** The output of a Wald tests, based on TLQR(0.5) in Table 8.5.

2) However, based on the TLHQR(0.1) and TLHQR(0.9), the *log(X1)* and *log(X2)* have insignificant effects in some of the cells *(A = i, B = j)*.

3) Figure 8.32 presents a Wald test based on the TLHQR(0.5). Then, at the10% level, we can conclude that both *log(X1)* and *log(X2)* have insignificant difference joint effects on *log(Y1)* between the four cells of *(A,B)*, based on the *Chi-square* statistic of 8.530372 with *df* =6 and *p*-value = 0.2018.

4) The function of TLHQR(0.5) within each cell *(A = i, B = j)* can be transformed to a multiplicative or nonlinear function. For instance, within *(A = 1, B = 1)*, we have one of the following equations:

$$log\,(Y1) = 1.668 + 0.202 log\,(X1) + 0.179 log\,(X2)$$

and

$$Y1 = Exp\,(1.668) \times X1^{0.202} \times X2^{0.179}$$

## 8.4 The HLQR of *Y1* on *(X1,X2,X3)* by a Cell-Factor

Referring to the path diagram we saw earlier in Figure 6.1, which are representing alternative additive and interaction linear QRs of *Y1* on *(X1,X2,X3)*, this section presents their extensions to additive-HLQR and interaction-HLQR of an endogenous variable *Y1* on tri-variate exogenous variable *(X1,X2,X3)*, by a *CF*, which can be one or more categorical factors.

### 8.4.1 An Additive HLQR of *Y1* on *(X1,X2,X3)* by *CF*

The additive HLQR (AHLQR) of *Y1* on *X1, X2,* and *X3*, by *CF*, has the following alternative ESs:

$$Y1\,C\,@Expand\,(CF, @Dropfirst)\,X1*@Expand\,(CF)\,X2*@Expand\,(CF)$$
$$X3*@Expand\,(CF) \qquad\qquad (8.16a)$$
$$Y1\,C\,X1\,X2\,X3@Expand\,(CF, @Dropfirst)\,X1*@Expand\,(CF, @Dropfirst)$$
$$X2*@Expand\,(CF, @Dropfirst)\,X3*@Expand\,(CF, @Dropfirst) \qquad (8.16b)$$

The ES (8.16b) should be applied if we want to conduct the test of various hypotheses on the effect differences of *X1, X2,* or *X3* between the levels of *CF*, which can be tested directly using the *t*-statistic in the output. Furthermore, the RVT can be applied, in addition to the Wald test.

### 8.4.2 A Full Two-Way Interaction HQR of *Y1* on *(X1,X2,X3)* by *CF*

As an extension of the F2WI-QR in (6.3), we have a full two-way interaction HQR (F2WI-HQR) with the following alternative ESs:

$$Y1\,C@Expand\,(CF, @Dropfirst)\,X1*X2*@Expand\,(CF)$$
$$X1*X3*@Expand\,(CF)\,X2*X3*@Expand\,(CF)\,X1*@Expand\,(CF)$$
$$X2*@Expand\,(CF)\,X3*@Expand\,(CF) \qquad\qquad (8.17a)$$

$$Y1 \; C \; X1^*X2 \; X^*X3 \; X2^*X3 \; X1 \; X2 \; X3 \, @ \, Expand \, (CF, \, @Dropfirst)$$

$$X1^*X2^* \, @ \, Expand \, (CF, \, @Dropfirst) \; X1^*X3^* \, @ \, Expand \, (CF, \, @Dropfirst)$$

$$X2^*X3^* \, @ \, Expand \, (CF, \, @Dropfirst) \; X1^* \, @ \, Expand \, (CF, \, @Dropfirst)$$

$$X2^* \, @ \, Expand \, (CF, \, @Dropfirst) \; X3^* \, @ \, Expand \, (CF, \, @Dropfirst) \tag{8.17b}$$

l?As with the ES (8.16b), I recommend applying the ES (8.17b) if we want to conduct the test of various hypotheses on the effect differences of the two-way interaction, as well as the main variables $X1, X2,$ or $X3$ between the levels of $CF$, which can be tested directly using the $t$-statistic in the output. In addition, the RVT and the Wald test can be applied for testing other hypotheses.

### 8.4.3 A Full Three-Way Interaction HQR of *Y1* on *(X1,X2,X3)* by *CF*

As an extension of the F3WI-QR in (6.4), we have a full thee-way interaction HQR(F3WI-HDR) with the following ESs:

$$Y1 \; C \, @ \, Expand \, (CF, \, @Dropfirst) \; X1^*X2^*X2^* \, @ \, Expand \, (CF)$$

$$X1^*X2^* \, @ \, Expand \, (CF) \; X1^*X3^* \, @ \, Expand \, (CF) \; X2^*X3^* \, @ \, Expand \, (CF)$$

$$X1^* \, @ \, Expand \, (CF) \; X2^* \, @ \, Expand \, (CF) \; X3^* \, @ \, Expand \, (CF) \tag{8.18a}$$

$$Y1 \; C \; X1^*X2^*X3 \; X1^*X2 \; X1^*X3 \; X2^*X3 \; X1 \; X2 \; X3 \, @ \, Expand \, (CF, \, @Dropfirst)$$

$$X1^*X2^*X3^* \, @ \, Expand \, (CF, \, @Dropfirst) \; X1^*X2^* \, @ \, Expand \, (CF, \, @Dropfirst)$$

$$X1^*X3^* \, @ \, Expand \, (CF, \, @Dropfirst) \; X2^*X3^* \, @ \, Expand \, (CF, \, @Dropfirst)$$

$$X1^* \, @ \, Expand \, (CF, \, @Dropfirst) \; X2^* \, @ \, Expand \, (CF, \, @Dropfirst)$$

$$X3^* \, @ \, Expand \, (CF, \, @Dropfirst) \tag{8.18b}$$

As with the ESs (8.16b) and (8.17b), I recommend applying the ES (8.17b) if we want to conduct the test of various hypotheses on the effect differences of the three-way interaction, two-way interaction, and the main variables $X1, X2,$ or $X3$ between the levels of CF, which can be tested directly using the $t$-statistic in the output. In addition, the RVT and the Wald test can be applied for testing other hypotheses on their effect differences.

**Example 8.13** *An Application of (8.18a) for* **CF = A**
Figure 8.33 presents the output of the QR(0.5) in (8.18a) for $CF = A$. Based on this output, the following findings and notes are presented:

1) It is important to note that if we want to test the differential of one or more numerical IVs between the levels of the factor $A$, then the QRs within the two levels of $A$ should have exactly the same set of numerical IVs. Otherwise, the effect of an IV will have different meaning from one level to the other. See the output of a reduced model presented in Table 8.6 below.
2) Based on the full model, the differential effects between any IVs can be tested using the Wald test. Such as, the interaction $X1^*X2^*X3$ has insignificant different effects on $Y1$ between the levels of $A$, based on the *Chi-square* statistic of 0.465 851 with $df = 1$ and $p$-value = 0.4949.

Dependent Variable: Y1
Method: Quantile Regression (Median)
Date: 11/28/18   Time: 15:02
Sample: 1 300
Included observations: 300
Huber Sandwich Standard Errors and Covariance
Sparsity method: Kernel (Epanechnikov) using residuals
Bandwidth method: Hall-Sheather, bw = 0.14513
Estimation successfully identifies unique optimal solution

| Variable | Coefficient | Std. Error | t-Statistic | Prob. |
|---|---|---|---|---|
| C | 3.499767 | 1.266094 | 2.764223 | 0.0061 |
| X1*X3*(A = 1) | 0.488615 | 0.255597 | 1.911662 | 0.0569 |
| X2*X3*(A = 1) | 0.107739 | 0.075382 | 1.429248 | 0.1540 |
| A = 2 | 1.294066 | 1.371852 | 0.943298 | 0.3463 |
| X1*X2*X3*(A = 1) | −0.114331 | 0.066850 | −1.710268 | **0.0883** |
| X1*X2*X3*(A = 2) | 0.177726 | 0.028352 | 6.268463 | **0.0000** |
| X1*X2*(A = 1) | 0.365780 | 0.587960 | 0.622118 | 0.5344 |
| X1*X2*(A = 2) | −0.809516 | 0.149033 | −5.431807 | 0.0000 |
| X1*(A = 1) | −1.690387 | 2.322387 | −0.727866 | 0.4673 |
| X1*(A = 2) | 0.239060 | 0.363379 | 0.657880 | 0.5111 |

| | | | | |
|---|---|---|---|---|
| Pseudo R-squared | 0.328742 | Mean dependent var | | 5.751100 |
| Adjusted R-squared | 0.307910 | S.D. dependent var | | 0.945077 |
| S.E. of regression | 0.669171 | Objective | | 74.49283 |
| Quantile dependent var | 5.670000 | Restr. objective | | 110.9750 |
| Sparsity | 1.493357 | Quasi-LR statistic | | 195.4371 |
| Prob(Quasi-LR stat) | 0.000000 | | | |

**Figure 8.33**   The output of the F3WI-HQR(0.5) in (8.19) for CF = G4.

3) As an illustration, the reduced model in Figure 8.34 is obtained using the trial-and-error method and shows the interaction has a significant effect in both QRs. However, the QRs have different sets of IVs. So, based on this reduced model, we can't test differential effects any of the numerical IVs on Y1 between the levels of the factor A, because the interaction X1*X2*X3 has different characteristics in the two QRs.

4) Note that the QRs within each level of the factor A do not have the same set of numerical IVs. The output of the QR in A = 1 has X1*X3 and X2*X3 as IVs, indicated by the IVs X1*X3*(A = 1) and X2*X3*(A = 1) in the output, whereas the QR in A = 2 does not. So the effect of X1*X2*X3 on Y1 within A = 1 is an effect adjusted for X1*X2, X1*X3, X2*X3 and X1, but within the level A = 2, its effect is adjusted for the IVs X1*X2 and X1 only.

**Table 8.6** The output of a manual multistage selection method of the ES (8.19).

| Parameter | Stage-1 | | | Stage-2 | | | Final-QR | | |
|---|---|---|---|---|---|---|---|---|---|
| | Coef. | t-Stat. | Prob. | Coef. | t-Stat. | Prob. | Coef. | t-Stat. | Prob. |
| C(10) | 4.536 | 21.077 | 0.000 | 4.925 | 12.402 | 0.000 | 3.500 | 2.818 | 0.005 |
| **C(11)** | **0.056** | **8.022** | **0.000** | **0.104** | **3.312** | **0.001** | **−0.114** | **−1.747** | **0.082** |
| C(12) | | | | −0.542 | −3.345 | 0.001 | 0.366 | 0.632 | 0.528 |
| C(13) | | | | 0.057 | 0.806 | 0.421 | 0.489 | 1.929 | 0.055 |
| C(14) | | | | | | | 0.108 | 1.454 | 0.147 |
| C(15) | | | | | | | −1.690 | −0.737 | 0.462 |
| C(16) | | | | | | | | | |
| C(17) | | | | | | | | | |
| C(20) | 4.183 | 16.667 | 0.000 | 4.519 | 8.943 | 0.000 | 1.704 | 2.511 | 0.013 |
| **C(21)** | **0.076** | **5.811** | **0.000** | **0.140** | **3.692** | **0.000** | **−0.197** | **−3.064** | **0.002** |
| C(22) | | | | −0.619 | −3.483 | 0.001 | 0.273 | 0.670 | 0.503 |
| C(23) | | | | 0.084 | 1.250 | 0.212 | 0.486 | 2.535 | 0.012 |
| C(24) | | | | | | | 0.254 | 5.093 | 0.000 |
| C(25) | | | | | | | −0.270 | −0.194 | 0.846 |
| C(26) | | | | | | | | | |
| C(27) | | | | | | | | | |
| Pseudo R-sq | 0.253 | | | 0.318 | | | 0.387 | | |
| Adj. R-sq | 0.245 | | | 0.302 | | | 0.363 | | |
| S.E. of reg. | 0.709 | | | 0.667 | | | 0.626 | | |
| Quantile DV | 5.670 | | | 5.670 | | | 5.670 | | |
| Sparsity | 1.829 | | | 1.582 | | | 1.300 | | |
| Objective | 82.937 | | | 75.693 | | | 68.045 | | |
| Restr. obj. | 110.98 | | | 110.98 | | | 110.98 | | |

**Example 8.14** *An Alternative Reduced Model of (8.18a) for CF = A*

As an additional illustration, Table 8.6 presents the summary of the manual multistage selection methods, using the full model which is presented using the following ES:

$$Y1 = (C(10) + C(11)*X1*X2*X3 + C(12)*X1*X2 + C(13)*X1*X3$$
$$+ C(14)*X2*X3 + C(15)*X1 + C(16)*X2 + C(17)*X3)*(A = 1)$$
$$+ (C(20) + C(21)*X1*X2*X3 + C(22)*X1*X2 + C(23)*X1*X3$$
$$+ C(24)*X2*X3 + C(25)*X1 + C(26)*X2 + C(27)*X3)*(A = 2) \qquad (8.19)$$

Based on this summary, the following findings and notes are presented:

1) In this case, we try to obtain an acceptable reduced model so that the QRs within the two levels of A have the same set of numerical IVs and the interaction X1*X2*X3 has a significant effect on Y1, as the Final-QR in Table 8.6.

Dependent Variable: Y1
Method: Quantile Regression (Median)
Date: 11/28/18   Time: 15:02
Sample: 1 300
Included observations: 300
Huber Sandwich Standard Errors & Covariance
Sparsity method: Kernel (Epanechnikov) using residuals
Bandwidth method: Hall-Sheather, bw = 0.14513
Estimation successfully identifies unique optimal solution

| Variable | Coefficient | Std. Error | t-Statistic | Prob. |
|---|---|---|---|---|
| C | 3.499767 | 1.266094 | 2.764223 | 0.0061 |
| X1*X3*(A = 1) | 0.488615 | 0.255597 | 1.911662 | 0.0569 |
| X2*X3*(A = 1) | 0.107739 | 0.075382 | 1.429248 | 0.1540 |
| A = 2 | 1.294066 | 1.371852 | 0.943298 | 0.3463 |
| X1*X2*X3*(A = 1) | −0.114331 | 0.066850 | −1.710268 | **0.0883** |
| X1*X2*X3*(A = 2) | 0.177726 | 0.028352 | 6.268463 | **0.0000** |
| X1*X2*(A = 1) | 0.365780 | 0.587960 | 0.622118 | 0.5344 |
| X1*X2*(A = 2) | −0.809516 | 0.149033 | −5.431807 | 0.0000 |
| X1*(A = 1) | −1.690387 | 2.322387 | −0.727866 | 0.4673 |
| X1*(A = 2) | 0.239060 | 0.363379 | 0.657880 | 0.5111 |

| | | | | |
|---|---|---|---|---|
| Pseudo R-squared | 0.328742 | Mean dependent var | | 5.751100 |
| Adjusted R-squared | 0.307910 | S.D. dependent var | | 0.945077 |
| S.E. of regression | 0.669171 | Objective | | 74.49283 |
| Quantile dependent var | 5.670000 | Restr. objective | | 110.9750 |
| Sparsity | 1.493357 | Quasi-LR statistic | | 195.4371 |
| Prob(Quasi-LR stat) | 0.000000 | | | |

**Figure 8.34** The output of an acceptable reduced model of the QR in Figure 8.33.

2) So, the effect of the interaction *X1*X2*X3* on *Y1* is adjusted for the same set of variables in both QRs. Hence, we can test its effect difference between the two levels of the factor *A*, using the Wald test. We find that the effects have an insignificant effect difference, based on the *Chi-square* statistic of 0.814 616 with $df = 1$ and p-value = 0.2668.

3) In addition, Figure 8.35 shows that the joint effects of all IVs have insignificant differences between the two levels of *A*, based on the *Chi-square* statistic of 5.244 614 with $df = 5$ and p-value = 0.3868. So, in the statistical sense, the reduced 3WI-HQR(0.5) could be reduced to a 3WI-Homogeneous QR(0.5).

4) However, I prefer to present the Final-QR as an acceptable reduced 3WI-HQR, because in general or in the theoretical sense, an IV should have different adjusted effects on the corresponding dependent variable (DV) between the groups of the individuals (research objects).

Wald test:
Equation: EQ35_FINAL

| Test statistic | Value | df | Probability |
|---|---|---|---|
| F-statistic | 1.048923 | (5, 288) | 0.3891 |
| Chi-square | 5.244614 | 5 | 0.3868 |

Null Hypothesis: $C(11) = C(21)$, $C(12) = C(22)$, $C(13) = C(23)$,
    $C(14) = C(24)$, $C(15) = C(25)$
Null hypothesis summary:

| Normalized Restriction (= 0) | Value | Std. Err. |
|---|---|---|
| $C(11) - C(21)$ | 0.082835 | 0.091777 |
| $C(12) - C(22)$ | 0.092760 | 0.707819 |
| $C(13) - C(23)$ | 0.003101 | 0.317574 |
| $C(14) - C(24)$ | -0.146727 | 0.089377 |
| $C(15) - C(25)$ | -1.420446 | 2.683435 |

Restrictions are linear in coefficients.

**Figure 8.35** The Wald test for the joint effect differences of all numerical IVs.

**Example 8.15** *An Application the F3WI-QR(Median) in (8.18b) for CF = G4*
As an illustration, Figure 8.36 present the output of a F3WI-QR(Median) in (8.18b) for $CF = G4$, and Figure 8.37 present its Representations' output. Based on these outputs, the following findings and notes are presented:

1) Based on the *QLR* statistic of $QLR_0 = 316.9843$ with $df = 31$ and *p*-value $= 0.000\,000$, we can conclude all IVs are jointly significant.
2) Based on the output of its representations, the following equation of the QR(Median) in the level $G4 = 1$, as the referent group, can easily be written using the block or select the first eight terms in the *"Substituted Coefficients"* and copy-paste to the text. Then inserting the symbol $\widehat{Y1}$, and using the coefficiets in three decimals, we have the following regression in $G4 = 1$.

$$\widehat{Y1} = -2.183 - 0.499*X1*X2*X3 + 2.998*X1*X2 - 0.949*X1*X3$$
$$+ 0.585*X2*X3 + 4.283*X1 - 3.325*X2 + 1.522*X3$$

3) The *t*-statistic starting from the row of $X1*X2*X3*(G4 = 2)$ up to $X3*(G4 = 4)$, with the model parameters $C(12)$ to $C(32)$, can be used to tests one- or two-sided hypotheses on the effect differences of the corresponding numerical variables between a pair of $G4$ levels, with $G4 = 1$ as the referent level. For instance, at the 5% level, $X1*X2*X3$ has

Dependent Variable: Y1
Method: Quantile Regression (Median)
Date: 08/31/19   Time: 19:36
Sample: 1 300
Included observations: 300
Huber Sandwich Standard Errors & Covariance
Sparsity method: Kernel (Epanechnikov) using residuals
Bandwidth method: Hall-Sheather, bw = 0.14513
Estimation successfully identifies unique optimal solution

| Variable | Coefficient | Std. Error | t-Statistic | Prob. |
|---|---|---|---|---|
| C | −2.182824 | 15.41753 | −0.141581 | 0.8875 |
| X1*X2*X3 | −0.499035 | 1.585625 | −0.314724 | 0.7532 |
| X1*X2 | 2.997522 | 7.004571 | 0.427938 | 0.6690 |
| X1*X3 | −0.949049 | 3.714910 | −0.255470 | 0.7986 |
| X2*X3 | 0.584775 | 1.578177 | 0.370538 | 0.7113 |
| X1 | 4.282567 | 16.09775 | 0.266035 | 0.7904 |
| X2 | −3.324819 | 6.732877 | −0.493818 | 0.6218 |
| X3 | 1.521843 | 3.716688 | 0.409462 | 0.6825 |
| G4 = 2 | 104.4786 | 52.57357 | 1.987284 | 0.0479 |
| G4 = 3 | −112.3243 | 70.13467 | −1.601551 | 0.1104 |
| G4 = 4 | 41.61558 | 62.66092 | 0.664139 | 0.5072 |
| X1*X2*X3*(G4 = 2) | −6.178425 | 3.174549 | −1.946237 | 0.0527 |
| X1*X2*X3*(G4 = 3) | 6.253580 | 3.387740 | 1.845945 | 0.0660 |
| X1*X2*X3*(G4 = 4) | −0.204114 | 2.449692 | −0.083322 | 0.9337 |
| X1*X2*(G4 = 2) | 33.55382 | 16.74346 | 2.003996 | 0.0461 |
| X1*X2*(G4 = 3) | −34.12841 | 18.97423 | −1.798671 | 0.0732 |
| X1*X2*(G4 = 4) | −2.387099 | 14.03353 | −0.170100 | 0.8651 |
| X1*X3*(G4 = 2) | 15.75295 | 8.204715 | 1.919988 | 0.0559 |
| X1*X3*(G4 = 3) | −13.31668 | 8.796066 | −1.513937 | 0.1312 |
| X1*X3*(G4 = 4) | 5.832164 | 6.480467 | 0.899960 | 0.3689 |
| X2*X3*(G4 = 2) | 7.971982 | 3.932123 | 2.027399 | 0.0436 |
| X2*X3*(G4 = 3) | −9.142398 | 4.700605 | −1.944941 | 0.0528 |
| X2*X3*(G4 = 4) | 0.668764 | 3.631086 | 0.184177 | 0.8540 |
| X1*(G4 = 2) | −82.69335 | 41.68102 | −1.983957 | 0.0483 |
| X1*(G4 = 3) | 74.75583 | 49.18545 | 1.519877 | 0.1297 |
| X1*(G4 = 4) | −24.69889 | 38.55135 | −0.640675 | 0.5223 |
| X2*(G4 = 2) | −43.22343 | 20.95481 | −2.062697 | 0.0401 |
| X2*(G4 = 3) | 49.52644 | 26.90119 | 1.841050 | 0.0667 |
| X2*(G4 = 4) | 2.357279 | 22.23393 | 0.106022 | 0.9156 |
| X3*(G4 = 2) | −19.92851 | 10.29292 | −1.936139 | 0.0539 |
| X3*(G4 = 3) | 20.21736 | 12.31451 | 1.641751 | 0.1018 |
| X3*(G4 = 4) | −9.780488 | 9.974777 | −0.980522 | 0.3277 |

**Figure 8.36**   The output of the F3WI-QR(Median) in (8.18b).

| | | | |
|---|---|---|---|
| Pseudo R-squared | 0.450689 | Mean dependent var | 5.751100 |
| Adjusted R-squared | 0.387149 | S.D. dependent var | 0.945077 |
| S.E. of regression | 0.586219 | Objective | 60.95978 |
| Quantile dependent var | 5.670000 | Restr. objective | 110.9750 |
| Sparsity | 1.262276 | Quasi-LR statistic | 316.9843 |
| Prob(Quasi-LR stat) | 0.000000 | | |

**Figure 8.36** (*continued*)

a significant smaller effect in the level $G4 = 2$ than $G4 = 1$, based on the $t$-statistic of $t_0 = -1.946\,237$ with $df = 268$ and $p$-value $= 0.0527/2 = 0.026\,35$.

4) The list of variables can easily be used to conduct various RVTs using the block copy-paste method, such as follows:

4.1 To do the RVT for the numerical IVs in the level $G4 = 1$. With the output of the representation on-screen, *block select* the list of the seven variables $X1*X2*X3\ X1*X2\ X1*X3\ X2*X3\ X1\ X2\ X3$, in the Estimation Command of QREG then click *copy*. Then select *View/Coefficient Diagnostics/Redundant Variables – Likelihood Ration*, click *paste*, and click *OK* to obtain the output in Figure 8.38, which shows the seven IVs are jointly significant at the 1% level.

4.2 To do the RVT of the effect differences of $X1*X2*X3$ between the four levels of $G4$. With the output of the representation on-screen, *block select* the IV $X1*X2*X3*@$ *Expand(G4,@Dropfirst)* in the Estimation Command of QREG then click *copy*. Then select . *View/Coefficient Diagnostics/Redundant Variables – Likelihood Ration*, click *paste*, and then click *OK* to obtain the output in Figure 8.39, which shows $X1*X2*X3$ has significant differences between the four levels of $G4$ at the 1% level of significance.

4.3 Next, we want to test the hypothesis that the effect of $X1$ on $Y1$, depends on $X2$ and $X3$, conditional for the level $G4 = 1$. We should apply the Wald test for the null hypothesis, $H_0$: $C(2) = C(3) = C(4) = C(6) = 0$, with the output presented in Figure 8.40, which shows the effect of $X1$ on $Y1$ is significantly dependent on $X2*X3, X2$ and $X3$ at the 5% level.

4.4 Considering the hypotheses presented in point (4.2), and (4.3) let's text the complex hypotheses as follows

4.4.1 To test a complex hypothesis on the differences of *"the effect of X1 on Y1 depends on X2 and X3"* between the levels of $G4$. With the output of the representations on-screen, *block select* all the IVs with the variable $X1$ in the Estimation Command of QREG then click *copy*. Then select *View/Coefficient Diagnostics/Redundant Variables – Likelihood Ratio*, click *paste*, and then click *OK* to obtain the output in Figure 8.41. It shows the redundant variables are jointly significant at the 1% level, based on the $QLR$ $L$-statistic of 36.863 29 with $df = 12$ and $p$-value $= 0.0002$.

Estimation Command:

=========================

QREG Y1 C X1*X2*X3 X1*X2 X1*X3 X2*X3 X1 X2 X3 @EXPAND(G4,@DROPFIRST)

    X1*X2*X3*@EXPAND(G4,@DROPFIRST) X1*X2*@EXPAND(G4,@DROPFIRST)

    X1*X3*@EXPAND(G4,@DROPFIRST) X2*X3*@EXPAND(G4,@DROPFIRST)

    X1*@EXPAND(G4,@DROPFIRST) X2*@EXPAND(G4,@DROPFIRST)

    X3*@EXPAND(G4,@DROPFIRST)

Estimation Equation:

=========================

Y1 = C(1) + C(2)*X1*X2*X3 + C(3)*X1*X2 + C(4)*X1*X3 + C(5)*X2*X3 + C(6)*X1 + C(7)*X2 +
C(8)*X3 + C(9)*(G4=2) + C(10)*(G4=3) + C(11)*(G4=4) + C(12)*X1*X2*X3*(G4=2) +
C(13)*X1*X2*X3*(G4=3) + C(14)*X1*X2*X3*(G4=4) + C(15)*X1*X2*(G4=2) +
C(16)*X1*X2*(G4=3) + C(17)*X1*X2*(G4=4) + C(18)*X1*X3*(G4=2) + C(19)*X1*X3*(G4=3) +
C(20)*X1*X3*(G4=4) + C(21)*X2*X3*(G4=2) + C(22)*X2*X3*(G4=3) + C(23)*X2*X3*(G4=4) +
C(24)*X1*(G4=2) + C(25)*X1*(G4=3) + C(26)*X1*(G4=4) + C(27)*X2*(G4=2) +
C(28)*X2*(G4=3) + C(29)*X2*(G4=4) + C(30)*X3*(G4=2) + C(31)*X3*(G4=3) +
C(32)*X3*(G4=4)

Forecasting Equation:

=========================

Y1 = C(1) + C(2)*X1*X2*X3 + C(3)*X1*X2 + C(4)*X1*X3 + C(5)*X2*X3 + C(6)*X1 + C(7)*X2 +
C(8)*X3 + C(9)*(G4=2) + C(10)*(G4=3) + C(11)*(G4=4) + C(12)*X1*X2*X3*(G4=2) +
C(13)*X1*X2*X3*(G4=3) + C(14)*X1*X2*X3*(G4=4) + C(15)*X1*X2*(G4=2) +
C(16)*X1*X2*(G4=3) + C(17)*X1*X2*(G4=4) + C(18)*X1*X3*(G4=2) + C(19)*X1*X3*(G4=3) +
C(20)*X1*X3*(G4=4) + C(21)*X2*X3*(G4=2) + C(22)*X2*X3*(G4=3) + C(23)*X2*X3*(G4=4) +
C(24)*X1*(G4=2) + C(25)*X1*(G4=3) + C(26)*X1*(G4=4) + C(27)*X2*(G4=2) +
C(28)*X2*(G4=3) + C(29)*X2*(G4=4) + C(30)*X3*(G4=2) + C(31)*X3*(G4=3) +
C(32)*X3*(G4=4)

Substituted Coefficients:

=========================

Y1 = -2.18282373414 - 0.499034804656*X1*X2*X3 + 2.99752240616*X1*X2 -
0.949048624498*X1*X3 + 0.58477518186*X2*X3 + 4.28256687431*X1 - 3.3248191684*X2 +
1.52184256011*X3 + 104.47864861*(G4=2) - 112.324255824*(G4=3) + 41.6155835254*(G4=4) -
6.17842482141*X1*X2*X3*(G4=2) + 6.25357970736*X1*X2*X3*(G4=3) -
0.204113576632*X1*X2*X3*(G4=4) + 33.5538232012*X1*X2*(G4=2) -
34.128406676*X1*X2*(G4=3) - 2.38709899092*X1*X2*(G4=4) + 15.7529540024*X1*X3*(G4=2) -
13.3166849776*X1*X3*(G4=3) + 5.83216421349*X1*X3*(G4=4) + 7.97198188861*X2*X3*(G4=2)
- 9.14239790273*X2*X3*(G4=3) + 0.668763780704*X2*X3*(G4=4) - 82.6933508492*X1*(G4=2) +
74.7558299631*X1*(G4=3) - 24.6988944118*X1*(G4=4) - 43.2234313024*X2*(G4=2) +
49.5264354255*X2*(G4=3) + 2.35727897749*X2*(G4=4) - 19.9285105183*X3*(G4=2) +
20.2173603037*X3*(G4=3) - 9.78048779626*X3*(G4=4)

**Figure 8.37** The output of the representations of the F3WI-QR(Median) in (8.18b).

Redundant variables test
Equation: EQ1_F3WI_QR
Redundant variables: X1*X2*X3 X1*X2 X1*X3 X2*X3 X1 X2 X3
Specification: Y1 C X1*X2*X3 X1*X2 X1*X3 X2*X3 X1 X2 X3
    @EXPAND(G4,@DROPFIRST) X1*X2*X3 *@EXPAND(G4,@DROPFIRST)
    X1*X2*@EXPAND(G4,@DROPFIRST) X1*X3*@EXPAND(G4,@DROPFIRST)
    X2 *X3*@EXPAND(G4,@DROPFIRST) X1 *@EXPAND(G4,@DROPFIRST)
    X2*@EXPAND(G4,@DROPFIRST) X3*@EXPAND(G4,@DROPFIRST)
Null hypothesis: X1*X2*X3 X1*X2 X1*X3 X2*X3 X1 X2 X3 are jointly insignificant

|  | Value | df | Probability |
|---|---|---|---|
| QLR L-statistic | 90.62011 | 7 | 0.0000 |
| QLR Lambda-statistic | 81.40783 | 7 | 0.0000 |

**Figure 8.38** The output of the RVT of the numerical IVs in $G4 = 1$.

Redundant variables test
Equation: EQ1_F3WI_QR
Redundant variables: X1*X2*X3*@EXPAND(G4,@DROPFIRST)
Specification: Y1 C X1*X2*X3 X1*X2 X1*X3 X2*X3 X1 X2 X3@
    EXPAND(G4,@DROPFIRST) X1*X2*X3*@EXPAND(G4,@DROPFIRST)
    X1*X2*@EXPAND(G4,@DROPFIRST) X1*X3*@EXPAND(G4,@DROPFIRST)
    X2*X3*@EXPAND(G4,@DROPFIRST)
    X1*@EXPAND(G4,@DROPFIRST) X2*@EXPAND(G4,@DROPFIRST)
    X3*@EXPAND(G4,@DROPFIRST)
Null hypothesis: X1*X2*X3*@EXPAND(G4,@DROPFIRST) are jointly insignificant

|  | Value | df | Probability |
|---|---|---|---|
| QLR L-statistic | 12.52503 | 3 | 0.0058 |
| QLR Lambda-statistic | 12.32629 | 3 | 0.0063 |

**Figure 8.39** The output of the RVT of the effect differences of *X1*X2*X3* between the four levels of *G4*.

Wald test:
Equation: EQ1_F3WI_QR

| Test statistic | Value | df | Probability |
|---|---|---|---|
| F-statistic | 2.455366 | (4, 268) | 0.0461 |
| Chi-square | 9.821464 | 4 | 0.0435 |

Null Hypothesis: $C(2) = C(3) = C(4) = C(6) = 0$

**Figure 8.40**   The output of a Wald test for testing the hypothesis on the effect of *X1* on *Y1* depending on *X2* and *X3*.

Redundant variables test
Equation: EQ1_F3WI_QR
Redundant variables: X1*X2*X3*@EXPAND(G4,@DROPFIRST)
    X1*X2*@EXPAND(G4,@DROPFIRST) X1*X3*@EXPAND(G4,@DROPFIRST)
    X1*@EXPAND(G4,@DROPFIRST)
Specification: Y1 C X1*X2*X3 X1*X2 X1*X3 X2*X3 X1 X2 X3
    @EXPAND(G4,@DROPFIRST) X1*X2*X3*@EXPAND(G4,@DROPFIRST)
    X1*X2*@EXPAND(G4,@DROPFIRST) X1*X3*@EXPAND(G4,@DROPFIRST)
    X2*X3*@EXPAND(G4,@DROPFIRST) X1*@EXPAND(G4,@DROPFIRST)
    X2*@EXPAND(G4,@DROPFIRST) X3*@EXPAND(G4,@DROPFIRST)
Null hypothesis: X1*X2*X3*@EXPAND(G4,@DROPFIRST)
    X1*X2*@EXPAND(G4,@DROPFIRST) X1*X3*@EXPAND(G4,@DROPFIRST)
    X1*@EXPAND(G4,@DROPFIRST) are jointly insignificant

| | Value | df | Probability |
|---|---|---|---|
| QLR L-statistic | 36.86329 | 12 | 0.0002 |
| QLR Lambda-statistic | 35.20907 | 12 | 0.0004 |

**Figure 8.41**   The output of the RVT of the effect differences of all numerical variables having *X1* between the four levels of *G4*.

4.4.2  In addition, note that the set of redundant variables presented in the output consist of $(3 \times 4) = 12$ IVs, since each *V*@Expand(G4,@Dropfirst)* represents three variables, *V*(G4 = 2), V*(G4 = 3)* and *V*(G4 = 4)*.

# 9

# Quantile Regressions Based on CPS88.wf1

## 9.1 Introduction

In this chapter, our quantile regressions (QRs) will be based on an EViews workfile called CPS88.wfi. The file includes *construction professional service* (CPS) data related to architectural engineering, project management, and planning. This file contains the variables presented in Figure 9.1.

These variables include several numerical variables:

- *LWAGE*: Log of hourly wages
- *AGE*: Worker age
- *GRADE*: Highest educational grade completed
- *POTEXP*: Years of potential experience

It also contains another ordinal variable *OCC1* (an occupational category indicator), a nominal variable *IND1* (an industrial category indicator), and four dichotomous variables:

- *HIGH*: An indicator for a highly unionized industry
- *MARRIED*: An indicator for marital status
- *UNION*: A union member indicator
- *PARTT*: An indicator for part-time workers

We can apply the QRs of *LWAGE* on alternative sets on exogenous variables similar to the QRs presented in the eight previous chapters. In this case, the variable *AGE,* as well as the ordinal variables *GRADE, POTEXP,* and *OCC1* can be used as numerical, rank, or categorical independent variables (IVs), as well as the four dichotomous variables. In addition, we also could have the QRs of *OCC1* on selected set of IVs. So, we can present a lot of possible QRs based on the data in CPS88.wf1.

However, in this chapter, we'll focus on only a few selected QRs for each type of models.

*Quantile Regression: Applications on Experimental and Cross Section Data Using EViews,* First Edition.
I Gusti Ngurah Agung.
© 2021 John Wiley & Sons Ltd. Published 2021 by John Wiley & Sons Ltd.

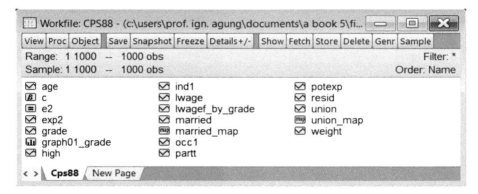

**Figure 9.1** The list of variables in CPS88.wf1.

## 9.2 Applications of an ANOVA Quantile Regression

### 9.2.1 One-Way ANOVA-QR

**Example 9.1** *An Application of a One-Way ANOVA-QR*
Figure 9.2 presents the output of a one-way ANOVA-QR of *LWAGE* by an ordinal categorical variable *GRADE*. We'll use the following equation specification (ES), where *GRADE* is treated as a nominal categorical variable:

$$LWAGE \ C \ @Expand \ (GRADE, @Dropfirst) \tag{9.1}$$

Based on these outputs, the following findings and notes are presented:

1) The 18 medians of *LWAGE* have significant differences between the 18 *GRADE* levels, based on the *QLR* statistic: $QLR_0 = 142.4658$ with $df = 17$ and $p$-value = 0.0000.
2) The $t$-statistic in the output can be used to test one- or two-sided hypotheses between the medians of *LWAGE* for the level *GRADE* = 1 and each of the other levels. Do these as exercises.
3) With the output in Figure 9.2 on-screen, select the *Forecast* button and click *OK* to obtain the statistical results in Figure 9.3a.. A new forecast variable *LWAGEF* is added to the file. Based on these results, the findings and notes are as follows:
   3.1 The Theil inequality coefficient (TIC) of the forecast value variable is TIC = 0.109 610, which is a relative measure, where TIC = 0 represents the perfect forecast and TIC = 1 is the worst. Note that the *LWAGEF* variable in fact is equal to the fitted variable *(FV)* of the model, namely *FV_ONEWAY*.
   3.2 In addition, Figure 9.3b presents the scatter graphs of *LWAGEF* and *LWAGE* on *GRADE*, which show *LWAGE* has a polynomial relationship with *GRADE*. See the section 9.3.1.1.
4) As an additional illustration, Figure 9.4 presents the output of a quantile slope equality test (QSET). Based on this output, the following findings and notes are presented:
   4.1 The output is obtained based on the QR(0.1), indicated by "Estimated equation quantile *tau* = 0.1." So, we can test any pairs of quantiles for *tau* ≥ 0.1.

Dependent Variable: LWAGE

Method: Quantile Regression (Median)

Included observations: 1000

Huber Sandwich Standard Errors & Covariance

Sparsity method: Kernel (Epanechnikov) using residuals

Bandwidth method: Hall-Sheather, bw=0.097156

Estimation successful but solution may not be unique

| Variable | Coefficient | Std. Error | t-Statistic | Prob. |
|---|---|---|---|---|
| C | 1.208960 | 0.167321 | 7.225403 | 0.0000 |
| GRADE = 2 | 0.736950 | 0.236627 | 3.114391 | 0.0019 |
| GRADE = 3 | 0.495788 | 0.236627 | 2.095227 | 0.0364 |
| GRADE = 4 | 0.495788 | 0.289808 | 1.710746 | 0.0874 |
| GRADE = 5 | 1.165946 | 0.224935 | 5.183479 | 0.0000 |
| GRADE = 6 | 0.295117 | 0.199507 | 1.479233 | 0.1394 |
| GRADE = 7 | 0.825746 | 0.215958 | 3.823642 | 0.0001 |
| GRADE = 8 | 0.933456 | 0.198208 | 4.709468 | 0.0000 |
| GRADE = 9 | 1.006977 | 0.184167 | 5.467749 | 0.0000 |
| GRADE = 10 | 0.736950 | 0.203678 | 3.618216 | 0.0003 |
| GRADE = 11 | 0.540240 | 0.256190 | 2.108747 | 0.0352 |
| GRADE = 12 | 0.988265 | 0.170734 | 5.788327 | 0.0000 |
| GRADE = 13 | 0.895174 | 0.183811 | 4.870085 | 0.0000 |
| GRADE = 14 | 1.093625 | 0.185304 | 5.901778 | 0.0000 |
| GRADE = 15 | 1.209184 | 0.207986 | 5.813775 | 0.0000 |
| GRADE = 16 | 1.412079 | 0.180735 | 7.812975 | 0.0000 |
| GRADE = 17 | 1.316769 | 0.193310 | 6.811710 | 0.0000 |
| GRADE = 18 | 1.691462 | 0.183633 | 9.211106 | 0.0000 |

| | | | | |
|---|---|---|---|---|
| Pseudo R-squared | 0.104522 | Mean dependent var | | 2.275496 |
| Adjusted R-squared | 0.089020 | S.D. dependent var | | 0.563464 |
| S.E. of regression | 0.514578 | Objective | | 204.4032 |
| Quantile dependent var | 2.302585 | Restr. objective | | 228.2616 |
| Sparsity | 1.339736 | Quasi-LR statistic | | 142.4658 |
| Prob(Quasi-LR stat) | 0.000000 | | | |

**Figure 9.2** The output of the one-way ANOVA-QR(Median) in (9.1).

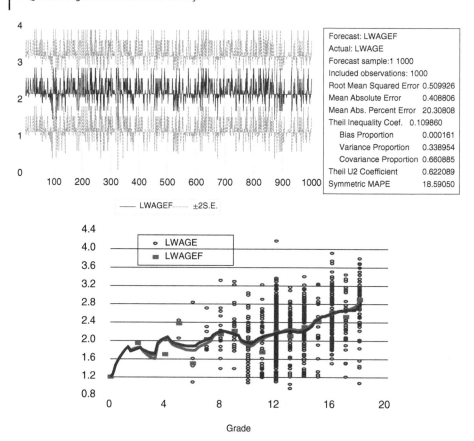

**Figure 9.3** (a) The graphs of LWAGEF and its forecast evaluation. (b) The scatter graphs of *LWAGE* and *LWAGEF* on *GRADE* with their kernel polynomial fit.

4.2 Using the pair of quantiles (0.25, 0.75), the output unexpectedly presents two pairs of quantiles (0.1, 0.25) and (0.25, 0.75). If we want to obtain the output for only the pair of quantiles (0.25, 0.75), the test should be done based on the QR(0.25).

4.3 The output presents only the three selected parameters C(4), C(10), and C(16). For the conclusion of the test, please refer to the QSET presented in previous example 8.8, Figure 8.9.

5) As an additional illustration, Figure 9.5 presents the scatter graphs of the three forecast variables for tau = 0,1, 0.5, and 0.9, on *GRADE,* which clearly shows each forecast variable has a polynomial relationship with *GRADE*.

6) As additional one-way ANOVA-QRs, we can have the QRs of *LWAGE* on each of the variables *AGE, OCC1, IND1,* and *POTEXP,* and the QRs of *OCC1* on each of the variables *AGE, GRADE, IND1,* and *POTEXP*. Do these as exercises.

Quantile slope Equality test

Equation: EQ02

Specification: LWAGE C @EXPAND(GRADE,@DROPFIRST)

Estimated equation quantile tau = 0.1

User-specified test quantiles: 0.25 0.75

Test statistic compares user-specified coefficients: C(4) C(10) C(16)

| Test Summary | Chi-Sq. Statistic | Chi-Sq. d.f. | Prob. |
|---|---|---|---|
| Wald test | 66.07790 | 6 | 0.0000 |

Restriction detail: b(tau_h) − b(tau_k) = 0

| Quantiles | Variable | Restr. value | Std. Error | Prob. |
|---|---|---|---|---|
| 0.1, 0.25 | GRADE = 4 | 0.000000 | 0.180396 | 1.0000 |
| | GRADE = 10 | −0.356675 | 0.138571 | 0.0101 |
| | GRADE = 16 | −0.447555 | 0.127931 | 0.0005 |
| 0.25, 0.75 | GRADE = 4 | −0.310155 | 0.254242 | 0.2225 |
| | GRADE = 10 | −0.741937 | 0.187637 | 0.0001 |
| | GRADE = 16 | −0.789458 | 0.160063 | 0.0000 |

**Figure 9.4**   The output of a quantile slope equality test.

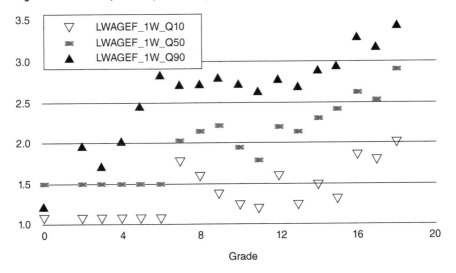

**Figure 9.5**   Scatter graphs of forecast variables *LWAGEF_1W* for quantiles = 0.1, 0.5, and 0.9.

### 9.2.2 Two-Way ANOVA Quantile Regression

#### 9.2.2.1 The Simplest Equation of Two-Way ANOVA-QR

The simplest ES of a two-way ANOVA-QR (2W-ANOVA-QR) of *LWAGE* by any two categorical factors *F1* and *F2*, in CPS88.wf1, can be presented as follows:

$$LWAGE \ @Expand \ (F1, F2) \tag{9.2}$$

The main objective of applying this model is to identify various quantiles of *LWAGE* for each cell or group generated by the two factors. See the following example.

**Example 9.2** *An Application of a Two-Way ANOVA-QR*
Figure 9.6 presents the output from using the following ES:

$$LWAGE \ @Expand \ (GRADE, PARTT) \tag{9.3}$$

| Dependent Variable: LWAGE | | | | |
|---|---|---|---|---|
| Method: Quantile Regression (median) | | | | |
| Sample: 1 1000 IF GRADE > 9 | | | | |
| Included observations: 931 | | | | |
| Estimation successful but solution may not be unique | | | | |

| Variable | Coefficient | Std. Error | t-Statistic | Prob. |
|---|---|---|---|---|
| PARTT = 0,GRADE = 10 | 2.079442 | 0.111081 | 18.72007 | 0.0000 |
| PARTT = 0,GRADE = 11 | 2.055725 | 0.161526 | 12.72688 | 0.0000 |
| PARTT = 0,GRADE = 12 | 2.197225 | 0.033292 | 65.99820 | 0.0000 |
| PARTT = 0,GRADE = 13 | 2.148434 | 0.072480 | 29.64182 | 0.0000 |
| PARTT = 0,GRADE = 14 | 2.420368 | 0.066621 | 36.33057 | 0.0000 |
| PARTT = 0,GRADE = 15 | 2.525729 | 0.143839 | 17.55939 | 0.0000 |
| PARTT = 0,GRADE = 16 | 2.621039 | 0.069694 | 37.60785 | 0.0000 |
| PARTT = 0,GRADE = 17 | 2.525729 | 0.094194 | 26.81407 | 0.0000 |
| PARTT = 0,GRADE = 18 | 2.944439 | 0.069546 | 42.33826 | 0.0000 |
| PARTT = 1,GRADE = 10 | 1.568616 | 0.167158 | 9.384037 | 0.0000 |
| PARTT = 1,GRADE = 11 | 1.294727 | 0.083377 | 15.52851 | 0.0000 |
| PARTT = 1,GRADE = 12 | 1.791759 | 0.114941 | 15.58845 | 0.0000 |
| PARTT = 1,GRADE = 13 | 1.386294 | 0.106566 | 13.00883 | 0.0000 |
| PARTT = 1,GRADE = 14 | 1.348073 | 0.090547 | 14.88814 | 0.0000 |
| PARTT = 1,GRADE = 15 | 1.321756 | 0.149714 | 8.828535 | 0.0000 |
| PARTT = 1,GRADE = 16 | 1.504077 | 0.234986 | 6.400715 | 0.0000 |
| PARTT = 1,GRADE = 17 | 1.504077 | 0.234986 | 6.400715 | 0.0000 |
| PARTT = 1,GRADE = 18 | 2.302585 | 0.158018 | 14.57169 | 0.0000 |

| | | | | |
|---|---|---|---|---|
| Pseudo R-squared | 0.150886 | Mean dependent var | | 2.288417 |
| Adjusted R-squared | 0.135075 | S.D. dependent var | | 0.567753 |
| S.E. of regression | 0.499124 | Objective | | 182.1274 |
| Quantile dependent var | 2.302585 | Restr. objective | | 214.4910 |
| Sparsity | 1.236659 | | | |

**Figure 9.6** The output of the QR(Median) in (9.3), for the subsample *{GRADE > 9}*.

Based on this output, the findings and notes are as follows:

1) The Coefficient column presents the values of the median of *LWAGE* for each group of respondents generated by the factor *PARTT* and an ordinal variable *GRADE,* which are the estimates of the parameters C(1) to C(18).
2) The estimate of $\widehat{C(1)}$ = 2.079 442 indicates 50% of the group *(PARTT = 0, GRADE = 10)* has the log of hourly wages below 2.079 442.
3) As an additional illustration, Figure 9.7 presents the QPE based on the 2W-ANOVA-QR (0.5) in (9.3), specific for C(1) and C(10), which represent the medians of *LWAGE* in *(PARTT = 0, GRADE = 10)* and *(PARTT = 1, GRADE = 10),* respectively, for the 10 quantiles.

### 9.2.2.2 A Special Equation of the Two-Way ANOVA-QR

As a modification of the ES in (9.3), a special equation specification of a 2W-ANOVA-QR of *Y* by specific categorical characteristics of the research objects, namely *F1,* such as *GRADE,*

---

Quantile Process Estimates

Equation: EQ07

Specification: LWAGE @EXPAND(PARTT,GRADE)

Estimated equation quantile tau = 0.5

Number of process quantiles: 10

Display user-specified coefficients: PARTT = 0,GRADE = 10 PARTT = 1,GRADE = 10

| | Quantile | Coefficient | Std. Error | t-Statistic | Prob. |
|---|---|---|---|---|---|
| PARTT = 0,GRADE = 10 | 0.100 | 1.321756 | 0.120468 | 10.97188 | 0.0000 |
| | 0.200 | 1.609438 | 0.119734 | 13.44177 | 0.0000 |
| | 0.300 | 1.791759 | 0.136781 | 13.09951 | 0.0000 |
| | 0.400 | 1.945910 | 0.116908 | 16.64475 | 0.0000 |
| | 0.500 | 2.079442 | 0.111081 | 18.72007 | 0.0000 |
| | 0.600 | 2.207275 | 0.105485 | 20.92507 | 0.0000 |
| | 0.700 | 2.302585 | 0.109028 | 21.11918 | 0.0000 |
| | 0.800 | 2.484907 | 0.161222 | 15.41292 | 0.0000 |
| | 0.900 | 2.708050 | 0.120346 | 22.50221 | 0.0000 |
| PARTT = 1,GRADE = 10 | 0.100 | 1.078810 | 0.146654 | 7.356143 | 0.0000 |
| | 0.200 | 1.208960 | 0.196007 | 6.167928 | 0.0000 |
| | 0.300 | 1.435084 | 0.236261 | 6.074152 | 0.0000 |
| | 0.400 | 1.568616 | 0.167119 | 9.386214 | 0.0000 |
| | 0.500 | 1.568616 | 0.167158 | 9.384037 | 0.0000 |
| | 0.600 | 1.704748 | 0.130156 | 13.09771 | 0.0000 |
| | 0.700 | 1.791759 | 0.135708 | 13.20309 | 0.0000 |
| | 0.800 | 1.791759 | 0.112350 | 15.94807 | 0.0000 |
| | 0.900 | 2.484907 | 0.232525 | 10.68661 | 0.0000 |

**Figure 9.7** The quantile process estimates specific for the parameters C(1) and C(10).

*AGE,* and *OCC1,* and any factor *F2,* can be presented using the following alternative ESs:

$$Y @ Expand\,(F1) @ Expand\,(F1) * @ Expand\,(F2, @ Droplast) \tag{9.4a}$$

$$Y\,C @ Expand\,(F1, @ Droplast) @ Expand\,(F1) * @ Expand\,(F2, @ Droplast) \tag{9.4b}$$

The main objectives of applying these models are to identify and test the quantiles differences of *Y* between the levels of *F2* for each level of *F1.* See the following example.

### Example 9.3 *An Application of a Special Two-Way ANOVA-QR*

Figure 9.8 presents the output of the specific QR(Median) in (9.4a) using the variables in CPS88.wf1:

$$LWAGE\ @ Expand\,(GRADE) @ Expand\,(GRADE) * @ Expand\,(PARTT, @ Droplast)$$

$$\tag{9.5}$$

Based on this output, the following findings and notes can be presented:

1) We can develop the table of the model parameters by *GRADE* and *PARTT,* as presented in Table 9.1, using the following steps:
    1.1 As the coefficient of *GRADE* = 10, the parameter C(1) is inserted in the row of *GRADE* = 10 in both levels of *PARTT.* Similarly, parameters C(2) to C(9) are inserted in the rows of *GRADE* = 11 to *GRADE* = 18, respectively.
    1.2 The parameter C(10) as the coefficient of the dummy interaction or cross-product *(GRADE* = 10)*(PARTT* = 0) is added to C(1) only within the level *PARTT* = 0. Similarly, the parameters C(11) to C(18) are added to C(2) to C(19), respectively, only within the level *PARTT* = 0.
2) The estimates of the parameters C(10) to C(18) represent the quantile differences *PARTT*(0 −1) for each group generated by *GRADE* and *PARTT.* The output shows that all of their estimates are positive. So, the medians of *LWAGE* for *PARTT* = 0 are greater than those for *PARTT* = 1, for all *GRADE* levels. For instance, the estimate $\widehat{C}(10) = 0.510\,826$ indicates that specific for *GRADE* = 10, the median of *LWAGE* for *PARTT* = 0 is 0.510 826 greater than for *PARTT* = 1.
    2.1 Each of the *t*-statistic in the output can be used to test the one- or two-sided hypotheses of the median differences of *LWAGE,* between *PARTT* = 0 and *PARTT* = 1, for each level of *GRADE.* Do these as an exercise.
    2.2 Also an exercise, develop a tabulation of parameters based on the model in (9.4b).
3) Many other hypotheses on the quantile differences can easily be tested using the Wald test, including a hypothesis of the interaction effect of *GRADE*PARTT* on *LWAGE,* with the following statistical hypothesis, and its statistical result presented in Figure 9.9:

$$H_0 : C\,(10) = C\,(11) = \dots = C\,(18)\ vs\ H_1 : Otherwise$$

Based on this statistical result, we can conclude that the null hypothesis is rejected at the 1% level of significance. Hence, we can say that the effect of *GRADE* on *LWAGE* is significantly dependent on *PARTT,* or the median *LWAGE* differences between the levels of *PARTT* are significantly dependent on *GRADE.*

Dependent variable: LWAGE

Method: Quantile Regression (median)

Date: 12/02/18   Time: 11:11

Sample: 1 1000 IF GRADE > 9

Included observations: 931

Huber Sandwich Standard Errors & Covariance

Sparsity method: Kernel (Epanechnikov) using residuals

Bandwidth method: Hall-Sheather, bw = 0.099499

Estimation successful but solution may not be unique

| Variable | Coefficient | Std. Error | t-Statistic | Prob. |
|---|---|---|---|---|
| GRADE = 10 | 1.568616 | 0.167394 | 9.370811 | 0.0000 |
| GRADE = 11 | 1.294727 | 0.083310 | 15.54100 | 0.0000 |
| GRADE = 12 | 1.791759 | 0.114922 | 15.59110 | 0.0000 |
| GRADE = 13 | 1.386294 | 0.106488 | 13.01828 | 0.0000 |
| GRADE = 14 | 1.386294 | 0.096903 | 14.30600 | 0.0000 |
| GRADE = 15 | 1.906575 | 0.405990 | 4.696110 | 0.0000 |
| GRADE = 16 | 2.484907 | 0.234399 | 10.60120 | 0.0000 |
| GRADE = 17 | 1.504077 | 0.234399 | 6.416749 | 0.0000 |
| GRADE = 18 | 2.302585 | 0.157700 | 14.60100 | 0.0000 |
| (GRADE = 10)*(PARTT = 0) | 0.510826 | 0.200910 | 2.542566 | 0.0112 |
| (GRADE = 11)*(PARTT = 0) | 0.760998 | 0.181798 | 4.185948 | 0.0000 |
| (GRADE = 12)*(PARTT = 0) | 0.405466 | 0.119644 | 3.388941 | 0.0007 |
| (GRADE = 13)*(PARTT = 0) | 0.762140 | 0.128762 | 5.918975 | 0.0000 |
| (GRADE = 14)*(PARTT = 0) | 1.034074 | 0.117599 | 8.793216 | 0.0000 |
| (GRADE = 15)*(PARTT = 0) | 0.619154 | 0.430626 | 1.437799 | 0.1508 |
| (GRADE = 16)*(PARTT = 0) | 0.136132 | 0.244547 | 0.556670 | 0.5779 |
| (GRADE = 17)*(PARTT = 0) | 1.021652 | 0.252623 | 4.044176 | 0.0001 |
| (GRADE = 18)*(PARTT = 0) | 0.641854 | 0.172334 | 3.724486 | 0.0002 |

| | | | | |
|---|---|---|---|---|
| Pseudo R-squared | 0.150886 | Mean dependent var | | 2.288417 |
| Adjusted R-squared | 0.135075 | S.D. dependent var | | 0.567753 |
| S.E. of regression | 0.497779 | Objective | | 182.1274 |
| Quantile dependent var | 2.302585 | Restr. objective | | 214.4910 |
| Sparsity | 1.246909 | | | |

**Figure 9.8**   The output of the special ANOVA-QR(Median) in (9.5).

**Table 9.1** Parameters of the special QR in (9.5).

| GRADE | PARTT = 0 | PARTT = 1 | PARTT(0 − 1) |
|-------|-----------|-----------|--------------|
| 10 | C(1) + C(10) | C(1) | C(10) |
| 11 | C(2) + C(11) | C(2) | C(11) |
| 12 | C(3) + C(12) | C(3) | C(12) |
| 13 | C(4) + C(13) | C(4) | C(13) |
| 14 | C(5) + C(14) | C(5) | C(14) |
| 15 | C(6) + C(15) | C(6) | C(15) |
| 16 | C(7) + C(16) | C(7) | C(16) |
| 17 | C(8) + C(17) | C(8) | C(17) |
| 18 | C(9) + C(18) | C(9) | C(18) |

Wald Test:
Equation: EQ_(9.5)

| Test statistic | Value | df | Probability |
|----------------|-------|-----|-------------|
| F-statistic | 2.804148 | (8, 913) | 0.0045 |
| Chi-square | 22.43318 | 8 | 0.0042 |

Null hypothesis: C(10) = C(11) = C(12) = C(13) = C(14) = C(15)

**Figure 9.9** The statistical results of a Wald test.

4) As an additional illustration, based on the two outputs in Figures 9.6 and 9.8, we can develop a statistical results summary as shown in Figure 9.10. Note that both QRs in Figures 9.6 and 9.8 in fact are the same QR, indicated by the same values of the five statistics from pseudo $R$-squared to sparsity.

However, both statistical results present the note "Estimation successful but solution may not be unique," which causes their corresponding parameters' estimates to have different values. Why do they have different values? In fact, the Eqs. (9.3) and (9.8) present the same QR. For these different findings, my EViews provider gave the following explanation:

> *The estimation you are performing does not have a closed form solution and the output is indicating that there may be multiple solutions which will lead to the same value for the objective function.*

(Email August 21, 2019).

| | Output in Figure 9.6 | | | | | | Output in Figure 9.8 | | |
|---|---|---|---|---|---|---|---|---|---|
| | PARTT = 0 | | | PARTT = 1 | | | PARTT(0 − 1) | | |
| Variable | Coef. | t-Stat. | Prob. | Coef. | t-Stat. | Prob. | Coef. | t-Stat. | Prob. |
| GRADE = 10 | 2.079 | 18.720 | 0.000 | 1.569 | 9.384 | 0.000 | 0.511 | 2.543 | 0.011 |
| GRADE = 11 | 2.056 | 12.727 | 0.000 | 1.295 | 15.529 | 0.000 | 0.761 | 4.186 | 0.000 |
| GRADE = 12 | 2.197 | 65.998 | 0.000 | 1.792 | 15.588 | 0.000 | 0.405 | 3.389 | 0.001 |
| GRADE = 13 | 2.148 | 29.642 | 0.000 | 1.386 | 13.009 | 0.000 | 0.762 | 5.919 | 0.000 |
| GRADE = 14 | 2.420 | 36.331 | 0.000 | 1.348 | 14.888 | 0.000 | 1.034 | 8.793 | 0.000 |
| GRADE = 15 | 2.526 | 17.559 | 0.000 | 1.322 | 8.829 | 0.000 | 0.619 | 1.438 | 0.151 |
| GRADE = 16 | 2.621 | 37.608 | 0.000 | 1.504 | 6.401 | 0.000 | 0.136 | 0.557 | 0.578 |
| GRADE = 17 | 2.526 | 26.814 | 0.000 | 1.504 | 6.401 | 0.000 | 1.022 | 4.044 | 0.000 |
| GRADE = 18 | 2.944 | 42.338 | 0.000 | 2.303 | 14.572 | 0.000 | 0.642 | 3.724 | 0.000 |
| Pse. R-sq | 0.150886 | | | | | | 0.150886 | | |
| Adj. R-sq | 0.135075 | | | | | | 0.135075 | | |
| S.E. of reg | 0.499124 | | | | | | 0.497779 | | |
| Quantile DV | 2.302585 | | | | | | 2.302585 | | |
| Sparsity | 1.236659 | | | | | | 1.246909 | | |

**Figure 9.10** Statistical results summary based on the outputs in Figures 9.6 and 9.8.

5) As an alternative statistical results summary, we can develop a summary using the two outputs of the following ES, specific for each of the two subsamples *{PARTT = 0}* and *{PARTT = 1}*:

$$LWAGE @ Expand (GRADE) \tag{9.6}$$

The result would be two different sets of the five statistics, from pseudo *R*-squared to sparsity, instead of only one set based on the output in Figure 9.6. Do these as exercises.

### 9.2.2.3 An Additive Two-Way ANOVA-QR

As an illustration, this section presents an additive 2W-ANOVA-QR. I consider this to be at best not-recommended or even the worst ANOVA-QR, with the following general ESs, for any factors *F1* and *F2*:

$$Y @ Expand (F1) @ Expand (F2, @ Droplast) \tag{9.7a}$$

$$Y \ C @ Expand (F1, @ Droplast) @ Expand (F2, @ Droplast) \tag{9.7b}$$

See the following example.

**Example 9.4    *An Application of the ES (9.7a)***
As a modification of the QR in Figure 9.8, Figure 9.11 presents the output of a specific QR(Median) using the following ES:

$$Y @ Expand\,(GRADE)\,@ Expand\,(PART, @ Droplast) \tag{9.8}$$

Based on this output, the following findings and notes are presented:

1) We can develop the tabulation of the model parameters as presented in Table 9.2.
2) Compared with Table 9.1, this table shows the quantiles different between $PARTT = 0$ and $PARTT = 1$ are the same for all $GRADE$ levels, which never happens in reality. For this reason, I consider an additive 2W-ANOVA-QR for any factors $F1$ and $F2$ to be the worst QR. The case is similar for additive N-Way ANOVA-QRs.
3) As an additional illustration, refer to the additive ANOVA-QR presented in Section 2.5, Chapter 2.

---

Dependent Variable: LWAGE

Method: Quantile Regression (median)

Sample: 1 1000 IF GRADE > 9

Sparsity method: Kernel (Epanechnikov) using residuals

Bandwidth method: Hall-Sheather, bw = 0.099499

Estimation successful but solution may not be unique

| Variable | Coefficient | Std. Error | t-Statistic | Prob. |
|---|---|---|---|---|
| GRADE = 10 | 1.437588 | 0.122011 | 11.78243 | 0.0000 |
| GRADE = 11 | 1.398717 | 0.094972 | 14.72768 | 0.0000 |
| GRADE = 12 | 1.582769 | 0.083832 | 18.88031 | 0.0000 |
| GRADE = 13 | 1.498212 | 0.102951 | 14.55262 | 0.0000 |
| GRADE = 14 | 1.756041 | 0.098088 | 17.90274 | 0.0000 |
| GRADE = 15 | 1.883875 | 0.157403 | 11.96847 | 0.0000 |
| GRADE = 16 | 1.979185 | 0.108309 | 18.27348 | 0.0000 |
| GRADE = 17 | 1.883875 | 0.113910 | 16.53829 | 0.0000 |
| GRADE = 18 | 2.302585 | 0.096302 | 23.91007 | 0.0000 |
| PARTT = 0 | 0.641854 | 0.081050 | 7.919249 | 0.0000 |

| | | | |
|---|---|---|---|
| Pseudo R-squared | 0.145592 | Mean dependent var | 2.288417 |
| Adjusted R-squared | 0.137243 | S.D. dependent var | 0.567753 |
| S.E. of regression | 0.495849 | Objective | 183.2629 |
| Quantile dependent var | 2.302585 | Restr. objective | 214.4910 |
| Sparsity | 1.313492 | | |

**Figure 9.11**   The output of the QR(Median) in (9.8).

**Table 9.2** Parameters of the special QR in (9.8).

| GRADE | PARTT = 0 | PARTT = 1 | PARTT(0 − 1) |
|-------|-----------|-----------|--------------|
| 10 | C(1) + C(10) | C(1) | C(10) |
| 11 | C(2) + C(10) | C(2) | C(10) |
| 12 | C(3) + C(10) | C(3) | C(10) |
| 13 | C(4) + C(10) | C(4) | C(10) |
| 14 | C(5) + C(10) | C(5) | C(10) |
| 15 | C(6) + C(10) | C(6) | C(10) |
| 16 | C(7) + C(10) | C(7) | C(10) |
| 17 | C(8) + C(10) | C(8) | C(10) |
| 18 | C(9) + C(10) | C(9) | C(10) |

### 9.2.3 Three-Way ANOVA-QRs

As an extension of the ESs of the two-way interaction QRs (2WI-QRs) we can represent the 3WI-QR of $Y$, by the three factors $F1, F2$, and $F3$, using one of the following alternative ESs:

$$Y @ Expand\,(F1, F2, F2) \tag{9.9a}$$

$$Y\ C @ Expand\,(F1, F2, F3, @Dropfirst) \tag{9.9b}$$

$$Y\ C @ Expand\,(F1, F2, @Droplast)\,@Expand\,(F1, F2)*@Expand\,(F3, @Droplast) \tag{9.9c}$$

**Example 9.5** *Application of a Specific QR in (9.9c)*
Figure 9.12 presents the output of a specific $2 \times 2 \times 2$ factorial QR of *LWAGE* in (9.9c) with the following ES:

$$LWAGE\ C @ Expand\,(HIGH, PARTT, @DRopfirst)\,@Expand\,(HIGH, PARTT)*$$
$$@Expand\,(MARRIED, @Dropfirst) \tag{9.10}$$

Based on this output, the following findings and notes are presented.

1) Based on the output in Figure (9.12), the Table 9.3 of the model parameters is developed as follows:

 1.1 As the intercept, C(1) should be inserted in the eight cells generated by the three factors. As the coefficient of the dummy variable *(HIGH = 0,PARTT = 0)*, C(2) is added to both levels of *MARRIED*, within the row of *(HIGH = 0,PARTT = 0)*.

 1.2 Similarly, the coefficients C(3), C(4), and C(5), respectively, are added to both levels of *MARRIED*, in the rows of cells generated by *HIGH* and *PARTT*.

 1.3 The parameters C(5), C(6), and C(7) are added to the cells *(HIGH,PART, MARRIED)* = (0,1,S); (1,0,S); and (1,1,S), respectively.

 1.4 Finally, the table is completed, and the bold lines in the table indicate the difference between the specific pairs of the quantiles. For instance, C(5) is the *parameter quantile difference* of *LWAGE* between the two cells (0, 0, S) and (0, 0, M), which can be tested using the $t$-statistic in the output.

Dependent Variable: LWAGE

Method: Quantile Regression (median)

Included observations: 1000

Bandwidth method: Hall-Sheather, bw = 0.097156

Estimation successful but solution may not be unique

| Variable | Coefficient | Std. Error | t-Statistic | Prob. |
|---|---|---|---|---|
| C | 1.945910 | 0.128428 | 15.15173 | 0.0000 |
| HIGH = 0,PARTT = 0 | 0.451985 | 0.137047 | 3.298027 | 0.0010 |
| HIGH = 0,PARTT = 1 | −0.064919 | 0.190753 | −0.340329 | 0.7337 |
| HIGH = 1,PARTT = 0 | 0.451985 | 0.132081 | 3.422020 | 0.0006 |
| (HIGH = 0 AND PARTT = 0)*(MARRIED = "single") | −0.382992 | 0.098388 | −3.892652 | 0.0001 |
| (HIGH = 0 AND PARTT = 1)*(MARRIED = "single") | −0.494697 | 0.152448 | −3.245016 | 0.0012 |
| (HIGH = 1 AND PARTT = 0)*(MARRIED = "single") | −0.200670 | 0.069197 | −2.900001 | 0.0038 |
| (HIGH = 1 AND PARTT = 1)*(MARRIED = "single") | −0.441833 | 0.164214 | −2.690587 | 0.0073 |

| | | | |
|---|---|---|---|
| Pseudo R-squared | 0.096877 | Mean dependent var | 2.275496 |
| Adjusted R-squared | 0.090505 | S.D. dependent var | 0.563464 |
| S.E. of regression | 0.522676 | Objective | 206.1482 |
| Quantile dependent var | 2.302585 | Restr. objective | 228.2616 |
| Sparsity | 1.336484 | Quasi-LR statistic | 132.3676 |
| Prob(quasi-LR stat) | 0.000000 | | |

**Figure 9.12** The output of the QR(Median) in (9.10).

**Table 9.3** Parameters of QR(Median) in (9.10) by the factors *HIGH*, *PARTT*, and *MARRIED*.

| | | "single" = S | "married" = M | S − M |
|---|---|---|---|---|
| HIGH = 0 | PARTT = 0 | C(1) + C(2) + C(5) | C(1) + C(2) | *C(5)* |
| HIGH = 0 | PARTT = 1 | C(1) + C(3) + C(6) | C(1) + C(3) | *C(6)* |
| HIGH = 1 | PARTT = 0 | C(1) + C(4) + C(7) | C(1) + C(4) | *C(7)* |
| HIGH = 1 | PARTT = 1 | C(1) + C(8) | C(1) | *C(8)* |
| **HIGH = 0** | **PARTT(0 − 1)** | **C(2) − C(3) + C(5) − C(6)** | **C(2) − C(3)** | **C(5) − C(6)** |
| **HIGH = 1** | **PARTT(0 − 1)** | **C(4) − C(7) − C(8)** | **C(4)** | **C(7) − C(8)** |
| **HIGH(0 − 1)** | **PARTT = 0** | **C(2) − C(4) + C(5) − C(7)** | **C(2) − C(4)** | **C(5) − C(7)** |
| **HIGH(0 − 1)** | **PARTT = 1** | **C(3) + C(6) − C(8)** | **C(3)** | **C(6) − C(8)** |

2) Based on the $QLR$ statistic of $QLR_0 = 132.3676$ with a $p$-value 0.0000, we can conclude all IVs have a joint significant effect on $LWAGE$. In other words, the medians of $LWAGE$ have significant differences between the eight cells generated by the three factors.

3) Then, based on each of the parameters C(3) to C(8), one- or two-sided hypotheses can directly be tested using the $t$-statistic in the output. For instance, $H_0$: $C(3) = 0$ is accepted based on the $t$-statistic of $t_0 = -0.340329$ with $df = 992 = (1000 - 8)$ with $p$-value $= 0.7337$. So, we can conclude the medians of $LWAGE$ have insignificant differences between the levels of $PARTT$, conditional for $HIGH = 1$ and $MARRIED = M$.

4) Other hypotheses can be tested using the Wald test, especially the hypotheses on the interaction factors effects, such as follows:

4.1 Based on the conditional two-way interaction effects, which reflect the well-known *conditional DID* (difference-in-differences) in the common two-way ANOVA model, or in a LS-regression for the 3WI-ANOVA-QR in (9.10), we have the following conditional two-way interaction Quantile-DID:

$$IE\,(PARTT^*MARRIED \mid HIHG = 0) = C\,(5) - C\,(6)$$

$$IE\,(PARTT^*MARRIED \mid HIHG = 1) = C\,(7) - C\,(8)$$

$$IE\,(HIGH^*MARRIED \mid PARTT = 0) = C\,(5) - C\,(7)$$

$$IE\,(HIGH^*MARRIED \mid PARTT = 1) = C\,(6) - C\,(8)$$

$$IE\,(HIGH^*PARTT \mid \text{``}single\text{``}) = (C\,(2) - C\,(3) + C\,(5)$$
$$-C\,(6)) - (C\,(4) - C\,(7) - C\,(8)$$
$$IE\,(HIGH^*PARTT \mid \text{``}married\text{''}) = C\,(2) - C\,(3) - C\,(4)$$

4.2 Referring to the three-way interaction effect presented in Example 3.1, Chapter 3, based on each pair of the previous conditional two-way interactions, we have the three-way interaction effect as follows:

$$IE\,(HIGH^*PARTT^*MARRIED) = C\,(5) - C\,(6) - C\,(7) + C\,(8)$$

5) Each interaction can be tested as a two-sided hypothesis. For the three-way interaction, we have the following statistical hypothesis:

$$H_0 : C\,(5) - C\,(6) - C\,(7) + C\,(8) = 0 \, vs \, H_1 : Otherwise$$

We find that the null hypothesis is accepted based on the *Chi-square* statistic of $\chi_0^2 = 0.259131$ with $df = 1$ and $p$-value $= 0.6107$. So, we can conclude that the interaction of the three factors has an insignificant effect on $LWAGE$.

**Example 9.6** *An Application of a Specific QR in (9.9b)*

Figure 9.13 presents the output of a specific $2 \times 14 \times 2$ factorial QR of $LWAGE$ in (9.9b) with the following ES:

$$LWAGE \, C@Expand\,(HIGH, IND1, PARTT, @DRopfirst) \tag{9.11}$$

Dependent Variable: LWAGE

Method: Quantile Regression (median)

Sample: 1 1000

Included observations: 1000

Huber Sandwich Standard Errors & Covariance

Sparsity method: Kernel (Epanechnikov) using residuals

Bandwidth method: Hall-Sheather, bw = 0.097156

Estimation successful but solution may not be unique

| Variable | Coefficient | Std. Error | t-Statistic | Prob. |
|---|---|---|---|---|
| C | 2.668616 | 0.046594 | 57.27375 | 0.0000 |
| HIGH = 0,IND1 = 6,PARTT = 1 | −0.248248 | 0.165926 | −1.496134 | 0.1349 |
| HIGH = 0,IND1 = 7,PARTT = 0 | −0.366031 | 0.077547 | −4.720104 | 0.0000 |
| HIGH = 0,IND1 = 7,PARTT = 1 | −1.059178 | 0.165577 | −6.396881 | 0.0000 |
| HIGH = 0,IND1 = 8,PARTT = 0 | −0.729874 | 0.074527 | −9.793412 | 0.0000 |
| HIGH = 0,IND1 = 8,PARTT = 1 | −1.401668 | 0.066746 | −20.99991 | 0.0000 |
| HIGH = 0,IND1 = 9,PARTT = 0 | 0.039434 | 0.114723 | 0.343733 | 0.7311 |
| HIGH = 0,IND1 = 9,PARTT = 1 | 0.039434 | 0.279737 | 0.140968 | 0.8879 |
| HIGH = 0,IND1 = 12,PARTT = 0 | −0.589174 | 0.198061 | −2.974705 | 0.0030 |
| HIGH = 0,IND1 = 12,PARTT = 1 | −0.836034 | 0.169133 | −4.943050 | 0.0000 |
| HIGH = 0,IND1 = 13,PARTT = 0 | −0.598594 | 0.137498 | −4.353486 | 0.0000 |
| HIGH = 0,IND1 = 13,PARTT = 1 | −0.876857 | 0.359129 | −2.441621 | 0.0148 |
| HIGH = 1,IND1 = 1,PARTT = 0 | −0.260670 | 0.146273 | −1.782080 | 0.0750 |
| HIGH = 1,IND1 = 1,PARTT = 1 | −0.620923 | 0.229983 | −2.699868 | 0.0071 |
| HIGH = 1,IND1 = 2,PARTT = 0 | −0.307762 | 0.061942 | −4.968588 | 0.0000 |
| HIGH = 1,IND1 = 2,PARTT = 1 | −0.876857 | 0.250700 | −3.497641 | 0.0005 |
| HIGH = 1,IND1 = 3,PARTT = 0 | −0.391349 | 0.117408 | −3.333240 | 0.0009 |
| HIGH = 1,IND1 = 3,PARTT = 1 | −0.963868 | 0.279045 | −3.454168 | 0.0006 |
| HIGH = 1,IND1 = 4,PARTT = 0 | −0.366031 | 0.080014 | −4.574570 | 0.0000 |
| HIGH = 1,IND1 = 4,PARTT = 1 | −0.876857 | 0.108026 | −8.117085 | 0.0000 |
| HIGH = 1,IND1 = 5,PARTT = 0 | −0.226269 | 0.098937 | −2.287004 | 0.0224 |
| HIGH = 1,IND1 = 5,PARTT = 1 | −0.366031 | 0.200096 | −1.829273 | 0.0677 |
| HIGH = 1,IND1 = 10,PARTT = 0 | −0.305406 | 0.106277 | −2.873671 | 0.0041 |
| HIGH = 1,IND1 = 10,PARTT = 1 | −1.360283 | 0.086428 | −15.73883 | 0.0000 |
| HIGH = 1,IND1 = 11,PARTT = 0 | −0.366031 | 0.227591 | −1.608283 | 0.1081 |
| HIGH = 1,IND1 = 11,PARTT = 1 | −0.653713 | 0.160628 | −4.069742 | 0.0001 |
| HIGH = 1,IND1 = 14,PARTT = 0 | −0.165360 | 0.096297 | −1.717191 | 0.0863 |
| HIGH = 1,IND1 = 14,PARTT = 1 | −1.257629 | 0.127547 | −9.860127 | 0.0000 |

| | | | | |
|---|---|---|---|---|
| Pseudo R-squared | 0.154418 | Mean dependent var | | 2.275496 |
| Adjusted R-squared | 0.130929 | S.D. dependent var | | 0.563464 |
| S.E. of regression | 0.501934 | Objective | | 193.0140 |
| Quantile dependent var | 2.302585 | Restr. objective | | 228.2616 |
| Sparsity | 1.135545 | Quasi-LR statistic | | 248.3220 |
| Prob(quasi-LR stat) | 0.000000 | | | |

**Figure 9.13** The output of the QR(Median) in (9.11).

Based on this output, the following findings and notes are presented:

1) Note that the ordering of the three variables *HIGH, IND1,* and *PARTT* in the function *@Expand(\*)* is one of six possible orderings. It has the specific objective of easily identifying the median differences of *LWAGE* between *PARTT* = 0 and *PARTT* = 1, for each group of the firms classified by *HIGH* and *IND1*. For instance, the intercept parameter "C" represents the median of the first group, namely *(HIGH = 0,IND1 = 6,PARTT = 0)*, which is greater than the median of the second group *(HIGH = 0,IND1 = 6,PARTT = 1)*. And the median of *LWAGE* in the third group, *(HIGH = 0,IND1 = 7,PARTT = 0)* = (2.668 616 − 0.366 031), is greater than the median in fourth group *(HIGH = 0,IND1 = 7, PARTT = 1)* = (2.668 616 − 1.059 179).

2) Note that the 14 levels of *IND1* in fact are classified into two groups of *HIGH* = 0 and *HIGH* = 1. So the model (9.11) represents only $14 \times 2 = 28$ groups of industries, instead of $2 \times 14 \times 2 = 56$. Hence, at the 1% level, the medians of *LWAGE* have significant differences between the 28 groups of workers based on the *QLR* statistic of $QLR_0 = 248.322$ with *p*-value = 0.0000.

3) Based on the *t*-statistic in the output, we can directly test the median difference of *LWAGE* between each group of the dummy IV and the first group, namely *(HIGH = 0, IND1 = 6,PARTT = 0)*. For instance, at the 10% level of significance, based on the *t*-statistic of $t_0 = -1.496\,134$ with $df = 944$ with *p*-value = 0.1349, the medians of *LWAGE* have an insignificant difference between the first two groups. However, at the 10% level, the median of the first group is significantly greater than the second group based on *t*-statistic of $t_0 = -1.496\,134$ with $df = 944$ with *p*-value = 0.1349/2 = 0.067 45.

4) The other hypotheses on the median differences within a set of groups can be tested using the Wald test. For instance, between the 12 firms, conditional for *HIGH* = 0, the null hypothesis $H_0$: $C(i) = 0$, for $i = 2$ to $i = 12$, is rejected based on the *Chi-square* statistic of $\chi_0^2 = 524.9477$ with $df = 11$ and *p*-value = 0.0000.

5) As exercises, test other hypotheses, based on alternative QR(*tau*), and more advanced analyses using EViews' *Quantile Process* in Example 2.11 and *Forecast* in section 2.7.2, Chapter 2.

## 9.3 Quantile Regressions with Numerical Predictors

### 9.3.1 QR of *LWAGE* on *GRADE*

In this case, we can develope various QRs of *LWAGE* on *GRADE*, which are similar to alternative QRs presented in Chapter 4. In this section, we'll look at two example QRs of *LWAGE* on *GRADE*.

#### 9.3.1.1 A Polynomial QR of *LWAGE* on *GRADE*
Observing the scatter graphs in Figure 9.5, I propose that a polynomial QR of *LWAGE* on the numerical variable *GRADE* would be a good fit QR. However, the polynomial degree should be obtained using the trial-and-error method. See the following example.

**Example 9.7** *A Third-Degree Polynomial QR of LWAGE on GRADE*
Figure 9.14 presents the statistical results of the third-degree polynomial QR of *LWAGE* on *GRADE,* using the following ES, with its representations:

$$LWAGE\ C\ GRADE\ GRADE^2\ GRADE^3 \tag{9.12}$$

Dependent Variable: LWAGE

Method: Quantile Regression (median)

Included observations: 1000

Estimation successfully identifies unique optimal solution

| Variable | Coefficient | Std. Error | t-Statistic | Prob. |
|---|---|---|---|---|
| C | 1.208960 | 0.182382 | 6.628727 | 0.0000 |
| GRADE | 0.237788 | 0.066419 | 3.580108 | 0.0004 |
| GRADE^2 | −0.023359 | 0.007433 | −3.142522 | 0.0017 |
| GRADE^3 | 0.000854 | 0.000242 | 3.528984 | 0.0004 |

| | | | | |
|---|---|---|---|---|
| Pseudo R-squared | 0.084391 | Mean dependent var | | 2.275496 |
| Adjusted R-squared | 0.081633 | S.D. dependent var | | 0.563464 |
| S.E. of regression | 0.520338 | Objective | | 208.9983 |
| Quantile dependent var | 2.302585 | Restr. objective | | 228.2616 |
| Sparsity | 1.336118 | Quasi-LR statistic | | 115.3387 |
| Prob(Quasi-LR stat) | 0.000000 | | | |

Estimation Command:

=========================

QR(K = E, NGRID = 100) LWAGE C GRADE GRADE^2 GRADE^3

Estimation Equation:

=========================

LWAGE = C(1) + C(2)*GRADE + C(3)*GRADE^2 + C(4)*GRADE^3

Substituted Coefficients:

=========================

LWAGE = 1.20896 + 0.237788454762*GRADE − 0.0233585881393*GRADE^2 + 0.000853815564374*GRADE^3

**Figure 9.14** Statistical results based on the QR(Median) in (9.12), with its representations.

Based on the output, the findings and notes are as follows:

1) The IVs have significant joint effects on *LWAGE*, based on the *QLR* statistic of $QLR_0 = 15.3387$ with $df = 3$ and p-value = 0.000 000, and each IV has a significant adjusted effect at the 1% level.

2) As an additional analysis, Figure 9.15 presents the scatter graphs of *LWAGE* and its three forecast variables for the quantiles 0.10, 0.50, and 0.90, on *GRADE*, with their forecast evaluations presented in Figure 9.16. Note that the forecast variable of an QR is the same as its fitted variable. The advantage of using the forecast variable is that the output presents its forecast evaluation, as shown in Figure 9.15.

3) Figure 9.17 presents the output of a QSET specific for $X1^3$, which shows that $X1^3$ has insignificant effects on *LWAGE* between the pairs of quantiles (0.1, 0.5) and (0.5, 0.9) that correspond to the graphs of the three forecast variables in Figure 9.15, which are almost parallel.

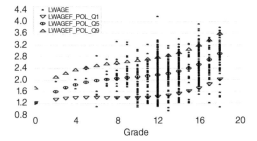

**Figure 9.15**  Scatter graphs of *LWAGE* and its forecast variables on *GRADE*.

| Forecast: LWAGEF_POL_Q1 | |
| --- | --- |
| Actual: LWAGE | |
| Forecast sample: 1 1000 | |
| Included observations: 1000 | |
| Root Mean Squared Error | 0.909537 |
| Mean Absolute Error | 0.782597 |
| Mean Abs. Percent Error | 31.65568 |
| Theil Inequality Coef. 0.234153 | |
| Bias Proportion | 0.671534 |
| Variance Proportion | 0.181465 |
| Covariance Proportion | 0.147000 |
| Theil U2 Coefficient | 1.086726 |
| Symmetric MAPE | 38.76313 |

| Forecast: LWAGEF_POL_Q5 | |
| --- | --- |
| Actual: LWAGE | |
| Forecast sample: 1 1000 | |
| Included observations: 1000 | |
| Root Mean Squared Error | 0.519296 |
| Mean Absolute Error | 0.417997 |
| Mean Abs. Percent Error | 21.04832 |
| Theil Inequality Coef. 0.111689 | |
| Bias Proportion | 0.001270 |
| Variance Proportion | 0.415488 |
| Covariance Proportion | 0.583242 |
| Theil U2 Coefficient | 0.630162 |
| Symmetric MAPE | 19.02183 |

| Forecast: LWAGEF_POL_Q9 | |
| --- | --- |
| Actual: LWAGE | |
| Forecast sample: 1 1000 | |
| Included observations: 1000 | |
| Root Mean Squared Error | 0.828316 |
| Mean Absolute Error | 0.686146 |
| Mean Abs. Percent Error | 37.63461 |
| Theil Inequality Coef. 0.157005 | |
| Bias Proportion | 0.604989 |
| Variance Proportion | 0.131599 |
| Covariance Proportion | 0.263412 |
| Theil U2 Coefficient | 0.988485 |
| Symmetric MAPE | 28.54150 |

**Figure 9.16**  The three forecast evaluations of *LWAGE* presented in Figure 9.14.

Quantile Slope Equality Test

Equation: EQ10_POL_3

Specification: LWAGE C GRADE GRADE^2 GRADE^3

Estimated equation quantile tau = 0.5

User-specified test quantiles: 0.1 0.9

Test statistic compares user-specified coefficients: C(4)

| Test summary | Chi-Sq. statistic | Chi-Sq. d.f. | Prob. |
| --- | --- | --- | --- |
| Wald test | 0.708565 | 2 | 0.7017 |

Restriction detail: b(tau_h) − b(tau_k) = 0

| Quantiles | Variable | Restr. value | Std. Error | Prob. |
| --- | --- | --- | --- | --- |
| 0.1, 0.5 | GRADE^3 | −0.000307 | 0.000367 | 0.4022 |
| 0.5, 0.9 | | 5.98E−05 | 0.000614 | 0.9224 |

**Figure 9.17**  A quantile slope equality test specific for *X1^3*.

Quantile Process Estimates
Equation: UNTITLED
Specification: LWAGE C GRADE GRADE^2 GRADE^3
Estimated equation quantile tau = 0.5
Number of process quantiles: 10
Display all coefficients

|  | Quantile | Coefficient | Std. Error | t-Statistic | Prob. |
|---|---|---|---|---|---|
| C | 0.100 | 1.208960 | 0.113668 | 10.63588 | 0.0000 |
|  | 0.200 | 1.208960 | 0.151661 | 7.971456 | 0.0000 |
|  | 0.300 | 1.318421 | 0.195951 | 6.728316 | 0.0000 |
|  | 0.400 | 1.208960 | 0.181056 | 6.677271 | 0.0000 |
|  | 0.500 | 1.208960 | 0.182382 | 6.628727 | 0.0000 |
|  | 0.600 | 1.736389 | 0.521242 | 3.331254 | 0.0009 |
|  | 0.700 | 1.790761 | 0.459003 | 3.901413 | 0.0001 |
|  | 0.800 | 1.733785 | 0.399943 | 4.335086 | 0.0000 |
|  | 0.900 | 1.735669 | 0.708136 | 2.451038 | 0.0144 |
| GRADE | 0.100 | 0.083165 | 0.068231 | 1.218883 | 0.2232 |
|  | 0.200 | 0.088127 | 0.071361 | 1.234958 | 0.2171 |
|  | 0.300 | 0.160020 | 0.080007 | 2.000091 | 0.0458 |
|  | 0.400 | 0.188490 | 0.070936 | 2.657174 | 0.0080 |
|  | 0.500 | 0.237788 | 0.066419 | 3.580108 | 0.0004 |
|  | 0.600 | 0.104224 | 0.152339 | 0.684159 | 0.4940 |
|  | 0.700 | 0.089148 | 0.135979 | 0.655600 | 0.5122 |
|  | 0.800 | 0.121428 | 0.125673 | 0.966222 | 0.3342 |
|  | 0.900 | 0.227471 | 0.223366 | 1.018381 | 0.3087 |
| GRADE^2 | 0.100 | −0.011970 | 0.009760 | −1.226420 | 0.2203 |
|  | 0.200 | −0.009079 | 0.009093 | −0.998471 | 0.3183 |
|  | 0.300 | −0.019043 | 0.009306 | −2.046284 | 0.0410 |
|  | 0.400 | −0.018783 | 0.008339 | −2.252398 | 0.0245 |
|  | 0.500 | −0.023359 | 0.007433 | −3.142522 | 0.0017 |
|  | 0.600 | −0.010614 | 0.014001 | −0.758064 | 0.4486 |
|  | 0.700 | −0.006415 | 0.012641 | −0.507505 | 0.6119 |
|  | 0.800 | −0.008401 | 0.012222 | −0.687357 | 0.4920 |
|  | 0.900 | −0.021197 | 0.021525 | −0.984774 | 0.3250 |
| GRADE^3 | 0.100 | 0.000546 | 0.000360 | 1.519947 | 0.1288 |
|  | 0.200 | 0.000432 | 0.000309 | 1.395214 | 0.1633 |
|  | 0.300 | 0.000796 | 0.000308 | 2.585194 | 0.0099 |
|  | 0.400 | 0.000730 | 0.000279 | 2.613902 | 0.0091 |
|  | 0.500 | 0.000854 | 0.000242 | 3.528984 | 0.0004 |
|  | 0.600 | 0.000491 | 0.000402 | 1.221821 | 0.2221 |
|  | 0.700 | 0.000314 | 0.000366 | 0.859524 | 0.3903 |
|  | 0.800 | 0.000359 | 0.000366 | 0.981715 | 0.3265 |
|  | 0.900 | 0.000794 | 0.000643 | 1.235619 | 0.2169 |

**Figure 9.18** The table of the quantile process estimates of the QR(Median) in (9.12) for 10 quantiles.

Quantile process estimates

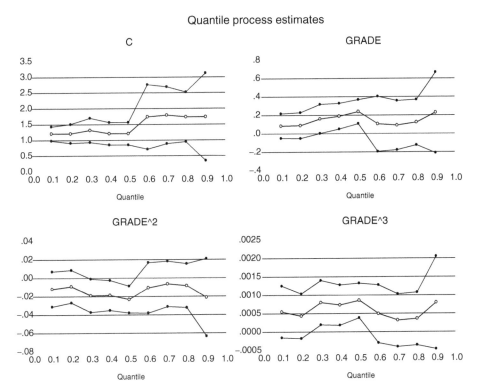

**Figure 9.19** The graphs of quantile process estimates of the QR(Median) in (9.11), for 10 quantiles.

4) In addition, Figures 9.18 and 9.19 presents the table and graphs of quantile process estimates of the QR (9.11) for 10 quantiles, from 0.1 to 0.9, for the intercept C, and each of the IVs, with their 95% confidence intervals (quantiles ±2 se).

5) Based on the output in Figure 9.18, we could easily write the QR function for each quantile from 0.1 to 0.9. However, that takes time. So, if you want to write each of the third-degree polynomial QRs, I recommend running QR(tau) for each value of tau. Then we can copy the QR function from the output of the representations, as presented in Figure 9.14, for the quantile = 0.5.

#### 9.3.1.2 The Simplest Linear QR of *Y1* on a Numerical *X1*

The simplest linear QR (SLQR) of *Y1* on a numerical variable *X1* always can be applied, even though *Y1* and *X1* do not have a linear relationship, because we want to study the trend of the quantiles of *Y1* with respect to *X1*. We'll do so with the following ES:

$$Y1 \ C \ X1 \tag{9.13}$$

#### Example 9.8  *An Application of the SLQR of LWAGE on GRADE*

Figure 9.20a present the quantile process estimates of the SLQR of *LWAGE* on *GRADE*, which show that *GRADE* has a significant linear effect on *LWAGE* for each of the five quantiles at the 1% level. And Figure 9.20b presents a quantile Process Estimates of the QR(0.5) in Figure 9.20a

(a)
Dependent variable: LWAGE
Method: quantile regression (median)
Date: 01/04/19 Time: 13:42
**Sample: 1 1000**
Included observations: 1000
Huber sandwich standard errors and covariance
Sparsity method: Kernel (Epanechnikov) using residuals
Bandwidth method: Hall-Sheather, bw = 0.097156
Estimation successfully identifies unique optimal solution

| Variable | Coefficient | Std. Error | t-Statistic | Prob. |
|---|---|---|---|---|
| C | 1.208960 | 0.109587 | 11.03192 | 0.0000 |
| GRADE | 0.083536 | 0.008263 | 10.11002 | 0.0000 |
| Pseudo R-squared | 0.071092 | Mean dependent var | | 2.275496 |
| Adjusted R-squared | 0.070161 | S.D. dependent var | | 0.563464 |
| S.E. of regression | 0.524745 | Objective | | 212.0341 |
| Quantile dependent var | 2.302585 | Restr. objective | | 228.2616 |
| Sparsity | 1.401185 | Quasi-LR statistic | | 92.65021 |
| Prob(quasi-LR stat) | 0.000000 | | | |

(b)
Quantile process estimates
Equation: (9.12)
Specification: LWAGE C GRADE
Estimated equation quantile tau = 0.5
Number of process quantiles: 5
Display all coefficients

| | Quantile | Coefficient | Std. Error | t-Statistic | Prob. |
|---|---|---|---|---|---|
| C | 0.200 | 0.786363 | 0.205056 | 3.834869 | 0.0001 |
| | 0.400 | 1.172962 | 0.116367 | 10.07982 | 0.0000 |
| | 0.500 | 1.208960 | 0.109587 | 11.03192 | 0.0000 |
| | 0.600 | 1.373716 | 0.102385 | 13.41712 | 0.0000 |
| | 0.800 | 1.561063 | 0.086828 | 17.97870 | 0.0000 |
| GRADE | 0.200 | 0.078225 | 0.016069 | 4.868001 | 0.0000 |
| | 0.400 | 0.075540 | 0.008850 | 8.535852 | 0.0000 |
| | 0.500 | 0.083536 | 0.008263 | 10.11002 | 0.0000 |
| | 0.600 | 0.082287 | 0.007831 | 10.50736 | 0.0000 |
| | 0.800 | 0.089833 | 0.006694 | 13.42011 | 0.0000 |

**Figure 9.20** (a) The output of the simplest linear QR(Median). (b) The quantile process estimates of the QR(Median) in Figure 9.20a.

**Example 9.9** *Alternative QRs of LWAGE on GRADE*

As an extension of the third-degree polynomial QR in (9.11), Figure 9.21 presents the output of a fourth-degree polynomial QR(Median) *LWAGE* on *GRADE* and one of its acceptable in the statistical sense, a reduced model,. The ESs applied are as follows:

$$LWAGE\ C\ GRADE\ GRADE^{\wedge}2\ GRADE^{\wedge}3\ Grade^{\wedge}4 \tag{9.14}$$

$$LWAGE\ C\ GRADE\ GRADE^{\wedge}2\ Grade^{\wedge}4 \tag{9.15}$$

Figure 9.20 presents the scatter graphs of *LWAGE* and five of its forecast value variables (*LWAGEF_Cat, LWAGEF_Lin, LWAGEF_Pol3, LWAGEF_Pol4,* and *LWAGEF_Red _Pol4*) as the results of the QR(Median)s in (9.1, 9.11–9.14).

Based on Figures 9.20 and 9.21, the following findings and notes are presented:

1) Two of the IVs of the fourth-degree polynomial QR(Median) have insignificant effects at the 10% level. So, to keep the fourth-degree polynomial, then by using the trial-and-error method, we have to delete either one of *GRADE, GRADE^2,* and *GRADE^3*. We then obtain an acceptable reduced QR(Median), as presented in Figure 9.21.
2) Each IV of the reduced QR(Median) has either a positive or negative significant effect on *LWAGE* at the 1% level of significance. For instance, at the 5% level, *GRADE^2* has a significant effect based on the *t*-statistic of $t_0 = -2.3629$ with $df = (1000 - 4)$ and *p*-value $= 0.0183/2 = 0.009\,15 < 0.01$.
3) The graphs in Figure 9.22 show that the three polynomial forecast value variables coincide or almost coincide. Although two IVs of the QR(Median) in (9.13) have insignificant effects at the 10% level, it can be considered the best QR among the three polynomial QRs.

| Variable | Full QR(Median) | | | A reduced QR(Median) | | |
|---|---|---|---|---|---|---|
| | Coef. | t-Stat. | Prob. | Coef. | t-Stat. | Prob. |
| C | 1.2090 | 6.6564 | 0.0000 | 1.3109 | 6.3794 | 0.0000 |
| GRADE | 0.3205 | 2.4292 | 0.0153 | 0.1695 | 2.8575 | 0.0044 |
| GRADE^2 | −0.0442 | −1.5712 | 0.1165 | −0.0110 | −2.3629 | 0.0183 |
| GRADE^3 | 0.0025 | 1.1603 | 0.2462 | | | |
| GRADE^4 | 0.0000 | −0.7574 | 0.4490 | 0.0000 | 3.1558 | 0.0016 |
| Pse. R-sq | 0.0849 | | | 0.0838 | | |
| Adj. R-sq | 0.0813 | | | 0.0811 | | |
| S.E. of reg | 0.5198 | | | 0.5208 | | |
| Quant DV | 2.3026 | | | 2.3026 | | |
| Sparsity | 1.3495 | | | 1.3349 | | |
| QLR-stat | 114.93 | | | 114.66 | | |
| Prob | 0.0000 | | | 0.0000 | | |

**Figure 9.21** The outputs of the QR(Median) in (9.13) and (9.14).

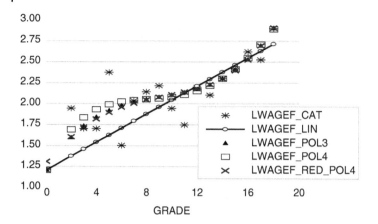

**Figure 9.22** The scatter graphs of *LWAGE* and four of its forecast value variables.

4) However, among the five QRs, the best QR is the QR(Median) of *LWAGE* with the dummy variables in (9.6) because it has the largest pseudo *R*-squared and adjusted *R*-squared, and the worst is the SLQR(Median) in (9.13).

### 9.3.2 Quantile Regressions of *Y1* on *(X1,X2)*

Referring to all alternative QRs presented in Section 5.2, we can conduct the analysis based on the same QRs of *Y1* = *LWAGE*, on *X1* = *GRADE*, and *X2* = *POTEXP*. For a better illustration, similar to the graphs in Figure 5.1, Figure 9.23 presents only two alternative causal relationships based on numerical variables *(X1,X2,Y1)*, where the solid arrows represent that each pair of the variables has a true causal relationship, and the arrow with dotted line in Figure 9.23b represents that *X2* at least is an upper variable of *Y1* because it has an indirect effect on *Y1* through *X1*.

The following subsections present two common two-way interaction QRs: a hierarchical 2WI-QR with its two possible nonhierarchical reduced models, as presented in the path diagram in Figure 9.23, and a special polynomial 2WI-QR.

#### 9.3.2.1 Hierarchical and Nonhierarchical Two-Way Interaction QRs

Based on Figures 9.23a,b, we have a hierarchical 2WI-QR and two nonhierarchical 2WI-QRs with the following ESs:

$$Y1\ C\ X1*X2\ X1\ X2 \tag{9.16}$$

$$Y1\ C\ X1*X2\ X1 \tag{9.17}$$

$$Y1\ C\ X1*X2\ X2 \tag{9.18}$$

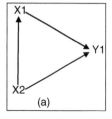 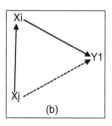

**Figure 9.23** Two alternative causal effects based on the numerical variables *(X1,X2,Y1)*.

**Example 9.10**   *An Application of the QR in (9.16)*

Figure 9.24 presents the statistical results of the QR(Median) in (9.12) for $Y1 = LWAGE$, $X1 = GRADE$, $X2 = POTEXP$, and its reduced QR. Based on these results, the following findings and notes are presented:

1) Even though $X1*X2$ has a large $p$-value as an IV of the QR(Median) in (9.15), it should be kept in the model because it has these special characteristics:

   1.1 The effect of $X1 = GRADE$ decreases with an increasing score of $X2 = POTEXP$. This can be explained using the following QR function, which shows $X2*X1$ has a negative coefficient. Hence, the values of $(0.124\,13 - 0.000\,45*X2)$ will decrease with an increasing value of $X2 = POTEXP$.

$$Y1 = 0.273\,51 + 0.027\,16*X2 + (0.124\,13 - 0.000\,45*X2)*X1 \tag{9.19}$$

   1.2 In addition, $X1*X2$ and $X1$ have significant joint effects on $Y1 = LWAGE$, based on the redundant variables test (RVT) with the output presented in Figure 9.25. So, there is a reason to keep $X1*X2$ as an IV.

2) Note that the interaction $X1*X2$ has a large $p$-value. In fact, it has the greatest correlation with $Y1$, compared with $X1$ and $X2$, as presented in Figure 9.26, and each pairs of the IVs are significantly correlated $p$-values $= 0.0000$. These findings show the unpredicted impacts of the bivariate correlations or multicollinearity between the IVs.

3) In addition, we also find that $X1*X2$ and $X2$ have significant joint effects on $Y1$, based on the *Chi-square* statistic of $\chi_0^2 = 213.7919$ with $df = 2$ and $p$-value $= 0.0000$.

4) Based on these two findings, I prefer not to reduce the model. However, as an illustration, Figure 9.24 shows two reduced QRs. The reduced QR-1 is better than the reduced QR-2 because QR-1 has a greater pseudo $R$-squared and adjusted $R$-squared. Based on QR-1 and QR-2, we can conclude only that the effect of $X1$ is significantly dependent on $X2$, or the effect of $X2$ is significantly dependent on $X1$.

| Variable | Q R(Median) in (9.15) | | | Reduced QR-1 in (9.16) | | | Reduced QR-2 in (9.17) | | |
|---|---|---|---|---|---|---|---|---|---|
| | Coef. | t-Stat. | Prob. | Coef. | t-Stat. | Prob. | Coef. | t-Stat. | Prob. |
| C | 0.2735 | 1.5206 | 0.1287 | 0.9267 | 9.4840 | 0.0000 | 1.8587 | 45.837 | 0.0000 |
| X1*X2 | −0.0005 | −0.7870 | **0.4315** | 0.0017 | 12.938 | 0.0000 | 0.0033 | 10.100 | 0.0000 |
| X1 | 0.1241 | 8.7879 | 0.0000 | 0.0740 | 9.7313 | 0.0000 | | | |
| X2 | 0.0272 | 3.9527 | 0.0001 | | | | −0.0195 | −5.1445 | 0.0000 |
| Pseudo R-sq | 0.1948 | | | 0.1836 | | | 0.1463 | | |
| Adj. R-sq. | 0.1924 | | | 0.1820 | | | 0.1446 | | |
| S.E. of reg | 0.4730 | | | 0.4796 | | | 0.4992 | | |
| Quantile DV | 2.3026 | | | 2.3026 | | | 2.3026 | | |
| Sparsity | 1.1414 | | | 1.1701 | | | 1.2326 | | |
| QLR stat | 311.62 | | | 286.51 | | | 216.77 | | |
| Prob | 0.0000 | | | 0.0000 | | | 0.0000 | | |

**Figure 9.24**   The output of the QR(Median) in (9.16) and its reduced QRs.

Redundant variables test
Null hypothesis: X1*X2 X1
Equation: UNTITLED
Specification: Y1 C X1*X2 X1 X2

|  | Value | df | Probability |
|---|---|---|---|
| QLR L-statistic | 219.8728 | 2 | 0.0000 |
| QLR Lambda-statistic | 203.0045 | 2 | 0.0000 |

**Figure 9.25** The RVT of *X1*X2* and *X1*.

Covariance analysis: ordinary
Date: 12/04/18 Time: 17:41
*Sample: 1 1000*
Included observations: 1000

Correlation
t-Statistic

| Probability | Y1 | X1*X2 | X1 | X2 |
|---|---|---|---|---|
| Y1 | 1.000000 |  |  |  |
|  | — |  |  |  |
|  | — |  |  |  |
| X1*X2 | 0.432306 | 1.000000 |  |  |
|  | 15.14543 | — |  |  |
|  | 0.0000 | — |  |  |
| X1 | 0.367766 | 0.124448 | 1.000000 |  |
|  | 12.49372 | 3.962259 | — |  |
|  | 0.0000 | 0.0001 | — |  |
| X2 | 0.308757 | 0.908137 | −0.241762 | 1.000000 |
|  | 10.25502 | 68.52401 | −7.871026 | — |
|  | 0.0000 | 0.0000 | 0.0000 | — |

**Figure 9.26** The output of bivariate correlations between *Y1*, *X1*X2*, *X1*, and *X2*.

### 9.3.2.2 A Special Polynomial Interaction QR

**Example 9.11** *A Third-Degree Polynomial Interaction QR of Y1 on (X1,X2)*
Figure 9.27 presents the statistical results of a special third-degree polynomial interaction QR of *Y1* on *(X1,X2)*, using the following ES and two of its reduced models:

$$Y1\ C\ X1^*X2\ X1^\wedge 2^*X2\ X1^\wedge 3^*X2\ X1\ X1^\wedge 2\ X1^\wedge 3\ X2 \tag{9.20}$$

Based on these results, the following findings and notes are presented:

1) Although each of the IVs, *X1*, *X1^2*, and *X1^3*, has an insignificant effect on *Y1* at the 10% level, the QR(Median) in (9.19) is an acceptable QR because each of their

| Variable | Full QR(Median) in (9.13) | | | RM-1 | | | RM-2 | | |
|---|---|---|---|---|---|---|---|---|---|
| | Coef. | t-Stat. | Prob. | Coef. | t-Stat. | Prob. | Coef. | t-Stat. | Prob. |
| C | 0.477 | 1.343 | 0.180 | 1.046 | 2.392 | 0.017 | 1.315 | 19.396 | 0.000 |
| X1*X2 | −0.010 | −1.840 | 0.066 | −0.005 | −1.828 | 0.068 | −0.005 | −1.654 | 0.099 |
| X1^2*X2 | 0.001 | 2.103 | 0.036 | 0.001 | 2.382 | 0.017 | 0.001 | 2.664 | 0.008 |
| X1^3*X2 | 0.000 | −2.291 | 0.022 | 0.000 | −2.593 | 0.010 | 0.000 | −3.384 | 0.001 |
| X1 | 0.210 | 1.625 | 0.105 | 0.030 | 0.592 | **0.554** | | | |
| X1^2 | −0.017 | −1.188 | 0.235 | | | | | | |
| X1^3 | 0.001 | 1.465 | 0.143 | 0.000 | 1.880 | 0.060 | 0.000 | 8.636 | 0.000 |
| X2 | 0.041 | 2.789 | 0.005 | 0.027 | 2.385 | 0.017 | 0.021 | 2.569 | 0.010 |
| Pse. R-sq | 0.201 | | | 0.201 | | | 0.201 | | |
| Adj. R-sq | 0.196 | | | 0.196 | | | 0.197 | | |
| S.E. of reg | 0.474 | | | 0.473 | | | 0.473 | | |
| Quantile DV | 2.303 | | | 2.303 | | | 2.303 | | |
| Sparsity | 1.096 | | | 1.098 | | | 1.103 | | |
| QLR stat. | 335.12 | | | 333.99 | | | 332.23 | | |
| Prob. | 0.000 | | | 0.000 | | | 0.000 | | |

**Figure 9.27** Statistical results of the full QR(Median) in (9.20) and its reduced models.

interactions, $X1*X2$, $X1^2*X2$, and $X1^3*X3$, has either a negative or positive significant effect on $Y1$ at the 5% level. For instance, $X1*X2$ has a negative significant effect, based on the $t$-statistic of $t_0 = -1.840$ with $df = 992$ and $p$-value $= 0.066/2 = 0.033$.

2) Note that, at the 10% level, both $X1$ and $X1^3$ have a positive significant effect on $Y1$, based on the $t$-statistic with $p$-values of 0.105/2 and 0.143/2, respectively. And, at the 15% level, $X1^2$ has a negative significant effect, with $p$-value $= 0.235/2 < 0.15$.

3) However, as an illustration, consider the reduced QR RM-1. Deleting $X1^2$, which has the largest probability in QR (9.13), shows that $X1*X2$ has a negative significant effect, based on the $t$-statistic of $t_0 = -1.828$ with $df = 993$ and a slight greater $p$-value $= 0.068/2 = 0.034$. However, $X1$ has a large probability of 0.554. So, I would say RM-1 is worse than the full QR, in the statistical sense.

4) Then, by deleting $X1$ from RM-1, RM-2 is obtained, which shows that $X1*X2$ still has a negative significant effect at the 5% level, based on the $t$-statistic of $t_0 = -1.654$ with $df = 994$ and a slightly greater $p$-value $= 0.099/2 = 0.0495$, and each of the other IVs has a significant effect at the 1% level of significance.

5) As a more advanced analysis, Figure 9.28 presents the graphs of sorted $Y1$ and two of its fitted variables, $FV\_EQ915$ and $FV\_RM2$. The figure shows that the graphs of the two fitted variables coincide, with the correlation of 0.998 810. In the statistical sense, RM-2 is the best, even though it has a slightly greater adjusted $R$-squared.

6) In addition, Figure 9.29 presents a special QSET for the IVs of RM-2, between the pair of quantiles (0.25, 0.75), using the RM-2 ES, for tau $= 0.25$, as shown in the line "*Estimated equation quantile tau $= 0.25$.*" The steps of the analysis are as follows:

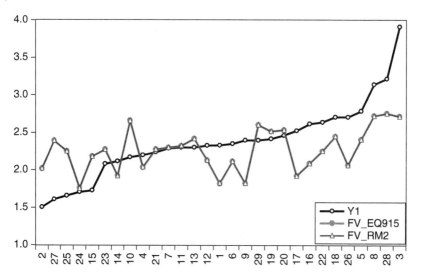

**Figure 9.28** The graphs of sorted *Y1* and its fitted variables, *FV_EQ915* and *FV_RM2*.

Quantile slope equality test
Equation: EQ_RM-2
Specification: Y1 C X1*X2 X1^2*X2 X1^3*X2 X1^3 X2
Estimated equation quantile tau = 0.25
User-specified test quantiles: 0.25, 0.75
Test statistic compares all coefficients

| Test summary | Chi-Sq. statistic | Chi-Sq. d.f. | Prob. |
|---|---|---|---|
| Wald test | 9.961364 | 5 | 0.0763 |

Restriction detail: b(tau_h) − b(tau_k) = 0

| Quantiles | Variable | Restr. value | Std. Error | Prob. |
|---|---|---|---|---|
| 0.25, 0.75 | X1*X2 | 0.001407 | 0.003984 | 0.7239 |
| | X1^2*X2 | −0.000190 | 0.000441 | 0.6665 |
| | X1^3*X2 | 7.42E−06 | 1.48E−05 | 0.6151 |
| | X1^3 | −7.06E−05 | 3.90E−05 | 0.0704 |
| | X2 | −0.004725 | 0.010576 | 0.6550 |

**Figure 9.29** A quantile slope equality test using the ES of RM-2, for tau = 0.25, between the pair of quantiles (0.25, 0.75).

6.1 Run the QR(0.25), using the RM-2 ES.

6.2 With the output on-screen, select *View/Quantile Process/Slope Equality Test* to display the Quantile Slope Equality Test dialog with options for the quantiles and coefficient specifications.

6.3 Inserting 0.25, 0.75 as the *User-specified quantiles-pairs* and clicking *OK* obtains the output in Figure 9.29.

7) Based on the output in Figure 9.29, the following findings and notes are presented:

7.1 At the 10% level, the five IVs have significantly different joint effects on $Y1 = LWAGE$ between QR(0.25) and QR(0.75), based on the *Chi-square* statistic of $\chi_0^2 = 9.961\ 364$ with $df = 5$ and $p$-value = 0.0763.

7.2 For each of the five IVs, only $X1\hat{}\ 3$ has a significant slope difference.

7.3 Note that if the QSET is done based on the QR(0.5), then the output of the test will present two pairs of quantiles (0.25, 0.5) and (0.5, 0.75). As exercise, do this based on QR(0.1), QR(0.75), and QR(0.9).

### 9.3.2.3 A Double Polynomial Interaction QR of Y1 on (X1,X2)

In this section, I present a double polynomial interaction QR (DPI-QR) of *Y1* on *(X1,X2)* because I found the QR(0.5) of *Y1* and *X2* to have polynomial relationships, as shown in Figure 9.30.

**Figure 9.30** A polynomial QR(0.5) of *Y1 = LWAGE* on *X2 = POTEXP*.

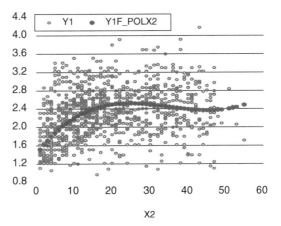

### Example 9.12 An Application of the DPI-QR(Median) of Y1 on (X1,X2)

Figure 9.31 presents the summary of three DPI-QRs of *Y1* on *(X1,X2)*. Based on this summary, the following findings and notes are presented:

1) Each of the first three IVs of the QR-1 has an insignificant effect at the 10% level, and *X1\*X2* has a very large *p*-value, so reducing the model results in QR-2, with each IV having a significant effect at the 1% level.

2) Then using the trial-and error method to insert the four variables, $X1\hat{}2$, $X1\hat{}3$, $X2\hat{}2$, and $X2\hat{}3$, one by one as additional IVs of the QR-2, obtains QR(Median)-3 as a final acceptable model. It is the best fit model among the three because it has the largest pseudo *R*-squared and adjusted *R*-squared.

| Variable | QR(Median)-1 | | | QR(Median)-2 | | | QR(Median)-3 | | |
|---|---|---|---|---|---|---|---|---|---|
| | Coef. | t-Stat. | Prob. | Coef. | t-Stat. | Prob. | Coef. | t-Stat. | Prob. |
| C | 0.3602 | 2.0964 | 0.0363 | 0.3394 | 2.0794 | 0.0378 | 0.6775 | 3.0047 | 0.0027 |
| X1*X2 | 0.0011 | 0.3994 | **0.6897** | | | | | | |
| X1^2*X2 | 0.0003 | 1.2318 | 0.2183 | 0.0004 | 3.5744 | 0.0004 | 0.0005 | 3.5773 | 0.0004 |
| X1^3*X2 | 0.0000 | −1.092 | 0.2750 | 0.0000 | −2.923 | 0.0035 | 0.0000 | −2.865 | 0.0043 |
| X1*X2^2 | −0.0002 | −3.965 | 0.0001 | −0.0002 | −4.004 | 0.0001 | −0.0002 | −3.122 | 0.0018 |
| X1*X2^3 | 0.0000 | 2.8866 | 0.0040 | 0.0000 | 2.8135 | 0.0050 | 0.0000 | 1.9337 | 0.0534 |
| X1 | 0.0909 | 5.8867 | 0.0000 | 0.0929 | 6.4441 | 0.0000 | 0.0490 | 1.8357 | 0.0667 |
| X2 | 0.0390 | 4.8013 | 0.0000 | 0.0422 | 8.5449 | 0.0000 | 0.0338 | 5.5034 | 0.0000 |
| X1^3 | | | | | | | 0.0001 | 1.8928 | 0.0587 |
| Pse. R-sq | 0.2304 | | | 0.2303 | | | 0.2324 | | |
| Adj. R-sq | 0.2250 | | | 0.2256 | | | 0.2270 | | |
| S.E. of reg | 0.4550 | | | 0.4553 | | | 0.4545 | | |
| Quantile DV | 2.3026 | | | 2.3026 | | | 2.3026 | | |
| Sparsity | 1.0603 | | | 1.0571 | | | 1.0459 | | |
| QLR stat | 396.85 | | | 397.79 | | | 405.75 | | |
| Prob. | 0.0000 | | | 0.0000 | | | 0.0000 | | |

**Figure 9.31** Statistical results summary of three DPI-QRs of Y1 on (X1,X2).

3) More advanced analysis could easily be done based on the QR-3, as presented in previous examples, such as the *Quantile Process Estimates* (QPE), *Quantile Sole Equality Test* (QSET)*t*, and *Symmetric Quantile Test* (SQT), as in Appendix A. Do as exercises.

### 9.3.3 QRs of Y1 on Numerical Variables (X1,X2,X3)

Chapter 6 presented examples of various QRs of Y1 on (X1,X2,X3). This section presents two additional QRs of Y1 = *LWAGE* on X1 = *GRADE*, X2 = *POTEXP*, and X3 = *OCC1.X1, X2, and X3* are ordinal variables, which will be used as numerical predictors. For the three predictors, we would have X1 and X2 have direct effects on X3, and X1 has direct effect on X2, since higher X1 = *GRADE* tends to have less years of *POTEXP* = X2. So, referring to the path diagram in Figure 6.1(c), we will have the a full 2-way and 3-way interaction QRs, having the same ESs as the equations (6.3) and (6.4).

#### 9.3.3.1 A Full Two-Way Interaction QR

**Example 9.13** *An Application of a F2WI-QR of Y1 on (X1,X2,X3)*
Figure 9.32 presents the statistical results summary of a manual three-stage selection method of a F2WI-QR of Y1 on (X1,X2,X3), with the following ES:

$$Y1\ C\ X1{*}X2\ X1{*}X3\ X2{*}X3\ X1\ X2\ X3 \tag{9.21}$$

| Variable | Stage-1 | | | Stage-2 | | | Stage-3 | | |
|---|---|---|---|---|---|---|---|---|---|
| | Coef. | t-Stat. | Prob. | Coef. | t-Stat. | Prob. | Coef. | t-Stat. | Prob. |
| C | 1.916 | 24.083 | 0.000 | 0.970 | 8.393 | 0.000 | 0.598 | 2.001 | 0.046 |
| X1*X2 | 0.002 | 12.190 | 0.000 | 0.001 | 4.889 | 0.000 | 0.001 | 5.170 | 0.000 |
| X1*X3 | −0.001 | −0.957 | **0.339** | −0.004 | −5.562 | 0.000 | −0.009 | −2.593 | 0.010 |
| X2*X3 | −0.001 | −5.042 | 0.000 | 0.001 | 3.048 | 0.002 | 0.001 | 1.563 | 0.118 |
| X1 | | | | 0.093 | 9.363 | 0.000 | 0.113 | 5.899 | 0.000 |
| X3 | | | | | | | 0.077 | 1.475 | 0.140 |
| Pse. R-sq | 0.149 | | | 0.200 | | | 0.202 | | |
| Ad. R-sq | 0.146 | | | 0.197 | | | 0.198 | | |
| S.E. of reg. | 0.495 | | | 0.471 | | | 0.471 | | |
| Quantile DV | 2.303 | | | 2.303 | | | 2.303 | | |
| Sparsity | 1.238 | | | 1.106 | | | 1.123 | | |
| QLR stat | 219.3 | | | 330.4 | | | 327.8 | | |
| Prob. | 0.000 | | | 0.000 | | | 0.000 | | |

**Figure 9.32** The statistical results summary based on the F2WI-QR(Median) in (9.21).

Based on this summary, the following findings and notes are presented:

1) By using the F2WI-QR(Median), we obtain two of the two-way interactions. So, applying the manual three-stage selection method obtains the output summary presented in Figure 9.32.
2) At Stage-1, the QR(Median) has only the three two-way interactions as the IVs. Although one has an insignificant effect, we still keep it as an IV since we never can predict the impacts of inserting additional IVs. However, another way to present the analysis is to delete the X1*X3 from the first stage model. Do that as an exercise.
3) Each IV of the QR at Stage-2 has a significant effect at the 1% level. So, in the statistical sense, we can consider this QR to be an acceptable nonhierarchical reduced model of the F2WI-QR in (9.21).
4) Finally, inserting the variables X2 or X3 at Stage-3 obtains an acceptable nonhierarchical QR(Median), since at the 10%, both X2*X3 and X3 have a positive significant effect, with p-values of $0.118/1 = 0.059$, and $0.140/2 = 0.07$, respectively.

### 9.3.3.2 A Full-Three-Way-Interaction QR
Referring to the ES (6.4), the F3WI-QR of Y1 on (X1,X2,X3) would have the following ES:

$$Y1 \ C \ X1^*X2^*X3 \ X1^*X2 \ X1^*X3 \ X2^*X3 \ X1 \ X2 \ X3 \qquad (9.22)$$

In this case, the interaction X1*X2*X3 is defined as the most important predictor of $Y = LWAGE = log(Wage)$. So, we have to find a final QR, which generally will be a nonhierarchical 3WI-QR.

**Example 9.14   *Application of the F3WI-QR(Median)***
It is really unexpected that the output of the F3WI-QR(Median) in (9.22) shows that each IV has an significant effect at the 1% level, as we can see in Figure 9.33, and all its IVs have significant joint effects, based on the *QLR*-statistic of $QLR_0 = 362.8720$ with $df = 7$ and *p*-value $= 0.000\,000$. This case shows a very special impact of bivariate correlations or multicollinearity of the IVs, since all pairs have significant correlations, based on the *t*-statistic with *p*-values smaller than 0.0001, and a pair has a *p*-value of 0.0024, namely *X2* and *X1\*X3*.

As an additional illustration, Figure 9.34 presents the output of F3WI-QR(tau) for tau $= 0.1$ and tau $= 0.9$, which also shows that each IV of the QR(0.9) has a significant effect at the 1% level. However, four of the seven IVs of the QR(0.1) have insignificant effects at the 10%, which should be accepted as a F3WI-QR(0.1). But, we could reduce the QR model to obtain a better QR. By deleting *X2*, we find the QR(0.1) with each of the interaction IVs having a significant effect at the 1%, 5%, or 10% level.

As a more advanced analysis, Figure 9.35 presents the output of a QSET between a pair of quantiles (0.1, 0.9), based on the QR(0.1). With this output, the findings and notes are as follows:

1)  The slopes of the seven IVs have significant differences, based on the *Chi-Square* statistic of $\chi_0^2 = 27.554\,44$ with $df = 7$ and *p*-value $= 0.0003$.

---

Dependent variable: Y1
Method: quantile regression (median)
Included observations: 1000
Bandwidth method: Hall-Sheather, bw = 0.097156
Estimation successfully identifies unique optimal solution

| Variable | Coefficient | Std. Error | t-Statistic | Prob. |
|---|---|---|---|---|
| C | −1.147800 | 0.471166 | −2.436085 | 0.0150 |
| X1*X2*X3 | 0.000702 | 0.000217 | 3.240122 | 0.0012 |
| X1*X2 | −0.003662 | 0.001184 | −3.092568 | 0.0020 |
| X1*X3 | −0.025165 | 0.005997 | −4.196311 | 0.0000 |
| X2*X3 | −0.009605 | 0.002776 | −3.460759 | 0.0006 |
| X1 | 0.229678 | 0.031984 | 7.181066 | 0.0000 |
| X2 | 0.073642 | 0.016231 | 4.537260 | 0.0000 |
| X3 | 0.322851 | 0.082503 | 3.913205 | 0.0001 |

| | | | |
|---|---|---|---|
| Pseudo R-squared | 0.212061 | Mean dependent var | 2.275496 |
| Adjusted R-squared | 0.206501 | S.D. dependent var | 0.563464 |
| S.E. of regression | 0.465881 | Objective | 179.8562 |
| Quantile dependent var | 2.302585 | Restr. objective | 228.2616 |
| Sparsity | 1.097403 | Quasi-LR statistic | 352.8720 |
| Prob(quasi-LR stat) | 0.000000 | | |

**Figure 9.33**   The output of the F3WI-QR(Median) in (9.22).

| Variable | QR(0.1) Coef. | t-Stat. | Prob. | QR(0.9) Coef. | t-Stat. | Prob. |
|---|---|---|---|---|---|---|
| C | −1.7884 | −1.7905 | 0.0737 | −1.5190 | −2.1598 | 0.0310 |
| X1*X2*X3 | 0.0006 | 1.1127 | 0.2661 | 0.0008 | 3.7275 | 0.0002 |
| X1*X2 | −0.0037 | −1.0112 | 0.3122 | −0.0059 | −3.7614 | 0.0002 |
| X1*X3 | −0.0296 | −2.6767 | 0.0076 | −0.0291 | −3.9471 | 0.0001 |
| X2*X3 | −0.0100 | −1.3613 | 0.1737 | −0.0129 | −4.3071 | 0.0000 |
| X1 | 0.2184 | 3.0649 | 0.0022 | 0.2898 | 5.8952 | 0.0000 |
| X2 | 0.0762 | 1.5535 | 0.1206 | 0.1078 | 4.9833 | 0.0000 |
| X3 | 0.4230 | 2.7794 | 0.0055 | 0.4011 | 3.9483 | 0.0001 |
| Pse.R-sq | 0.1349 | | | 0.2539 | | |
| Adj. R-sq | 0.1288 | | | 0.2486 | | |
| S.E. of reg | 0.7843 | | | 0.7173 | | |
| Quant DV | 1.5041 | | | 3.0093 | | |
| Sparsity | 2.5157 | | | 1.8738 | | |
| QLR-stat | 115.33 | | | 299.60 | | |
| Prob | 0.0000 | | | 0.0000 | | |

**Figure 9.34** An outputs summary of two F3WI-QR(tau) for tau = 0.1 and 0.9.

Quantile slope equality test
Equation: EQ02_F3WI
Specification: Y1 C X1*X2*X3 X1*X2 X1*X3 X2*X3 X1 X2 X3
Estimated equation quantile tau = 0.1
User-specified test quantiles: 0.1 0.9
Test statistic compares all coefficients

| Test summary | Chi-Sq. statistic | Chi-Sq. d.f. | Prob. |
|---|---|---|---|
| Wald test | 27.55444 | 7 | 0.0003 |

Restriction detail: b(tau_h) − b(tau_k) = 0

| Quantiles | Variable | Restr. value | Std. Error | Prob. |
|---|---|---|---|---|
| 0.1, 0.9 | X1*X2*X3 | −0.000231 | 0.000571 | 0.6854 |
| | X1*X2 | 0.002232 | 0.003802 | 0.5572 |
| | X1*X3 | −0.000477 | 0.012626 | 0.9698 |
| | X2*X3 | 0.002891 | 0.007649 | 0.7055 |
| | X1 | −0.071370 | 0.082195 | 0.3852 |
| | X2 | −0.031579 | 0.051795 | 0.5421 |
| | X3 | 0.021885 | 0.173782 | 0.8998 |

**Figure 9.35** The output of a quantile slope equality test based on the F3WI-QR(0.1).

2) However, unexpectedly, the pair of quantiles (0.1, 0.9) for each of the seven IVs has an insignificant difference, with minimum and maximum *p*-values of 0.3852 and 0.9698, respectively.

## 9.4 Heterogeneous Quantile-Regressions

Various heterogeneous linear QRs of *LWAGE* on ordinal variables can be applied based on the data in CPS88.wf1, by single or multiple factors.

### 9.4.1 Heterogeneous Quantile Regressions by a Factor

For the heterogeneous linear QR (HLQR) of *Y1* on *X1* by a factor *F1,* I recommend applying the following alternative ESs:

$$Y1\ C@Expand\,(F1,\,@Dropfirst)\ X1*@Expand\,(F1) \tag{9.23a}$$

$$Y1\ C\ X1@Expand\,(F1,\,@Dropfirst)\ X1*@Expand\,(F1,\,@Dropfirst) \tag{9.23b}$$

#### 9.4.1.1 A Heterogeneous Linear QR of *LWAGE* on *POTEXP* by *IND1*

**Example 9.15** *An Application of the QR (9.23a)*
Figure 9.36 presents the output of a specific QR(Median) in (9.23a) for the sample {*IND1* > 9}, using the following ES:

$$LWAGE\ C@Expand\,(IND1,\,@Dropfirst)\ POTEXP*@Expand\,(IND1) \tag{9.24}$$

Based on this output, the following findings and additional analyses are presented:

1) The output of the HLQR(Median) shows *POTEXP* has significant effect on *LWAGE* for only two of the five industries, namely *IND1* = 10 and *IND1* = 13.
2) As an advanced analysis, Figure 9.37 presents the output of a QSET of a selected pairs of quantiles, based on the QR(Median), for only its numerical IV, *POTEXP*, for each *IND1*. The conclusions are as follows:
   2.1 At the 10% level, the 10 pairs of quantiles have significant differences, based on the *Chi-Square* statistic of $\chi_0^2 = 16.311\,53$ with *df*-10 and *p*-value = 0.0911.
   2.2 However, two of the10 pairs of quantiles have significant difference, at the 1% and 5% levels, respectively, for *IND1* = 13 only.
3) As an additional analysis, Figure 9.38 presents the output of the Wald test for the linear effects differences of *POTEXP* on *LWAGE* between the five industries. Based on this output, the following findings and notes are presented:
   3.1 The linear effects of *POTEXP* on *LWAGE* have insignificant differences between the five industries. So, in the statistical sense, the HLQR(Median) could be reduced to an ANCOVA-QR(Median), with the following ES.

$$LWAGE\ C\ POTEXP\ @Expand\,(IND1,\,@Dropfirst) \tag{9.25}$$

Dependent variable: LWAGE
Method: quantile regression (median)
Date: 01/06/19 Time: 18:28
*Sample: 1 1000 IF IND1 > 9*
Included observations: 241
Huber sandwich standard errors and covariance
Sparsity method: Kernel (Epanechnikov) using residuals
Bandwidth method: Hall-Sheather, bw = 0.15612
Estimation successfully identifies unique optimal solution

| Variable | Coefficient | Std. Error | t-Statistic | Prob. |
|---|---|---|---|---|
| C | 1.686400 | 0.252009 | 6.691817 | 0.0000 |
| IND1 = 11 | 0.219102 | 0.394234 | 0.555767 | 0.5789 |
| IND1 = 12 | 0.465206 | 0.422690 | 1.100585 | 0.2722 |
| IND1 = 13 | 0.102997 | 0.290956 | 0.353996 | 0.7237 |
| IND1 = 14 | 0.705662 | 0.344329 | 2.049382 | 0.0416 |
| POTEXP*(IND1 = 10) | 0.030049 | 0.012668 | 2.371941 | **0.0185** |
| POTEXP*(IND1 = 11) | 0.021880 | 0.023445 | 0.933238 | 0.3517 |
| POTEXP*(IND1 = 12) | −0.006014 | 0.012901 | −0.466158 | 0.6415 |
| POTEXP*(IND1 = 13) | 0.020601 | 0.007208 | 2.858126 | **0.0047** |
| POTEXP*(IND1 = 14) | 0.005569 | 0.010230 | 0.544420 | 0.5867 |

| | | | |
|---|---|---|---|
| Pseudo R-squared | 0.083578 | Mean dependent var | 2.236418 |
| Adjusted R-squared | 0.047873 | S.D. dependent var | 0.579965 |
| S.E. of regression | 0.570553 | Objective | 53.85368 |
| Quantile dependent var | 2.251292 | Restr. objective | 58.76516 |
| Sparsity | 1.567234 | Quasi-LR statistic | 25.07083 |
| Prob(quasi-LR stat) | 0.002894 | | |

**Figure 9.36** The output of the HLQR(Median) in (9.24) for the sample {*IND1* > 9}.

3.2 However, I prefer not to present the ANCOVA-QR(Median) because the insignificant differences of the linear effects does not directly mean the five parameters, C(6) to C(10), have the same values in the corresponding population. In addition, the output in Figure 9.37 clearly shows the difference between their effects; two have significant effects, and two have very large $p$-values $> 0.58$.

Quantile slope equality test
Equation: UNTITLED
Specification: LWAGE C @EXPAND(IND1,@DROPFIRST) POTEXP*
@EXPAND(IND1)
Estimated equation quantile tau = 0.5
User-specified test quantiles: 0.1 0.9
Test statistic compares user-specified coefficients: C(6) C(7) C(8) C(9) C(10)

| Test summary | Chi-Sq. statistic | Chi-Sq. d.f. | Prob. |
|---|---|---|---|
| Wald test | 16.31153 | 10 | 0.0911 |

Restriction detail: b(tau_h) − b(tau_k) = 0

| Quantiles | Variable | Restr. value | Std. Error | Prob. |
|---|---|---|---|---|
| 0.1, 0.5 | POTEXP*(IND1 = 10) | −0.007728 | 0.012474 | 0.5355 |
| | POTEXP*(IND1 = 11) | 0.001733 | 0.022491 | 0.9386 |
| | POTEXP*(IND1 = 12) | 0.009239 | 0.012300 | 0.4526 |
| | POTEXP*(IND1 = 13) | −0.025527 | 0.008782 | **0.0037** |
| | POTEXP*(IND1 = 14) | −0.011940 | 0.009884 | 0.2270 |
| 0.5, 0.9 | POTEXP*(IND1 = 10) | 0.007441 | 0.016130 | 0.6445 |
| | POTEXP*(IND1 = 11) | 0.019271 | 0.022810 | 0.3982 |
| | POTEXP*(IND1 = 12) | −0.000954 | 0.019866 | 0.9617 |
| | POTEXP*(IND1 = 13) | 0.020891 | 0.009048 | **0.0209** |
| | POTEXP*(IND1 = 14) | 0.006148 | 0.010320 | 0.5514 |

**Figure 9.37** The quantile slope equality specific for the numerical IVs of the QR(0.5) in Figure 9.36.

Wald test:
Equation: untitled

| Test statistic | Value | df | Probability |
|---|---|---|---|
| F-statistic | 1.401013 | (4, 231) | 0.2344 |
| Chi-square | 5.604054 | 4 | 0.2307 |

Null hypothesis: C(6) = C(7) = C(8) = C(9) = C(10)

**Figure 9.38** The output of the Wald test for testing the linear effects differences of *POTEXP* on *LWAGE* between the five industries, based on the QR in Figure (9.36).

#### 9.4.1.2 A Heterogeneous Third-Degree Polynomial QR of *LWAGE* on *GRADE*

Referring to the third-degree polynomial QR of *LWAGE* on *GRADE* in (9.11), we can present an alternative heterogeneous third-degree polynomial QR of *LWAGE* on one or two selected categorical variables in CPS88.wf1.

#### Example 9.16 *An Application of a Polynomial QR of LWAGE on GRADE*

Figure 9.39 presents the output of a third-degree polynomial QR(0.25) of *LWAGE* on *GRADE* by *PARTT*, using the following ES:

$$LWAGE \; C@Expand\,(PARTT,\,@Dropfirst) \; GRADE*@Expand\,(PARTT)$$

$$GRADE\hat{\;}2*@Expand\,(PARTT) \; GRADE\hat{\;}3*@Expand\,(PARTT) \tag{9.26}$$

Dependent variable: LWAGE
Method: quantile regression (median)
Date: 01/09/19 Time: 05:19
*Sample: 1 1000*
Included observations: 1000
Huber sandwich standard errors and covariance
Sparsity method: Kernel (Epanechnikov) using residuals
Bandwidth method: Hall-Sheather, bw = 0.097156
Estimation successfully identifies unique optimal solution

| Variable | Coefficient | Std. Error | t-Statistic | Prob. |
|---|---|---|---|---|
| C | 1.208960 | 0.164236 | 7.361108 | 0.0000 |
| PARTT = 0 | 0.659399 | 0.393428 | 1.676037 | 0.0940 |
| GRADE*(PARTT = 0) | 0.049781 | 0.112094 | 0.444100 | 0.6571 |
| GRADE*(PARTT = 1) | 0.491167 | 0.178557 | 2.750756 | 0.0061 |
| GRADE^2*(PARTT = 0) | −0.006239 | 0.010966 | −0.568950 | 0.5695 |
| GRADE^2*(PARTT = 1) | −0.069370 | 0.024847 | −2.791924 | 0.0053 |
| GRADE^3*(PARTT = 0) | 0.000368 | 0.000330 | 1.117432 | 0.2641 |
| GRADE^3*(PARTT = 1) | 0.002521 | 0.000850 | 2.965939 | 0.0031 |

| | | | |
|---|---|---|---|
| Pseudo R-squared | 0.139029 | Mean dependent var | 2.275496 |
| Adjusted R-squared | 0.132954 | S.D. dependent var | 0.563464 |
| S.E. of regression | 0.498891 | Objective | 196.5265 |
| Quantile dependent var | 2.302585 | Restr. objective | 228.2616 |
| Sparsity | 1.246633 | Quasi-LR statistic | 203.6527 |
| Prob(quasi-LR stat) | 0.000000 | | |

**Figure 9.39** The output of the QR(Median) in (9.26).

Based on this output, the findings and notes are as follows:

1) Specific for the level $PARTT = 1$, each of the numerical IVs has either a positive or negative significant effect at the 1% level.
2) However, specific for $PARTT = 0$ at the 10% level, each numerical IV has an insignificant effect on $LWAGE$. Even though, I do not reduce the model to test the joint effects difference of the third-degree polynomial QR(Medians)s between the two levels of $PARTT$. Because Figure 9.40 shows the joint effects of the numerical IVs have a significant difference at the 10% level, based on the Wald test ($F$-statistic) of $F_0 = 2.248\,037$ with $df = (3992)$ and $p$-value $= 0.0812$. So, in the statistical sense, the pairs of the third-degree polynomial QR(Median)s can't be reduced to a single third-degree polynomial QR because the dummy variable ($PARTT = 0$) also has a significant effect at the 10% level.
3) As an additional analysis, I try to reduce the third-degree polynomial QR by deleting either $GRADE$ or $GRADE\hat{\ }2$, specific for the level $PARTT = 0$. Deleting $GRADE\hat{\ }2$ obtains a better fit reduced third-degree polynomial QR(Median) since it has a larger adjusted $R$-squared. The ES applied is as follows, with the output presented in Figure 9.41:

$$LWAGE\ C@Expand\,(PARTT,\,@Dropfirst)\ GRADE^*\,(\textbf{\textit{PARTT}} = \textbf{\textit{1}})$$
$$GRADE\hat{\ }2^*@Expand\,(PARTT)\ GRADE\hat{\ }3^*@Expand\,(PARTT) \tag{9.27}$$

4) Even though $GRADE\hat{\ }2^*(PARTT = 0)$ has a large $p$-value, it can be kept as an IV because $GRADE\hat{\ }3^*(PARTT = 0)$ has a significant effect at the 5% level. So, this reduced third-degree polynomial QR(Median) is an acceptable QR.
5) We also can do a more advanced analysis by using the object "*Quantile Process…*". As an illustration, Figure 9.42 presents the output of specific QSETs, based on the QR(0.5) and QR(0.1) in (9.24). From this output, we have the following conclusions:

| Wald test: | | | |
|---|---|---|---|
| Equation: EQ08 | | | |
| Test statistic | Value | df | Probability |
| F-statistic | 2.248037 | (3, 992) | 0.0812 |
| Chi-square | 6.744112 | 3 | 0.0805 |
| Null hypothesis: $C(3) = C(4), C(5) = C(6), C(7) = C(8)$ | | | |

**Figure 9.40** The output of a Wald test for the joint effects difference, based on Eq. (9.27).

Dependent variable: LWAGE
Method: quantile regression (median)
Included observations: 1000
Estimation successfully identifies unique optimal solution

| Variable | Coefficient | Std. Error | t-Statistic | Prob. |
|---|---|---|---|---|
| C | 1.208960 | 0.164324 | 7.357193 | 0.0000 |
| PARTT = 0 | 0.815800 | 0.210029 | 3.884234 | 0.0001 |
| GRADE*(PARTT = 1) | 0.491167 | 0.178595 | 2.750166 | 0.0061 |
| GRADE^ 2*(PARTT = 0) | −0.001368 | 0.002200 | −0.621941 | 0.5341 |
| GRADE^ 2*(PARTT = 1) | −0.069370 | 0.024852 | −2.791371 | 0.0053 |
| GRADE^ 3*(PARTT = 0) | 0.000224 | 0.000107 | 2.101682 | 0.0358 |
| GRADE^ 3*(PARTT = 1) | 0.002521 | 0.000850 | 2.965359 | 0.0031 |

| | | | | |
|---|---|---|---|---|
| Pseudo R-squared | 0.138691 | Mean dependent var | | 2.275496 |
| Adjusted R-squared | 0.133487 | S.D. dependent var | | 0.563464 |
| S.E. of regression | 0.498757 | Objective | | 196.6038 |
| Quantile dependent var | 2.302585 | Restr. objective | | 228.2616 |
| Sparsity | 1.256583 | Quasi-LR statistic | | 201.5484 |
| Prob(quasi-LR stat) | 0.000000 | | | |

**Figure 9.41**   The output of the QR(Median) in (9.27).

5.1 Even though the test quantiles 0.1 and 0.9 are applied, the output in Figure 9.42a shows 12 pairs of slopes between the two pairs of quantiles (0.1, 0.5) and (0.5, 0.9). Based on the *Chi-Square* statistic of $\chi_0^2 = 164.2875$ with $df = 12$ and a $p$-value $= 0.0000$, we can conclude the 12 pairs of slopes have significant differences. However, at the 1% level, only three of the numerical variables— *GRADE\*(PARTT = 1)*, *GRADE^ 2\*(PARTT = 1)*, and *GRADE^ 3\*(PARTT = 1)*—have significant effect differences between the quantiles 0.1 and 0.5.

5.2 Compared with the QSET in Figure 9.42a, Figure 9.42b presents a specific QSET for only the quantiles 0.1 and 0.9, based on QR(0.1) in (9.26). Based on the *Chi-Square* statistic of $\chi_0^2 = 124.6221$ with $df = 6$ and a $p$-value $= 0.0000$, we can conclude that the slopes of the six IVs have significant differences between the quantiles 0.1 and 0.9. However, at the 10% level of significance, each numerical IV has an insignificant slopes difference between the pairs of quantiles.

(a)

Quantile slope equality test

Equation: EQ08_RD1

Specification: LWAGE C (PARTT = 0) GRADE*(PARTT = 1)
  GRADE^2*@EXPAND(PARTT) GRADE^3*@EXPAND(PARTT)

Estimated equation quantile tau = 0.5

User-specified test quantiles: 0.1 0.9

Test statistic compares all coefficients

| Test summary | Chi-Sq. statistic | Chi-Sq. d.f. | Prob. |
|---|---|---|---|
| Wald test | 164.2875 | 12 | 0.0000 |

Restriction detail: b(tau_h) − b(tau_k) = 0

| Quantiles | Variable | Restr. value | Std. Error | Prob. |
|---|---|---|---|---|
| 0.1, 0.5 | PARTT = 0 | −0.631045 | 0.259628 | 0.0151 |
| | GRADE*(PARTT = 1) | −0.451163 | 0.173266 | **0.0092** |
| | GRADE^2*(PARTT = 0) | 0.001249 | 0.003653 | 0.7325 |
| | GRADE^2*(PARTT = 1) | 0.062531 | 0.024190 | **0.0097** |
| | GRADE^3*(PARTT = 0) | −0.000102 | 0.000182 | 0.5750 |
| | GRADE^3*(PARTT = 1) | −0.002230 | 0.000832 | **0.0073** |
| 0.5, 0.9 | PARTT = 0 | −0.654234 | 0.379680 | 0.0849 |
| | GRADE*(PARTT = 1) | 0.177496 | 0.259614 | 0.4942 |
| | GRADE^2*(PARTT = 0) | 0.001989 | 0.005575 | 0.7213 |
| | GRADE^2*(PARTT = 1) | −0.044117 | 0.036956 | 0.2326 |
| | GRADE^3*(PARTT = 0) | −0.000117 | 0.000264 | 0.6575 |
| | GRADE^3*(PARTT = 1) | 0.001857 | 0.001289 | 0.1498 |

**Figure 9.42** (a) The output of a QSET of the quantile 0.1 and 0.9, based on QR(0.5). (b) The output of a QSET of the quantile 0.1 and 0.9, based on QR(0.1).

(b)
Quantile slope equality test
Equation: EQ08_RD2
Specification: LWAGE C (PARTT = 0) GRADE*@EXPAND(PARTT)
GRADE^2*(PARTT = 1) GRADE^3*@EXPAND(PARTT)
Estimated equation quantile tau = 0.1
User-specified test quantiles: 0.1 0.9
Test statistic compares all coefficients

| Test summary | Chi-Sq. statistic | Chi-Sq. d.f. | Prob. |
|---|---|---|---|
| Wald test | 124.6221 | 6 | 0.0000 |

Restriction detail: b(tau_h) − b(tau_k) = 0

| Quantiles | Variable | Restr. value | Std. Error | Prob. |
|---|---|---|---|---|
| 0.1, 0.9 | PARTT = 0 | −1.322306 | 0.760991 | 0.0823 |
| | GRADE^2*(PARTT = 1) | 0.018414 | 0.037819 | 0.6263 |
| | GRADE*(PARTT = 0) | 0.024477 | 0.084936 | 0.7732 |
| | GRADE*(PARTT = 1) | −0.273667 | 0.263710 | 0.2994 |
| | GRADE^3*(PARTT = 0) | −0.000100 | 0.000152 | 0.5076 |
| | GRADE^3*(PARTT = 1) | −0.000373 | 0.001327 | 0.7784 |

**Figure 9.42** *(continued)*

### 9.4.1.3 An Application of QR for a Large Number of Groups

**Example 9.17** *An HLQR(Median) of LWAGE on POTEXP by Two Factors*
Figure 9.43 presents the output summary of a heterogeneous linear QR(Median), or
HLQR(Median), of *LWAGE* on *POTEXP* by two categorical factors *HIGH* and *IND1*,
with $2 \times 14$ groups of firms, using following ES:

$$lwage\ c\ potexp\ @expand\,(high, ind1, @drop\,(1, 1))$$
$$potexp*@expand\,(high, ind1, @drop\,(1, 1)) \tag{9.28}$$

Because the 14 levels of *IND1* are classified as the two levels of *HIGH*, the output presents
only 14 groups of industries. Based on this summary, the following findings and notes are
presented:

1) The output in Figure 9.43 shows *LWAGE* = C(1) + C(2)*POTEXP is the QR(Median) in
   the first group (*HIGH* = 0, *IND1* = 1) or the level *IND1* = 1, corresponding to the used
   of the function *@Drop(1,1)* in the ES (9.28). Hence, the *t*-statistic in the output and the
   RVT can easily be applied for testing specific hypotheses, such as follows.

| Variable | Coef. | t-Stat. | Prob. | Variable | Coef. | t-Stat. | Prob. |
|---|---|---|---|---|---|---|---|
| C | 2.310 | 5.917 | 0.000 | | | | |
| POTEXP | 0.006 | 0.395 | 0.693 | | | | |
| HIGH = 0,IND1 = 6 | 0.253 | 0.614 | 0.539 | POTEXP*(HIGH = 0,IND1 = 6) | −0.001 | −0.084 | 0.933 |
| HIGH = 0,IND1 = 7 | −0.513 | −1.203 | 0.229 | POTEXP*(HIGH = 0,IND1 = 7) | 0.019 | 1.092 | 0.275 |
| HIGH = 0,IND1 = 8 | −0.858 | −2.163 | 0.031 | POTEXP*(HIGH = 0,IND1 = 8) | 0.023 | 1.396 | **0.163** |
| HIGH = 0,IND1 = 9 | 0.298 | 0.680 | 0.497 | POTEXP*(HIGH = 0,IND1 = 9) | −0.002 | −0.131 | 0.896 |
| HIGH = 0,IND1 = 12 | −0.158 | −0.327 | 0.744 | POTEXP*(HIGH = 0,IND1 = 12) | −0.012 | −0.662 | 0.508 |
| HIGH = 0,IND1 = 13 | −0.520 | −1.250 | 0.212 | POTEXP*(HIGH = 0,IND1 = 13) | 0.014 | 0.866 | 0.387 |
| HIGH = 1,IND1 = 2 | 0.128 | 0.320 | 0.749 | POTEXP*(HIGH = 1,IND1 = 2) | −0.009 | −0.580 | 0.562 |
| HIGH = 1,IND1 = 3 | −0.315 | −0.733 | 0.464 | POTEXP*(HIGH = 1,IND1 = 3) | 0.009 | 0.545 | 0.586 |
| HIGH = 1,IND1 = 4 | −0.407 | −1.013 | 0.311 | POTEXP*(HIGH = 1,IND1 = 4) | 0.013 | 0.765 | 0.445 |
| HIGH = 1,IND1 = 5 | −0.266 | −0.623 | 0.533 | POTEXP*(HIGH = 1,IND1 = 5) | 0.008 | 0.455 | 0.649 |
| HIGH = 1,IND1 = 10 | −0.623 | −1.257 | 0.209 | POTEXP*(HIGH = 1,IND1 = 10) | 0.024 | 1.170 | 0.242 |
| HIGH = 1,IND1 = 11 | −0.404 | −0.881 | 0.379 | POTEXP*(HIGH = 1,IND1 = 11) | 0.016 | 0.594 | 0.553 |
| HIGH = 1,IND1 = 14 | 0.082 | 0.181 | 0.856 | POTEXP*(HIGH = 1,IND1 = 14) | −0.001 | −0.029 | 0.977 |
| Pseudo R-sq | 0.148 | | | Mean dependent var | 2.275 | | |
| Adjusted R-sq | 0.124 | | | S.D. dependent var | 0.563 | | |
| S.E. of reg | 0.510 | | | Objective | 194.6 | | |
| Quantile DV | 2.303 | | | Restr. objective | 228.3 | | |
| Sparsity | 1.199 | | | Quasi-LR statistic | 224.8 | | |
| Prob(QLR stat) | 0.000 | | | | | | |

**Figure 9.43** The output summary of the HLQR(Median) in (9.28).

2) The output shows that the linear effect of *POTEXP* between each of last 13 groups and the first group *(HIGH = 0,IND1 = 1)* has insignificant difference, based on the *t*-statistic with *p*-value = Prob. > 0.10, shown in the last column in Figure 9.43.

3) However, at the 1% level, the output of the RVT in Figure 9.44 shows that *POTEXP* has significant different effects between the 14 groups of firms, based on the *QLR L*-statistic of $QLR_0 = 37.934\,63$ with $df = 13$ and *p*-value = 0.0003. So, the QR(Median) (9.28) can't be reduced to an ANCOVA-QR(Median) with the following ES:

$$lwage\ c\ potexp\ @expand\ (high, ind1, @drop\ (1, 1)) \tag{9.29}$$

4) The output of the RVT can be obtained easily with the following steps:

4.1 With the output of the QR(Median) in (9.28) on-screen, select *View/ Coefficient Diagnostics/Redundant Variables-Likelihood Ratio* and then select the block shown on-screen for the RVT.

4.2 Next, we can insert the series or function *potexp*@expand(high,ind1,@drop(1,1))* using the *block-copy-paste* method and clicking *OK* to obtain the complete RVT output.

Redundant variables test
Equation: (9.28)
Redundant variables: POTEXP*@EXPAND(HIGH,IND1,
@DROP(1,1))
Specification: LWAGE C POTEXP @EXPAND(HIGH,IND1,
@DROP(1,1)) POTEXP*@EXPAND(HIGH,IND1,@DROP(1,1))
Null hypothesis: POTEXP*@EXPAND(HIGH,IND1,
@DROP(1,1)) are jointly insignificant

|                    | Value    | df | Probability |
|--------------------|----------|----|-------------|
| QLR L-statistic    | 37.93463 | 13 | 0.0003      |
| QLR Lambda-statistic | 37.39113 | 13 | 0.0004    |

**Figure 9.44** The part of the RVT's output based of the HLQR(Median) in (9.27).

4.3 Note that the output presents a specific null hypothesis, by using EViews 10, as presented in Figure 9.44. But by using EViews 8, the output presents different type of null hypothesis, as presented in Figures 9.46.

**Example 9.18** *A Heterogeneous Interaction QR(Median) of LWAGE on POTEXP and GRADE by HIGH and IND1*

As an extension of the QR(Median) in Figure 9.43, Figure 9.45 presents the output summary of a heterogeneous interaction QR(Median), or HIQR(Median), of *LWAGE* on *POTEXP* and *GRADE* by *HIGH* and *IND1,* using the following ES:

$$lwage\ c\ potexp\ grade\ potexp*grade@expand\ (high, ind1, @drop\ (1, 1))\ potexp*$$
$$@expand\ (high, ind1, @drop\ (1, 1))\ grade*@expand\ (high, ind1, @drop\ (1, 1))$$
$$potexp*grade*@expand\ (high, ind1, @drop\ (1, 1)) \tag{9.30}$$

Based this summary, the findings and notes are as follows:

1) Similar to the previous example, the QR in the first group *(HIGH = 0,IND1 = 1)*, as the reference QR has the following regression function, which show that the interaction *POTEXP*GRADE* has significant effect at the 5% level. Hence this regression is an acceptable two-way interaction QR(Median), even though the main variable *GRADE* has a large p-value.

$$\widehat{LWAGE} = 2.718 - 0.106*POTEXP - 0.052*GRADE + 0.010*POTEXP*GRADE$$

2) The t-test statistic of each numerical IV in the output can be easily used to test a one- or two-sided hypothesis against the reference QR. For instance, at the 5% level, it can be concluded that the interaction *P*G*(HIGH = 0,IND1 = 6)* has a significant difference with *P*G*(HIGH = 0,IND1 = 1)*, based on the *t*-statistic of $t_0 = -2.154$ with *p*-value = 0.032.

| Variable | Coef. | t-Stat. | Prob. | Varible | Coef. | t-Stat. | Prob. |
|---|---|---|---|---|---|---|---|
| C | 2.718 | 1.883 | 0.060 | | | | |
| POTEXP | −0.106 | −1.669 | 0.096 | | | | |
| GRADE | −0.052 | −0.528 | 0.598 | | | | |
| POTEXP*GRADE | 0.010 | 2.289 | 0.022 | | | | |
| HIGH = 0,IND1 = 6 | −1.376 | −0.827 | 0.408 | G*(HIGH = 0,IND1 = 6) | 0.135 | 1.157 | 0.248 |
| HIGH = 0,IND1 = 7 | −2.019 | −1.114 | 0.266 | G*(HIGH = 0,IND1 = 7) | 0.128 | 0.960 | 0.338 |
| HIGH = 0,IND1 = 8 | −1.589 | −1.066 | 0.287 | G*(HIGH = 0,IND1 = 8) | 0.082 | 0.779 | 0.436 |
| HIGH = 0,IND1 = 9 | −3.240 | −1.971 | 0.049 | G*(HIGH = 0,IND1 = 9) | 0.244 | 2.133 | 0.033 |
| HIGH = 0,IND1 = 12 | −3.974 | −2.222 | 0.027 | G*(HIGH = 0,IND1 = 12) | 0.298 | 2.299 | 0.022 |
| HIGH = 0,IND1 = 13 | −3.605 | −2.380 | 0.018 | G*(HIGH = 0,IND1 = 13) | 0.267 | 2.540 | 0.011 |
| HIGH = 1,IND1 = 2 | −2.582 | −1.703 | 0.089 | G*(HIGH = 1,IND1 = 2) | 0.203 | 1.921 | 0.055 |
| HIGH = 1,IND1 = 3 | −2.380 | −1.569 | 0.117 | G*(HIGH = 1,IND1 = 3) | 0.176 | 1.686 | 0.092 |
| HIGH = 1,IND1 = 4 | −1.199 | −0.687 | 0.492 | G*(HIGH = 1,IND1 = 4) | 0.084 | 0.650 | 0.516 |
| HIGH = 1,IND1 = 5 | −1.271 | −0.787 | 0.431 | G*(HIGH = 1,IND1 = 5) | 0.107 | 0.952 | 0.341 |
| HIGH = 1,IND1 = 10 | −2.619 | −1.315 | 0.189 | G*(HIGH = 1,IND1 = 10) | 0.155 | 1.193 | 0.233 |
| HIGH = 1,IND1 = 11 | −2.823 | −1.395 | 0.163 | G*(HIGH = 1,IND1 = 11) | 0.195 | 1.283 | 0.200 |
| HIGH = 1,IND1 = 14 | −2.152 | −1.290 | 0.197 | G*(HIGH = 1,IND1 = 14) | 0.171 | 1.506 | 0.132 |
| P*(HIGH = 0,IND1 = 6) | 0.117 | 1.743 | 0.082 | P*G*(HIGH = 0,IND1 = 6) | −0.010 | −2.154 | 0.032 |
| P*(HIGH = 0,IND1 = 7) | 0.096 | 1.188 | 0.235 | P*G*(HIGH = 0,IND1 = 7) | −0.007 | −1.174 | 0.241 |
| P*(HIGH = 0,IND1 = 8) | 0.099 | 1.458 | 0.145 | P*G*(HIGH = 0,IND1 = 8) | −0.007 | −1.478 | 0.140 |
| P*(HIGH = 0,IND1 = 9) | 0.140 | 2.016 | 0.044 | P*G*(HIGH = 0,IND1 = 9) | −0.011 | −2.250 | 0.025 |
| P*(HIGH = 0,IND1 = 12) | 0.139 | 1.971 | 0.049 | P*G*(HIGH = 0,IND1 = 12) | −0.013 | −2.484 | 0.013 |
| P*(HIGH = 0,IND1 = 13) | 0.234 | 3.298 | 0.001 | P*G*(HIGH = 0,IND1 = 13) | −0.019 | −3.752 | 0.000 |
| P*(HIGH = 1,IND1 = 2) | 0.138 | 2.125 | 0.034 | P*G*(HIGH = 1,IND1 = 2) | −0.011 | −2.521 | 0.012 |
| P*(HIGH = 1,IND1 = 3) | 0.120 | 1.863 | 0.063 | P*G*(HIGH = 1,IND1 = 3) | −0.009 | −2.128 | 0.034 |
| P*(HIGH = 1,IND1 = 4) | 0.100 | 1.424 | 0.155 | P*G*(HIGH = 1,IND1 = 4) | −0.008 | −1.542 | 0.123 |
| P*(HIGH = 1,IND1 = 5) | 0.121 | 1.837 | 0.067 | P*G*(HIGH = 1,IND1 = 5) | −0.010 | −2.229 | 0.026 |
| P*(HIGH = 1,IND1 = 10) | 0.128 | 1.751 | 0.080 | P*G*(HIGH = 1,IND1 = 10) | −0.009 | −1.913 | 0.056 |
| P*(HIGH = 1,IND1 = 11) | 0.136 | 1.828 | 0.068 | P*G*(HIGH = 1,IND1 = 11) | −0.010 | −1.895 | 0.058 |
| P*(HIGH = 1,IND1 = 14) | 0.131 | 1.728 | 0.084 | P*G*(HIGH = 1,IND1 = 14) | −0.011 | −2.140 | 0.033 |
| Pseudo R-squared | 0.297 | | | Mean dependent var | 2.275 | | |
| Adjusted R-squared | 0.256 | | | S.D. dependent var | 0.563 | | |
| S.E. of regression | 0.434 | | | Objective | 160.57 | | |
| Quantile dv | 2.303 | | | Restr. objective | 228.26 | | |
| Sparsity | 0.922 | | | Quasi-LR statistic | 587.30 | | |
| Prob | 0.000 | | | | | | |

(*) P = POTEXP and G = GRADE

**Figure 9.45** Presents the output summary of the QR(Median) in (9.30).

Redundant variables test
Null hypothesis:
POTEXP*GRADE*@EXPAND(HIGH,IND1,@DROP(1,1))
Equation: (9.29)
Specification: LWAGE C POTEXP GRADE POTEXP*GRADE
 @EXPAND(HIGH,IND1,@DROP(1,1)) POTEXP
 *@EXPAND(HIGH,IND1,@DROP(1,1)) GRADE
 *@EXPAND(HIGH,IND1,@DROP(1,1)) POTEXP*GRADE
 *@EXPAND(HIGH,IND1,@DROP(1,1))

|  | Value | df | Probability |
| --- | --- | --- | --- |
| QLR L-statistic | 24.72197 | 13 | 0.0251 |
| QLR Lambda-statistic | 24.50519 | 13 | 0.0268 |

**Figure 9.46** A part of the RVT for testing the interaction *POTEXP*GRADE* between the 14 firm groups, using EViews 8.

3) Figure 9.46 presents the output of a RVT for testing the interaction effects differences between the 14 firm groups, with the null hypothesis as presented in the output. We can conclude that the interaction has significant differences between the 14 first groups, based on the *QLR L*-statistic of $QLR_0 = 24.721\,97$ with $df = 13$ and p-value = 0.0251.
4) As an additional illustration, with the output of the QR(Median) in (9.30) on-screen, select the Forecast button, insert the name: *LWAGEF_Eq28,* and click *OK* to make the forecast variable directly available in the file.
5) Finally, based the two variables *LWAGE* and *LWAGEF_EQ28,* we can develop their line and symbol graphs of the 1000 observations. However, Figure 9.47 presents their graphs for only the last 29 observations, sorted by the *LWAGE.* So, we can clearly see their

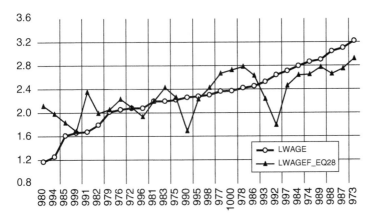

**Figure 9.47** A part of graphs of the sorted *LWAGE* and its forecast variable *LWAGE_EQ28.*

deviations. Note that, based on each observed or measured score of *LWAGE,* we have a single predicted median, which is an abstract score because it can be measured or estimated only by using a specific statistical method.

#### 9.4.1.4 Comparison Between Selected Heterogeneous QR(Median)

As an comparison, Table 9.4 presents four good-of-fit (GOF) measures for the five selected heterogeneous QR(Median)s, with the Eqs. (9.24), (9.26)–(9.30), which show the QR in (9.30) is the best, in the statistical sense, since it has the largest adjusted *R*-squared and the smallest TIC among the five heterogeneous QR(Median)s. As an additional note, a TIC of 0 indicates the forecast is the perfect forecast, and a TIC of 1(one) is the worst.

The deviation scores of each forecast variable from the observed scores of *LWAGE* can be computed easily. However, as an illustration, Figure 9.48 presents only the line and symbol graphs of the sorted *LWAGE* and its five forecast variables for the last 29 observations. Based on these graphs, the following findings and notes are presented:

1) These graphs clearly show the deviations of forecast values from the observed scores of *LWAGE* compared with the graphs for the 1000 observations.
2) Figure 9.48a presents the graphs of sorted *LWAGE,* with the worst and the best forecast value variables, namely *LWAGEF_Eq22* and *LWAGEF_Eq28,* among the five

**Table 9.4** Good-of-fit measures for five selected QR(Median)s.

| GOF measure | Eq. (9.24) | Eq. (9.26) | Eq. (9.27) | Eq. (9.28) | Eq. (9.30) |
|---|---|---|---|---|---|
| Pseudo R-squared | 0.083578 | 0.139029 | 0.138691 | 0.147557 | 0.298543 |
| Adjusted R-squared | 0.047873 | 0.132954 | 0.133487 | 0.123578 | 0.255558 |
| Root mean square error | 0.502856 | 0.496891 | 0.497086 | 0.493542 | 0.42154 |
| Theil inequality coefficient | 0.108248 | 0.107344 | 0.107082 | 0.106464 | 0.090448 |

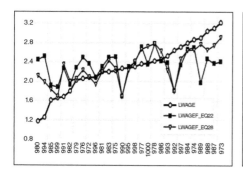 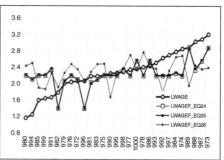

**Figure 9.48** Graphical presentation of *LWAGE* and its five forecast variables for the last 29 observations.

heterogeneous QRs. The scores of *LWAGEF_Eq28* have smaller deviations from the observed scores of *LWAGE* than the score of *LWAGEF_Eq22*.

3) Figure 9.48b shows the scores of *LWAGEF_Eq24* and *LWAGEF_Eq25* coincide. In fact, they are the forecast values variables of the third-degree polynomial heterogeneous QR(Median) and its reduced QR in (9.26) and (9.27), respectively. Hence, these findings show a reduced QR is as good as its full model.

# 10

# Quantile Regressions of a Latent Variable

## 10.1   Introduction

The data used in this chapter is in BBM.wf1, which contains several baby birth measurements—such as *BBW* (baby birth weight), *FUNDUS*, and *MUAC* (mid-upper arm circumference)—and mother indicators or variables. The data points are listed in Figure 10.1, and include ordinal, dummy, and numerical variables. The data file is provided by Dr. Lilis Heri Miscicih (graduated from the School of Public Health, University of Indonesia, and one of my advisories), and after graduated she is a part-time lecturer in the School of Public Health.

Similar to all types of quantile regressions (QRs) presented in previous chapters, we can easily develop QRs using each baby birth indicator, namely *BBW*, *FUNDUS*, and *MUAC*, as a dependent variable (DV). So we can have three sets of QRs of each baby birth indicator as a DV having various alternative independent variables (IVs) in BBW.wf1. For instance, we can easily develop a set of QRs of *FUNDUS* on various selected sets of the categorical or/and numerical variables of the mother indicators.

In this case, we will develop a baby latent variable (*BLV*) based on the three baby birth indicators, *BBW*, *FUNDUS*, and *MUAC*, and apply various QRs of *BLV* on selected mother indicators. We will take this approach instead of having three sets of QRs, each for *BBW*, *FUNDUS*, and *MUAC*, since there is no a multivariate QR. In addition, we will also develop a mother latent variable (*MLV*) based on the two ordinal variables *ED* (education level) and *SE* (socioeconomic) and two dummy variables, *SMOKING* (defined as *SMOKING* = 1 for the smoking mothers) and *NO_ANC* (defined as *NO_ANC* = 1 for the mothers who don't have antenatal care, or *ANC*).

## 10.2   Spearman-rank Correlation

As a preliminary analysis, Figure 10.2 presents the Spearman correlation matrix (*View/Covarince Analysis, Method: Spearman rank-order*) of *BBM* and the nine mother indicators or variables. Based on this matrix, the following findings and notes are presented:

1. In the parametric statistic, the *t*-statistic presented in the output can be used to test a hypothesis on the causal or *up-and-downward relationships* between pairs of the corresponding variables.

*Quantile Regression: Applications on Experimental and Cross Section Data Using EViews,* First Edition.
I Gusti Ngurah Agung.

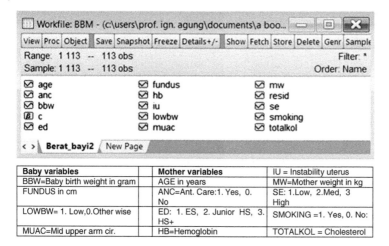

**Figure 10.1** The list of variables in BBM.wf1.

2. The first column of the matrix correlation shows the effect of each of the nine mother variables or indicators on *BLV*. Based on the outputs, I'll present only the following selected findings, in the parametric sense. Do the others as exercises.

    2.1 At the 10% level of significance, *AGE* has a significant negative effect on *BLV* based on the *t*-statistic of $t_0 = -1.441\,799$ with $df = (113 - 2)$ and *p*-value $= 0.1522/2 = 0.0761$. This case shows that increasing scores of *AGE* correspond with decreasing scores of *BLV*. On the other hand, we find that *AGE* has a significant negative effect on *BBW* based on the *t*-statistic of $t_0 = -2.047\,20$ with $df = (113 - 2)$ and *p*-value $= 0.0430/2 = 0.0215$.

    2.2 Similarly, instability uterus (*IU*) has a significant negative effect on *BLV* based on the *t*-statistic of $t_0 = -1.488\,944$ with $df = (113 - 2)$ and *p*-value $= 0.1393/2 = 0.069\,65$.

    2.3 The correlation (*BLV,ANC*) is a biserial correlation since *ANC* is a zero–one or dummy variable. Based on the *t*-statistic of $t_0 = 0.209\,454$ with $df = 111$ and *p*-value 0.8345, we can conclude that the means of *BLV* have insignificant difference between *ANC* $= 1$ (the group of mothers having antenatal care, or *ANC*), and *ANC* $= 0$. However, for the quantiles of *BLV*, we can arrive at different conclusions. See the following examples.

    2.4 The dummy variable *SMOKING,* with *SMOKING* $= 1$ for the smoking mothers, has a significant positive correlation with the variables *ED* and *SE*, with *p*-values of 0.0001 and 0.0039, respectively. Hence, we can conclude that the group of smoking mothers tends to have higher education and socioeconomic conditions.

3. The steps of the analysis to obtain the Spearman Rank Order matrix correlation are as follows:

    3.1 Having the data of BBM.wf1 on screen, show BBW and the nine mother indicators on-screen.

    3.2 Select *View/Covaraince Analysis …,* the dialog for doing the covariance analysis appears on screen.

    3.3 Select the *Spearman order rank method, Correlation, t-statistic,* and *Probability, … OK,* the matrix correlation in Figure 10.2 obtained.

Covariance Analysis: Spearman rank-order
Date: 03/15/19 Time: 10:17
Sample: 1 113
Included observations: 113

| Correlation<br>t-Statistic<br>Probability | BLV | AGE | ANC | ED | HB | IU | MW | SE | SMOKING |
|---|---|---|---|---|---|---|---|---|---|
| AGE | -0.135586<br>-1.441799<br>0.1522 | 1.000000<br>-----<br>----- | | | | | | | |
| ANC | 0.019877<br>0.209454<br>0.8345 | -0.106855<br>-1.119405<br>0.2654 | 1.000000<br>-----<br>----- | | | | | | |
| ED | 0.214447<br>2.313153<br>0.0228 | 0.116552<br>1.236375<br>0.2189 | -0.217022<br>-2.342298<br>0.0209 | 1.000000<br>-----<br>----- | | | | | |
| HB | 0.195868<br>2.104359<br>0.0378 | -0.068468<br>-0.723053<br>0.4712 | -0.172854<br>-1.848964<br>0.0671 | 0.335261<br>3.749172<br>0.0003 | 1.000000<br>-----<br>----- | | | | |
| IU | -0.139934<br>-1.489044<br>0.1393 | -0.010475<br>-0.110371<br>0.9123 | 0.106279<br>1.126068<br>0.2626 | -0.121942<br>-1.294399<br>0.1982 | -0.051682<br>-0.545229<br>0.5867 | 1.000000<br>-----<br>----- | | | |
| MW | 0.348470<br>3.916949<br>0.0002 | -0.048841<br>-0.494042<br>0.6223 | 0.230443<br>2.495025<br>0.0141 | -0.099768<br>-1.058375<br>0.2931 | -0.197221<br>-2.116480<br>0.0383 | -0.076559<br>-0.840873<br>0.4022 | 1.000000<br>-----<br>----- | | |
| SE | 0.138202<br>1.470151<br>0.1443 | 0.042336<br>0.446439<br>0.6561 | -0.219573<br>-2.371217<br>0.0195 | 0.296000<br>3.264860<br>0.0015 | 0.199972<br>2.150267<br>0.0337 | -0.019452<br>-0.204975<br>0.8380 | -0.151833<br>-1.618425<br>0.1084 | 1.000000<br>-----<br>----- | |
| SMOKING | 0.167785<br>1.793140<br>0.0757 | 0.146619<br>1.561808<br>0.1212 | -0.319789<br>-3.555912<br>0.0006 | 0.366380<br>4.148513<br>0.0001 | 0.445997<br>5.249928<br>0.0000 | -0.006791<br>-0.071548<br>0.9431 | -0.689362<br>-10.02583<br>0.0000 | 0.269507<br>2.948537<br>0.0039 | 1.000000<br>-----<br>----- |
| TOTALKOL | 0.281118<br>3.086216<br>0.0026 | -0.065833<br>-0.695103<br>0.4884 | 0.220821<br>2.385382<br>0.0188 | 0.073166<br>0.772924<br>0.4412 | 0.071064<br>0.750607<br>0.4545 | -0.466928<br>-5.563063<br>0.0000 | 0.190074<br>2.046416<br>0.0431 | 0.028187<br>0.275996<br>0.7831 | -0.087665<br>-0.927177<br>0.3558 |

**Figure 10.2** The output of covariance analysis, Spearman rank-order.

## 10.3  Applications of ANOVA-QR($\tau$)

Many alternative ANOVA-QR($\tau$)s, for each $\tau \in (0, 1)$, can easily be applied for the *BLV*, using all categorical mother indicators and dummy variables. However, for the one-way ANOVA-QR, I propose a specific one-way ANOVA-QR, using the transformed numerical mother indicators.

### 10.3.1  One-way ANOVA-QR of *BLV*

**Example 10.1**  *ANOVA-QR of BLV on AGE as Categorical Variable*
Figure 10.3 presents the output summary of the one-way ANOVA-QR($\tau$) for $\tau = 0.1$; 0.5; and 0.9. The outputs are obtained using the following equation specification (ES), in which

| Variable | One-way ANOVA-Q(0.1) Coef. | t-Stat | Prob. | One-way ANOVA-Q(0.5) Coef. | t-Stat | Prob. | One-way ANOVA-Q(0.9) Coef. | t-Stat | Prob. |
|---|---|---|---|---|---|---|---|---|---|
| C | −0.6676 | −3.4289 | 0.001 | 0.0448 | 0.3097 | 0.758 | 0.4959 | 2.9419 | 0.004 |
| AGE=19 | 0.3420 | 1.4667 | 0.146 | −0.3096 | −1.1883 | 0.238 | −0.4939 | −2.1661 | 0.033 |
| AGE=20 | 0.3659 | 1.4050 | 0.163 | 0.0479 | 0.1628 | 0.871 | 0.8788 | 2.2546 | 0.027 |
| AGE=21 | −0.2924 | −0.7865 | 0.434 | 0.1809 | 0.3113 | 0.756 | 0.2272 | 0.6074 | 0.545 |
| AGE=22 | −0.0933 | −0.1790 | 0.858 | 0.1835 | 0.6359 | 0.527 | 0.6035 | 1.6563 | 0.101 |
| AGE=23 | 0.0916 | 0.3780 | 0.706 | −0.1768 | −0.6820 | 0.497 | −0.1146 | −0.5086 | 0.612 |
| AGE=24 | 0.6522 | 1.9086 | 0.060 | 0.3668 | 0.7709 | 0.443 | 0.5215 | 1.3192 | 0.190 |
| AGE=25 | 0.0857 | 0.1892 | 0.850 | 0.4601 | 1.5013 | 0.137 | 0.4865 | 1.5777 | 0.118 |
| AGE=26 | −0.0924 | −0.2802 | 0.780 | −0.0942 | −0.4034 | 0.688 | −0.4855 | −2.3896 | 0.019 |
| AGE=27 | −0.0632 | −0.2556 | 0.799 | −0.1918 | −0.2384 | 0.812 | 0.0613 | 0.1922 | 0.848 |
| AGE=28 | −0.1210 | −0.2318 | 0.817 | −0.1770 | −0.7416 | 0.460 | −0.1409 | −0.6080 | 0.545 |
| AGE=29 | 0.0577 | 0.2377 | 0.813 | −0.3732 | −1.2594 | 0.211 | 0.2356 | 0.5283 | 0.599 |
| AGE=30 | 0.4675 | 0.9572 | 0.341 | 0.4118 | 1.5861 | 0.116 | 0.4530 | 1.4205 | 0.159 |
| AGE=31 | −0.2113 | −0.5510 | 0.583 | −0.0782 | −0.2160 | 0.830 | 0.7311 | 1.3714 | 0.174 |
| AGE=32 | −0.4438 | −1.0711 | 0.287 | −0.4559 | −1.5267 | 0.130 | −0.6526 | −2.4366 | 0.017 |
| AGE=33 | 0.5607 | 2.4050 | 0.018 | −0.0180 | −0.0755 | 0.940 | −0.1798 | −0.8080 | 0.421 |
| AGE=34 | 0.6276 | 2.3499 | 0.021 | −0.0847 | −0.2296 | 0.819 | −0.5358 | −2.0103 | 0.047 |
| AGE=36 | −0.0962 | −0.3603 | 0.719 | −0.8086 | −2.1925 | 0.031 | −1.2597 | −4.7262 | 0.000 |
| AGE=37 | −0.6752 | −2.5279 | 0.013 | −1.3875 | −3.7624 | 0.000 | −1.8387 | −6.8983 | 0.000 |
| AGE=38 | 0.5561 | 2.0820 | 0.040 | −0.1562 | −0.4236 | 0.673 | −0.6074 | −2.2788 | 0.025 |
| AGE=41 | 1.3316 | 4.9855 | 0.000 | 0.6193 | 1.6792 | 0.097 | 0.1681 | 0.6308 | 0.530 |
| AGE=42 | 0.2033 | 0.6281 | 0.532 | 0.3088 | 0.6161 | 0.539 | −0.1424 | −0.4224 | 0.674 |
| Pse. R^2 | 0.2851 | | | 0.2065 | | | 0.3471 | | |
| Adj. R^2 | 0.1201 | | | 0.0234 | | | 0.1965 | | |
| S.E. of reg | 0.8846 | | | 0.5357 | | | 0.8854 | | |
| Quant DV | −0.6676 | | | −0.0129 | | | 0.6745 | | |
| Sparsity | 1.7847 | | | 1.2099 | | | 1.7259 | | |
| QLR stat | 36.774 | | | 33.317 | | | 51.576 | | |
| Prob | 0.0179 | | | 0.0428 | | | 0.0002 | | |

**Figure 10.3**  The output summary of one-way ANOVA-QR($\tau$) in (10.1) for $\tau = 0.1$; 0.5; and 0.9.

Quantile Slope Equality Test

Equation: EQ01_AGE

Specification: BLV C @EXPAND(AGE,@DROPFIRST)

Estimated equation quantile tau = 0.1

User-specified test quantiles: 0.1, 0.9

Test statistic compares user-specified coefficients: C(5) C(20)

| Test summary | Chi-Sq. statistic | Chi-Sq. d.f. | Prob. |
|---|---|---|---|
| Wald test | 15.26236 | 2 | 0.0005 |

Restriction detail: b(tau_h) − b(tau_k) = 0

| Quantiles | Variable | Restr. Value | Std. Error | Prob. |
|---|---|---|---|---|
| 0.1, 0.9 | AGE = 22 | −0.696796 | 0.602251 | 0.2473 |
| | AGE = 38 | 1.163485 | 0.355952 | 0.0011 |

**Figure 10.4** The output of a QSET for special selected quantiles and coefficients.

the numerical variable *AGE* is transformed to a set of dummy variables:

$$BLV\ C\ @Expand\ (AGE, @Dropfirst) \tag{10.1}$$

Based on these outputs, the findings and notes are as follows:

1. At the 1% or 5% level of significance and based on the *Quasi-LR* statistic, we can conclude that the categorical variable *AGE* of each QR has a significant effect on the corresponding *BLV* quantile.
2. The *t*-statistic in the outputs can be used directly to test either a two-sided or one-sided hypothesis on the quantiles difference between pairs of age levels, with *AGE* = 18 as the referent group. For the other hypotheses on the quantiles differences between the age levels can be tested using the Wald test or redundant variables test (RVT). Do these as exercises. Since the QR has large number dummy IVs, I recommend conducting the analysis using the quantile process for only for some selected quantiles, as well as coefficients.

For instance, Figure 10.4 presents the output for a quantile slope equality test (QSET) for the specific quantiles 0.1, 0.9 and specific coefficients C(5), C(20), which represent the levels *AGE* = 22 and *AGE* = 38, respectively.

**Example 10.2** *Forecasting Based on the QR (10.1)*

With each QR($\tau$) of *BLV* in (10.1) for $\tau = 0.1, 0.5$, and 0.9 on-screen, we can easily obtain its forecast value variables, *BLVF_Q10*, *BLVF_Q50*, and *BLVF_Q90*. Figure 10.5 presents

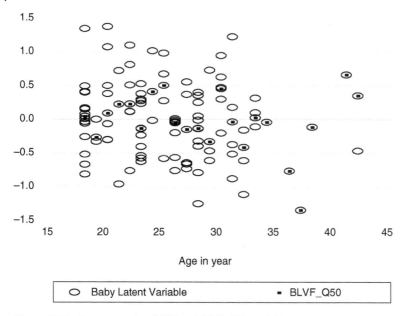

**Figure 10.5** Scatter graphs of *BLV* and *BLVF_Q50* on *AGE*.

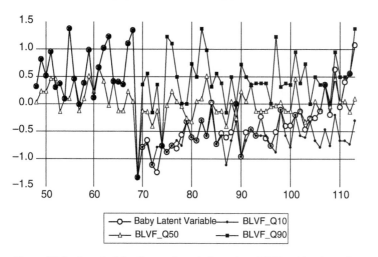

**Figure 10.6** A part of the line and symbol graphs of *BLV* and its three forecast value variables.

the scatter graphs of *BLV* and *BLVF_Q50* on *AGE*, and Figure 10.6 presents a part of the line and symbol graphs of *BLV* and its three forecast variables. Based on these graphs, the following findings and notes are presented:

1. The output of the scatter graphs in Figure 10.5 do not have the lines between two consecutive points. But the output of the graphs in Figure 10.6 have the lines already.
2. Unexpectedly, the graphs in Figure 10.5 present a much smaller number of points compared with the graphs in Figure 10.6 because a point in Figure 10.5 represents several observed scores of *AGE*. For instance, there are 18 observations for *AGE* = 18.

3. Figure 10.6 shows that many of the forecast variables for *BLVF_Q50* and *BLV* have coinciding scores.

### Example 10.3   A Special One-way ANOVA-QR of BLV

Figure 10.7 presents the output of a special one-way ANOVA-QR(Median) of *BLV* on *@Round(MW)*, using the following ES:

$$BLV\ C\ @Expand\ (@Round\ (MW)\ ,\ @Dropfirst) \tag{10.2}$$

*@Round(MW)* is an integer value variable, which is a transformed variable of the numerical variable *MW* (mother weight). Based on this output, the following findings and notes are presented:

1. At the 1% level of significance, we can conclude that the categorical variable *@Round(MW)* has a significant effect on *BLV*-median, based on the *QLR* statistic of $QLR_0 = 77.258\,58$ with $df = 28$ and *p*-value $= 0.000\,002$.
2. The *t*-statistic in the output can be used directly to test either a two-sided or one-sided hypothesis on the medians difference between pairs of *@Round(MW)* levels with the first level as the referent group. The other hypotheses on the medians differences between the scores of *@Round(MW)* can be tested using the Wald test.
3. I recommend doing other additional analyses, similar to the analyses presented in previous Example (10.1), as exercises.

### Example 10.4   An Alternative One-way ANOVA-QR of BLV

Figure 10.8 present the output of an alternative one-way ANOVA-QR(Median), without an intercept, that uses the following ES:

$$BLV\ C\ @Expand\ (OMW) \tag{10.3}$$

where the *OMW* (ordinal mother weight) is an ordinal variable having 10 levels, which is generated using the following equation:

$$omw = 1 + 1 * (mw >= @quantile\ (mw, 0.1)) + 1 * (mw >= @quantile\ (mw, 0.2))$$
$$+1 * (mw >= @quantile\ (mw, 0.3)) + 1 * (mw >= @quantile\ (mw, 0.4))$$
$$+1 * (mw >= @quantile\ (mw, 0.5)) + 1 * (mw >= @quantile\ (mw, 0.6))$$
$$+1 * (mw >= @quantile\ (mw, 0.7)) + 1 * (mw >= @quantile\ (mw, 0.8))$$
$$+1 * (mw >= @quantile\ (mw, 0.9)) \tag{10.4}$$

Based on this output, the following findings and notes are presented:

1. Compared with to the models in (10.1, 10.2), this is a model without an intercept because it has a special objective to present the graphs of 10 QR($\tau$) for $\tau = 0.1, 0.2, \ldots, 0.9$, which are the output of quantile process estimates (QPEs) of the one-way ANOVA-Q($\tau$) in (10.3), as shown in Figure 10.9.
2. Note that for each level of *OMW*, the values of quantile($\tau$) increase with $\tau$. But we can't test the quantiles differences between quantile($\tau_1$) and quantile($\tau_2$) for ($\tau_1 \neq \tau_2$). Based on the output in Figure 10.8, the quantile(0.25) differences between the levels of OMW can be tested easily using the Wald test.

Dependent Variable: BLV
Method: Quantile Regression (Median)
Sample: 1 113
Included observations: 113
Huber Sandwich Standard Errors & Covariance
Sparsity method: Kernel (Epanechnikov) using residuals
Bandwidth method: Hall-Sheather, bw = 0.20096
Estimation successful but solution may not be unique

| Variable | Coefficient | Std. Error | $t$-statistic | Prob. |
|---|---|---|---|---|
| C | −0.959943 | 0.277346 | −3.461176 | 0.0008 |
| @ROUND(MW) = 21 | 0.422904 | 0.341472 | 1.238474 | 0.2190 |
| @ROUND(MW) = 23 | 0.703609 | 0.346279 | 2.031917 | 0.0453 |
| @ROUND(MW) = 24 | 1.360679 | 0.348511 | 3.904270 | 0.0002 |
| @ROUND(MW) = 25 | 0.495552 | 0.340421 | 1.455703 | 0.1492 |
| @ROUND(MW) = 26 | 1.073775 | 0.371273 | 2.892144 | 0.0049 |
| @ROUND(MW) = 27 | 0.910542 | 0.345922 | 2.632214 | 0.0101 |
| @ROUND(MW) = 28 | 1.314995 | 0.554692 | 2.370677 | 0.0200 |
| @ROUND(MW) = 29 | 0.893555 | 0.392226 | 2.278161 | 0.0253 |
| @ROUND(MW) = 30 | 1.213490 | 0.348705 | 3.479987 | 0.0008 |
| @ROUND(MW) = 31 | 1.262077 | 0.679356 | 1.857754 | 0.0667 |
| @ROUND(MW) = 32 | 0.229194 | 0.350320 | 0.654243 | 0.5147 |
| @ROUND(MW) = 33 | 0.727821 | 0.522908 | 1.391872 | 0.1676 |
| @ROUND(MW) = 34 | 0.658238 | 0.309712 | 2.125320 | 0.0365 |
| @ROUND(MW) = 35 | 0.634368 | 0.356153 | 1.781169 | 0.0785 |
| @ROUND(MW) = 36 | 0.661118 | 0.341212 | 1.937558 | 0.0560 |
| @ROUND(MW) = 37 | 1.129521 | 0.309224 | 3.652758 | 0.0005 |
| @ROUND(MW) = 38 | 0.944587 | 0.340806 | 2.771624 | 0.0069 |
| @ROUND(MW) = 39 | 0.812924 | 0.333149 | 2.440119 | 0.0168 |
| @ROUND(MW) = 40 | 1.005554 | 0.325169 | 3.092400 | 0.0027 |
| @ROUND(MW) = 41 | 0.848490 | 0.390136 | 2.174860 | 0.0325 |
| @ROUND(MW) = 42 | 1.188180 | 0.309909 | 3.833971 | 0.0002 |
| @ROUND(MW) = 43 | 1.061289 | 0.312835 | 3.392491 | 0.0011 |
| @ROUND(MW) = 44 | 1.195362 | 0.480377 | 2.488381 | 0.0148 |
| @ROUND(MW) = 45 | 1.683096 | 0.392226 | 4.291134 | 0.0000 |
| @ROUND(MW) = 46 | 1.634486 | 0.392226 | 4.167200 | 0.0001 |
| @ROUND(MW) = 47 | 1.258715 | 0.355543 | 3.540264 | 0.0007 |
| @ROUND(MW) = 48 | 0.970401 | 0.458759 | 2.115271 | 0.0374 |
| @ROUND(MW) = 49 | 1.455872 | 0.392226 | 3.711817 | 0.0004 |
| Pseudo R-squRed | 0.346734 | Mean dependent var | | 9.04E−17 |
| Adjusted R-squRed | 0.128979 | S.D. dependent var | | 0.550900 |
| S.E. of regression | 0.527854 | Objective | | 15.94105 |
| Quantile dependent var | −0.012850 | Restr. objective | | 24.40210 |
| Sparsity | 0.875822 | Quasi-LR statistic | | 77.28558 |
| Prob(Quasi-LR stat) | 0.000002 | | | |

**Figure 10.7** The output of a special one-way ANOVA-QR(Median) in (10.2).

Dependent Variable: BLV

Method: Quantile Regression (tau = 0.25)

Date: 03/12/19   Time: 12:03

Sample: 1 113

Included observations: 113

Huber Sandwich Standard Errors & Covariance

Sparsity method: Kernel (Epanechnikov) using residuals

Bandwidth method: Hall-Sheather, bw = 0.13918

Estimation successful but solution may not be unique

| Variable | Coefficient | Std. Error | t-statistic | Prob. |
|---|---|---|---|---|
| OMW = 1 | −0.537039 | 0.127085 | −4.225816 | 0.0001 |
| OMW = 2 | −0.066388 | 0.166424 | −0.398911 | 0.6908 |
| OMW = 3 | −0.760891 | 0.098631 | −7.714479 | 0.0000 |
| OMW = 4 | −0.564834 | 0.148689 | −3.798756 | 0.0002 |
| OMW = 5 | −0.517339 | 0.228625 | −2.262829 | 0.0257 |
| OMW = 6 | −0.106848 | 0.161267 | −0.662554 | 0.5091 |
| OMW = 7 | −0.370333 | 0.227069 | −1.630928 | 0.1060 |
| OMW = 8 | −0.147019 | 0.107733 | −1.364666 | 0.1753 |
| OMW = 9 | −0.032696 | 0.135903 | −0.240586 | 0.8104 |
| OMW = 10 | 0.235419 | 0.113241 | 2.078920 | 0.0401 |

| | | | | |
|---|---|---|---|---|
| Pseudo R-squRed | 0.248608 | Mean dependent var | | 9.04E−17 |
| Adjusted R-squRed | 0.182953 | S.D. dependent var | | 0.550900 |
| S.E. of regression | 0.605242 | Objective | | 14.74445 |
| Quantile dependent var | −0.393213 | Restr. objective | | 19.62286 |
| Sparsity | 1.091885 | | | |

**Figure 10.8** The output of a one-way ANOVA_Q(0.25) in (10.3).

3. If we want to test the quantiles differences, we should be running the model using an intercept, as presented in Figures 10.3 and 10.7.

**Example 10.5** *An Additional One-Way ANOVA Quantile($\tau$)*

As an additional illustration, Figure 10.10 presents a summary of testing hypotheses on the quantiles($\tau$) differences for each $\tau = 0.1, 0.5$, and $0.9$, between the levels of the categorical mother variables *ANC* and *SMOKING*, with two levels each, and *SE* and *ED*, with three levels each. Based on this summary, the following findings and notes are presented:

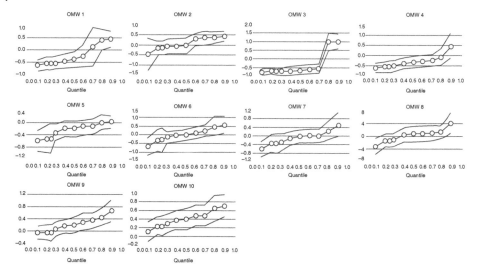

**Figure 10.9** The graphs of the QPE of the one-way ANOVA quantile($\tau$) in (10.3).

| QR(0.1) | ANC(2) | SMOKING(2) | SE(3) | ED(3) |
|---|---|---|---|---|
| QUASI-LR Stat | 1.617789 | 2.097617 | 3.029984 | 1.316437 |
| Prob. | 0.2034 | 0.147529 | 0.21981 | 0.517773 |
| QR(0.5) | ANC(2) | SMOKING(2) | SE(3) | ED(3) |
| QUASI-LR Stat | 0.186360 | 2.177175 | 0.334699 | 4.819868 |
| Prob. | 0.665963 | 0.140071 | 0.845904 | *0.089821* |
| QR(0.9) | ANC(2) | SMOKING(2) | SE(3) | ED(3) |
| QUASI-LR Stat | 0.260435 | 2.973721 | 0.63427 | 9.44269 |
| Prob. | 0.609821 | *0.084627* | 0.728232 | *0.008903* |

**Figure 10.10** The summary of the testing of quantiles differences between levels of categorical mother variables.

1. At the 10% level and based on the output in Figure 10.10, we can conclude that the quantiles 0.1, 0.5, and 0.9 of *BLV* each have insignificant differences between the levels of *ANC* and *SE*.

2. As a comparison, Figure 10.11 presents the output of nonparametric tests for equality of medians of *BLV* by *ANC*. Based on this output, the findings and notes are as follows:

   2.1 The output shows 84 of 113 mothers (74.34%) have antenatal care. This can be considered as good health care already. However, unexpectedly, the median(*BLV*) = −0.004 375 for the level *ANC* = 1 shows that more than 50% of the 84 mothers having the antenatal care have negative *BLV* scores.

Test for Equality of Medians of BLV

Categorized by values of ANC

Date: 03/16/19   Time: 07:51

Sample: 1 113

Included observations: 113

| Method | df | Value | Probability |
|---|---|---|---|
| Wilcoxon/Mann-Whitney | | 0.207067 | 0.8360 |
| Wilcoxon/Mann-Whitney (tie-adj.) | | 0.207067 | 0.8360 |
| Med. Chi-square | 1 | 0.349142 | 0.5546 |
| Adj. Med. Chi-square | 1 | 0.140997 | 0.7073 |
| Kruskal-Wallis | 1 | 0.044249 | 0.8334 |
| Kruskal-Wallis (tie-adj.) | 1 | 0.044249 | 0.8334 |
| van der Waerden | 1 | 0.004681 | 0.9455 |

Category statistics

| ANC | Count | Median | > Overall Median | Mean rank | Mean score |
|---|---|---|---|---|---|
| 0 | 29 | −0.039908 | 13 | 55.89655 | −0.010608 |
| 1 | 84 | −0.004375 | 43 | 57.38095 | 0.003662 |
| All | 113 | −0.012850 | 56 | 57.00000 | −1.08E-16 |

**Figure 10.11**   The output of a test for equality of medians of *BLV* by *ANC*.

2.2 Each of the nonparametric tests shows the medians of *BLV* have an insignificant difference between the levels of *ANC*. And we find the medians of BBW also have significant difference between the levels between the levels of ANC

3. Specific for the *ED* indicator, at the 10% level of significance, the quantiles(0.5) and quantiles(0.9) have significant differences between the three levels, based on the *QLR*-statistic with *p*-values of 0.089 821 and 0.008 903, respectively.

4. However, specific for the *SMOKING* indicator, at the 10% level, only the quantiles(0.9) has a significant difference between the two levels.

## 10.3.2   A Two-Way ANOVA-QR of *BLV*

For the analysis of a two-way ANOVA-QR of *BLV* on two categorical factors or predicators *A* and *B*, I propose to apply the alternative general ESs in the three following sections. Two important hypotheses will be tested based on a two-way ANOVA-QR, as follows:

**Table 10.1** Model parameters of a $2 \times 2$ factorial ANOVA-QR in (10.5).

|            | B = 1          | B = 2          | B(1 − 2)                     |
|------------|----------------|----------------|------------------------------|
| **A = 1**  | C(1)           | C(2)           | C(2) + C(3) − C(4)           |
| **A = 2**  | C(3)           | C(4)           | C(2)                         |
| **A(1 − 2)** | **C(1) − C(3)** | **C(2) − C(4)** | **C(1) − C(2) + C(3) − C(4)** |

(i) The hypothesis that the quantiles difference between the levels of a factor is *conditional for* the other factor, and

(ii) The hypothesis that the quantiles differences between all levels of a factor *depend on* the other factor. In other words, we test the quantile DID (difference-in-differences) between the cells/groups generated by the two factors, or the effect of the interaction of the two factors on the dependent variable (DV) quantiles.

### 10.3.2.1 The Simplest Equation of a Two-Way ANOVA-QR of *BLV*

The simplest equation of a two-way ANOVA-QR of *BLV* is an equation without an intercept, which is a $I \times J$ factorial ANOVA-QR with the following general ES:

$$BLV @ Expand (A, B) \tag{10.5}$$

As an illustration, Table 10.1 presents the model parameters of a $2 \times 2$ factorial ANOVA-QR, with the statistical hypothesis of its interaction effect as follows:

$H_0: C(1) - C(3) = C(2) - C(4)$ vs $H_1$: *Otherwise, or*

$H_0: C(1) - C(2) - C(3) + C(4) = 0$ vs $H_1$: *Otherwise*

### 10.3.2.2 A Two-way ANOVA-QR of *BLV* with an Intercept

The equation of an $I \times J$ factorial ANOVA-QR with an intercept has the following general ES, where the function *@Droplast* can be replaced by *@Dropfirst* or *@Drop(i,j)* by dropping the $(i,j)$th cell/group of the cells generated by two factors $A$ and $B$:

$$BLV \, C @ Expand (A, B, @Droplast) \tag{10.6}$$

As an illustration, Table 10.2 presents the model parameters of a $2 \times 2$ factorial ANOVA-QR, with the statistical hypothesis on its interaction effect as follows:

$H_0: C(2) - C(4) = C(3)$ vs $H_1$: *Otherwise, or*

$H_0: C(2) - C(3) - C(4) = 0$ vs $H_1$: *Otherwise*

**Table 10.2** Model parameters of a $2 \times 2$ factorial ANOVA-QR in (10.6).

|            | B = 1          | B = 2          | A(1 − 2)                     |
|------------|----------------|----------------|------------------------------|
| **A = 1**  | C(1) + C(2)    | C(1) + C(3)    | C(2) − C(3)                  |
| **A = 2**  | C(1) + C(4)    | C(1)           | C(4)                         |
| **A(1 − 2)** | **C(2) − C(4)** | **C(3)**       | **C(2) − C(3) − C(4)**       |

**Table 10.3** Model parameters of a $3 \times 2$ factorial ANOVA-QR in (10.7).

|  | B = 1 | B = 2 | B(1 − 2) |
|---|---|---|---|
| A = 1 | C(1) + C(2) + C(4) | C(1) + C(2) | C(4) |
| A = 2 | C(1) + C(3) + C(5) | C(1) + C(3) | C(5) |
| A = 3 | C(1) + C(6) | C(1) | C(6) |
| A(1 − 3) | C(2) + C(4) − C(6) | C(2) | C(4) − C(6) |
| A(2 − 3) | C(3) + C(5) − C(6) | C(3) | C(5) − C(6) |

### 10.3.2.3 A Special Equation of Two-Way ANOVA-QR of BLV

The special equation of an $I \times J$ factorial ANOVA-QR with an intercept has the following general ES:

$$BLV\ C\ @Expand\,(A, @Droplast)\,@Expand\,(A) * @Expand\,(B, @DropLast) \quad (10.7)$$

As an illustration, Table 10.3 presents the model parameters of a $3 \times 2$ factorial ANOVA-QR in (10.7), with the statistical hypothesis of its interaction effect as follows:

$$H_0 : C\,(4) = C\,(5) = C\,(6)\,;\,vs\,H_1 :\ Otherwise,\,or$$
$$H_0 : C\,(4) - C\,(6) = C\,(5) - C\,(6) = 0;\,vs\,H_1 :\ Otherwise$$

In addition, it is important to note that for each coefficient from C(2) to C(6), we can define either one-sided or two-sided hypotheses on the quantiles difference between pairs of a factor's levels, conditional for the level of the other factor. We then can test is easily using the $t$-statistic presented in the output. See the following example.

### Example 10.6   *An Application of a $3 \times 2$ ANOVA-QR(τ) of BLV*

As an empirical result, Figure 10.12 presents the summary of the following QR(τ) for $τ = 0.1$, 0.5, and 0.9:

$$BLV\ C\ @Expand\,(ED, @Droplast)$$
$$@Expand\,(ED) * @Expand\,(ANC, @Droplast) \quad (10.8)$$

Based on this summary, the following findings and notes are presented:

1. Only one of the coefficient parameters, C(3), in the three QR(τ) for $τ = 0.1$, 0.5, and 0.9, has small $p$-values of 0.014, 0.050, and 0.019, respectively. Referring to the model parameters in Table 10.3, we can conclude that the quantiles of *BLV* have significant differences between the two levels, $A = ED = 2$ and $A = ED = 3$, conditional for $ANC(B = 2) = 1$, because *ANC* is a dummy variable.
2. Note each parameter C(2) to C(6) is representing a quantiles difference, between a pair of cell or group. Although a pair of quantiles have insignificant differences, the two-way ANOVA can't be reduced, by deleting the corresponding parameter. This is because each QR should have six coefficient parameters, corresponding to the six cells of a $3 \times 2$ factorial ANOVA.

| Variable | QR(0.1) | | | QR(0.5) | | | QR(0.9) | | |
|---|---|---|---|---|---|---|---|---|---|
| | Coef. | t-stat. | Prob. | Coef. | t-stat. | Prob. | Coef. | t-stat. | Prob. |
| C(1) | −0.8129 | −5.1077 | 0.000 | −0.1959 | −1.0600 | 0.292 | 0.5110 | 4.5570 | 0.000 |
| C(2) | 0.1453 | 0.6779 | 0.499 | 0.1627 | 0.7669 | 0.445 | −0.0062 | −0.0276 | 0.978 |
| C(3) | 0.5807 | 2.5093 | **0.014** | 0.4730 | 1.9837 | **0.050** | 0.5065 | 2.3852 | **0.019** |
| C(4) | −0.0924 | −0.3683 | 0.713 | 0.3494 | 0.9409 | 0.349 | 0.1592 | 0.5644 | 0.574 |
| C(5) | 0.2758 | 1.2627 | 0.210 | 0.1465 | 0.5558 | 0.580 | −0.1305 | −0.5390 | 0.591 |
| C(6) | −0.378 | −1.338 | 0.184 | −0.1939 | −0.791 | 0.431 | −0.035 | −0.088 | 0.930 |
| Pse. R^2 | 0.0442 | | | 0.0483 | | | 0.1182 | | |
| Adj. R^2 | −0.0005 | | | 0.0038 | | | 0.0770 | | |
| S.E. Reg. | 0.8274 | | | 0.5492 | | | 0.8716 | | |
| Quant DV | −0.6676 | | | −0.0129 | | | 0.6745 | | |
| Sparsity | 2.4998 | | | 1.4238 | | | 2.7037 | | |
| QLR stat | 4.0681 | | | 6.6221 | | | 11.215 | | |
| Prob. | 0.5397 | | | 0.2503 | | | **0.0473** | | |

**Figure 10.12** The summary of outputs of the $3 \times 2$ ANOVA-QR($\tau$) in (10.8), for $\tau = 0.1$, 0.5, and 0.9.

| Wald test: | | QT(0.1) | | QT(0.5) | | QT(0.9) | |
|---|---|---|---|---|---|---|---|
| Statistic | df | Value | Prob. | Value | Prob. | Value | Prob. |
| F-statistic | (2, 107) | 1.7580 | 0.1773 | 0.8842 | 0.4161 | 0.3056 | 0.7373 |
| Chi-square | 2 | 3.5159 | 0.1724 | 1.7683 | 0.4131 | 0.6113 | 0.7367 |
| Null hypothesis: C(4) = C(5) = C(6) | | | | | | | |

**Figure 10.13** The summary of the testing hypothesis on the interaction effects of *ED*ANC*.

3. Only for the QR(0.9), at the 5% level, do the six quantiles of *BLV* have significant differences, based on the *QLR* statistic of $QLR_0 = 11.215$ with $p$-value $= 0.0473$.
4. In addition, Figure 10.13 presents the summary of the testing hypotheses on the effect of the interaction *ED*ANC* on *BLV* for each QR, which shows the interaction has an insignificant effect.

## 10.4   Three-way ANOVA-QR of *BLV*

So far, I have found only two of my advisories writing their theses using a $2 \times 2 \times 2$ factorial ANOVA parametric regression, and one of them tested the three-way interaction effect of

**Table 10.4** The model parameters of a $2 \times 2 \times 2$ factorial ANOVA-QR.

|  |  | G = 1 | G = 2 | G(1 – 2) |
|---|---|---|---|---|
| A = 1 | B = 1 | C(1) + C(2) + C(5) | C(1) + C(2) | C(5) |
|  | B = 2 | C(1) + C(3) + C(6) | C(1) + C(3) | C(6) |
| **IE(B\*G\|A = 1)** |  |  |  | **C(5) – C(6)** |
| A = 2 | B = 1 | C(1) + C(4) + C(7) | C(1) + C(4) | C(7) |
|  | B = 2 | C(1) + C(8) | C(1) | C(8) |
| **IE(B\*G\|A = 2)** |  |  |  | **C(7) – C(8)** |

the factors. To make easier testing hypotheses, based on an $I \times J \times K$ factorial ANOVA-QR of *BLV*, on three factors $A$, $B$, and $G$, I recommend applying the only general specific equation, as follows:

$$BLV \ C \ @Expand \ (A, B, @Droplast)$$
$$@Expand \ (A, B) * @Expand \ (G, @Droplast) \tag{10.9}$$

### Example 10.7 An Application of a $2 \times 2 \times 2$ ANOVA-QR($\tau$) of BLV

As an illustration, specific for dichotomous factors $A$, $B$, and $G$, its model parameters can be presented as in Table 10.4. The following notes are important:

1. The table presents two blocks of a $2 \times 2$ table of $B$ and $G$, conditional for $A = 1$ and $A = 2$, respectively.
2. The symbol *IE(B\*G|A = 1) = C(5) – C(6)* represents a two-way interaction (2WI) effect $B*G$ on *BLV*, conditional for $A = 1$, which can be tested using the Wald test.
3. Similarly, the symbol *IE(B\*G|A = 2) = C(7) – C(8)* represents a 2WI effect $B*G$ on *BLV*, conditional for $A = 2$.
4. Then, the statistical hypothesis on the effect of the three-way interaction $A*B*G$ on *BLV* can be presented as follows and be tested using the Wald test:

$$H_0: C(5) - C(6) = C(7) - C(8) ; vs \ H_1 : Otherwise$$
$$H_0: C(5) - C(6) - (7) + C(8) = 0; vs \ H_1 : Otherwise$$

5. In addition, based on each parameter, C(2) to C(8), either a one-sided or a two-sided hypothesis can be defined and easily tested using the $t$-statistic presented in the output. Refer to the previous example.

### Example 10.8 An Application of a $3 \times 2 \times 2$ ANOVA-QR($\tau$) of BLV

As an extension of the $2 \times 2 \times 2$ ANOVA-QR in (10.9) with its model parameters presented in Table 10.4, Table 10.5 presents the model parameters of a $3 \times 2 \times 2$ ANOVA-QR in Figure 10.14. Then we have three conditional 2WIs, namely *IE(B\*G|A = 1)*, *IE(B\*G|A = 2)*, and *IE(B\*G|A = 3)*, as presented in the Table 10.5, with the statistical hypothesis on the

**Table 10.5** The model parameters of a $3 \times 2 \times 2$ factorial ANOVA-QR in (10.9).

|  |  | G = 1 | G = 2 | G(1 − 2) |
|---|---|---|---|---|
| A = 1 | B = 1 | C(1) + C(2) + C(7) | C(1) + C(2) | C(7) |
|  | B = 2 | C(1) + C(3) + C(8) | C(1) + C(3) | C(8) |
| **IE(B\*G\|A = 1)** |  |  |  | **C(7) − C(8)** |
| A = 2 | B = 1 | C(1) + C(4) + C(9) | C(1) + C(4) | C(9) |
|  | B = 2 | C(1) + C(5) + C(10) | C(1) + C(5) | C(10) |
| **IE(B\*G\|A = 2)** |  |  |  | **C(9) − C(10)** |
| A = 3 | B = 1 | C(1) + C(6) + C(11) | C(1) + C(6) | C(11) |
|  | B = 2 | C(1) + C(12) | C(1) | C(12) |
| **IE(B\*G\|A = 3)** |  |  |  | **C(11) − C(12)** |

three-way interaction $A*B*G$ defined as follows. And Figure 10.15 presents its outputs for the three QRs in Figure 10.14.

$$H_0: C(7) - C(8) = C(9) - C(10) = C(11) - C(12) \; ; vs \; H_1: Otherwise$$

As an empirical results, Figure 10.14 presents the summary of the $3 \times 2 \times 2$ QR in (10.9), using three factors $A$, $B$, and $G$, which are the transformed factors from the numerical variables $AGE$, $MW$, and $HB$. The factors are generated using the following equation:

$$A = 1 + 1 * (AGE > 19) + 1 * (AGE > 29)$$
$$B = 1 + 1 * (MW >= @Quntile (MW, 0.5))$$
$$G = 1 + 1 * (HB >= @Qunatile (HB, 0.5))$$

## 10.5 QRs of *BLV* on Numerical Predictors

For these types of QRs, we have the numerical mother variables, $AGE$, $MW$, $HB$, and $TOTALKOL$. We can easily develop or define QRs of $BLV$ with one or more of these numerical mother variables or indicators. See the following examples.

### 10.5.1 QRs of *BLV* on MW

As the preliminary analysis, Figure 10.16 presents the scatter graphs of $BLV$ on $MW$, with their linear fit and Lowess polynomial-2 fit, which clearly show a polynomial LS-regression (LSR) is better than the linear LSR.

#### 10.5.1.1 The Simplest Linear Regression of *BLV* on MW
The simplest linear regression of $BLV$ on $MW$, for either an LSR or mean regression (MR), and QR($\tau$) for $\tau \epsilon$ (0, 1) can be represented using the following ES:

$$BLV \, C \, MW \tag{10.10}$$

| Variable | 3W-ANOVA-QR(0.1) | | | 3W-ANOVA-QR(0.5) | | | 3W-ANOVA-QR(0.9) | | |
|---|---|---|---|---|---|---|---|---|---|
| | Coef. | t-stat | Prob. | Coef. | t-stat | Prob. | Coef. | t-stat | Prob. |
| C(1) | −0.764 | −2.600 | 0.011 | 0.101 | 0.371 | 0.712 | 0.316 | 1.929 | 0.057 |
| C(2) | −0.049 | −0.124 | 0.902 | −0.427 | −1.185 | 0.239 | −0.305 | −1.264 | 0.209 |
| C(3) | 0.809 | 2.636 | 0.010 | 0.052 | 0.166 | 0.868 | 0.095 | 0.464 | 0.644 |
| C(4) | 0.199 | 0.618 | 0.538 | −0.151 | −0.442 | 0.660 | 0.701 | 2.528 | 0.013 |
| C(5) | 0.435 | 1.266 | 0.208 | 0.124 | 0.395 | 0.694 | 0.407 | 1.833 | 0.070 |
| C(6) | −0.115 | −0.264 | 0.792 | −0.135 | −0.366 | 0.715 | 0.140 | 0.551 | 0.583 |
| C(7) | 0.145 | 0.445 | 0.657 | 0.061 | 0.190 | 0.850 | 0.390 | 0.973 | 0.333 |
| C(8) | −0.077 | −0.226 | 0.822 | 0.343 | 0.554 | 0.581 | 0.934 | 2.530 | 0.013 |
| C(9) | −0.196 | −1.138 | 0.258 | −0.560 | −2.143 | 0.035 | −1.150 | −3.748 | 0.000 |
| C(10) | −0.315 | −0.859 | 0.392 | −0.107 | −0.519 | 0.605 | 0.089 | 0.250 | 0.803 |
| C(11) | −0.464 | −1.153 | 0.252 | −0.478 | −1.040 | 0.301 | 0.770 | 1.571 | 0.119 |
| C(12) | 0.393 | 1.147 | 0.254 | −0.075 | −0.209 | 0.835 | 0.633 | 2.103 | 0.038 |
| Pse.$R^2$ | 0.232 | | | 0.150 | | | 0.242 | | |
| Adj. $R^2$ | 0.148 | | | 0.057 | | | 0.159 | | |
| S.E. reg. | 0.829 | | | 0.525 | | | 0.921 | | |
| Quant DV | −0.668 | | | −0.013 | | | 0.675 | | |
| Sparsity | 1.911 | | | 1.266 | | | 2.277 | | |
| QLR stat | 27.904 | | | 23.098 | | | 27.218 | | |
| Prob. | 0.003 | | | 0.017 | | | 0.004 | | |

**Figure 10.14** The output summary of the $3 \times 2 \times 2$ ANOVA-QR($\tau$) for $\tau = 0.1, 0.5$, and $0.9$.

| Wald test: | | QR(0.1) | | QR(0.5) | | QR(0.9) | |
|---|---|---|---|---|---|---|---|
| Test statistic | df | Value | Prob. | Value | Prob. | Value | Prob. |
| F-statistic | (2, 101) | 1.4042 | 0.2503 | 0.0252 | 0.9751 | 1.7387 | 0.1810 |
| Chi-square | 2 | 2.8085 | 0.2456 | 0.0504 | 0.9751 | 3.4773 | 0.1758 |
| Test for 3W-Intercation: Null Hypothesis: C(7) − C(8) = C(9) − C(10) = C(11) − C(12) | | | | | | | |

**Figure 10.15** The output summary of testing hypotheses on the three-way interaction $A^*B^*G$.

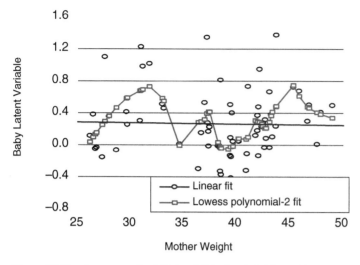

**Figure 10.16** Scatter graph of BLV on MW, with their Linear Fit and Lowess Polynomial-2 Fit.

| Variable | LS-regression | | | | Variable | QR(0.5) | | |
|---|---|---|---|---|---|---|---|---|
| | Coef. | t-Stat. | Prob. | | | Coef. | t-Stat. | Prob. |
| C | −0.9549 | −3.4932 | 0.0007 | | C | −1.2678 | −4.2115 | 0.0001 |
| MW | 0.0269 | 3.5515 | 0.0006 | | MW | 0.0336 | 4.3120 | 0.0000 |
| R^2 | 0.1020 | | | | Pse. R^2 | 0.0877 | | |
| Adj. R^2 | 0.0939 | | | | Adj. R^2 | 0.0795 | | |
| S.E. reg. | 0.5244 | | | | S.E. reg. | 0.5316 | | |
| SSR | 30.523 | | | | Quant. DV | −0.0129 | | |
| LL | −86.386 | | | | Sparsity | 1.2513 | | |
| F-stat. | 12.613 | | | | QLR stat | 13.6875 | | |
| Prob. | 0.001 | | | | Prob | 0.0002 | | |

**Figure 10.17** The outputs of the simplest LS-regression and QR(Median) in 10.10).

### Example 10.9 *An Application of the ES in (10.10)*

Figure 10.17 presents the outputs of the LSR and QR(Median) using the ES in (10.10). Based on these outputs, the following findings and notes are presented:

1. At the 1% level of significance, we can conclude that *MW* has a positive significant effect based both LSR and QR(0.5).
2. Figure 10.18 presents (a) the outputs table of 10 Quantile Process Estimates of the QR, and (b) the outputs graph of 10 Quantile Process Estimates of the QR. And Figure 10.18a shows that *MW* has insignificant effects on *BLV* only in the QR(0.8) and QR(0.9).
3. The slopes of *MW* are decreasing from $\tau = 0.1$ to $\tau = 0.8$ (Figure 10.19).

Quantile Process Estimates
Equation: UNTITLED
Specification: BLV C MW
Estimated equation quantile tau = 0.5
Number of process quantiles: 10
Display all coefficients

|  | Quantile | Coefficient | Std. Error | t-Statistic | Prob. |
|---|---|---|---|---|---|
| C | 0.100 | -2.446888 | 0.708268 | -3.454747 | 0.0008 |
|  | 0.200 | -2.155854 | 0.685346 | -3.145645 | 0.0021 |
|  | 0.300 | -1.557807 | 0.331064 | -4.705449 | 0.0000 |
|  | 0.400 | -1.349346 | 0.297515 | -4.535387 | 0.0000 |
|  | 0.500 | -1.267785 | 0.301029 | -4.211511 | 0.0001 |
|  | 0.600 | -0.857209 | 0.343083 | -2.498550 | 0.0139 |
|  | 0.700 | -0.652289 | 0.397002 | -1.643036 | 0.1032 |
|  | 0.800 | 0.220729 | 0.431203 | 0.511892 | 0.6097 |
|  | 0.900 | 0.367906 | 0.425925 | 0.863780 | 0.3896 |
| MW | 0.100 | 0.051689 | 0.018708 | 2.762956 | 0.0067 |
|  | 0.200 | 0.048045 | 0.017641 | 2.723506 | 0.0075 |
|  | 0.300 | 0.036182 | 0.008664 | 4.176411 | 0.0001 |
|  | 0.400 | 0.034039 | 0.007731 | 4.402976 | 0.0000 |
|  | 0.500 | 0.033588 | 0.007789 | 4.312038 | 0.0000 |
|  | 0.600 | 0.025617 | 0.008760 | 2.924465 | 0.0042 |
|  | 0.700 | 0.023438 | 0.009990 | 2.346199 | 0.0207 |
|  | 0.800 | 0.006052 | 0.010683 | 0.566551 | **0.5722** |
|  | 0.900 | 0.007841 | 0.010747 | 0.729574 | **0.4672** |

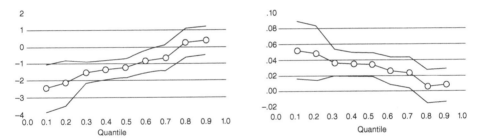

**Figure 10.18** (a) The output table of 10 quantile process estimates of the QR in (10.10). (b) The output graph of 10 quantile process estimates of the QR in (10.10).

4. In addition, Figure 10.19 presents the output of a QSET, and Figure 10.20 presents the output of a symmetric quantile test, which could easily be used to derive the conclusions of each test.

### 10.5.1.2 Polynomial Regression of *BLV* on *MW*

Because the lowess polynomial-2 fit in Figure 10.20 has five relative maximum and minimum points, I propose applying a sixth-degree polynomial regression with the following ES:

$$BLV\ C\ MW\ MW^{\wedge}2\ MW^{\wedge}3\ MW^{\wedge}4\ MW^{\wedge}5\ MW^{\wedge}6 \tag{10.11}$$

In general, the six IVs are highly correlated. In cases when the QR's IVs have large or very large p-values, we should modify the QR using the trail-and-error method. So, the acceptable empirical regression, in the statistical sense, does not have all $MW^{\wedge}k's$ as the IVs. See the following example.

Quantile Slope Equality Test

Equation: UNTITLED

Specification: BLV C MW

Estimated equation quantile tau = 0.5

Number of test quantiles: 4

Test statistic compares all coefficients

| Test summary | Chi-Sq. Statistic | Chi-Sq. d.f. | Prob. |
| --- | --- | --- | --- |
| Wald test | 7.424053 | 2 | 0.0244 |

Restriction Detail: b(tau_h) − b(tau_k) = 0

| Quantiles | Variable | Restr. Value | Std. Error | Prob. |
| --- | --- | --- | --- | --- |
| 0.25, 0.5 | MW | 0.006040 | 0.008399 | 0.4721 |
| 0.5, 0.75 | | 0.022882 | 0.008835 | 0.0096 |

**Figure 10.19** The output of four quantile slope equality test based on the QR(0.5).

Symmetric Quantiles Test

Equation: UNTITLED

Specification: BLV C MW

Estimated equation quantile tau = 0.5

Number of test quantiles: 4

Test statistic compares all coefficients

| Test summary | Chi-Sq. Statistic | Chi-Sq. d.f. | Prob. |
| --- | --- | --- | --- |
| Wald test | 3.832106 | 2 | 0.1472 |

Restriction Detail: b(tau) + b(1-tau) − 2*b(0.5) = 0

| Quantiles | Variable | Restr. Value | Std. Error | Prob. |
| --- | --- | --- | --- | --- |
| 0.25, 0.75 | C | 0.766580 | 0.485856 | 0.1146 |
| | MW | −0.016842 | 0.012483 | 0.1773 |

**Figure 10.20** The output of a symmetric quantile test based on the QR(0.5).

**Example 10.10**   *An Application of the ES in (10.11)*

Figure 10.21 presents the output of a reduced LSR and a reduced QR(Median) of the ES in (10.11), which are acceptable regressions, in the statistical sense. Based on these outputs, the following findings and notes are presented:

1. Since each IV of the QR(Median) in (10.11) has an insignificant effect on *BLV* at the 10% level of significance, I do the analysis for the LSR by using the STEPLS-Regression: Combinatorial and the trial-and-error method. The output of a reduced LSR, presented in Figure 10.22, shows each IV has a significant effect on *BLV* at the 1% level.
2. Then the output of the reduced QR(Median) is obtained using the LSR's ES. Note that each IV of the QR(Median) has a significant effect on *BLV* at the 5% level.
3. Since the coefficients of $MW^{\wedge}5$ and $MW^{\wedge}6$ are very close to zero, the QR(Median) can be modified by using $(MW/10)^{\wedge}k$'s as its IVs, with the alternative output in Figure 10.22.
4. As additional illustration, Figure 10.23 shows the output of two QSETs for the test quantiles "0.1, 0.9," based on the reduced QR(0.5) and QR(0.1), respectively. Based on these outputs, the following findings and notes are presented:
   4.1 The output based on the reduced QR(0.5) presents the test for eight pairs of slopes of the four IVs for the test quantiles of 0.1, 0.5 and 0.5, 0.9. At the 5% level, we can conclude the eight slopes have significant differences, based on the *Chi-Square* statistic of $\chi_0^2 = 17.044\,57$ with $df = 8$ and p-value $= 0.0290$. However, the pairs of each IV have an insignificant difference at the 10% Level.

| Variable | Reduced LS-regression | | | | Variable | Reduced QR(Median) | | |
|---|---|---|---|---|---|---|---|---|
| | Coef. | t-Stat. | Prob. | | | Coef. | t-Stat. | Prob. |
| C | −5.80551 | −2.7858 | 0.0063 | | C | −4.37281 | −2.4308 | 0.0167 |
| MW^5 | 8.43E-07 | 2.7929 | 0.0062 | | MW^5 | 6.19E-07 | 2.1821 | 0.0313 |
| MW^2 | 0.04175 | 2.7825 | 0.0064 | | MW^2 | 0.030477 | 2.2014 | 0.0298 |
| MW^6 | −9.78E-09 | −2.7592 | 0.0068 | | MW^6 | −7.16E-09 | −2.1547 | 0.0334 |
| MW^3 | −0.00168 | −2.7978 | 0.0061 | | MW^3 | −0.00123 | −2.1917 | 0.0305 |
| R^2 | 0.1786 | | | | Pse. R^2 | 0.1420 | | |
| Adj. R^2 | 0.1482 | | | | Adj. R^2 | 0.1102 | | |
| S.E. reg. | 0.5084 | | | | S.E. reg. | 0.5140 | | |
| SSR | 27.920 | | | | Quant. DV | −0.0129 | | |
| LL | −81.350 | | | | Sparsity | 1.2230 | | |
| F-stat. | 5.8712 | | | | QLR stat | 22.659 | | |
| Prob | 0.0003 | | | | Prob | 0.0001 | | |

**Figure 10.21**   The outputs of the reduced LS-regression and QR(Median) in (10.11).

Dependent Variable: BLV
Method: Quantile Regression (Median)
Included observations: 113
Huber Sandwich Standard Errors and Covariance
Sparsity method: Kernel (Epanechnikov) using residuals
Bandwidth method: Hall-Sheather, bw = 0.20096
Estimation successfully identifies unique optimal solution

| Variable | Coefficient | Std. Error | t-Statistic | Prob. |
|---|---|---|---|---|
| C | −4.372809 | 1.798909 | −2.430812 | 0.0167 |
| (MW/10)^2 | 3.047670 | 1.384406 | 2.201428 | 0.0298 |
| (MW/10)^3 | −1.224568 | 0.558734 | −2.191682 | 0.0305 |
| (MW/10)^5 | 0.061880 | 0.028358 | 2.182087 | 0.0313 |
| (MW/10)^6 | −0.007165 | 0.003325 | −2.154714 | 0.0334 |

| | | | | |
|---|---|---|---|---|
| Pseudo R-squRed | 0.141958 | Mean dependent var | | 9.04E-17 |
| Adjusted R-squRed | 0.110179 | S.D. dependent var | | 0.550900 |
| S.E. of regression | 0.513979 | Objective | | 20.93803 |
| Quantile dependent var | −0.012850 | Restr. objective | | 24.40210 |
| Sparsity | 1.223020 | Quasi-LR statistic | | 22.65914 |
| Prob(Quasi-LR stat) | 0.000148 | | | |

**Figure 10.22** An alternative output of the QR(median) in Figure 10.21.

4.2 The output based on the reduced QR(0.1) presents the test for only four pairs of slopes for the test quantiles "0.1, 0.9," which shows they have significant differences, based on the *Chi-Square* statistic of $\chi_0^2 = 12.050\,53$ with $df = 8$ and $p$-value $= 0.0170$. But the pairs of each IV have an insignificant difference at the 10% level.

### 10.5.2 QRs of *BLV* on Two Numerical Predictors

#### 10.5.2.1 An Additive QR of *BLV*

Based on the four numerical mother variables, *AGE*, *MW*, *HB*, and *TOTALKOL*, we can easily develop six additive QRs of *BLV* on two of the numerical variables. Do these as exercises.

#### 10.5.2.2 A Two-Way Interaction QR of BLV on MW and AGE

Similarly, we also can easily develop six two-way interaction QRs (2WI-QRs) of *BLV* on the four numerical mother variables, *AGE*, *MW*, *HB*, and *TOTALKOL*. In this section, I present

Quantile Slope Equality Test

Equation: EQ02

Specification: BLV C (MW/10)^2 (MW/10)^3 (MW/10)^5 (MW/10)^6

Estimated equation quantile tau = 0.5

User-specified test quantiles: 0.1 0.9

Test statistic compares all coefficients

| Test Summary | Chi-Sq. Statistic | Chi-Sq. d.f. | Prob. |
| --- | --- | --- | --- |
| Wald test | 17.04457 | 8 | 0.0296 |

Restriction Detail: b(tau_h)-b(tau_k) = 0

| Quantiles | Variable | Restr. Value | Std. Error | Prob. |
| --- | --- | --- | --- | --- |
| 0.1, 0.5 | (MW/10)^2 | 1.517512 | 1.837757 | 0.4090 |
| | (MW/10)^3 | −0.702832 | 0.747881 | 0.3473 |
| | (MW/10)^5 | 0.042588 | 0.038701 | 0.2711 |
| | (MW/10)^6 | −0.005280 | 0.004590 | 0.2500 |
| 0.5, 0.9 | (MW/10)^2 | −0.991231 | 1.739864 | 0.5689 |
| | (MW/10)^3 | 0.269231 | 0.725150 | 0.7104 |
| | (MW/10)^5 | −0.003634 | 0.038194 | 0.9242 |
| | (MW/10)^6 | 5.19E-05 | 0.004514 | 0.9908 |

Quantile Slope Equality Test

Equation: EQ02

Specification: BLV C (MW/10)^2 (MW/10)^3 (MW/10)^5 (MW/10)^6

Estimated equation quantile tau = 0.1

User-specified test quantiles: 0.1 0.9

Test statistic compares all coefficients

| Test Summary | Chi-Sq. Statistic | Chi-Sq. d.f. | Prob. |
| --- | --- | --- | --- |
| Wald Test | 12.05053 | 4 | 0.0170 |

Restriction Detail: b(tau_h)-b(tau_k) = 0

| Quantiles | Variable | Restr. Value | Std. Error | Prob. |
| --- | --- | --- | --- | --- |
| 0.1, 0.9 | (MW/10)^2 | 0.526281 | 2.209096 | 0.8117 |
| | (MW/10)^3 | −0.433602 | 0.920039 | 0.6374 |
| | (MW/10)^5 | 0.038954 | 0.048770 | 0.4244 |
| | (MW/10)^6 | −0.005228 | 0.005803 | 0.3677 |

**Figure 10.23** The outputs of the QSET, for the test quantiles 0.1 and 0.9 based on the Reduced QR(0.5) and Q(0.1), respectively.

one of them, with the following ES and the main objective of studying the effect of *MW* on *BLV*, dependent on *AGE*:

$$BLV\ C\ MW * AGE\ MW\ AGE \tag{10.12}$$

### Example 10.11   *An Application of the ES in (10.12)*
Figure 10.24 presents the output of the full two-way interaction (F2WI) QR(0.5) in (10.12) with two of its reduced models. Based on these outputs, the following findings and notes are presented:

1. Since the interaction *MW*AGE* in the full model has a very large *p*-value, it looks like the full model is not a good fit 2WI model. In fact, its IVs have significant joint effects on *BLV*, based on the *Quasi-LR* statistic of $QLR_0 = 14.85$ with *p*-value = 0.002.
2. Even though the interaction *MW*AGE* has a large *p*-value, the two IVs, *MW*AGE and MW,* have significant joint effects on *BLV,* based on the *QLR* L-statistic of $QLRL_0 = 13.464\,553$ with $df = 2$ and *p*-value 0.0012, as presented in Figure 10.25. Hence, we can conclude the effect of *MW* on *BLV* is significantly dependent on *AGE*, with the following equation:

$$\widehat{BLV} = (-1.170\,69 - 0.002\,14 \times AGE) + (0.037\,13 - 0.000\,19 \times AGE) \times MW$$

Based on this equation, we can conclude that the effect of *MW* is positive for all *AGE* scores because $(0.037\,13 - 0.000\,19 \times AGE) > +0.02\,915$, as *Max(AGE)* = 42, and its effect decreases with the increasing scores of *AGE*.

3. On the other hand, we find that the IVs *MW*AGE* and *AGE* have insignificant effects on *BLV*, based on the *QLR* L-statistic of $QLRL_0 = 1.890\,813$ with $df = 2$ and *p*-value 0.3885. So, we can conclude that the effect of *AGE* on *BLV* is insignificantly dependent on *MW*,

| Variable | F2WI-QR(0.5) | | | Reduced1-QR(0.5) | | | Reduced2-QR(0.5) | | |
|---|---|---|---|---|---|---|---|---|---|
| | Coef. | t-Stat. | Prob. | Coef. | t-stat. | Prob. | Coef. | t-stat. | Prob. |
| C | −1.170 | −0.800 | 0.426 | −1.220 | −3.715 | 0.000 | 0.194 | 0.650 | 0.517 |
| MW*AGE | 0.000 | −0.124 | 0.902 | 0.000 | −0.807 | 0.421 | 0.001 | 2.918 | 0.004 |
| MW | 0.037 | 0.960 | 0.339 | 0.039 | 3.788 | 0.000 | | | |
| AGE | −0.002 | −0.038 | 0.970 | | | | −0.047 | −3.060 | 0.003 |
| Pse. R^2 | 0.101 | | | 0.100 | | | 0.087 | | |
| Adj. R^2 | 0.076 | | | 0.084 | | | 0.070 | | |
| S.E. Reg. | 0.532 | | | 0.529 | | | 0.529 | | |
| Quant. DV | −0.013 | | | −0.013 | | | −0.013 | | |
| Sparsity | 1.322 | | | 1.317 | | | 1.364 | | |
| QLR Stat | 14.85 | | | 14.86 | | | 12.45 | | |
| Prob. | 0.002 | | | 0.001 | | | 0.002 | | |

**Figure 10.24**   The outputs of the 2WI-QR(0.5) in (10.12), and its reduced models.

Redundant Variables test

Equation: UNTITLED

Redundant variables: MW*AGE MW

Specification: BLV C MW*AGE MW AGE

Null hypothesis: MW*AGE MW are jointly insignificant

| | Value | df | Probability |
|---|---|---|---|
| QLR L-statistic | 13.46553 | 2 | 0.0012 |
| QLR Lambda-statistic | 12.82595 | 2 | 0.0016 |

**Figure 10.25** The RVT of the two IVs *MW\*AGE* and *MW* of the F2WI-QR(Median).

with the following equation:

$$\widehat{BLV} = (-1.170\,44 + 0.037\,13 \times MW) - (0.002\,14 + 0.000\,19 \times MW) \times AGE$$

Note that this equation shows that *AGE* has an insignificant negative effect on *BLV* for all scores of the positive variable *MW*.

4. Since each IV of the F2WI-QR(0.5) has a large *p*-value, we generally might want to see its reduced models. For this reason, two of its reduced models, Reduced1-QR(0.5) and Reduced2-QR(0.5), are included in Figure 10.24. Based on these reduced models, the following findings and notes are presented:

   4.1 Even though the interaction *MW\*AGE* in the Reduced1 has a large *p*-value, it has the greatest adjusted *R*-squared. And the joint effects of the IVs have a significant effect with a *p*-value of 0.001. So, we can considere it to be the best fit model among the three models, in the statistical sense.

   4.2 On the other hand, at the 1% level, the interaction *MW\*AGE* in the Reduced2 has a significant effect, based on the *t*-statistic of $t_0 = 2.918$ with *p*-value $= 0.004$, but it has the smallest adjusted *R*-squared. So, in the statistical sense, we can say the Reduced2 is the worst fit model.

   4.3 Both reduced regressions are more restricted regression functions compared with the F2WI-QR.

### 10.5.2.3 A Two-way Interaction Polynomial QR of BLV on *MW* and *AGE*

As an extension of the reduced polynomial QR(0.5) presented in Figure 10.21, this section presents the reduced polynomial 2WI-QR(0.5) of *BLV* with the following ES:

$$BLV \; C \; AGE^*(MW/10)^\wedge 2 \; AGE^*(MW/10)^\wedge 3 \; AGE^*(MW/10)^\wedge 5$$
$$AGE^*(MW/10)^\wedge 6 \; AGE \tag{10.13}$$

**Example 10.12** *The Output of the QR(Median) in (10.13)*

Figure 10.26 presents the output of the QR(Median) in (10.13). Based on this output, the following findings and notes are presented:

1. It is very unexpected that each IV has either a positive or negative significant adjusted effect at the 1% or 5% level of significance. For instance, $AGE^*(MW/10)^5$ has a significant positive effect, based on the $t$-statistic of $t_0 = 2.413\,643$ with $p$-value $= 0.0175/2 = 0.008\,75 < 0.01$; and, $AGE^*(MW/10)^6$ has a significant negative effect, based on the $t$-statistic of $t_0 = -2.297\,951$ with $p$-value $= 0.0235/2 = 0.011\,75 < 0.05$.

2. As an additional illustration, Figure 10.27 presents the output of a QSET of the QR(0.1) in (10.13). Based on this output, the following findings and notes are presented:

   2.1 Based on the Wald test (*Chi-Square* statistic) of $\chi_0^2 = 24.366\,01$ with $df = 5$ and $p$-value $= 0.0002$, we can conclude that the five pairs slopes of the IVs between the quantiles 0.1 and 0.9 have significant differences.

---

Dependent Variable: BLV

Method: Quantile Regression (Median)

Date: 03/19/19   Time: 15:16

Sample: 1 113

Included observations: 113

Huber Sandwich Standard Errors & Covariance

Sparsity method: Kernel (Epanechnikov) using residuals

Bandwidth method: Hall-Sheather, bw $= 0.20096$

Estimation successfully identifies unique optimal solution

| Variable | Coefficient | Std. Error | t-Statistic | Prob. |
|---|---|---|---|---|
| C | 0.234263 | 0.275226 | 0.851167 | 0.3966 |
| AGE*(MW/10)^2 | 0.163477 | 0.059429 | 2.750772 | 0.0070 |
| AGE*(MW/10)^3 | −0.064155 | 0.024385 | −2.630946 | 0.0098 |
| AGE*(MW/10)^5 | 0.003112 | 0.001289 | 2.413643 | 0.0175 |
| AGE*(MW/10)^6 | −0.000355 | 0.000154 | −2.297951 | 0.0235 |
| AGE | −0.250312 | 0.081126 | −3.085490 | 0.0026 |

| | | | | |
|---|---|---|---|---|
| Pseudo R-squRed | 0.155875 | Mean dependent var | | 9.04E-17 |
| Adjusted R-squRed | 0.116430 | S.D. dependent var | | 0.550900 |
| S.E. of Regression | 0.502648 | Objective | | 20.59843 |
| Quantile dependent var | −0.012850 | Restr. objective | | 24.40210 |
| Sparsity | 1.189356 | Quasi-LR statistic | | 25.58475 |
| Prob(Quasi-LR stat) | 0.000107 | | | |

**Figure 10.26** The output of the 2WI-polynomial QR(0.5) in (10.13).

Quantile Slope Equality test

Equation: UNTITLED

Specification: BLV C AGE*(MW/10)^2 AGE*(MW/10)^3 AGE*(MW/10)^5 AGE*(MW/10)^6 AGE

Estimated equation quantile tau = 0.1

User-specified test quantiles: 0.1 0.9

Test statistic compares all coefficients

| Test summary | Chi-Sq. Statistic | Chi-Sq. d.f. | Prob. |
|---|---|---|---|
| Wald test | 24.38601 | 5 | 0.0002 |

Restriction Detail: b(tau_h) − b(tau_k) = 0

| Quantiles | Variable | Restr. Value | Std. Error | Prob. |
|---|---|---|---|---|
| 0.1, 0.9 | AGE*(MW/10)^2 | −0.089752 | 0.103797 | 0.3872 |
| | AGE*(MW/10)^3 | 0.027583 | 0.043521 | 0.5262 |
| | AGE*(MW/10)^5 | −0.000777 | 0.002327 | 0.7385 |
| | AGE*(MW/10)^6 | 6.82E-05 | 0.000278 | 0.8061 |
| | AGE | 0.197148 | 0.136553 | 0.1488 |

**Figure 10.27**   The output of a QSET based on the 2WI-polynomial QR(0.1).

    2.2  However, the pairs of each slope of the five IVs unexpectedly have an insignificant difference at the 10% level.

3. As another additional statistical result, Figure 10.28 presents the graph of a QPE, based on the QR in (10.13)

4. As an alternative 2WI-polynomial QR(0.5), if we replace *AGE* with *HB* or *TOTALKOL* in (10.13), we find that each of the five IVs also has a significant effect at the 1%, 5%, or 10% level, and their IVs have significant joint effects based on the *QLR* statistic with $p$-values of 0.000 022, and 0.000 502, respectively.

### 10.5.3   QRs of BLV on Three Numerical Variables

Based on the four numerical mother variables, *AGE*, *HB*, *MW*, and *TOTALKOL*, we can easily select four possible sets of three numerical IVs to present the QRs of *BLV* on three numerical IVs, as shown in the next three examples.

#### 10.5.3.1   Additive QR of BLV on MW, AGE, and HB

As an illustration, consider an additive QR of *BLV* on *MW*, *AGE*, and *HB*, with the following ES:

    *BLV C MW AGE HB*                                        (10.14)

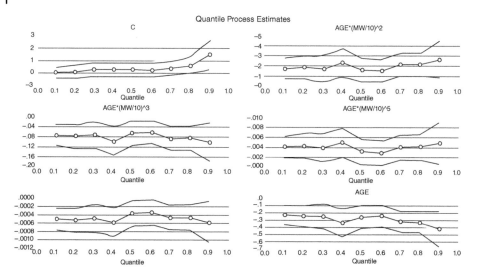

**Figure 10.28** The graphs of a quantile process estimate of the QR in (10.14).

| Variable | QR(0.1) | | | QR(0.5) | | | QR(0.9) | | |
|---|---|---|---|---|---|---|---|---|---|
| | Coef. | t-Stat. | Prob. | Coef. | t-Stat. | Prob. | Coef. | t-Stat. | Prob. |
| C | −3.058 | −3.620 | 0.000 | −2.508 | −2.684 | 0.008 | −1.807 | −0.991 | 0.324 |
| MW | 0.055 | 5.356 | 0.000 | 0.036 | 4.038 | 0.000 | 0.024 | 0.890 | 0.375 |
| AGE | −0.027 | −2.351 | 0.021 | −0.005 | −0.431 | 0.667 | −0.014 | −0.758 | 0.450 |
| HB | 0.099 | 1.694 | 0.093 | 0.107 | 1.799 | 0.075 | 0.170 | 1.435 | 0.154 |
| Pse. R^2 | 0.268 | | | 0.141 | | | 0.037 | | |
| Adj. R^2 | 0.247 | | | 0.117 | | | 0.011 | | |
| S.E.reg | 0.796 | | | 0.505 | | | 0.907 | | |
| Quant DV | −0.668 | | | −0.013 | | | 0.675 | | |
| Sparsity | 1.706 | | | 1.236 | | | 4.131 | | |
| QLR-stat | 36.12 | | | 22.20 | | | 2.306 | | |
| Prob. | 0.000 | | | 0.000 | | | 0.511 | | |

**Figure 10.29** The output summary of the QR($\tau$) in (10.14) for $\tau = 0.1, 0.5$, and 0.9.

### Example 10.13 *Outputs of QR($\tau$) in (10.14)*

Figure 10.29 presents the output summary of the QR($\tau$) in (10.14) for $\tau = 0.1, 0.5$, and 0.9. Based on this summary, the findings and notes are as follows:

1. Each IV, *MW, AGE,* and *HB,* of the QR(0.1) has either a positive or negative significant effect on *BLV* at the 1% or 5% level. For instance, *AGE* has a negative significant effect based on the *t*-statistic of $t_0 = 2.351$ with p-value $= 0.021/2 = 0.0105 < 0.05$. On the

| Variable | QR(0.1) | | | QR(0.5) | | | QR(0.9) | | |
|---|---|---|---|---|---|---|---|---|---|
| | Coef. | t-Stat. | Prob. | Coef. | t-Stat. | Prob. | Coef. | t-Stat. | Prob. |
| C | −6.660 | −1.877 | 0.063 | −7.153 | −1.078 | 0.283 | −9.172 | −0.920 | 0.360 |
| MW*AGE | −0.003 | −1.840 | 0.069 | 0.000 | −0.035 | 0.972 | 0.001 | 0.275 | 0.784 |
| MW*HB | −0.007 | −1.416 | 0.160 | −0.012 | −1.002 | 0.318 | −0.019 | −0.337 | 0.737 |
| AGE*HB | 0.008 | 1.403 | 0.163 | 0.003 | 0.324 | 0.747 | −0.004 | −0.042 | 0.966 |
| MW | 0.198 | 2.738 | 0.007 | 0.185 | 1.222 | 0.224 | 0.226 | 0.334 | 0.739 |
| AGE | −0.010 | −0.108 | **0.915** | −0.042 | −0.318 | 0.751 | 0.010 | 0.008 | 0.993 |
| HB | 0.180 | 0.724 | **0.471** | 0.480 | 0.874 | 0.384 | 0.853 | 1.043 | 0.299 |
| Pse. R^2 | 0.311 | | | 0.157 | | | 0.053 | | |
| Adj. R^2 | 0.272 | | | 0.110 | | | 0.000 | | |
| S.E. reg | 0.739 | | | 0.506 | | | 0.947 | | |
| Quant DV | −0.668 | | | −0.013 | | | 0.675 | | |
| Sparsity | 1.554 | | | 1.229 | | | 3.804 | | |
| QLR-stat | 46.07 | | | 24.98 | | | 3.592 | | |
| Prob. | 0.000 | | | 0.000 | | | 0.732 | | |

**Figure 10.30** The output summary of the QR($\tau$) in (10.15) for $\tau = 0.1, 0.5,$ and 0.9.

other hand, *HB* has a positive significant effect, based on the *t*-statistic of $t_0 = 1.694$ with *p*-value $= 0.093/2 = 0.0465 < 0.05$.
2. At the 10% level, the IV *AGE* of the QR(0.5) has a negative insignificant effect, based on the *t*-statistic of $t_0 = -0.431$ with *p*-value $= 0.667/2 = 0.3335$. But its IVs have a joint significant effect, based on the *QLR* statistic of $QLR_0 = 22.20$ with $df = 3$ and *p*-value $= 0.000$.
3. Each IV of the QR(0.9) has an insignificant effect at the 10% level of significance. And its IVs also have insignificant joint effects, based on the *QLR* statistic of $QLR_0 = 2.306$ with $df = 3$ and *p*-value $= 0.511$. However, the model can't be reduced because the three QRs should have the same set of IVs for the QSET.

#### 10.5.3.2 A Full Two-Way Interaction QR of *BLV* on *MW, GE,* and *HB*
As an extension of the additive QR in (10.14), I present a F2WI-QR of *BLV* on *MW, AGE,* and *HB*, with the following ES. The main objective of this model is to study the effect and test the hypothesis that the effect of an IV on *BLV* depends on the other two IVs.

$$BLV \; C \; MW * AGE \; MW * HB \; AGE * HB \; MW \; AGE \; HB \tag{10.15}$$

**Example 10.14** *An Application of F2WI-QR(0.1) in (10.15)*
Figure 10.30 presents the output summary of the QR($\tau$) in (10.15) for $\tau = 0.1, 0.5,$ and 0.9. Based on this summary, the following findings and notes are presented:

1. All IVs of the F2WI-QR(0.1) and -QR(0.5) have significant joint effects on *BLV* with *p*-value = 0.000. But the IVs of F2WI-QR(0.9) are jointly insignificant, based on the *QLR* statistic of $QLR_0 = 3.592$ with $df = 6$ and *p*-value = 0.732.
2. Specific for the QR(0.1), each 2WI IV has either a positive or negative significant effect at the 5% or 10% level. For instance, *AGE\*HB* has a positive significant effect at the 10% level, based on the *t*-statistic of $t_0 = 1.403$ with *p*-value = $0.163/2 < 0.10$. So, even though the main variables *AGE* and *HB* have large *p*-values, they can be kept as the IVs.
3. The F2WI-QR(0.5) and F2WI-QR(0.9) are presented for comparison and testing the QSET between the three QR($\tau$)s, which can be done easily. Refer to the statistical results in Figure 10.23 and do these as exercises.
4. If the QR(0.1) must be reduced, then QR(0.5) and QR(0.9) have to be reduced so the three QRs have the same set of IVs. See the following example.
5. As an additional analysis, the three IVs, *MW\*AGE*, *AGE\*HB*, and *AGE* of the F2WI-QR(0.1), have a significant joint effects at the 5% level, based on the *QLR* statistic of $QLR_0 = 9.931\,200$ with $df = 3$ and *p*-value = 0.0192. Hence, we can conclude the effect of *AGE* on *BLV* is significantly dependent on *MW* and *HB*. However, based on both QR(0.5) and QR(0.9), the effects of *AGE* on *BLV* are insignificantly dependent on *MW* and *HB*.

### 10.5.3.3 A Full Three-Way Interaction QR of *BLV* on *AGE*, *HB*, and *MW*

As an extension of the F2WI-QR in (10.15), I present a full three-way interaction QR (F3WI-QR) of *BLV* on *MW, AGE*, and *HB*, with the following ES:

$$BLV \ C \ MW^*AGE^*HB \ MW^*AGE \ MW^*HB \ AGE^*HB \ MW \ AGE \ HB \qquad (10.16)$$

**Example 10.15** *An Illustration of the QR in (10.16)*
Figure 10.31 presents a summary of the acceptable reduced 3WI-QR($\tau$)s, for $\tau = 0.1, 0.5$, and 0.9 of the F3WI-QR, with ES (10.16). Based on these outputs, the following findings and notes are presented.

1. The IVs of each of the three reduced QRs, R3WI-QR(0.1), R3WI-QR(0.5), and R3WI-QR(0.9), have joint significant effects on *BLV,* based on the *QLR* statistic, with the *p*-values of 0.000, 0.000, and 0.002, respectively.
2. At the 5% level, the 3WI *MV\*AGE\*HB* of the R3WI-QR(0.25) has a negative significant effect, based on the *t*-statistic of $t_0 = -1.806$ with *p*-value = $0.074/2 = 0.037 < 0.05$. And at the 10% level, the interaction of the R3WI-QR(0.5) also has a negative significant effect, based on the *t*-statistic of $t_0 = -1.527$ with *p*-value = $0.130/2 = 0.065 < 0.10$.
3. But the 3WI IV of the R3WI-QR(0.75) has an insignificant effect, based on the *t*-statistic with *p*-value = 0.469. However, all of its IVs are jointly significant, based on the *QLR* statistic of $QLR_0 = 11.45$ with $df = 4$ and *p*-value 0.022.
4. Hence, based on each of the three QRs, the effect on *AGE* on *BLV* is significantly dependent on a function of *MW\*HB, MW*, and *HB*.

| Variable | R3WI-QR(0.25) | | | R3WI-QR(0.50) | | | R3WI-QR(0.75) | | |
|---|---|---|---|---|---|---|---|---|---|
| | Coef. | t-stat. | Prob. | Coef. | t-stat. | Prob. | Coef. | t-stat. | Prob. |
| C | −0.061 | −0.225 | 0.822 | 0.168 | 0.581 | 0.562 | 0.051 | 0.215 | 0.831 |
| MW*AGE*HB | 0.000 | −1.806 | 0.074 | −0.001 | −1.527 | 0.130 | 0.000 | −0.726 | 0.469 |
| MW*AGE | 0.007 | 2.321 | 0.022 | 0.008 | 1.834 | 0.069 | 0.005 | 0.968 | 0.335 |
| AGE*HB | 0.022 | 2.528 | 0.013 | 0.025 | 1.938 | 0.055 | 0.018 | 1.137 | 0.258 |
| AGE | −0.346 | −3.074 | 0.003 | −0.365 | −2.249 | 0.027 | −0.264 | −1.332 | 0.186 |
| Pse. R^2 | 0.227 | | | 0.153 | | | 0.072 | | |
| Adj. R^2 | 0.199 | | | 0.121 | | | 0.038 | | |
| S.E. reg. | 0.615 | | | 0.504 | | | 0.546 | | |
| Quant. DV | −0.393 | | | −0.013 | | | 0.370 | | |
| Sparsity | 1.284 | | | 1.213 | | | 1.335 | | |
| QLR stat | 37.08 | | | 24.56 | | | 11.45 | | |
| Prob. | 0.000 | | | 0.000 | | | 0.022 | | |

**Figure 10.31**  The outputs of the reduced 3WI-QR($\tau$) in (10.16), $\tau = 0.25, 0.50$, and 0.7.

## 10.6  Complete Latent Variables QRs

In all preceding sections, the QRs are the QRs of a baby latent variable, *BLV*, on the measured IVs or predictors. In this section, I present QRs of *BLV* with a mother latent IV, *MLV*. The latent variable *MLV* is developed using the ordinal variables, *ED and SE*, and the dummy variables *SMOKING* and *NO_ANC,* which I defined as *NO_ANC* = 1 if the observed score of *ANC* = 0, and *NO_ANC* = 0 otherwise. This transformation is applied to ensure the *MLV* is generated based on positively correlated measured variables. Then we obtain the *MLV* using the following equation,:

$$MLV = 0.530573ED + 0.491249SE + 0.582860SMOKING + 0.441722NO\_ANC$$

$$(10.17)$$

(See Appendix C, "Applications of Factor Analysis," for more details on generating the *MLV.*)

And the Z-scores variables of *ZAGE, ZHB, ZMV*, and *ZTKOL* = *ZTOTALKOL*, are generated using the following general formula/equation:

$$ZV = Mean(V)/@Stdev(V)$$

For the following examples, we use the data set in B_MLV.wf1, which is shown in Figure 10.32. In addition, to present more advanced latent variables QRs, the latent variables *BLV, MLV, ZAGE, ZHB, ZMW,* and *ZTKOL* are transformed to positive variables by adding five to each of the six variables. As a result, we have the positive latent variables

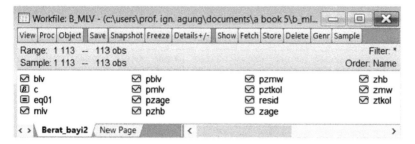

**Figure 10.32** The data set used to present the complete latent variables QRs.

| Statistic | PBLV | PMLV | PZGE | PZHB | PZMW | PZTKOL |
|---|---|---|---|---|---|---|
| Mean | 5 | 5 | 5 | 5 | 5 | 5 |
| Median | 4.9872 | 4.8642 | 4.9193 | 4.9269 | 5.0444 | 4.9391 |
| Maximum | 6.3747 | 6.9389 | 7.9012 | 7.5080 | 7.0565 | 7.5637 |
| Minimum | 3.6572 | 4.1771 | 3.6914 | 2.6039 | 2.0973 | 2.7174 |
| Std. Dev. | 0.5509 | 0.7540 | 1 | 1 | 1 | 1 |
| Skewness | 0.1498 | 0.8400 | 0.6286 | 0.2454 | −0.3671 | 0.2183 |
| Kurtosis | 2.9053 | 2.8465 | 3.1155 | 3.0792 | 2.7930 | 2.6533 |
| Jarque-Bera | 0.4650 | 13.3996 | 7.5034 | 1.1639 | 2.7391 | 1.4634 |
| Probability | 0.7925 | 0.0012 | 0.0235 | 0.5588 | 0.2542 | 0.4811 |
| Sum | 565 | 565 | 565 | 565 | 565 | 565 |
| Sum Sq. Dev. | 33.9910 | 63.6688 | 112 | 112 | 112 | 112 |
| Observations | 113 | 113 | 113 | 113 | 113 | 113 |

**Figure 10.33** The descriptive statistics of the positive latent variables and Z-scores.

*PBLV, PMLV, PZAGE, PZHB, PZMW, and PZTKOL,* as presented in Figure 10.32, with mean = 5, and their descriptive statistics which are shown in Figure 10.33. It is unexpected that the latent variables *PBLV* and *PMLV* don't have the standard deviations of one.

### 10.6.1 Additive Latent Variable QRs of *BLV*

We should accept that the mother ordinal variable *SE* is the most important causal factor of *BLV*, in addition to the *ED*, *SMOKING*, and *NO_ANC*, which are positively correlated with *SE* , then *MLV* should be the main causal factor of *BLV*. For this reason, an additive latent variable QR with the following ES is applied. Our main objectives are to study the main effect of *MLV* on *BLV* adjusted for the other IVs and also to test the hypothesis that *MLV* has a positive effect on *BLV,* adjusted for the other variables, *ZAGE, ZHB, ZMW,* and *ZTKOL.*

$$BLV\ C\ MLV\ ZAGE\ ZHB\ ZMW\ ZTKOL \tag{10.18}$$

| Variable | QR(0.1) | | | QR(0.5) | | | QR(0.9) | | |
|---|---|---|---|---|---|---|---|---|---|
| | Coef. | t-stat. | Prob. | Coef. | t-stat. | Prob. | Coef. | t-stat. | Prob. |
| C | −0.501 | −8.900 | 0.000 | −0.085 | −1.662 | 0.100 | 0.612 | 5.708 | 0.000 |
| MLV | 0.300 | 3.285 | 0.001 | 0.320 | 3.415 | 0.001 | 0.452 | 1.919 | 0.058 |
| ZAGE | −0.099 | −1.614 | 0.109 | −0.069 | −1.166 | 0.246 | −0.180 | −0.979 | 0.330 |
| ZHB | 0.126 | 2.281 | 0.025 | 0.117 | 1.878 | 0.063 | −0.056 | −0.543 | 0.588 |
| ZMW | 0.363 | 8.173 | 0.000 | 0.321 | 5.640 | 0.000 | 0.175 | 1.094 | 0.277 |
| ZTKOL | 0.019 | 0.231 | 0.818 | 0.107 | 1.860 | 0.066 | 0.200 | 1.853 | 0.067 |
| Pse. R^2 | 0.344 | | | 0.233 | | | 0.155 | | |
| Adj. R^2 | 0.313 | | | 0.197 | | | 0.115 | | |
| S.E. reg | 0.692 | | | 0.458 | | | 0.789 | | |
| Quant DV | −0.668 | | | −0.013 | | | 0.675 | | |
| Sparsity | 1.635 | | | 1.061 | | | 3.036 | | |
| QLR-stat | 48.41 | | | 42.80 | | | 13.05 | | |
| Prob. | 0.000 | | | 0.000 | | | 0.023 | | |

**Figure 10.34** The output summary of the QR($\tau$) in (10.18) for $\tau = 0.1, 0.5$, and 0.9.

### Example 10.16   *Outputs of the Latent Variables QRs in (10.18)*

Figure 10.34 presents the output summary of the latent variables QR($\tau$), for $\tau = 0.1, 0.5$, and 0.9. Based on this summary, the following findings and notes are presented:

1. The latent variable *MLV* of QR(0.1), QR(0.5), and QR(0.9), respectively, has a positive significant adjusted effect, based on the *t*-statistic with *p*-values of 0.001, 0.001, and 0.058. Hence, the outputs present the expected results. Or, in other words, the data supports the hypothesis.
2. The IVs of each of the QRs have significant joint effects on *BLV*, based on the *Quasi-LR* statistic, with *p*-values of 0.000, 0.000, and 0.020, respectively. Even though *ZTKOL* in QR(0.1) and *ZHB* in QR(0.9) have very large *p*-values, the QRs should not be reduced because all QRs should have the same set of IVs when doing an analysis of the quantile process. Do the analysis using the quantile process as exercises.

### Example 10.17   *An Application of the Stability Test*

As the output of an advanced analysis, Figure 10.35 presents the output of the stability test, specifically the Ramsey RESET Test, for the QR(Median) in (10.18). Based on this output, the following findings and notes are presented:

1. This test in fact is an omitted variable test (OVT) for the squares of fitted values, *FITTED^2*, which shows that it has a significant effect, based on the *QLR L*-statistic of $QLR_0 = 5.173\,521$ with $df = 1$ and *p*-value $= 0.0229$.

Ramsey RESET test
Equation: EQ06_Q
Specification: BLV C MLV ZAGE ZHB ZMW ZTKOL
**Omitted variables: squares of fitted values**

|  | Value | df | Probability |
|---|---|---|---|
| QLR L-statistic | 5.173521 | 1 | 0.0229 |
| QLR lambda-statistic | 5.080798 | 1 | 0.0242 |

| L-test summary: | Value |
|---|---|
| Restricted objective | 18.72468 |
| Unrestricted objective | 18.06147 |
| Scale | 0.256389 |

| Lambda-test summary: |  |
|---|---|
|  | Value |
| Restricted log obj. | 2.929843 |
| Unrestricted log obj. | 2.893781 |
| Scale | 0.014195 |

Unrestricted test equation:

Dependent variable: BLV

Method: quantile regression (Median)

Date: 04/02/19   Time: 06:27

Sample: 1 113

Included observations: 113

Huber Sandwich standard errors & covariance

Sparsity method: Kernel (Epanechnikov) using residuals

Bandwidth method: Hall-Sheather, bw = 0.20096

Estimation successfully identifies unique optimal solution

| Variable | Coefficient | Std. error | t-statistic | Prob. |
|---|---|---|---|---|
| C | 0.060595 | 0.065619 | 0.923432 | 0.3579 |
| MLV | 0.349110 | 0.094161 | 3.707595 | 0.0003 |
| ZAGE | −0.106092 | 0.055141 | −1.924013 | 0.0570 |
| ZHB | 0.027261 | 0.056678 | 0.480978 | 0.6315 |
| ZMW | 0.312925 | 0.060655 | 5.159083 | 0.0000 |
| ZTKOL | 0.082221 | 0.054093 | 1.519986 | 0.1315 |
| FITTED^2 | −0.840760 | 0.318487 | −2.639856 | 0.0095 |

| | | | | |
|---|---|---|---|---|
| Pseudo R-squared | 0.259840 | Mean dependent var | 9.04E-17 |
| Adjusted R-squared | 0.217944 | S.D. dependent var | 0.550900 |
| S.E. of regression | 0.443327 | Objective | 18.06147 |
| Quantile dependent var | −0.012850 | Restr. objective | 24.40210 |
| Sparsity | 1.025555 | Quasi-LR statistic | 49.46111 |
| Prob(Quasi-LR stat) | 0.000000 | | |

**Figure 10.35** The output of the stability test based on the QR(Median) in (10.18).

2. In the Unrestricted Test Equation, *FITTED^2* is presented as an additional IV of the QR(Median) in (10.18), which shows it has a negative significant adjusted effect on *BLV*, based on the *t*-statistic of $t_0 = -2.639\,856$ with $df = (113 - 7)$ and *p*-value $= 0.0096/2 = 0.0048$. In addition, the output also presents the statement "*Estimation successfully identifies unique optimal solution.*" Hence, the *FITTED^2* should be inserted as an additional IV of the QR(Median) to have a better fit QR(Median).

3. However, there is no statement about the stability condition of the QR(Median).

**Example 10.18**   *Outputs of Mean Regression in (10.18)*
As a comparison, Figure 10.36 presents the output of the LSR or MR of (10.18). Based on this output, the findings and notes are as follows:

1. Since the estimate of the intercept parameter is very close to zero, with a probability of 1.0000, then the MR could be modified to the model without the intercept. Do this as an exercise.

2. In contrast with the QR(0.5) in Figure 10.34, with *ZHB* having a significant effect at the 10% level, the IV *ZHB* of this MR has an insignificant effect with a *p*-value of 0.2399. But at the 15% level, it has a positive significant effect, based on the *t*-statistic of $t_0 = 1.181\,715$

Dependent variable: BLV

Method: least squares

Date: 04/02/19   Time: 06:00

Sample: 1 113

Included observations: 113

| Variable | Coefficient | Std. error | t-statistic | Prob. |
|---|---|---|---|---|
| C | 1.33E-16 | 0.041874 | 3.18E-15 | 1.0000 |
| MLV | 0.343925 | 0.072639 | 4.734730 | 0.0000 |
| ZAGE | -0.096481 | 0.043734 | -2.206085 | 0.0295 |
| ZHB | 0.057992 | 0.049074 | 1.181715 | 0.2399 |
| ZMW | 0.289498 | 0.049498 | 5.848735 | 0.0000 |
| ZTKOL | 0.122859 | 0.043274 | 2.839122 | 0.0054 |

| | | | |
|---|---|---|---|
| R-squRed | 0.376271 | Mean dependent var | 9.04E-17 |
| Adjusted R-squRed | 0.347125 | S.D. dependent var | 0.550900 |
| S.E. of regression | 0.445131 | Akaike info criterion | 1.270740 |
| Sum squRed resid | 21.20116 | Schwarz criterion | 1.415557 |
| Log likelihood | -65.79681 | Hannan-Quinn criter. | 1.329505 |
| F-statistic | 12.90980 | Durbin-Watson stat | 0.949652 |
| Prob(F-statistic) | 0.000000 | | |

**Figure 10.36**   The output of the mean regression in (10.18).

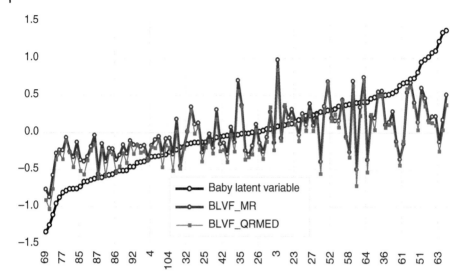

**Figure 10.37** The line and symbol graphs of *BLV*, *BLVF_MR*, and *BLVF_QRMed*.

with $df = (113 - 6)$ and $p$-value $= 0.2399/2 = 0.11995 < 0.15$. And all the IVs have significant joint effects, based on the $F$-statistic of $F_0 = 12.9098$ with $df = (5107)$ and $p$-value $= 0.000000$. For these reasons, the latent variables MR is an acceptable regression.

3. As an additional illustration, Figure 10.37 presents the graphs of the sorted *BLV* and its forecast value variables, *BLVF_MR* and *BLVF_QRMed*, of the MR and the QR(0.5) in (10.18), respectively. Based on these graphs, the following findings and notes are presented:

4. The forecast value variables *BLVF_MR* and *BLVF_QRMed* are the same as the fitted value variables of the MR and the QR(0.5) in (10.18).

5. The graphs clearly show that most values of *BLVF_MR* are greater than the values of *BLVF_QRMed*. However, they are significantly correlated, based on the $t$-statistic of $t_0 = 66.07$ with $df = 111$ and $p$-value $= 0.0000$, with $r = 0.987622$.

## 10.6.2 Advanced Latent Variables QRs

In order to develop advanced latent variables QRs, such as the interaction latent variables, polynomial-QRs (ILVP-QRs), and translog-linear QRs (TL-QRs), we should be using the positive latent variables, *PBLV, PMLV, PZAGE, PZHB, PZMW*, and *PZTKOL*, which were mentioned earlier in this section, because the interaction of two negative scores and the squared of a negative scores will be misleading. This is because the interaction *MLV\*ZAGE* for the negative scores of *MLV* and *ZAGE* can be greater than many or all of the multiplications of their positive scores. Similarly for each quadratic latent variable.

### 10.6.2.1 The Two-Way-Interaction QR of *PBLV* on *(PLV1,PLV2)*

The 2WI-QR of *PBLV* on two positive latent variables, *PLV1* and *PLV2*, has the following general ES.

$$PBLV \ C \ PLV1 * PLV2 \ PLV1 \ PLV2 \tag{10.19}$$

Based on the five positive latent variables in Figure 10.32, we could develop 10 possible 2WI-QRs of *PBLV*. The main objective of this model is to test the hypothesis that the effect of *PLV1* on *PBLV* depends on *PLV2*, or the effect of *PLV2* on *PBLV* depends on *PLV1*, which depency relationship is more suitable in the theoretical sense.

**Example 10.19**   *An Application of the 2WI-QR in (10.19)*
Figure 10.38 presents the output summary of the full model F2WI-QR(0.5) for *PLV1* = *PMLV* and *PLV2* = *PZMW* in (10.19) and two of its reduced models, Reduced QR1 and Reduced QR2. Based on this summary, the following findings and notes are presented:.

1. Each IV of the F2WI_QR has an insignificant effect at the 10% level, and the interaction *PMLV*PZMW* has very large probability of 0.535. However, all IVs have significant joint effects on *PBLV*, based on the *QLR* statistic of $QRL_0$ = 33.55 with *df* = 3 and *p*-value = 0.000. In addition, *PMLV*PZMW* and *PMLV* have significant joint effects, based on the Wald test *(Chi-square* statistic) of $\chi_0^2$ = 15.738 37 with *df* = 2 and *p*-value = 0.0004. So, the F2WI-QR is an acceptable QR in representing that the effect of *PMLV* on *BLV* is significantly dependent on *PZMW*.
2. Furthermore, *PMLV*PZMW* and *PZMW* also have significant joint effects, based on the Wald test *(Chi-square* statistic) of $\chi_0^2$ = 26.009 31 with *df* = 2 and *p*-value = 0.0000. So, the F2WI-QR is an acceptable QR in representing that the effect of *PZMW* on *BLV* is significantly dependent on *PMLV*.
3. As an additional illustration, the outputs of Reduced QR1 and Reduced QR2 are presented, which show that the interaction *PMLV*PZMW* has a significant effect on *PBLV* with the *p*-values of 0.000 and 0.001, respectively.

| Variable | F2WI-QR(0.5) | | | Reduced QR1 | | | Reduced QR2 | | |
|---|---|---|---|---|---|---|---|---|---|
| | Coef. | t-stat. | Prob. | Coef. | t-stat. | Prob. | Coef. | t-stat. | Prob. |
| C | 0.403 | 0.160 | 0.873 | 3.275 | 7.736 | 0.000 | 3.425 | 13.13 | 0.000 |
| PMLV*PZMW | −0.062 | −0.622 | 0.535 | 0.058 | 5.652 | 0.000 | 0.062 | 3.363 | 0.001 |
| PMLV | 0.591 | 1.299 | 0.197 | 0.048 | 0.754 | 0.453 | | | |
| PZMW | 0.625 | 1.174 | 0.243 | | | | −0.005 | −0.068 | 0.946 |
| Pse. R^2 | 0.196 | | | 0.173 | | | 0.171 | | |
| Adj. R^2 | 0.174 | | | 0.158 | | | 0.156 | | |
| S.E. reg. | 0.477 | | | 0.482 | | | 0.487 | | |
| Quant. DV | 4.987 | | | 4.987 | | | 4.987 | | |
| Sparsity | 1.139 | | | 1.127 | | | 1.135 | | |
| QLR stat | 33.55 | | | 29.90 | | | 29.37 | | |
| Prob. | 0.000 | | | 0.000 | | | 0.000 | | |

**Figure 10.38**   The output summary of the F2WI-QR(0.5) in (10.19) and its reduced QRs.

### 10.6.2.2 Two-Way-Interaction QRs of *PBLV* on (*PLV1,PLV2,PLV3*)
**Example 10.20** *An Application of a F2WI-QR(Median)*

Figure 10.39 presents the output of a F2WI-QR(Median) of *PBLV* on (*LV1,LV2,LV3*) with the following ES, with the main objective is to test the following hypothesis: The effect of the latent variable *PMLV* on *PBLV* depends on the linear combination of the *PZMW* and *PZHB*.

$$PBLV = C(1) + (C(2)*PZMW + C(3)*PZHB + C(4))*PMLV$$
$$+ C(5)*PZMW*PZHB + C(6)*PZMW + C(7)*PZHB \tag{10.20}$$

Dependent variable: PBLV

Method: quantile regression (Median)

Sample: 1 113

Included observations: 113

Huber Sandwich standard errors & covariance

Sparsity method: Kernel (Epanechnikov) using residuals

Bandwidth method: Hall-Sheather, bw = 0.20096

Estimation successfully identifies unique optimal solution

PBLV = C(1) + (C(2)*PZMW + C(3)*PZHB + C(4))*PMLV + C(5)*PZMW
 *PZHB + C(6)*PZMW + C(7)*PZHB

| | Coefficient | Std. error | t-statistic | Prob. |
|---|---|---|---|---|
| C(1) | 1.520569 | 4.060903 | 0.374441 | 0.7088 |
| C(2) | −0.007584 | 0.099486 | −0.076237 | 0.9394 |
| C(3) | 0.019724 | 0.081882 | 0.240879 | 0.8101 |
| C(4) | 0.217069 | 0.631277 | 0.343857 | 0.7316 |
| C(5) | −0.004561 | 0.063988 | −0.071281 | 0.9433 |
| C(6) | 0.418163 | 0.558672 | 0.748495 | 0.4558 |
| C(7) | 0.003977 | 0.663079 | 0.005997 | 0.9952 |

| | | | |
|---|---|---|---|
| Pseudo R-squRed | 0.211735 | Mean dependent var | 5.000000 |
| Adjusted R-squRed | 0.167116 | S.D. dependent var | 0.550900 |
| S.E. of regression | 0.483413 | Objective | 19.23532 |
| Quantile dependent var | 4.987150 | Restr. objective | 24.40210 |
| Sparsity | 1.105697 | Quasi-LR statistic | 37.38299 |
| Prob(Quasi-LR stat) | 0.000001 | | |

**Figure 10.39** The output of the F2WI-QR(Median) in (10.20).

| Wald test: | | | |
| --- | --- | --- | --- |
| Equation: QR(0.5) in Figure 10.38 | | | |
| Test statistic | Value | df | Probability |
| F-statistic | 4.221090 | (3, 106) | 0.0073 |
| Chi-square | 12.66327 | 3 | 0.0054 |
| Null hypothesis: C(2) = C(3) = C(4) = 0 | | | |

**Figure 10.40** The Wald test for the null hypothesis $H_0$: C(2) = C(3) = C(4) = 0.

Based on this output, the following findings and notes are presented:

1. Even though each IV has a large or very large p-value, all the IVs have significant joint effects on *PBLV*, based on the *QLR* statistic of $QLR_0 = 37.382\,99$ with $df = 2$ and p-value $= 0.0000$.

2. Based on the Wald test (*Chi-square* statistic) of $\chi_0^2 = 12.663\,27$ with $df = 3$ and p-value $= 0.0054$, we can see in Figure 10.40, that the effect of *PMLV* on *PBLV* is significantly dependent on the linear combination *(C(2)\*PZMW+ C(3)\*PZHB + C(4))*. Hence, the F2WI-QR(Median) in Figure 10.39 is an acceptable model in representing that the effect of *PMLV* on *PBLV* depends on *(C(2)\*PZMW+ C(3)\*PZHB + C(4))*.

3. Since each IV of the F2WI-QR(Median) in Figure (10.39) has large a p-value, I try to obtain its reduced model by using the trial-and-error method. The outputs of three alternative reduced QRs, Red-QR-1, Red-QR-2, and Red-QR-3, are presented in Figure 10.41. Based on these outputs, the following findings and notes are presented:

   3.1 The function Red-QR-1 has the following equation, which shows that the effect of *PMLV* on *PBLV* depends on (0.063*PZMW*+ 0.017*PZHB* − 0.08), and it can be tested using the Wald test. Do that as an exercise.

   $$PBLV = (0.063PZMW + 0.017PZHB - 0.088) * PMLV$$
   $$+ (3.398 + 0.001PZMW * PZHB)$$

   Note that the graphs of this function in a two-dimensional space *(PMLV, PBLV)* are regression lines of *PBLV* on *PMLV*, with various slopes and various intercepts.

   3.2 The function Red-QR-2 has the following equation, which shows that the linear effect of *PMLV* on *PBLV* depends on (0.063*PZMW* + 0.018*PZHB* − 0.099):

   $$PBLV = 3.446 + (0.063PZMW + 0.018PZHB - 0.099) * PMLV$$

   Note that the graphs of this function in a two-dimensional space *(PMLV,PBLV)* also are regression lines of *PBLV* on *PMLV*, with various slopes and a constant or fixed intercept of 3.446.

| Variable | Red-QR-1 | | | Red-QR-2 | | | Red-QR-3 | | |
|---|---|---|---|---|---|---|---|---|---|
| | Coef. | t-stat. | Prob. | Coef. | t-stat. | Prob. | Coef. | t-stat. | Prob. |
| C | 3.398 | 3.524 | 0.001 | 3.446 | 7.721 | 0.000 | 2.845 | 6.261 | 0.000 |
| PMLV*PZMW | 0.063 | 1.696 | 0.093 | 0.063 | 5.721 | 0.000 | 0.032 | 1.195 | 0.235 |
| PMLV*PZHB | 0.017 | 0.492 | 0.624 | 0.018 | 1.574 | 0.118 | 0.020 | 1.795 | 0.076 |
| PZMW*PZHB | 0.001 | 0.021 | 0.983 | | | | | | |
| PMLV | −0.088 | −0.243 | 0.808 | −0.099 | −0.877 | 0.382 | | | |
| PZMW | | | | | | | 0.162 | 1.236 | 0.219 |
| Pse. R^2 | 0.197 | | | 0.197 | | | 0.203 | | |
| Adj. R^2 | 0.167 | | | 0.175 | | | 0.181 | | |
| S.E. reg. | 0.477 | | | 0.474 | | | 0.474 | | |
| Quant. DV | 4.987 | | | 4.987 | | | 4.987 | | |
| Sparsity | 1.119 | | | 1.110 | | | 1.114 | | |
| QLR stat | 34.39 | | | 34.65 | | | 35.56 | | |
| Prob. | 0.000 | | | 0.000 | | | 0.000 | | |

**Figure 10.41** The outputs of alternative reduced QR(Median)s in (10.20).

3.3 The function Red-QR-3 has the following equation, which shows the effect of *PMLV* on *PBLV* depends on $(0.032PZMW + 0.020PZHB)$:

$$PBLV = (2.845 + 0.162PZMW) + (0.032PZMW + 0.020PZHB) * PMLV$$

Note that the graphs of this function in a two-dimensional space *(PMLV,PBLV)* are regression lines of *PBLV* on *PMLV*, with various slopes and various intercepts.

### 10.6.2.3 A Special Full-Two-Way-Interaction QR of *PBLV*

A special or specific F2WI-QR of *PBLV* on four predictors, *PZMW, PZAGE, PZHB*, and *PZTKOL*, has the following ES. The main objective of this QR is to test the hypothesis the effect of *PMLV* on *PBLV* depends on the linear combination of four variables: *PZMW, PZAGE, PZHB*, and *PZKOL*.

> PBLV  C PMLV * PZMW PMLV * PZAGE PMLV * PZHB PMLV * PZTKOL
>
> PMLV PZMW PZAGE PZHB PZTKOL                          (10.21)

**Example 10.21** *An Application of a F2WI-QR(Median) in (10.21)*
Figure 10.42 presents the output of the special F2WI-QR(Median) in (10.21). Based on this output, the following findings and notes are presented:

1. Even though six of nine IVs have very large *p*-values, all the IVs have significant joint effects on *PBLV*, based on the *QLR* statistic of $QLR_0 = 46.787\ 27$ with $df = 9$ and

Dependent Variable: PBLV

Method: Quantile Regression (Median)

Sample: 1 113

Included observations: 113

Huber Sandwich Standard Errors & Covariance

Sparsity method: Kernel (Epanechnikov) using residuals

Bandwidth method: Hall-Sheather, bw = 0.20096

Estimation successfully identifies unique optimal solution

| Variable | Coefficient | Std. Error | t-Statistic | Prob. |
|---|---|---|---|---|
| C | −1.602676 | 3.337844 | −0.480153 | 0.6321 |
| PMLV*PZMW | −0.145757 | 0.097293 | −1.498127 | 0.1372 |
| PMLV*PZAGE | 0.030445 | 0.074467 | 0.408832 | 0.6835 |
| PMLV*PZHB | −0.020263 | 0.059691 | −0.339463 | 0.7350 |
| PMLV*PZTKOL | 0.027012 | 0.090686 | 0.297863 | 0.7664 |
| PMLV | 0.850331 | 0.656003 | 1.296229 | 0.1978 |
| PZMW | 1.113346 | 0.493492 | 2.256057 | 0.0262 |
| PZAGE | −0.251251 | 0.362437 | −0.693225 | 0.4897 |
| PZHB | 0.156896 | 0.302490 | 0.518681 | 0.6051 |
| PZTKOL | −0.037622 | 0.446543 | −0.084251 | 0.9330 |

| | | | | |
|---|---|---|---|---|
| Pseudo R-squRed | 0.249302 | Mean dependent var | | 5.000000 |
| Adjusted R-squRed | 0.183707 | S.D. dependent var | | 0.550900 |
| S.E. of regression | 0.468152 | Objective | | 18.31862 |
| Quantile dependent var | 4.987150 | Restr. objective | | 24.40210 |
| Sparsity | 1.039973 | Quasi-LR statistic | | 46.79727 |
| Prob(Quasi-LR stat) | 0.000000 | | | |

**Figure 10.42**  The output of the F2WI-QR(Median) in (10.21).

p-value = 0.000 000. So the QR in (10.21) is an acceptable QR(Median) for testing whether the adjusted effect of *PMLV* on *PBLV* depends on a linear combination of *PZMW, PZAGE, PZHB,* and *PZTKOL*.

2. Based on the output in Figure 10.42, we have the following QR(Median) function, which shows that the effect of *PMLV* on *PBLV* depends on a linear function:

$$PBLV = -1.603 + PMLV * (-0.146PZMW + 0.0304PZAGE - 0.0202PZHB$$
$$+0.027PZTKOL + 0.850) + 1.113PZMW - 0.251PZAGE$$
$$+0.157PZHB - 0.039PZTKOL$$

3. And the output of the RVT in Figure 10.43 shows that the five IVs, *PMLV*PZMW, PMLV*PZAGE, PMLV*PZHB, PMLV*PZTKOL, and PMLV,* are jointly significant, based

Redundant Variables Test

Equation: (10.21)

Specification: PBLV C PMLV*PZMW PMLV*PZAGE PMLV*PZHB

PMLV*PZTKOL PMLV PZMW PZAGE PZHB PZTKOL

Redundant Variables: PMLV*PZMW PMLV*PZAGE PMLV*PZHB

PMLV *PZTKOL PMLV

Null hypothesis: PMLV*PZMW PMLV*PZAGE PMLV*PZHB

PMLV*PZTKOL PMLV are jointly insignificant

|                      | Value     | df | Probability |
|----------------------|-----------|----|-------------|
| QLR L-statistic      | 19.22428  | 5  | 0.0017      |
| QLR Lambda-statistic | 18.02122  | 5  | 0.0029      |

**Figure 10.43** The RVT for the null hypothesis in the output above.

on the *QLR L*-statistic of $QLR\_L_0 = 19.224\,28$ with $df = 5$ and $p$-value $= 0.0017$. So, we can conclude that *PMLV* on *PBLV* is significantly dependent on the linear combination of *PZMW, PZAGE, PZHB, and PZTKOL*. However, it is important to mention that the statements of the *null hypothesis* presented in Figure 10.43 is the output of a revised of updated software by the EViews' provider, since I found the original EViews 10 was misleading.

4. As an additional illustration, Figure 10.44 presents the output summary of three reduced QRs, Red-QR-1, Red-QR-2, and Red-QR-3, of the F2WI-QR(Median) in (10.21). Based on this summary, the findings and notes are as follows:

4.1 All IVs of each reduced QR have significant joint effects, based on the *QLR* statistic with a $p$-value of 0.000.

4.2 Only the Red-QR-3 has a 2WI IV, namely *PMLV*PZAGE*, which has a $p$-value $> 0.30$. However, we find that the five IVs, *PMLV*PZMW, PMLV*PZAGE, PMLV*PZHB, PMLV*PZTKOL,* and *PMLV,* are jointly significant, based on the *QRL L*-statistic of $QRL\text{-}L_0 = 27.406\,04$ with $df = 5$ and $p$-value $= 0.0000$. (Refer to the RVT output in Figure 10.43.)

### 10.6.2.4 An Application of a Nonlinear QR of PBLV

Referring to the nonlinear or multiplicative Cobb–Douglas production function, the following example presents the statistical results based on a TL-QR of *PBLV*, with the this ES:

$$\log(PBLV) \ C \ \log(PMLV) \ \log(PZAGE) \ \log(PZHB)$$
$$\log(PZMW) \ \log(PZTKOL) \tag{10.22}$$

**Example 10.22** *An Application of the QR(Median) in (10.22)*

Figure 10.45 presents the output of the translog-linear QR(Median) in (10.22). Based on this output, the following findings and notes are presented:

| Variable | Red-QR-1 | | | Red-QR-2 | | | Red-QR-3 | | |
|---|---|---|---|---|---|---|---|---|---|
| | Coef. | t-Stat. | Prob. | Coef. | t-Stat. | Prob. | Coef. | t-Stat. | Prob. |
| C | 3.097 | 8.265 | 0.000 | 5.596 | 3.538 | 0.001 | −0.175 | −0.065 | 0.948 |
| PMLV*PZMW | 0.060 | 5.177 | 0.000 | 0.059 | 4.944 | 0.000 | −0.120 | −1.286 | 0.201 |
| PMLV*PZAGE | −0.022 | −2.222 | 0.028 | 0.069 | 1.201 | 0.233 | 0.039 | 0.701 | **0.485** |
| PMLV*PZHB | 0.016 | 1.843 | 0.068 | 0.015 | 1.270 | 0.207 | 0.017 | 1.345 | 0.182 |
| PMLV*PZTKOL | 0.021 | 1.770 | 0.080 | 0.017 | 1.445 | 0.152 | 0.021 | 1.759 | 0.082 |
| PMLV | | | | −0.497 | −1.583 | 0.116 | 0.510 | 0.982 | 0.329 |
| PZAGE | | | | −0.431 | −1.413 | 0.161 | −0.284 | −0.996 | 0.322 |
| PZMW | | | | | | | 0.995 | 2.079 | 0.040 |
| Pse. R^2 | 0.219 | | | 0.224 | | | 0.247 | | |
| Adj. R^2 | 0.190 | | | 0.180 | | | 0.197 | | |
| S.E. reg. | 0.456 | | | 0.459 | | | 0.464 | | |
| Quant. DV | 4.987 | | | 4.987 | | | 4.987 | | |
| Sparsity | 1.102 | | | 1.086 | | | 1.049 | | |
| QLR stat | 38.72 | | | 40.21 | | | 45.93 | | |
| Prob. | 0.000 | | | 0.000 | | | 0.000 | | |

**Figure 10.44**   The outputs of three alternative reduced QRs of the QR(Median) in (10.21).

1. Based on the output in Figure 10.45, we have the following translog-linear function:

$$LOG\,(PBLV) = 0.406 + 0.377 * LOG\,(PMLV) - 0.103 * LOG\,(PZAGE)$$
$$+0.088 * LOG\,(PZHB) + 0.307 * LOG\,(PZMW) + 0.077 * LOG\,(PZTKOL)$$

which can be presented as the nonlinear or multiplicative function as follows:

$$PBLV = Exp\,(0.406) \times PMLV^{0.377} PZAGE^{0.103} PZHB^{0.088} \times$$
$$PZMW^{0.307} PZTKOL^{0.077}$$

2. As additional analysis, based on the translog linear QR(Median) in (10.22), we can obtain two alternative forecast value variables, namely *BLVF* and *LOGBLVF,* as presented in Figure 10.46a, b, respectively. Based on these outputs, the following findings and notes are presented.

   2.1 The forecast value variables *BLVF* and *LOGBLF* have the Theil inequality coefficient (TIC) of 0.044 445 and 0.027 267, respectively, where $TIC = 0$ indicates a perfect forecast and $TIC = 1$ indicates the worst forecast (Theil 1966).

   2.2 To generalize, by using *log(Y)* as a DV of various QRs, we can directly obtain two alternative forecast variables, *YF* and *LOGYF*.

Dependent Variable: LOG(PBLV)

Method: Quantile Regression (Median)

Included observations: 113

Huber Sandwich Standard Errors & Covariance

Sparsity method: Kernel (Epanechnikov) using residuals

Bandwidth method: Hall-Sheather, bw = 0.20096

Estimation successfully identifies unique optimal solution

| Variable | Coefficient | Std. Error | t-Statistic | Prob. |
|---|---|---|---|---|
| C | 0.406025 | 0.231658 | 1.752696 | 0.0825 |
| LOG(PMLV) | 0.377462 | 0.105112 | 3.591056 | 0.0005 |
| LOG(PZGE) | −0.102871 | 0.063042 | −1.631794 | 0.1057 |
| LOG(PZHB) | 0.087571 | 0.063760 | 1.373452 | 0.1725 |
| LOG(PZMW) | 0.307348 | 0.051951 | 5.916078 | 0.0000 |
| LOG(PZTKOL) | 0.077114 | 0.054572 | 1.413081 | 0.1605 |

| | | | | |
|---|---|---|---|---|
| Pseudo R-squRed | 0.232440 | Mean dependent var | | 1.603383 |
| Adjusted R-squRed | 0.196573 | S.D. dependent var | | 0.110877 |
| S.E. of regression | 0.089126 | Objective | | 3.766274 |
| Quantile dependent var | 1.606865 | Restr. objective | | 4.906814 |
| Sparsity | 0.218679 | Quasi-LR statistic | | 41.72467 |
| Prob(Quasi-LR stat) | 0.000000 | | | |

**Figure 10.45** The output of the QR(Median) in (10.22).

### 10.6.2.5 An Application of Semi-Log Polynomial QR of log(PBLV)

To learn the possibility for the application of a semi-log polynomial QR of *log(PBLV)*, that is a QR of *log(PBLV)* on positive latent variable IVs, I examined the scatter graphs of *log(PBLV)* on each of the four positive latent variables, *PMLV, PZGE, PZMW,* and *PZTKOL*, with their linear fit and linear polynomial fit, as presented in Figure 10.47. Hence, the semi-log polynomial QR would be an acceptable model, Corresponding to each of the graphs, I propose the semi-log polynomial QR of *log(PBLV)*, on a positive latent variable *PLV*, with the following general ES, where the power *k*, and its possible reduced model, will be obtain using the trial-and-error method:

$$log\,(PBLV)\ C\ PLuV\ PLV^2 \ldots PLV^k \tag{10.23}$$

**Example 10.23** *An Application of the QR($\tau$) in (10.23)*

Figure 10.48 presents the output summary of three QR($\tau$)s in (10.23). Based on this summary, the following findings and notes are presented:

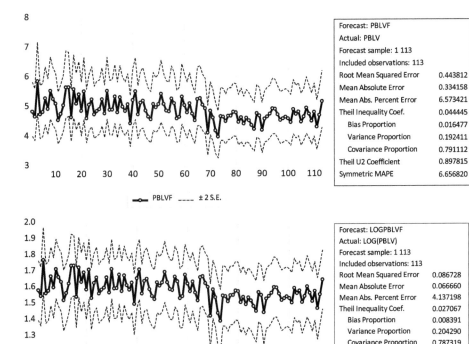

| Forecast: PBLVF | |
| --- | --- |
| Actual: PBLV | |
| Forecast sample: 1 113 | |
| Included observations: 113 | |
| Root Mean Squared Error | 0.443812 |
| Mean Absolute Error | 0.334158 |
| Mean Abs. Percent Error | 6.573421 |
| Theil Inequality Coef. | 0.044445 |
| Bias Proportion | 0.016477 |
| Variance Proportion | 0.192411 |
| Covariance Proportion | 0.791112 |
| Theil U2 Coefficient | 0.897815 |
| Symmetric MAPE | 6.656820 |

| Forecast: LOGPBLVF | |
| --- | --- |
| Actual: LOG(PBLV) | |
| Forecast sample: 1 113 | |
| Included observations: 113 | |
| Root Mean Squared Error | 0.086728 |
| Mean Absolute Error | 0.066660 |
| Mean Abs. Percent Error | 4.137198 |
| Theil Inequality Coef. | 0.027067 |
| Bias Proportion | 0.008391 |
| Variance Proportion | 0.204290 |
| Covariance Proportion | 0.787319 |
| Theil U2 Coefficient | 0.873075 |
| Symmetric MAPE | 4.162431 |

**Figure 10.46** (a) The forecast value variable, *BLVF*, based on the QR(Median) in (10.22). (b) The forecast value variable, *LOGBLVF*, based on the QR(Median) in (10.22).

1. Each IV of the QR(0.25) has either a positive or negative significant effect on *log(PBLV)* at the 5% or 10% level of significance. For instance, the IV *PMLV* has a negative significant effect, based on the *t*-statistic of $t_0 = -1.640$ with $df = (113-5)$ and *p*-value $= 0.104/2 = 0.052 < 0.10$. But, unexpectedly. all its IVs have an insignificant joint effect, based on the *QRL* statistic of $QLR_0 = 6.299$ with $df = 4$ and *p*-value $= 0.178$.

2. Similarly, each IV of the QR(0.50) has either a positive or negative significant effect on *log(PBLV)* at the 5% level. For instance, the IV *PMLV^4* has a positive significant effect, based on the *t*- statistic of $t_0 = 2.331$ with $df = (113-5)$ and *p*-value $= 0.022/2 = 0.011$. But, all its IVs all unexpectedly have an insignificant joint effect, based on the *QRL* statistic of $QLR_0 = 7.168$ with $df = 4$ and *p*-value $= 0.126$.

3. On the other hand, at the 10% level, each IV of the QR(0.75) has an insignificant effect on *log(PBLV)*, but each has either positive or negative significant effect at the 15% level (Lapin 1973; Agung 2009). However, it is also unexpected, at the 10% level, that all its IVs have a significant joint effect, based on the *QRL* statistic of $QLR_0 = 8.670$ with $df = 4$ and *p*-value $= 0.070$.

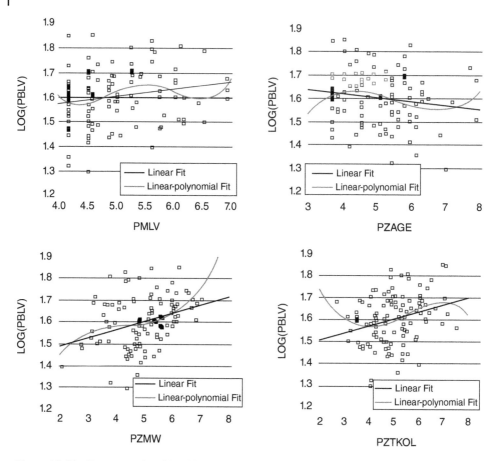

**Figure 10.47** Scatter graphs of *log(PBLV)* on each of the four positive latent variables, with their linear fit and linear-polynomial fit.

### Example 10.24 *An Extension of the QR(τ) in (10.23)*

As an extension of the additive semi-log the QR($\tau$) in (10.23), we also can have alternative interaction semi-log QR, such as the interaction latent-variables QRs presented earlier. This example presents a modification of only one of the special F2WI-QRs in (10.21), with the output presented in Figure 10.49. Its main objective is to test the hypothesis that the effect of *PMLV* on *log(PBLV)* depends on the linear combination of the positive latent variables, $a \times PZMW + b \times PZAGE + c \times PZHB + d \times PZTKOL$.

$$LOG\,(PBLV)\ C\ PMLV * PZMW\ PMLV * PZAGE\ PMLV * PZHB$$

$$PMLV * PZTKOL\ PMLV\ PZMW\ PZAGE\ PZHB\ PZTKOL \qquad (10.24)$$

Based on this output, the findings and notes are as follows:

1. Even though most of its IVs have very large $p$-values, they unexpectedly are all jointly significant, based on the QRL statistic of $QLR_0 = 44.64767$ with $df = 5$ and

| Variable | QR(0.25) | | | QR(0.75) | | | QR(0.75) | | |
|---|---|---|---|---|---|---|---|---|---|
| | Coef. | t-Stat. | Prob. | Coef. | t-Stat. | Prob. | Coef. | t-Stat. | Prob |
| C | 68.44 | 1.650 | 0.102 | 84.59 | 2.334 | 0.021 | 40.82 | 1.073 | 0.286 |
| PMLV | −51.05 | −1.640 | 0.104 | −62.92 | −2.306 | 0.023 | −30.35 | −1.055 | 0.294 |
| PMLV^2 | 14.43 | 1.667 | 0.098 | 17.66 | 2.318 | 0.022 | 8.680 | 1.076 | 0.284 |
| PMLV^3 | −1.792 | −1.694 | 0.093 | −2.177 | −2.326 | 0.022 | −1.086 | −1.092 | 0.277 |
| PMLV^4 | 0.083 | 1.722 | 0.088 | 0.099 | 2.331 | 0.022 | 0.050 | 1.104 | 0.272 |
| Pse. R^2 | 0.052 | | | 0.050 | | | 0.062 | | |
| Adj. R^2 | 0.017 | | | 0.014 | | | 0.027 | | |
| S.E. reg. | 0.130 | | | 0.111 | | | 0.125 | | |
| Quant. DV | 1.528 | | | 1.607 | | | 1.681 | | |
| Sparsity | 0.364 | | | 0.271 | | | 0.292 | | |
| QLR stat | 6.299 | | | 7.186 | | | 8.670 | | |
| Prob. | 0.178 | | | 0.126 | | | 0.070 | | |

**Figure 10.48**  The output summary of three QR($\tau$)s in (10.23).

$p$-value $= 0.000\,001$. And the output of the RVT in Figure 10.50 shows that the redundant variables are jointly significant, based on the $QRL$ L-statistic of $QLR - L_0 = 18.076\,41$ with $df = 5$ and $p$-value $= 0.0029$.

2. Hence, the semi-log F2WI-QR(Median) is an acceptable QR to show that the *PMLV* on *log(PBLV)* is significantly dependent on the linear combination of the positive latent variables *PZMW, PZAGE, PZHB*, and *PZTKOL*.

3. As an additional illustration, Figure 10.51 presents the output summary of three reduced QRs of the F2WI-QR(Median) in (10.24). Based on this summary, the following findings and notes are presented:

3.1 All IVs of each of reduced QR are jointly significant with a $p$-value of 0.000.

3.2 The interaction IV, *PMLV\*PZAGE*, of the reduced semi-log QR(0.5)-1 has a large $p$-value of 0.432.

3.3 We can delete either of the main IVs, *PMLV* or *PZAGE*, to obtain the outputs of the reduced QR(0.5)-2 and QR(0.5)-3, as presented in Figure 10.51. In addition, the simplest reduced QR(0.5) would be the QR without the main IVs, which shows each IV has either a positive or negative significant effect at the 1% or 5% level of significance. But the output is not presented.

3.4 Finally, by using the RVT, we can show that the effect of *PMLV* on *log(BLV)* is significantly dependent on the linear combination of *PZMW, PZAGE, PZHB*, and *PZTKOL* for each of the four reduced QRs. Do these as exercises.

Dependent Variable: LOG(PBLV)

Method: Quantile Regression (Median)

Date: 04/05/19   Time: 21 : 25

Sample: 1 113

Included observations: 113

Huber Sandwich Standard Errors & Covariance

Sparsity method: Kernel (Epanechnikov) using residuals

Bandwidth method: Hall-Sheather, bw = 0.20096

Estimation successfully identifies unique optimal solution

| Variable | Coefficient | Std. Error | t-Statistic | Prob. |
|---|---|---|---|---|
| C | 0.211953 | 0.712350 | 0.297541 | 0.7667 |
| PMLV*PZMW | −0.030495 | 0.020383 | −1.496141 | 0.1377 |
| PMLV*PZAGE | 0.005385 | 0.015738 | 0.342143 | 0.7329 |
| PMLV*PZHB | −0.004312 | 0.012973 | −0.332366 | 0.7403 |
| PMLV*PZTKOL | 0.004908 | 0.018701 | 0.262424 | 0.7935 |
| PMLV | 0.183912 | 0.139124 | 1.321929 | 0.1891 |
| PZMW | 0.229159 | 0.103878 | 2.206043 | 0.0296 |
| PZAGE | −0.046670 | 0.077785 | −0.599990 | 0.5498 |
| PZHB | 0.032197 | 0.066522 | 0.484006 | 0.6294 |
| PZTKOL | −0.003337 | 0.091941 | −0.036298 | 0.9711 |

| | | | | |
|---|---|---|---|---|
| Pseudo R-squRed | 0.247981 | Mean dependent var | | 1.603383 |
| Adjusted R-squRed | 0.182271 | S.D. dependent var | | 0.110877 |
| S.E. of regression | 0.091317 | Objective | | 3.690017 |
| Quantile dependent var | 1.606865 | Restr. objective | | 4.906814 |
| Sparsity | 0.218027 | Quasi-LR statistic | | 44.64767 |
| Prob(Quasi-LR stat) | 0.000001 | | | |

**Figure 10.49**   The output of the semi-log F2WI-QR(Median) in (10.24).

## 10.7   An Application of Heterogeneous Quantile-regressions

Each of the QRs with the numerical IVs in previous section can easily be modified to *heterogeneous quantile regressions* (HQRs) by one, two, or three categorical factors, using the dummy or categorical variables in BBM.wf1. In the following subsection, I present a few simple examples, which can easily be extended to HQRs with many numerical IVs.

The main objectives of an HQR are to present and test the effect differences of one or more numerical variables on the DV between the levels of the categorical factors.

Redundant Variables Test

Equation: QR(0.5) in Figure 10.48

Specification: LOG(PBLV) C PMLV*PZMW PMLV*PZAGE PMLV*PZHB

  PMLV*PZTKOL PMLV PZMW PZAGE PZHB PZTKOL

Redundant Variables: PMLV*PZMW PMLV*PZAGE PMLV*PZHB

  PMLV*PZTKOL PMLV

Null hypothesis: PMLV*PZMW PMLV*PZAGE PMLV*PZHB

  PMLV*PZTKOL PMLV are jointly insignificant

|  | Value | df | Probability |
|---|---|---|---|
| QLR L-statistic | 18.07641 | 5 | 0.0029 |
| QLR Lambda-statistic | 16.96743 | 5 | 0.0046 |

**Figure 10.50**  The output of a RVT based on semi-log F2WI-QR(Median) in (10.24).

| Variable | Red Semi-Log QR(0.5)-1 | | | Red Semi-Log QR(0.5)-2 | | | Red Semi-Log QR(0.5)-3 | | |
|---|---|---|---|---|---|---|---|---|---|
|  | Coef. | t-Stat. | Prob. | Coef. | t-Stat. | Prob. | Coef. | t-Stat. | Prob. |
| C | 1.627 | 4.995 | 0.000 | 1.244 | 11.69 | 0.000 | 1.218 | 11.67 | 0.000 |
| PMLV*PZMW | 0.012 | 4.809 | 0.000 | 0.012 | 5.119 | 0.000 | 0.012 | 5.061 | 0.000 |
| PMLV*PZAGE | 0.009 | 0.789 | 0.432 | −0.004 | −1.585 | 0.116 | −0.005 | −1.047 | 0.298 |
| PMLV*PZHB | 0.003 | 1.298 | 0.197 | 0.004 | 1.635 | 0.105 | 0.003 | 1.478 | 0.143 |
| PMLV*PZTKOL | 0.004 | 1.679 | 0.096 | 0.004 | 1.790 | 0.076 | 0.004 | 1.839 | 0.069 |
| PMLV | −0.079 | −1.228 | 0.222 | −0.008 | −0.281 | 0.779 |  |  |  |
| PZAGE | −0.066 | −1.041 | 0.300 |  |  |  | 0.003 | 0.091 | 0.928 |
| Pse. R^2 | 0.221 |  |  | 0.218 |  |  | 0.217 |  |  |
| Adj. R^2 | 0.177 |  |  | 0.181 |  |  | 0.181 |  |  |
| S.E. reg. | 0.091 |  |  | 0.091 |  |  | 0.091 |  |  |
| Quant. DV | 1.607 |  |  | 1.607 |  |  | 1.607 |  |  |
| Sparsity | 0.221 |  |  | 0.225 |  |  | 0.225 |  |  |
| QLR stat | 39.19 |  |  | 38.07 |  |  | 37.87 |  |  |
| Prob | 0.000 |  |  | 0.000 |  |  | 0.000 |  |  |

**Figure 10.51**  The output summary of three reduced semi-log QR(0.5)s in (10.24).

### 10.7.1 A Heterogeneous QR of *BLV* by a Categorical Factor

#### 10.7.1.1 A Two-level Heterogeneous QR of *BLV*
As an extension of each previous QR of *BLV* or *PBLV* with numerical IVs, we can easily have various HQRs by a single categorical factor/variable. This section presents one two-level additive HQR of *BLV* on *MLV, ZAGE, ZHB, ZMW, and ZTKOL*, by a dummy variable *LOWBW* as an extension of the additive QR of *BLV* in (10.18), using the following special ES. It is important to present the IVs in three groups. The first group is *(C(10) + C(20))\*(LOWBW)* to present the intercept of the whole model, where C(10) is the intercept of the QR for group *(LOWBW = 0)* and *(C(10) + C(20))* is the intercept of the QR for group *(LOWBW = 1)*. See the following example in more detail.

$$BLV = (C(10) + C(20) * (LOWBW = 1))$$
$$+ (C(11) * MLV + C(12) * ZAGE + C(13) * ZHB + C(14) * ZMW$$
$$+ C(15) * ZTKOL) * (LOWBW = 0)$$
$$+ (C(21) * MLV + C(22) * ZAGE + C(23) * ZHB + C(24) * ZMW$$
$$+ C(25) * ZTKOL) * (LOWBW = 1) \tag{10.25}$$

**Example 10.25** *The Output of HQR(Median) in (10.25)*
Figure 1.51 presents the output of the two-level additive HQR(Median) in (10.25), with its representations presented in Figure 10.52. Based on these outputs, the following findings and notes are presented:

1. All IVs of the HQR, which are represented by 11 parameters C(11) to C(15) and C(20) to C(25) are jointly significant, based on the QRL statistic of $QLR_0 = 72.237\,51$ with $df = 11$ and $p$-value $= 0.000\,000$.
2. Based on the output in Figure 10.52, we can easily derive the following QR functions:
$$BLV_{(Lowbw=0)} = 0.152 + 0.237 * MLV - 0.107 * ZAGE - 0.060 * ZHB$$
$$+0.057 * ZMW + 0.111 * ZTKOL$$
$$BLV_{(Lowbw=1)} = -0.406 + 0.441 * MLV - 0.158 * ZAGE + 0.047 * ZHB$$
$$+0.299 * ZMW + 0.149 * ZTKOL$$

3. All types of hypotheses on the effect differences of the numerical IVs between the two levels of *LOWBW* can easily be tested using the Wald test. Do them as exercises (Figure 10.53).
4. As an advanced analysis, Figure 10.54 presents a special line and symbol graph, which is obtained using the following steps:
   4.1 With the output in Figure 10.52 on-screen, click the Forecast button, enter the *BLVF_EQ1025* as the forecast name, and click *OK*. The forecast variable *BLVF_EQ1025* is inserted directly in the data file.
   4.2 Select the sample *LOWBW* = 1, which indicates low baby weight, and then generate the variable *BLVF_LOW = BLVF_EQ1025* as a new variable in the data file.
   4.3 Similarly, select the sample *LOWBW* = 0, which indicates high baby weight, and then generate the variable *BLVF_HIGH = BLVF_EQ1025* as a new variable in the data file.

Dependent Variable: BLV

Method: Quantile Regression (Median)

Date: 04/06/19   Time: 10:11

Sample: 1 113

Included observations: 113

Huber Sandwich Standard Errors & Covariance

Sparsity method: Kernel (Epanechnikov) using residuals

Bandwidth method: Hall-Sheather, bw = 0.20096

Estimation successfully identifies unique optimal solution

BLV = (C(10) + C(20)*(LOWBW = 1)) + (C(11)*MLV + C(12)*ZAGE + C(13)

   *ZHB + C(14)*ZMW + C(15)*ZTKOL)*(LOWBW = 0) + (C(21)*MLV

   +C(22)*ZAGE + C(23)*ZHB + C(24)*ZMW + C(25)*ZTKOL)

   *(LOWBW = 1)

|        | Coefficient | Std. Error | t-Statistic | Prob. |
|--------|-------------|------------|-------------|-------|
| C(10)  | 0.152499    | 0.076019   | 2.006068    | 0.0475 |
| C(20)  | −0.405813   | 0.178832   | −2.269247   | 0.0254 |
| C(11)  | 0.236649    | 0.121383   | 1.949616    | 0.0540 |
| C(12)  | −0.107003   | 0.089969   | −1.189327   | 0.2371 |
| C(13)  | −0.060376   | 0.077489   | −0.779152   | 0.4377 |
| C(14)  | 0.057212    | 0.096777   | 0.591172    | 0.5557 |
| C(15)  | 0.110878    | 0.069629   | 1.592419    | 0.1144 |
| C(21)  | 0.440507    | 0.221358   | 1.990020    | 0.0493 |
| C(22)  | −0.158454   | 0.073074   | −2.168412   | 0.0325 |
| C(23)  | 0.046545    | 0.098322   | 0.473395    | 0.6370 |
| C(24)  | 0.298550    | 0.193294   | 1.544536    | 0.1256 |
| C(25)  | 0.148623    | 0.073974   | 2.009127    | 0.0472 |

| | | | |
|---|---|---|---|
| Pseudo R-squRed | 0.346377 | Mean dependent var | 9.04E-17 |
| Adjusted R-squRed | 0.275190 | S.D. dependent var | 0.550900 |
| S.E. of regression | 0.422204 | Objective | 15.94978 |
| Quantile dependent var | −0.012850 | Restr. objective | 24.40210 |
| Sparsity | 0.936060 | Quasi-LR statistic | 72.23751 |
| Prob(Quasi-LR stat) | 0.000000 | | |

**Figure 10.52**   Output of the HQR(Median) in (10.25).

Estimation command:
==========================
QREG BLV = (C(10) + C(20)*(LOWBW = 1)) + (C(11)*MLV + C(12)*ZAGE + C(13)*ZHB +
C(14)*ZMW + C(15)*ZTKOL)*(LOWBW = 0) + (C(21)*MLV + C(22)*ZAGE + C(23)*ZHB +
C(24)*ZMW + C(25)*ZTKOL)*(LOWBW = 1)

Estimation equation:
==========================
BLV = (C(10) + C(20)*(LOWBW = 1)) + (C(11)*MLV + C(12)*ZAGE + C(13)*ZHB +
C(14)*ZMW + C(15)*ZTKOL)*(LOWBW = 0) + (C(21)*MLV + C(22)*ZAGE + C(23)*ZHB +
C(24)*ZMW + C(25)*ZTKOL)*(LOWBW = 1)

Substituted coefficients:
==========================
BLV = (0.152498869677 − 0.405813345158*(LOWBW = 1)) + (0.236649380705*MLV −
0.107002680177*ZAGE − 0.0603756650879*ZHB + 0.0572119252138*ZMW + 0.11087829777*
ZTKOL)*(LOWBW = 0) + (0.440506994137*MLV − 0.158454275736*ZAGE + 0.0465450841642*
ZHB + 0.298550083948*ZMW + 0.148623136107*ZTKOL)*(LOWBW = 1)

**Figure 10.53** The representations of the HQR(Median) in Figure 10.52.

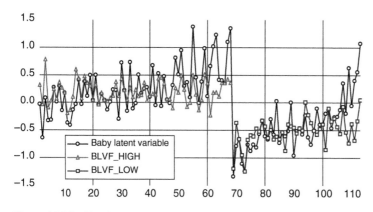

**Figure 10.54** The line and symbol graphs of the variable BLV, BLVF_HIGH, and BLVF_LOW.

4.4 Select the whole sample, and then you can develop the line and symbol graphs of the three variables *BLV, BLVF_HIGH*, and *BLVF_LOW,* with the output presented in Figure 10.54. And Figure 10.55 presents the graphs sorted by the variable *BLV*.

### 10.7.1.2 A Three-Level Heterogeneous QR of PBLV

Each QR of *BLV* or *PBLV* with numerical IVs presented previously can easily be extended to various three-level heterogeneous quantile regressions (3LH-QRs). This section presents a three-level heterogeneous interaction QR (3LHI-QR) of *PBLV* on *MLV* and *ZMW* by *SE* having three levels, with the following special *ES*, as an extension of the 2WI-QR in (10.19):

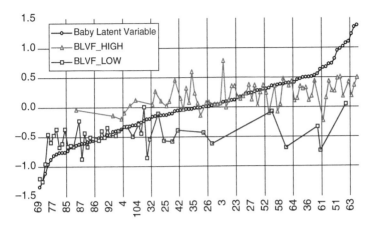

**Figure 10.55** The graphs of BLV, BLVF_HIGH, and BLVF_LOW, sorted by BLV.

$$PBLV = (C(10) + C(20) * (SE = 2) + C(30) * (SE = 3))$$
$$+ (C(11) * PMLV * PZMW + C(12) * PMLV + C(13) * PZMW) * (SE = 1)$$
$$+ (C(21) * PMLV * PZMW + C(22) * PMLV + C(23) * PZMW) * (SE = 2)$$
$$+ (C(31) * PMLV * PZMW + C(32) * PMLV + C(33) * PZMW) * (SE = 3)$$
$$(10.26)$$

**Example 10.26**  *The Output of 3LHI-QR($\tau$)s in (10.26)*

Figure 10.56 presents the output summary of three 3LHI-QR($\tau$) in (10.26). Based on this summary, the findings and notes are as follows:

1. All IVs of each QR are jointly significant, based on the *QRL* statistic at the 1% level of significance.
2. Based on the output of representations, we can easily derive the following three QR functions:

$$PBLV_{(SE=1)} = -9.970 - 0.487PMLV * PZMW + 2.759PMLV + 2.645PZMW$$
$$PBLV_{(SE=2)} = 11.266 - -0.033PMLV * PZMW + 0.434PMLV + 0.461PZMW$$
$$PBLV_{(SE=3)} = 6.702 - 0.099PMLV * PZMW + 0.941PMLV + 1.205PZMW$$

3. Similar to the graphs in Figures 10.52 and 10.54, we can develop the graphs of *PBLV* and its forecast value variable by *SE*, based on the 3LHI_QR(Median) in (10.26). Note that compared with the graphs in Figure 10.54, the graphs of the three forecast value variables in Figure 10.57 present many break points, or no lines between two consecutive points. However, its graphs sorted by *PBLV* in Figure 10.58 do have lines.

## 10.7.2 Heterogeneous QR of *BLV* by Two Categorical Factors

As an extension of each QR of *BLV* or *PBLV* with numerical IVs, or an extension of the HQR presented above, we can present various HQRs by two categorical factors. The examples in this section present a $2 \times 2$ factorial HQR of *BLV* or *PBLV*.

| Variable | 3LHI-QR(0.1) | | | 3LHI-QR(0.5) | | | 3LHI-QR(0.9) | | |
|---|---|---|---|---|---|---|---|---|---|
| | Coef. | t-Stat. | Prob. | Coef. | t-Stat. | Prob. | Coef. | t-Stat. | Prob |
| C(10) | −9.788 | −2.804 | 0.006 | −9.970 | −1.748 | 0.084 | −7.701 | −0.463 | 0.644 |
| C(20) | 9.145 | 2.188 | 0.031 | 11.266 | 1.228 | 0.222 | −8.714 | −0.484 | 0.629 |
| C(30) | −4.675 | −0.369 | 0.713 | 6.702 | 0.367 | 0.714 | −2.723 | −0.139 | 0.889 |
| C(11) | −0.409 | −3.504 | 0.001 | −0.487 | −2.001 | 0.048 | −0.418 | −0.646 | 0.520 |
| C(12) | 2.387 | 3.807 | 0.000 | 2.759 | 2.244 | 0.027 | 2.594 | 0.772 | 0.442 |
| C(13) | 2.484 | 3.948 | 0.000 | 2.645 | 2.347 | 0.021 | 2.152 | 0.675 | 0.501 |
| C(21) | −0.080 | −1.005 | 0.317 | −0.033 | −0.120 | 0.905 | −0.657 | −2.320 | 0.022 |
| C(22) | 0.674 | 1.741 | 0.085 | 0.434 | 0.340 | 0.735 | 3.620 | 2.794 | 0.006 |
| C(23) | 0.744 | 1.622 | 0.108 | 0.461 | 0.302 | 0.763 | 4.003 | 2.745 | 0.007 |
| C(31) | −0.672 | −1.092 | 0.277 | −0.099 | −0.140 | 0.889 | −0.366 | −0.878 | 0.382 |
| C(32) | 3.064 | 1.297 | 0.198 | 0.941 | 0.323 | 0.748 | 2.152 | 1.246 | 0.216 |
| C(33) | 4.189 | 1.351 | 0.180 | 1.205 | 0.293 | 0.770 | 2.772 | 1.141 | 0.257 |
| Pse. R^2 | 0.351 | | | 0.250 | | | 0.224 | | |
| Adj. R^2 | 0.280 | | | 0.169 | | | 0.139 | | |
| S.E. reg. | 0.753 | | | 0.485 | | | 0.800 | | |
| Quant. DV | 4.332 | | | 4.987 | | | 5.675 | | |
| Sparsity | 1.596 | | | 1.048 | | | 1.995 | | |
| QLR stat | 50.583 | | | 46.631 | | | 28.795 | | |
| Prob | 0.000 | | | 0.000 | | | 0.002 | | |

**Figure 10.56**   The outputs summary of the 3LHI-QR($\tau$) in (10.26) for $\tau = 0.1$, 0.5 and 0.9

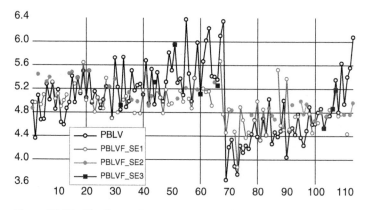

**Figure 10.57**   The line and symbol graphs of *PBLV* and its forecast value variables, *PBLVF_SE1*, *PBLVF_SE2*, and *PBLVF_SE3*, based on the 3LHI-QR(Median).

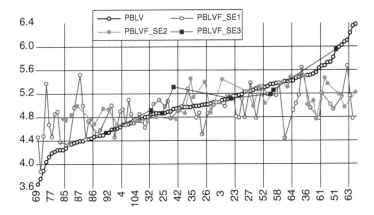

**Figure 10.58** The graphs of *PBLV*, *PBLVF_SE1*, *PBLVF_SE2*, and *PBLVF_SE3*, sorted by *PBLV*.

**Example 10.27** *A Heterogeneous Additive QR of BLV by LOWBW and ANC*

As an extension of the HQR of *BLV* in (10.25), Figure 10.59 presents the output of a $2 \times 2$ factorial HQR with the following special ES, which can be considered a four-level heterogeneous additive QR (4LHA-QR):

$$
\begin{aligned}
BLV = {}& (C(10) + C(20) * (LOWBW = 0) * (ANC = 1) \\
& + C(30) * (LOWBW = 1) * (ANC = 0) + C(40) * (LOWBW = 1) * (ANC = 1)) \\
& + (C(11) * MLV + C(12) * ZAGE + C(13) * ZHB + C(14) * ZMW \\
& + C(15) * ZTKOL) * (LOWBW = 0) * (ANC = 0) \\
& + (C(21) * MLV + C(22) * ZAGE + C(23) * ZHB + C(24) * ZMW \\
& + C(25) * ZTKOL) * (LOWBW = 0) * (ANC = 1) \\
& + (C(31) * MLV + C(32) * ZAGE + C(33) * ZHB + C(34) * ZMW \\
& + C(35) * ZTKOL) * (LOWBW = 1) * (ANC = 0) \\
& + (C(41) * MLV + C(42) * ZAGE + C(43) * ZHB \\
& + C(44) * ZMH \; CELL/GROUPW \\
& + C(45) * ZTKOL) * (LOWBW = 1) * (ANC = 1)
\end{aligned}
\tag{10.27}
$$

Based on this output, the following findings and notes are presented:

1. All the IVs are classified into five groups. The first is the group of an intercept and three dummy variables, and the other four groups consist of numerical IVs for the four cells/groups generated by the two dummy variables *LOWBW* and *ANC*.
2. All IVs, which are represented by the parameters C(10) to C(45), of the QR are jointly significant, based on the *QRL* statistic of $QLR_0 = 90.677\,171$ with $df = 23$ and *p*-value $= 0.000\,000$.
3. In addition, we can use the *t*-statistic in the output to test the effect of each numerical IV within each cell/group. All types of hypotheses on the effect differences of the numerical IVs between two or more groups that generated *LOWBW* and *ANC* can easily be tested using the Wald test. Do this as exercises.

Dependent variable: BLV
Method: Quantile Regression (Median)
Date: 02/03/21   Time: 12 : 53
Sample: 1 113
Included observations: 113
Huber Sandwich Standard Errors & Covariance
Sparsity method: Kernel (Epanechnikov) using residuals
Bandwidth method: Hall-Sheather, bw = 0.20096
Estimation successfully identifies unique optimal solution
BLV = (C(10) + C(20)*(LOWBW = 0)*(ANC = 1) + C(30)*(LOWBW = 1)
    *(ANC = 0) + C(40)*(LOWBW = 1)*(ANC = 1)) + (C(11)*MLV + C(12)
    *ZAGE + C(13)*ZHB + C(14)*ZMW + C(15)*ZTKOL)*(LOWBW = 0)
    *(ANC = 0) + (C(21)*MLV + C(22)*ZAGE + C(23)*ZHB + C(24)*ZMW
    +C(25)*ZTKOL)*(LOWBW = 0)*(ANC = 1) + (C(31)*MLV + C(32)
    *ZAGE + C(33)*ZHB + C(34)*ZMW + C(35)*ZTKOL)*(LOWBW = 1)
    *(ANC = 0) + (C(41)*MLV + C(42)*ZAGE + C(43)*ZHB + C(44)*ZMW
    +C(45)*ZTKOL)*(LOWBW = 1)*(ANC = 1)

|  | Coefficient | Std. error | t-statistic | Prob. |
|---|---|---|---|---|
| C(10) | 0.195586 | 0.288154 | 0.678756 | 0.4991 |
| C(20) | −0.054252 | 0.306201 | −0.177178 | 0.8598 |
| C(30) | −0.748758 | 0.376395 | −1.989287 | 0.0497 |
| C(40) | −0.433484 | 0.351103 | −1.234635 | 0.2202 |
| C(11) | −0.076587 | 0.234115 | −0.327134 | 0.7443 |
| C(12) | 0.035887 | 0.120215 | 0.298526 | 0.7660 |
| C(13) | 0.082250 | 0.134786 | 0.610226 | 0.5433 |
| C(14) | −0.067922 | 0.295748 | −0.229662 | 0.8189 |
| C(15) | 0.078240 | 0.160118 | 0.488639 | 0.6263 |
| *** |  |  |  |  |
| C(41) | 0.508783 | 0.333765 | 1.524376 | 0.1310 |
| C(42) | −0.219294 | 0.088156 | −2.487565 | 0.0147 |
| C(43) | 0.027324 | 0.118490 | 0.230602 | 0.8182 |
| C(44) | 0.397647 | 0.189537 | 2.097989 | 0.0387 |
| C(45) | 0.130265 | 0.085738 | 1.519341 | 0.1322 |

| | | | |
|---|---|---|---|
| Pseudo R-squred | 0.432585 | Mean dependent var | 9.04E-17 |
| Adjusted R-squred | 0.285950 | S.D. dependent var | 0.550900 |
| S.E. of regression | 0.425612 | Objective | 13.84612 |
| Quantile dependent var | −0.012850 | Restr. objective | 24.40210 |
| Sparsity | 0.815908 | Quasi-LR statistic | 103.5017 |
| Prob(Quasi-LR stat) | 0.000000 | | |

**Figure 10.59**   A part of the statistical result of the QR(Median) in (10.27).

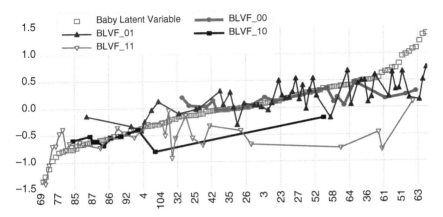

**Figure 10.60** The line an symbol graphs of *BLV* and four forecast variables *BLVF_00, BLVF_01, BLVF_10,* and *BLVF_11,* sort by *BLV*.

4. Based on the output of the representations of the QR(Median) in (10.27), we can easily derive the following four QR functions:

$$BLV\,(0,0) = 0.196 - 0.077^*MLV + 0.036^*ZAGE + 0.082^*ZHB$$
$$- 0.068^*ZMW + 0.078^*ZTKOL$$

$$BLV\,(0,1) = (0.196 - 0.054) + 0.365^*MLV - 0.200^*ZAGE - 0.094^*ZHB$$
$$+ 0.122^*ZMW + 0.168^*ZTKOL$$

$$BLV\,(1,0) = (0.196 - 0.749) + 0.260^*MLV - 0.046^*ZAGE + 0.076^*ZHB$$
$$+ 0.092^*ZMW + 0.082^*ZTKOL$$

$$BLV\,(1,1) = (0.196 - 0.433) + 0.509^*MLV - 0.219^*ZAGE - 0.027^*ZHB$$
$$+ 0.400^*ZMW + 0.130^*ZTKOL$$

5. Similar to the graphs in Figure 10.58, Figure 10.60 presents the line and symbol graphs of *BLV* and its forecast variables, *BLVF_00, BLVF_01, BLVF_10,* and *BLVF_11,* sorted by *BLV*. They are obtained using the same method as presented in Example 10.22, point-4.

6. As an additional analysis, Figure 10.61 presents the descriptive statistics of the five variables in Figure 10.60, which shows the Jarque-Bera (JB) test statistic support the normality distribution of each variables. As a comparison, Figure 10.62 presents the histogram of each variable with its normal density, which shows that the histogram of the forecast variable *BLVF_11* has a large deviation from normal density, which corresponds to the JB test having the smallest *p*-value $= 0.21 < 0.30$.

7. Although a theoretical concept of statistics, it is important to mention that all numerical variables, including the zero–one (dummy) variables, and ordinal variables have *theoretical normal distributions*. As an illustration, Figure 10.63 presents the theoretical distribution *of* an ordinal variable *SE* and the dummy variable *ANC*. For this reason, we don't need to conduct additional normality tests of any variables. Refer to the notes on the "misinterpretation of selected theoretical concepts of statistics" and the illustrative graphs, presented in Agung (2019).

| Statistic | BLV | BLVF_00 | BLVF_01 | BLVF_10 | BLVF_11 |
|---|---|---|---|---|---|
| Mean | 9.0E-17 | 1.6E-01 | 1.9E-01 | −5.4E-01 | −5.8E-01 |
| Median | −1.3E-02 | 1.5E-01 | 1.7E-01 | −5.4E-01 | −5.6E-01 |
| Maximum | 1.4E+00 | 4.7E-01 | 7.4E-01 | −1.9E-01 | 1.0E-01 |
| Minimum | −1.3E+00 | −4.0E-02 | −4.2E-01 | −8.0E-01 | −1.4E+00 |
| Std. Dev. | 5.5E-01 | 1.4E-01 | 2.8E-01 | 1.6E-01 | 3.1E-01 |
| Skewness | 1.5E-01 | 5.1E-01 | 2.2E-02 | 6.1E-01 | −4.7E-01 |
| Kurtosis | 2.9E+00 | 2.6E+00 | 2.4E+00 | 3.5E+00 | 4.1E+00 |
| Jarque-Bera | 4.7E-01 | 9.2E-01 | 7.0E-01 | 8.1E-01 | 3.1E+00 |
| Probability | 7.9E-01 | 6.3E-01 | 7.1E-01 | 6.7E-01 | 2.1E-01 |
| Sum | 1.3E-14 | 2.8E+00 | 9.5E+00 | −5.9E+00 | −2.0E+01 |
| Sum Sq. Dev. | 3.4E+01 | 3.4E-01 | 3.8E+00 | 2.6E-01 | 3.1E+00 |
| Observations | 113 | 18 | 50 | 11 | 34 |

**Figure 10.61** Descriptive statistics of the five variable in Figure 10.60.

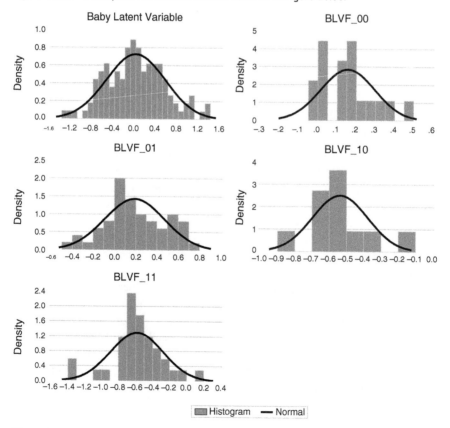

**Figure 10.62** Histogram of each variable in Figure 10.60, with its normal density.

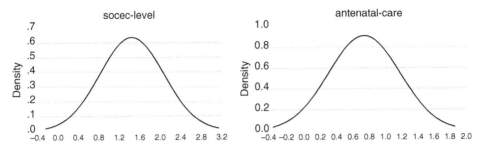

**Figure 10.63** Theoretical distributions of *BLVF_00*, and the dummy variable *ANC*.

**Example 10.28** *A Heterogeneous Interaction QR of log(BLV) by ANC and IU*
As a modification of the 3LHI-QR of *PBLV* in (10.26), Figure 10.64 presents the output of a $2 \times 2$ factorial heterogeneous interaction QR of *log(PBLV)* by two mother indicators, *ANC* and *IU*. Based on this output, the findings and notes are as follows:

1. As with the $2 \times 2$ factorial QR equation in (10.26), the IVs of this QR are classified into five groups of IVs. And the output presents the ES of the QR, which can be considered as a 4LHI-QR of *log(PBLV)*. Based on the output of its Representations, we can easily present the following four QR functions, for the groups generated by *ANC* and *IU*:

$$LOG(PBLV)_{(0,0)} = -0.728 - 0.061PMLV * PZMW + 0.356PMLV + 0.405PZMW$$

$$LOG(PBLV)_{(0,1)} = 1.863 - 0.031PMLV * PZMW + 0.191PMLV + 0.212PZMW$$

$$LOG(PBLV)_{(1,0)} = 1.034 + 0.014PMLV * PZMW + 0.026PMLV + 0.001PZMW$$

$$LOG(PBLV)_{(1,1)} = -0.105 - 0.069PMLV * PZMW + 0.416PMLV + 0.412PZMW$$

2. As an additional illustration, Figure 10.65 presents the Wald test for testing the hypothesis on the effect differences of the three numerical IVs on *log(PBLV)* between the four groups, which shows they have insignificant differences, based on the *Chi-square* statistic of $\chi_0^2 = 4.570\,954$ with $df = 9$ and p-value $= 0.8700$. Hence, in the statistical sense, the HQR(Median) can be reduced to an ANCOVA-QR(Median)with the following ES:

$$Log\,(PBLV)\,C\,@\,Expand\,(ANC, IU, @\,Dropfirst)$$
$$PMLV * PZMW\,PMLV\,PZMW \tag{10.28}$$

However, I do not recommend an ANCOVA model because having insignificant effect differences does not mean they have exactly the same effects in the corresponding population. Furthermore, an additive ANCOVA model of two or more categorical factors is considered to be the worst model, in the theoretical sense, as illustrated in Agung (2011, 2014). And in general, a numerical IV has different effects on the DV, in the theoretical sense, between groups of the objects. In fact, the output in above C(12), C(22), C(32), and C(33) shows different p-values. At the 5% level, it shows the interaction *PMLV*PZMW* has a positive significant effect on *log(BLV)*, specific within the group (*ANC* = 0, *IU* = 1), based on the t-statistic of $t_o = 1.732\,585$ with $df = (113 - 15)$ and p-value $= 0.0863/2 = 0/043\,15 < 0.05$; however, the others have insignificant effects.

Dependent Variable: LOG(PBLV)

Method: Quantile Regression (Median)

Sample: 1 113

Included observations: 113

Huber Sandwich standard errors & covariance

Sparsity method: Kernel (Epanechnikov) using residuals

Bandwidth method: Hall-Sheather, bw = 0.20096

Estimation successfully identifies unique optimal solution

LOG(PBLV) = (C(10) + C(20)*(IU = 0)*(ANC = 1) + C(30)*(IU = 1)*(ANC = 0)

  +C(40)*(IU)*(ANC = 1)) + (C(11)*PMLV*PZMW + C(12)*PMLV+C(13)

  *PZMW)*(ANC = 0)*(IU = 0) + (C(21)*PMLV*PZMW + C(22)*PMLV

  +C(23)*PZMW)*(ANC = 0)*(IU = 1) + (C(31)*PMLV*PZMW + C(32)

  *PMLV + C(33)*PZMW)*(ANC = 1)*(IU = 0) + (C(41)*PMLV*PZMW

  +C(42)*PMLV + C(43)*PZMW)*(ANC = 1)*(IU = 1)

| | Coefficient | Std. error | t-statistic | Prob. |
|---|---|---|---|---|
| C(10) | −0.728275 | 1.218185 | −0.597836 | 0.5513 |
| C(20) | 1.863196 | 1.428997 | 1.303849 | 0.1954 |
| C(30) | 1.034207 | 2.809271 | 0.368141 | 0.7136 |
| C(40) | −0.105444 | 1.968801 | −0.053557 | 0.9574 |
| C(11) | −0.060817 | 0.044430 | −1.368827 | 0.1742 |
| C(12) | 0.355763 | 0.205336 | 1.732585 | 0.0863 |
| C(13) | 0.405092 | 0.253027 | 1.600984 | 0.1126 |
| C(21) | −0.030655 | 0.100184 | −0.305987 | 0.7603 |
| C(22) | 0.191324 | 0.428706 | 0.446284 | 0.6564 |
| C(23) | 0.212272 | 0.582842 | 0.364202 | 0.7165 |
| C(31) | 0.013816 | 0.028447 | 0.485670 | 0.6283 |
| C(32) | 0.026147 | 0.143230 | 0.182556 | 0.8555 |
| C(33) | 0.001441 | 0.146388 | 0.009842 | 0.9922 |
| C(41) | −0.068667 | 0.061941 | −1.108595 | 0.2703 |
| C(42) | 0.416312 | 0.335781 | 1.239833 | 0.2180 |
| C(43) | 0.411884 | 0.286462 | 1.437830 | 0.1537 |

| | | | |
|---|---|---|---|
| Pseudo R-squRed | 0.247925 | Mean dependent var | 1.603383 |
| Adjusted R-squRed | 0.131625 | S.D. dependent var | 0.110877 |
| S.E. of regression | 0.097464 | Objective | 3.690290 |
| Quantile dependent var | 1.606865 | Restr. objective | 4.906814 |
| Sparsity | 0.216857 | Quasi-LR statistic | 44.87847 |
| Prob(Quasi-LR stat) | 0.000080 | | |

**Figure 10.64** The output of a 2 × 2 factorial HI-QR(Median) of *log(PBLV)*.

Wald test:

Equation: EQ15

| Test statistic | Value | df | Probability |
|---|---|---|---|
| F-statistic | 0.507884 | (9, 97) | 0.8656 |
| Chi-square | 4.570954 | 9 | 0.8700 |

Null hypothesis: C(11) = C(21) = C(31) = C(41),

  C(12) = C(22) = C(32) = C(42), C(13) = C(23) = C(33) = C(43)

**Figure 10.65**  The output of a Wald test for testing the effect differences of the numerical IVs between the four groups/cells.

## 10.8  Piecewise QRs

A *peacewise* (PW)-QR of *BLV* is an HQR of *BLV* having at least a numerical IV by an ordinal factor, which is generated based on a numerical IV of the model. For the examples, I generate ordinal factors based on the *MLV* with two, three, and four levels, *O2MLV, O3MLV*, and *O4MLV*, using the following equations:

$$O2MLV = 1 + 1 * (MLV >= @Quantile\,(MLV, 0.50))$$
$$O3MLV = 1 + 1 * (MLV >= @Quantile\,(MLV, 0.30))$$
$$+1 * (MLV >= @Quantile\,(MLV, 0.70))$$
$$O4MLV = 1 + 1 * (MLV >= @Quantile\,(MLV, 0.25))$$
$$+1 * (MLV >= @Quantile\,(MLV, 0.50))$$
$$+1 * (MLV >= @Quantile\,(MLV, 0.75))$$

### 10.8.1  The Simplest PW-QR of *BLV* on *MLV*

The following examples present two *simple peacewise* (SPW) QRs of *BLV* on *MLV* only by single factors, which obtain unexpected statistical results.

**Example 10.29**  *The Simplest-Pease-Wise (SPW)-QR of BLV*
Figure 10.63 presents the outputs summary of the SPW-QR($\tau$) of *BLV* on *MLV* only by *O2MLV*, for $\tau = 0.1$, 0.5 and 0.9. Note that a PW-QR in fact is a HQR, where the categorical IV is generated by one of the numerical variables. In this case, the factor O2MLV is generated from the numerical IV, *MLV*. Based on this summary the following findings and notes are presented:

1. All IVs of the SPW-QR(0.1) and SPW-QR(0.5) have insignificant joint effects with the *p*-values of 0.334 and 0.554, respectively. But the IVs of the SPW-QR(0.9) are jointly significant at the 10% level.
2. At the 5% level, the IV *MLV* of the SPW-QR(0.9) has a positive significant effect, based on the *t*-statistic of $t_0 = 1.671$ with $df = (113 - 4)$ and *p*-value $= 0.098/2 = 0.049$.

| Variable | SPW-QR(0.1) | | | SPW-QR(0.5) | | | SPW-QR(0.9) | | |
|---|---|---|---|---|---|---|---|---|---|
| | Coef. | t-stat. | Prob. | Coef. | t-stat. | Prob. | Coef. | t-stat. | Prob. |
| C | −0.768 | −2.106 | 0.038 | −0.242 | −0.679 | 0.499 | 0.916 | 2.649 | 0.009 |
| O2MLV = 2 | 0.048 | 0.112 | 0.911 | 0.344 | 0.917 | 0.361 | −0.181 | −0.461 | 0.646 |
| MLV | −0.008 | −0.014 | 0.989 | −0.254 | −0.494 | 0.622 | 0.777 | 1.671 | 0.098 |
| MLV*(O2MLV = 2) | 0.148 | 0.227 | 0.821 | 0.260 | 0.491 | 0.624 | −0.616 | −1.068 | 0.288 |
| Pse. R^2 | 0.042 | | | 0.015 | | | 0.060 | | |
| Adj. R^2 | 0.015 | | | −0.012 | | | 0.035 | | |
| S.E. reg. | 0.899 | | | 0.548 | | | 0.857 | | |
| Quant. DV | −0.668 | | | −0.013 | | | 0.675 | | |
| Sparsity | 2.825 | | | 1.435 | | | 2.468 | | |
| QLR stat | 3.401 | | | 2.088 | | | 6.284 | | |
| Prob. | 0.334 | | | 0.554 | | | 0.099 | | |

**Figure 10.66** The output summary of the SPW-QR($\tau$), $\tau$ = 0.1, 0.5, and 0.9.

3. As additional illustrations, Figures 10.67 and 10.68 present the line and symbol graphs of *MLV, BLV*, and the forecast variable (fitted variable) of *BLV* based on the SPW-QR(0.9) in Figure 10.66: that is, *BLVF_SPW_Q90*, for the levels *O2MLV = 1* and *O2MLV = 2*, respectively, sorted by *BLV*. These graphs show very unexpected results. However, they should be acceptable, in the statistical sense, as the results of the SPW-QR(0.9), since they are obtained using the valid and viable estimation method, under the assumption the model is a true population model.

**Example 10.30** *A Simple Piecewise QR of BLV by O3MLV*
Figure 10.69 presents the unexpected output of the SPW-QR(0.5) of *BLV* on *MLV* only by *O3MLV*, using the following ES:

$$BLV\ C@Expand\,(O3MLV, @Dropfirst)\ MLV * @Expand\,(O3MLV) \tag{10.29}$$

Note that this output presents a warning and the statement "*Singularity in sandwich covariance estimation*," with "NA" for the Std. Error, which should be highly dependent on the data. However, unexpectedly, additional analyses can be done, with their valid outputs, by deleting one of the IVs. Hence, a complete output would be obtained by deleting one of its IVs, which we will see in the following example. Based on this incomplete output, however, the following findings and notes can be presented:

1. The output presents a complete QR(Median) function, which can easily be written or presented based on the output of representations. Do this as an exercise, referring to Example 10.25.

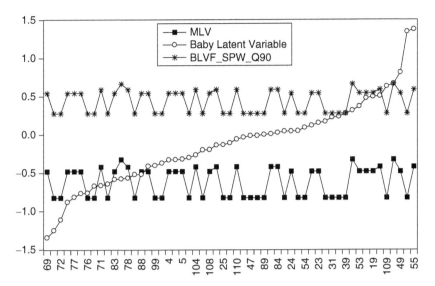

**Figure 10.67** The graphs of *MLV*, *BLV*, and *BLVF_SPW_Q90*, for *O2MLV = 1*, sorted by *BLV*.

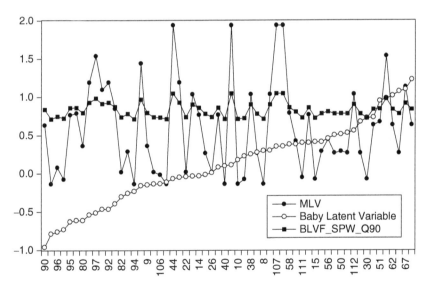

**Figure 10.68** The graphs of *MLV*, *BLV*, and BLVF_SPW_Q90, for *O2MLV = 2*, sorted by *BLV*.

2. Even though the output of the model (10.29) is not a complete output, we can conclude that its IVs have insignificant joint effects on *BLV*, based on the *QLR* statistic of $QLR_0 = 4.681\,91$ with $df = 5$ and $p$-value $= 0.320\,812$. We also can have a complete QR function, which we can see in the output of its representations.

3. In addition, even though the output is incomplete, it presents an acceptable and valid QR function. Do this as an exercise.

Dependent Variable: BLV

Method: Quantile Regression (Median)

Included observations: 113

Huber Sandwich Standard Errors & Covariance

Sparsity method: Kernel (Epanechnikov) using residuals

Bandwidth method: Hall-Sheather, bw = 0.20096

WARNING: linear dependencies found in regressors (some coefficients set to initial values)

*Singularity in sandwich covariance estimation*

| Variable | Coefficient | Std. error | t-Statistic | Prob. |
|---|---|---|---|---|
| C | 0.000000 | NA | NA | NA |
| O3MLV = 2 | 0.218810 | NA | NA | NA |
| O3MLV = 3 | 0.469295 | NA | NA | NA |
| MLV*(O3MLV = 1) | 0.040418 | NA | NA | NA |
| MLV*(O3MLV = 2) | 0.865247 | NA | NA | NA |
| MLV*(O3MLV = 3) | −0.208107 | NA | NA | NA |
| Pseudo R-squRed | 0.034381 | Mean dependent var | | 9.04E-17 |
| Adjusted R-squRed | −0.001382 | S.D. dependent var | | 0.550900 |
| S.E. of regression | 0.555362 | Objective | | 23.56312 |
| Quantile dependent var | −0.012850 | Restr. objective | | 24.40210 |
| Sparsity | 1.431649 | Quasi-LR statistic | | 4.688191 |
| Prob(Quasi-LR stat) | 0.320812 | | | |

**Figure 10.69** The output of the SPW-QR(Median) in (10.29).

4. Since it has a complete QR function, we can do its residual diagnostics, with the output presented in Figure 10.70, and obtain its forecast or fitted value variable, *BLVF_EQ29*, as shown in Figure 10.71.
5. We also can apply the RVT, with an output as presented in Figure 10.72. We can conclude that the effect of *MLV* on *BLV* has insignificant differences between the levels of *O3MLV*, based on the *QLR L*-statistic of $QLR\text{-}L_0 = 3.442\,543$ with $df = 2$ and $p$-value $= 0.1788$.
6. However, although we can't apply the quantile process, we can apply the stability test.

## 10.8.2 An Extension of the SPW-QR of *BLV* on *MLV* in (10.29)

Even though the SPW-QR in (10.29) has incomplete statistical results, as seen in Figure 10.69, the outputs of several statistical analyses are acceptable and valid, in the statistical and theoretical senses. For these reasons, the SPW-QR will be extended to a more advanced QR by inserting additional IVs, which can be learned by using the OMT. Then their ES can be modified to obtain complete output with a new ES, as shown in the following example.

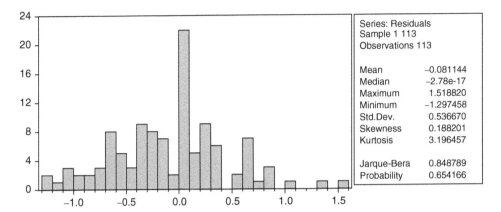

**Figure 10.70** The output of the residual diagnostics of the QR(Median) in (10.29).

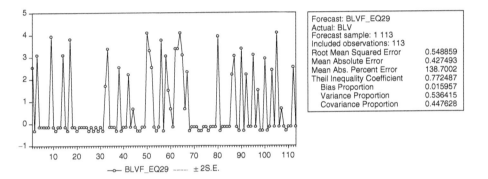

**Figure 10.71** The forecast evaluation based on the SQR(Median) in (10.29).

Redundant Variables Test

Equation: EQ in (10.29)

Specification: BLV C@EXPAND(O3MLV,@DROPFIRST)MLV MLV*@EXPAND (O3MLV,@DROP(2))

Redundant variables: MLV*@EXPAND(O3MLV,@DROP(2))

Null hypothesis: MLV*@EXPAND(O3MLV,@DROP(2)) are jointly insignificant

|  | Value | df | Probability |
|---|---|---|---|
| QLR L-statistic | 3.442543 | 2 | 0.1788 |
| QLR Lambda-statistic | 3.398309 | 2 | 0.1828 |

**Figure 10.72** The output of a RVT based on the SQR(Median) in (10.29).

**Example 10.31** *Application of OMT*

Referring to the problem of the output of F2WI-QR(Median) in Figure 10.69, we can extend the QR(Median) in (10.29) by selecting the variables *MLV\*ZMW\*@Expand(O3MLV)* and *ZMW\*@Expand(O3MLV)* for the OMT, with the output presented in Figure 10.73. Based on this output, the findings and notes are as follows:

1. The variables *MLV\*ZMW\*@Expand(O3MLV)* and *MZW\*@Expand(O3MLV)* are jointly significant, based on the *QLR* L-statistic of $QLR\text{-}L_0 = 37.23531$ with $df = 6$ and p-value = 0.0000. Hence, the variables need to be inserted as additional variables of the QR (10.29) to improve its good of fit.

2. All IVs of the unrestricted test equation are jointly significant, based on the *QLR*-statistic of $QLR_0 = 43.38602$ with $df = 11$ and p-value = 0.000 002. Hence, we have an acceptable PW-F2WI-HQR(Median) with the following ES:

$$BLV\ C@Expand\ (O3MLV, @dropfirst)\ MLV * ZMW * @Expand\ (O3MLV)$$

$$MLV * @Expand\ (O3MLV)\ ZMW * @Expand\ (O3MLV) \tag{10.30}$$

which should be modified to the following ES:

$$BLV\ C\ (O3MLV = 2)\ (O3MLV = 3)\ MLV * ZMW\ MLV\ ZMW$$

$$MLV * ZMW * @Expand\ (O3MLV, @Dropfirst)$$

$$MLV * @Expand\ (O3MLV, @Dropfirst)$$

$$ZMW * @Expand\ (O3MLV, @Dropfirst) \tag{10.31}$$

The modification enables us to use the RVT to make the hypotheses on the effect differences of each or all of the numerical IVs on *BLV* between the levels of *O3MLV* This is because the Wald test can't be applied for the QR (10.30) as the output of the Wald test will also present the *"NA"* Std. Error.

As an illustration, Figure 10.74 presents the output of the RVT for testing the joint effect differences of all numerical IVs on *BLV*, between the three levels of *O3MLV*, based on the QR(Median) in (10.31). The output shows that they are jointly insignificant, based on the *QLR* statistic of $QLR_0 = 4.464461$ with $df = 6$ and p-value = 0.6141.

3. Even though Figure 10.74 presents an incomplete output, the output shows a valid and acceptable QR(Median), which also can be obtained for each level of *O3MLV*. See how to present the functions in the following section.

## 10.8.3 Reduced Models of the Previous PW-QRs of *BLV*

### 10.8.3.1 A Reduced Model of the SPW-QR of *BLV* in (10.29)

Referring to the incomplete output of the previous SPW-QR of *BLV* by the ordinal variable *O3MLV* in (10.29), I should use the trial-and-error method to obtain a reduced model that gives complete output. At the first trial, by deleting the dummy variable *(O3MLV = 2)* from the QR (10.29), I unexpectedly obtain the complete output presented in Figure 10.75 from using the following ES:

$$BLV\ C\ (O3MLV = 3)\ MLV * @Expand\ (O3MLV) \tag{10.32}$$

Omitted variables test
Null hypothesis: MLV*ZMW*@EXPAND(O3MLV) ZMW*@EXPAND(O3MLV)
Equation: EQ20
Specification: BLV C @EXPAND(O3MLV,@DROPFIRST) MLV *@EXPAND(O3MLV)
Omitted Variables: MLV*ZMW*@EXPAND(O3MLV) ZMW*@EXPAND(O3MLV)
are jointly insignificant

| | Value | df | Probability |
|---|---|---|---|
| QLR L-statistic | 37.23531 | 6 | 0.0000 |
| QLR lambda-statistic | 32.89816 | 6 | 0.0000 |

L-test summary:

| | Value |
|---|---|
| Restricted objective | 23.56312 |
| Unrestricted objective | 18.48409 |
| Scale | 0.272808 |

Lambda-test summary:

| | Value |
|---|---|
| Restricted Log Obj. | 3.159683 |
| Unrestricted Log Obj. | 2.916910 |
| Scale | 0.014759 |

Unrestricted Test Equation:
Dependent variable: BLV
Method: Quantile Regression (Median)
Included observations: 113
Huber Sandwich Standard Errors & Covariance
Sparsity method: Kernel (Epanechnikov) using residuals
Bandwidth method: Hall-Sheather, bw = 0.20096
WARNING: linear dependencies found in regressors (some coefficients set to initial values)
Singularity in sandwich covariance estimation

| Variable | Coefficient | Std. Error | t-Statistic | Prob. |
|---|---|---|---|---|
| C | 0.000000 | NA | NA | NA |
| O3MLV = 2 | −0.075737 | NA | NA | NA |
| O3MLV = 3 | −0.241056 | NA | NA | NA |
| MLV*(O3MLV = 1) | 0.445889 | NA | NA | NA |
| MLV*(O3MLV = 2) | 0.459861 | NA | NA | NA |
| MLV*(O3MLV = 3) | 0.879026 | NA | NA | NA |
| MLV*ZMW*(O3MLV = 1) | −0.790684 | NA | NA | NA |
| MLV*ZMW*(O3MLV = 2) | −0.266531 | NA | NA | NA |
| MLV*ZMW*(O3MLV = 3) | 0.499379 | NA | NA | NA |
| ZMW*(O3MLV = 1) | 0.000000 | NA | NA | NA |
| ZMW*(O3MLV = 2) | 0.287732 | NA | NA | NA |
| ZMW*(O3MLV = 3) | 0.001728 | NA | NA | NA |

| | | | |
|---|---|---|---|
| Pseudo R-squRed | 0.242521 | Mean dependent var | 9.04E-17 |
| Adjusted R-squRed | 0.176333 | S.D. dependent var | 0.550900 |
| S.E. of regression | 0.477146 | Objective | 18.48409 |
| Quantile dependent var | −0.012850 | Restr. objective | 24.40210 |
| Sparsity | 1.091230 | Quasi-LR statistic | 43.38602 |
| Prob(Quasi-LR stat) | 0.000002 | | |

**Figure 10.73** The omitted variables test of *MLV*ZMW*@Expand(O3MLV)* and *ZMW*@Expand(O3MLV)*.

Redundant Variables Test

Equation: EQ22

Redundant variables: MLV*ZMW*@EXPAND(O3MLV,@DROPFIRST)

    MLV*@EXPAND(O3MLV,@DROPFIRST) ZMW

    *@EXPAND(O3MLV,@DROPFIRST)

Specification: BLV C (O3MLV = 2) (O3MLV = 3) MLV*ZMW MLV ZMW

    MLV*ZMW*@EXPAND(O3MLV,@DROPFIRST)MLV

    *@EXPAND(O3MLV,@DROPFIRST) ZMW

    *@EXPAND(O3MLV,@DROPFIRST)

Null hypothesis: MLV*ZMW*@EXPAND(O3MLV,@DROPFIRST)

    MLV*@EXPAND(O3MLV,@DROPFIRST)

    ZMW*@EXPAND(O3MLV,@DROPFIRST) are jointly insignificant

|  | Value | df | Probability |
|---|---|---|---|
| QLR L-statistic | 4.464461 | 6 | 0.6141 |
| QLR Lambda-statistic | 4.392495 | 6 | 0.6237 |

**Figure 10.74** The output of the RVT for the hypothesis of the joint effects of all numerical IVs of the QR(Median) in (10.31) between the levels of *O3MLV*.

Based on this output, the following findings and notes are presented:

1. We can derive the equations of the three conditional QR functions of *BLV*, as follows, which show that the first two functions have coinciding intercept of 0.218 810:

$$\widehat{BLV}_{(O3MLV=1)} = C\,(1) + C\,(3) \times MLV = 0.218\,810 + 0.306\,307 \times MLV$$

$$\widehat{BLV}_{(O3MLV=2)} = C\,(1) + C\,(4) \times MLV = 0.218\,810 + 0.865\,247 \times MLV$$

$$\widehat{BLV}_{(O3MLV=3)} = (C\,(1) + C\,(2)) + C\,(5) \times MLV$$

$$= (0.218\,810 + 0.250\,4844) - 0.208\,107 \times MLV$$

2. Even though all IVs of the QR(Median) are jointly insignificant, based on the *QLR* statistic of $QLR_0 = 4.688\,191$ with $df = 4$ and p-value $= 0.320\,812$, two IVs have a positive significant effect at the 10% level. The numerical IV *MLV\*(O3MLV = 1)* has a positive significant effect at the 10% level, based on the t-statistic of $t_0 = 2.397\,204$ with $df = (113 - 5)$ and p-value $= 0.0182/2 < 0.01$, and *MLV\*(O3MLV = 2)* also has a positive significant effect at the 10% level.

3. In addition, at the 5% level, we find that *MLV* has significant difference effects between the three levels of *O3MLV*, based on the Wald test (*Chi-square* statistic) of $\chi_0^2 = 6.570\,984$ with $df = 2$ and p-value $= 0.0374$.

### 10.8.3.2 A Reduced Model of the PW-2WI-QR of *BLV* in (10.31)

Referring to the incomplete output of the previous PW-2WI-QR of *BLV* by the ordinal variable *O3MLV* in (10.31), which is an advanced peace-wise QR, since the factor *O3MLV* is an

Dependent Variable: BLV

Method: Quantile Regression (Median)

Date: 05/19/19 Time: 09:51

Sample: 1 113

Included observations: 113

Huber Sandwich Standard Errors & Covariance

Sparsity method: Kernel (Epanechnikov) using residuals

Bandwidth method: Hall-Sheather, bw = 0.20096

Estimation successfully identifies unique optimal solution

| Variable | Coefficient | Std. Error | t-Statistic | Prob. |
|---|---|---|---|---|
| C | 0.218810 | 0.117835 | 1.856926 | 0.0660 |
| O3MLV = 3 | 0.250484 | 0.296509 | 0.844778 | 0.4001 |
| MLV*(O3MLV = 1) | 0.306307 | 0.208113 | 1.471826 | 0.1440 |
| MLV*(O3MLV = 2) | 0.865247 | 0.360940 | 2.397204 | 0.0182 |
| MLV*(O3MLV = 3) | −0.208107 | 0.215548 | −0.965478 | 0.3365 |

| | | | |
|---|---|---|---|
| Pseudo R-squRed | 0.034381 | Mean dependent var | 9.04E-17 |
| Adjusted R-squRed | −0.001382 | S.D. dependent var | 0.550900 |
| S.E. of regression | 0.552785 | Objective | 23.56312 |
| Quantile dependent var | −0.012850 | Restr. objective | 24.40210 |
| Sparsity | 1.431649 | Quasi-LR statistic | 4.688191 |
| Prob(Quasi-LR stat) | 0.320812 | | |

Estimation command:
=========================
QREG BLV C O3MLV = 3 MLV*@EXPAND(O3MLV)

Estimation equation:
=========================
BLV = C(1) + C(2)*(O3MLV = 3) + C(3)*MLV*(O3MLV = 1) + C(4)*MLV*(O3MLV = 2) +
C(5)*MLV*(O3MLV = 3)

Substituted coefficients:
=========================
BLV = 0.218810191668 + 0.250484400428*(O3MLV = 3) + 0.306306885037*MLV*(O3MLV = 1)
+ 0.865246586857*MLV*(O3MLV = 2) − 0.208106559655*MLV*(O3MLV = 3)

**Figure 10.75** The output of the QR(Median) in (10.32) and its representations.

Dependent Variable: BLV

Method: Quantile Regression (Median)

Date: 04/09/19   Time: 17:57

Sample: 1 113

Included observations: 113

Huber Sandwich Standard Errors & Covariance

Sparsity method: Kernel (Epanechnikov) using residuals

Bandwidth method: Hall-Sheather, bw = 0.20096

Estimation successfully identifies unique optimal solution

| Variable | Coefficient | Std. Error | t-Statistic | Prob. |
|---|---|---|---|---|
| C | −0.044858 | 0.189666 | −0.236512 | 0.8135 |
| O3MLV = 3 | 0.022516 | 0.487214 | 0.046214 | 0.9632 |
| MLV*ZMW | −0.325629 | 0.430132 | −0.757045 | 0.4507 |
| MLV | 0.526942 | 0.480058 | 1.097662 | 0.2749 |
| ZMW | 0.257547 | 0.163842 | 1.571924 | 0.1190 |
| MLV*ZMW*(O3MLV = 1) | −0.152095 | 0.360142 | −0.422318 | 0.6737 |
| MLV*ZMW*(O3MLV = 3) | 0.580320 | 0.646610 | 0.897481 | 0.3715 |
| MLV*(O3MLV = 1) | −0.135563 | 0.325103 | −0.416985 | 0.6775 |
| MLV*(O3MLV = 3) | 0.100948 | 0.448386 | 0.225137 | 0.8223 |

| | | | | |
|---|---|---|---|---|
| Pseudo R-squRed | 0.233142 | Mean dependent var | | 9.04E-17 |
| Adjusted R-squRed | 0.174153 | S.D. dependent var | | 0.550900 |
| S.E. of regression | 0.476745 | Objective | | 18.71295 |
| Quantile dependent var | −0.012850 | Restr. objective | | 24.40210 |
| Sparsity | 1.100323 | Quasi-LR statistic | | 41.36353 |
| Prob(Quasi-LR stat) | 0.000002 | | | |

**Figure 10.76**   A complete output of a reduced PW-QR(Median) in (10.31).

ordinal variable of the numerical IV, *MLV*, as an extension of the SPW-QR in the Example 10.29, we should obtain an acceptable reduced QR having the complete output. Then by using the trial-and-error method, I obtain a reduced model, with the following ES. Based on this output, the following finding and notes are presented.

$$BLV\ C\ (O3MLV = 3)\ MLV * ZMW\ MLV\ ZMW$$

$$MLV * ZMW * @EXPAND\,(O3MLV, @DROP\,(2))$$

$$MLV * @EXPAND\,(O3MLV, @DROP\,(2)) \tag{10.33}$$

Based this output, the following finding and notes are presented;

1. Note that different functions, *@Dropfirst* and *@Drop(2)*, are used in the ESs (10.31, 10.33), respectively. Also, the ES (10.33) does not have the interaction

*ZMW\*@Expand(O3MLV,@Dropfirst)*. These things indicate it takes time to obtain a complete output based on a reduced QR in (10.31) using the trial-and-error method.

2. Based on the output of its representations, we can derive easily the equations of the following three conditional QR functions of *BLV*:

$$\widehat{BLV}_{(O3MLV=1)} = C(1) + C(6) MLV * ZMW + C(8) MLV$$
$$= -0.045 - 0.152 MLV * ZMW - 0.136 MLV$$

$$\widehat{BLV}_{(O3MLV=2)} = C(1) + C(3) MLV * ZMW + C(4) MLV + C(5) ZMW$$
$$= -0.045 - 0.326 MLV * ZMW + 0.527 MLV + 0.258 ZMW$$

$$\widehat{BLV}_{(O3MLV=3)} = (C(1) + C(2)) + C(7) MLV * ZMW + C(9) MLV$$
$$= (-0.045 + 0.023) + 0.580 * MLV * ZMW + 0.101 * MLV$$

Note the regressions $\widehat{BLV}_{(O3MLV=1)}$ and $\widehat{BLV}_{(O3MLV=2)}$ have coinciding intercepts, and they are the reduced 2WI-QR because they do not have *ZMW* as their main IV.

3. Since the model has the complete output, all hypotheses can be tested using the Wald test, aside from the RVT. In addition, we also can do more advanced analyses, which have been presented earlier in this chapter, as well as in previous chapters.

# Appendix A

# Mean and Quantile Regressions

This appendix presents a comparison between the basic estimation methods of the mean regression and quantile regression. For more details about the mathematical concepts of quantile regression, I recommend the following books: Koenker (2005), Hao and Naiman (2007), and Davino et al. (2014).

## A.1  The Single Parameter Mean and Quantile Regressions

### A.1.1  The Single-Parameter Mean Regression

A mean regression (MR) based on a numerical random sample $Y_i$, $i = 1,2, ..., n$, has the following equation:

$$Y_i = \mu + \varepsilon_i \tag{A.1}$$

where $\mu$ is the mean parameter of the population, and $\varepsilon_i$, $i = 1, 2, ..., n$ are the random errors, which are assumed to have *independent identical normal distributions (IID) with zero mean and constant variance*, namely $IID -N(0, \sigma^2)$.

Then based on the Eq. (A.1), we have the following *sum of squared deviation (SSD)*, which is a function of the population parameter $\mu$, as follows:

$$SSD(\mu) = \sum_1^n \varepsilon_i^2 = \sum_1^n (Y_i - \mu)^2 \tag{A.2}$$

The necessary condition for obtaining the least squares estimate of the parameter $\mu$ is the partial derivative of $SSD(\mu)$ equals zero, that is:

$$\frac{\partial SSD(\mu)}{\partial \mu} = -2 \sum_1^n (Y_i - \mu) = 0 \tag{A.3}$$

$$\hat{\mu} = \frac{1}{n} \sum_1^n Y_i = \overline{Y} \tag{A.4}$$

*Quantile Regression: Applications on Experimental and Cross Section Data Using EViews*, First Edition.
I Gusti Ngurah Agung.
© 2021 John Wiley & Sons Ltd. Published 2021 by John Wiley & Sons Ltd.

## A.1.2 The Single-Parameter Quantile Regression

A quantile regression (QR) based on a numerical random sample $Y_i$, $i = 1, 2, \ldots, n$, has the following equation:

$$Y_i = q + \varepsilon_i(\tau) \tag{A.5}$$

where $q = q(\tau)$ is the quantile($\tau$) population parameter, and $\varepsilon_i(\tau)$, $i = 1, 2, \ldots, n$ are the random errors, which are assumed to have IID *with zero mean and constant variance only* for any fixed proportion $\tau \in (0, 1)$.

Then based on the Eq. (A.5), we have the following *sum of absolute deviation (SAD)*, which is a function of the population parameter $q$ for the fixed proportion $\tau \varepsilon (0, 1)$, as follows:

$$
\begin{aligned}
SAD(q) &= \sum_1^n |\varepsilon i(\tau)| = \sum_1^n |Yi - q| \\
&= \sum_{Y \le q} (\tau - 1)(Y - q) + \sum_{Y > q} \tau (Y - q) \\
&= (\tau - 1) \sum_{Y \le q} (Y - q) + \tau \sum_{Y > q} (Y - q)
\end{aligned} \tag{A.6}
$$

Note that in the preceding equation, a specific weighted sum is defined. Then the necessary condition for obtaining the least absolute estimate of the parameter $q$ is the partial derivative of $SAD(q)$ equals zero, that is:

$$\frac{\partial SAD(q)}{\partial q} = (\tau - 1) \sum_{Y \le q} (-1) + \tau \sum_{Y > q} (-1) = 0$$

Hence, we obtain the following equations:

$$(\tau - 1) \sum_{Y \le q} (1) + \tau \sum_{Y > q} (1) = 0$$

$$\tau \sum_1^n (1) - \sum_{Y \le q} (1) = 0$$

$$n\tau - nF_n(q) = 0$$

$$\tau - F_n(q) = 0$$

which give the estimate of the quantile($\tau$):

$$F_n(q) = P(Y \le q) = \tau \rightarrow F_n^{-1}(\tau) = q(\tau)$$

## A.2 The Simplest Conditional Mean and Quantile Regressions

### A.2.1 The Simplest Conditional Mean Regression

The simplest conditional MR is a two-parameter MR, based on a bivariate numerical random sample $(X_i, Y_i)$, $i = 1, 2, \ldots, n$, and has the following equation:

$$Y_i = \beta_0 + \beta_1 X_i + \varepsilon_i \tag{A.7}$$

Then based on the Eq. (A.7), we have the following *SSD*, conditional for the sample scores of the variable *X*, which is a function of the population parameter $(\beta_0,\beta_1)$ as follows:

$$SSD\left(\beta_0, \beta_1 \mid x\right) = \sum_1^n \varepsilon_i^2 = \sum \left(Y_i - \beta_0 - \beta_1 X_i\right)^2 \tag{A.8}$$

The least squares estimates of the parameters $(\beta_0,\beta_1)$ are obtained as the minimization of the $SSD(\beta_0,\beta_1)$. The necessary conditions are its partial derivatives equal to zero as its *normality equations*, as follows:

$$\frac{\partial SSD\left(\beta_0, \beta_1 \mid x\right)}{\partial \beta_0} = 0 \text{ and } \frac{\partial SSD\left(\beta_0, \beta_1 \mid x\right)}{\partial \beta_1} = 0 \tag{A.9}$$

Then we can derive the following results:

$$\hat{\beta}_0 = \overline{Y} + \hat{\beta}_1 \overline{X} \tag{A.10a}$$

$$\hat{\beta}_1 = \frac{n \sum X_i Y_i - \sum X_i \sum Y_i}{n \sum X_i^2 - \left(\sum X_i\right)^2} \tag{A.10b}$$

### A.2.2 The Simplest Conditional Quantile Regressions

The simplest conditional QR is a two-parameter QR, based on a bivariate numerical random sample $(X_i, Y_i)$, $i = 1, 2, ..., n$, and has the following equation:

$$Y_i = \alpha + \beta X_i + \varepsilon_i(\tau) \tag{A.11}$$

where $\alpha = \alpha(\tau)$ and $\beta = \beta(\tau)$ are the population parameters, and $\varepsilon_i(\tau)$, $i = 1, 2, ..., n$, are the random errors, which are assumed to have IID for a fixed proportion $\tau \in (0, 1)$.

Then based on the Eq. (A.11), we have the following *SAD*, conditional for the sample scores of the variable *X*, which is a function of the population parameters $\alpha(\tau)$ and $\beta(\tau)$ for any fixed proportion $\tau \varepsilon (0, 1)$, as follows:

$$SAD\left(\alpha, \beta \mid x\right) = \sum_1^n |\varepsilon_i(\tau)| = \sum_1^n |Y_i - \alpha - \beta X_i|$$

$$= (\tau - 1) \sum_{y \le \theta} \left(Y_i - \alpha - \beta X_i\right) + \tau + \tau \sum_{y > 0} \left(Y_i - \alpha - \beta X_i\right) \tag{A.12}$$

where subscript $y \le \theta$ indicates $Y_i \le \alpha + \beta X_i$.

The *least absolute estimates* of the parameters $(\alpha,\beta)$ are obtained as the minimization of the weighted sum as presented Eq. (A.12), which is a *linear programming problem* (Koenker 2005; Hao and Naiman 2007; Davino et al. 2014). So, we can't apply the partial derivation method as presented for the conditional MR earlier.

## A.3 The Estimation Process of the Quantile Regression

The basic estimation process of the QR is the following:

1) With the data file on-screen, selecting *Quick/Estimate Equation* opens the Equation Estimation dialog shown in Figure A.1a. The default method shown in the Estimation settings is LS – Least Squares (NLS and ARMA).

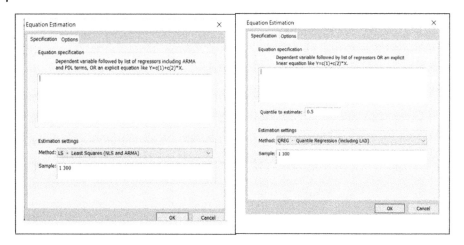

**Figure A.1** The Equation Estimation dialogs for (a) the LS – Least Squares (NLS and ARMA) method and (b) the QREG – Quantile Regression method.

2) Change the *Method* setting to *QREG – Quantile Regression (including LAD)* to display the dialog in Figure A.1b, which presents the default *Quantile to estimate* value 0.5. We can modify this value to any $\tau \ \varepsilon \ (0, 1)$.

3) Finally, by inserting the *Equation specification* of a QR and clicking *OK,* we obtain output of the QR($\tau$). The following subsections present more advanced QR analyses.

### A.3.1 Applications of the Quantile Process

With the output of the QR still on-screen, we select *View/Quantile Process* to find additional alternative analyses for the QR. The three options are *Process Coefficients, Slope Equality*

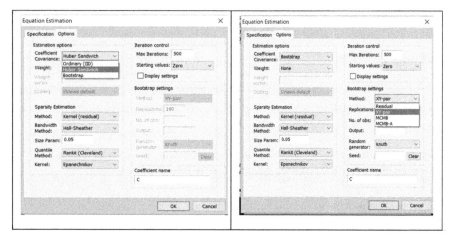

**Figure A.2** The dialog for the estimation options (a) The default options and (b) the coefficient covariance: bootstrap.

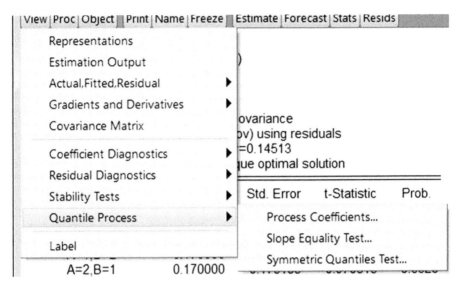

**Figure A.3** The alternative Quantile Process analyses.

*Test*, and *Symmetric Quantile Test,* as presented in Figure A.3. The other available analysis methods can easily be identified, and each can easily be run as exercises for alternative combinations of options, as discussed in Chapter 2, Section 2.8.

### A.3.1.1 An Application of Quantile Process/Process Coefficients

With the output of any QR($\tau$) on-screen, select *View/Quantile Process/Process Coefficients* to display the Quantile Process View dialog in Figure A.4.

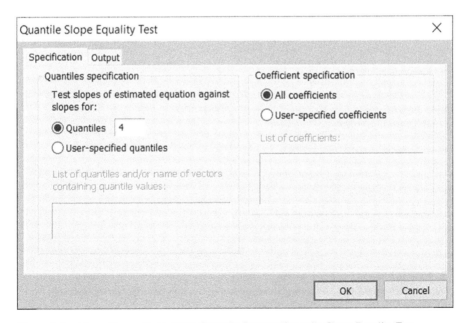

**Figure A.4** Alternative options of the Quantile Process/Quantile Slope Equality Test.

This figure shows (i) two alternatives for *Output*, *Table* or *Graph*; (ii) two possible *Quantiles specification* options, the number of *Quantiles* or *User-specified quantiles*; and (iii) three alternative coefficient specifications: *All coefficients*, *Intercept only*, and *User-specified coefficients*, which all are illustrated in the following notes:

1) This figure presents the default combination options. The output of this combination of options will present a tabulation of 10 quantiles, 0.1, 0.2, ..., 0.9, for all coefficients. An alternative output format is a graph.

2) For the *Quantiles specification,* we can have many possible alternative options, using either the number of quantiles or user-specified quantiles:
   - For the *Quantiles*, I recommend using 4, 5, 8, 10, or 20, and not 3, 6, 7, or 9, because the output of the latter will present incomplete decimal proportions, such as 0.333 ....
   - For *User-specified quantiles*, we can use either a single proportion or multiple proportions.

3) For the *Coefficient specification,* we have three possible choices. For the *User-specified coefficients*, we should enter selected model parameters, such as any number of coefficients from C(1) to C($k$), where C($k$) is last model parameter.

### A.3.1.2 An Application of Quantile Process/Slope Equality Test

With the output of a QR($\tau$) *with the intercept* on-screen, selecting *View/Quantile Process/Slope Equality Test* opens a Quantile Process View dialog as presented in Figure A.5, with two *Quantiles specification* options and two options for *Coefficient specification*, which should be selected from the slope parameters C(2) to C($k$).

### A.3.1.3 An Application of the Quantile Process/Symmetric Quantile Test

With the output of any QR($\tau$), *either with or without the intercept*, on-screen, we can select *View/Quantile Process/Symmetric Quantile Test* to open a dialog, as presented in Figure A.6 with two options of *Quantiles specification* and three options of *Coefficient specification*, which are the same as presented in Figure A.4.

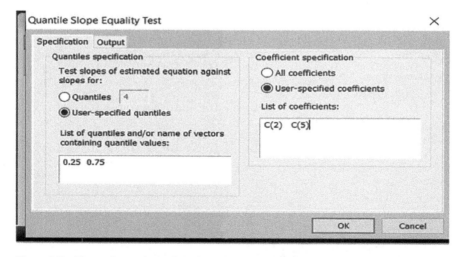

**Figure A.5** Alternative options of the Quantile Process/Quantile Slope Equality Test.

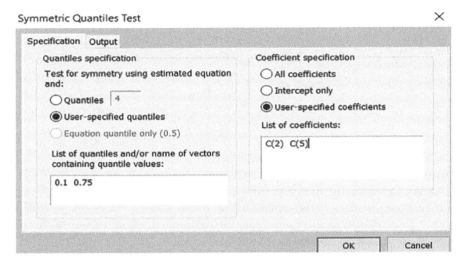

**Figure A.6**   Alternative options for Quantile Process/Symmetric Quantile Test.

### A.3.1.4   The Impacts of the Combinations of Options

In addition to the set of default options presented in Figures A.2a,b, many other combinations of options can be applied, which have been presented in Chapter 2, Section 2.8. We found that none of the alternative combinations of options have an impact on the parameters' estimates. In other words, we obtain the same QR function for all combinations of options. Hence, we will have the same residuals and outputs of all residual diagnostics.

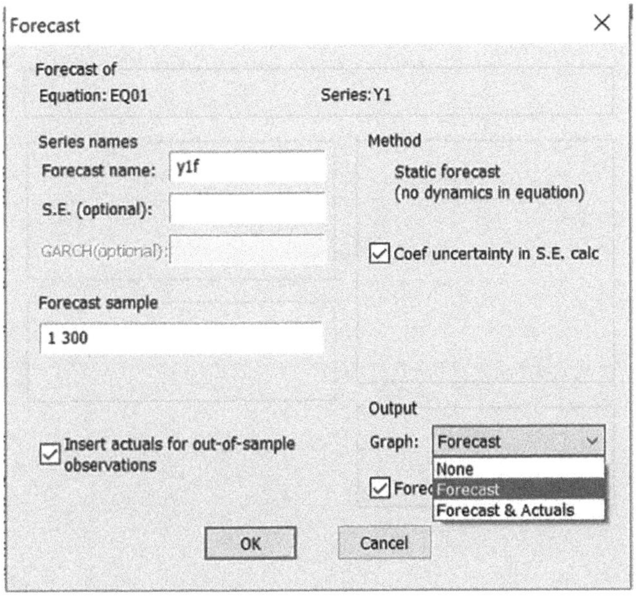

**Figure A.7**   The dialog for the forecasting.

## A.4   An Application of the Forecast Button

With the output of QR($\tau$) of any dependent variable *Y1* on-screen, clicking the *Forecast* button opens the Forecast dialog in Figure A.7.

The following notes are presented about this dialog:

1) The figure directly shows the forecast variable *Y1F*, which can be renamed, such as *Y1F_Eq01*, since the forecast is for the QR of *Y1* in Eq. 01.
2) The forecast sample 1 300 represents the whole sample of size 300. Hence, the forecast variable is equal to the fitted variable based on the QR. On the other hand, if *Y1* has missing values (NAs), the forecast variable *Y1F_Eq01* still has 300 scores. Refer to Chapter 6, Examples 6.11 and 6.12.
3) The output can include both the graph and forecast evaluation. I recommend selecting *Graph*: *Forecast and Actuals* so the output of the graph will show the deviation of the forecast scores from the observed scores of *Y1*.

# Appendix B

# Applications of the t-Test Statistic for Testing Alternative Hypotheses

This appendix presents applications of the $t$-test statistic for directly testing two-sided and one-sided hypotheses using the observed $t$-statistic and its probability presented in the corresponding statistical results. For an illustration, see the statistical results in Figure B.1 and its model parameters presented in Table B.1, which are taken from the results presented in Chapter 2, in Figure 2.11 and Table 2.4. Based on Figure B.1 and Table B.1, we will define and test several alternative hypotheses.

## B.1    Testing a Two-Sided Hypothesis

A two-sided statistical hypothesis can be presented as follows:

$$H_0 : C(k) = 0 \text{ vs } H_1 : C(k) \neq 0, \text{ for each } k = 2, 3, \ldots, 6 \tag{B.1}$$

with the $p$-value of each hypothesis equals to *Prob.* in the output.

Hence, at the $\alpha$ level of significance, the null hypothesis $H_0$ is rejected if the $p$-value = *Prob.* < $\alpha$. And it is accepted if the $p$-value $\geq \alpha$.

For instance, for $k = 2$, at the 5% level of significance, the null hypothesis is rejected based on the $t$-statistic of $t_0 = -2.523$ with $df = (300 - 6)$ and $p$-value = *Prob.*/2 = 0.0122 < 0.05. Hence, we can conclude that the medians of *Y1* in the two levels $G = 1$ and $G = 2$ have a significant difference, conditional for the level of $H = 2$.

## B.2    Testing a Right-Sided Hypothesis

A right-sided statistical hypothesis can be presented as follows:

$$H_0 : C(k) \leq 0 \text{ vs } H_1 : C(k) > 0, \text{ for each } k = 2, 3, \ldots, 6 \tag{B.2}$$

and the conclusion of the testing of each hypothesis depends on the sign of the estimate of the parameter $\widehat{C(k)}$, which is the same as the sign of the observed value of the $t$-statistic. And for the $p$-value of the hypothesis, we have the following criteria:

1) If the observed value of the $t$-statistic is positive, or $t_0 > 0$, then the $p$-value = *Prob.*/2.

*Quantile Regression: Applications on Experimental and Cross Section Data Using EViews*, First Edition.
I Gusti Ngurah Agung.
© 2021 John Wiley & Sons Ltd. Published 2021 by John Wiley & Sons Ltd.

```
Dependent Variable: Y1
Method: Quantile Regression (Median)
Date: 03/28/18   Time: 08:42
Sample: 1 300
Included observations: 300
Huber Sandwich Standard Errors & Covariance
Sparsity method: Kernel (Epanechnikov) using residuals
Bandwidth method: Hall-Sheather, bw=0.14513
Estimation successful but solution may not be unique
```

| Variable | Coefficient | Std. Error | t-Statistic | Prob. |
|---|---|---|---|---|
| C | 6.000000 | 0.142008 | 42.25110 | 0.0000 |
| G2=1 | -0.500000 | 0.198188 | -2.522856 | 0.0122 |
| (G2=1)*(H3=1) | -0.830000 | 0.217051 | -3.823989 | 0.0002 |
| (G2=1)*(H3=3) | 0.500000 | 0.206894 | 2.416694 | 0.0163 |
| (G2=2)*(H3=1) | -0.330000 | 0.207099 | -1.593444 | 0.1121 |
| (G2=2)*(H3=3) | 0.330000 | 0.203664 | 1.620315 | 0.1062 |

| | | | |
|---|---|---|---|
| Pseudo R-squared | 0.106330 | Mean dependent var | 5.751100 |
| Adjusted R-squared | 0.091132 | S.D. dependent var | 0.945077 |
| S.E. of regression | 0.849650 | Objective | 99.17500 |
| Quantile dependent var | 5.670000 | Restr. objective | 110.9750 |
| Sparsity | 2.119713 | Quasi-LR statistic | 44.53433 |
| Prob(Quasi-LR stat) | 0.000000 | | |

```
Estimation Command:
=========================
QREG(K=E,NGRID=100,Q=R) Y1 C @EXPAND(G2,@DROP(2))
@EXPAND(G2)*@EXPAND(H3,@DROP(2))

Estimation Equation:
=========================
Y1 = C(1) + C(2)*(G2=1) + C(3)*(G2=1)*(H3=1) + C(4)*(G2=1)*(H3=3) +
C(5)*(G2=2)*(H3=1) + C(6)*(G2=2)*(H3=3)

Substituted Coefficients:
=========================
Y1 = 6 - 0.5*(G2=1) - 0.83*(G2=1)*(H3=1) + 0.5*(G2=1)*(H3=3) - 0.33*
(G2=2)*(H3=1) + 0.33*(G2=2)*(H3=3)
```

**Figure B.1**  Statistical results of a $2 \times 3$ factorial QR in (2.8).

**Table B.1**  The parameters of the QR in Figure B.1.

| | $H = 1$ | $H = 2$ | $H = 3$ | $H(1 - 2)$ | $B(3 - 3)$ |
|---|---|---|---|---|---|
| $G = 1$ | $C(1) + C(2) + C(3)$ | $C(1) + C(2)$ | $C(1) + C(2) + C(4)$ | $C(3)$ | $C(4)$ |
| $G = 2$ | $C(1) + C(5)$ | $C(1)$ | $C(1) + C(6)$ | $C(5)$ | $C(6)$ |
| $G(1 - 2)$ | $C(2) + C(3) - C(5)$ | $C(2)$ | $C(2) + C(4) - C(6)$ | $C(3) - C(5)$ | $C(4) - C(6)$ |

2) If the observed value of the $t$-statistic is negative, or $t_0 < 0$, then the $p$-value is computed as $p$-value $= 1$- *Prob./2*.

### Example B.1  *Testing the Hypothesis (3.2) for $k = 6$*

In this case, the proposed hypothesis is *Conditional for the level $G = 2$, the median of Y1 in $H = 3$ is greater than in $H = 2$*, which is represented as $H_1: C(6) > 0$.

Since the output presents the estimate of the parameter $\widehat{C(6)} = 0.330 > 0$, and $t_0 = 1.620 > 0$, then the $p$-value of the hypothesis should be computed as $p$-value $=$ *Prob./2* $= 0.1062/2 = 0.0531$.

Hence, at the 10% level of significance, the null hypothesis is rejected based on the $t$-statistic of $t_0 = 1.620$ with $df = (300 - 6)$ and $p$-value $=$ *Prob./2* $= 0.0531 < 0.10$. We can conclude that the data supports the hypothesis.

### Example B.2  *Testing the Hypothesis (3.2) for $k = 2$*

In this case, the proposed hypothesis is *Conditional for the level $H = 2$, the median of Y1 in $G = 1$ is greater than in $G = 2$*, which is represented as $H_1: C(2) > 0$.

Since the output presents the estimate of the parameter $\widehat{C(2)} = -0.500 < 0$, and $t_0 = -2.523 < 0$, then the $p$-value $= 1 - (Prob./2) = 1 - (0.0122/2) = 1 - 0.0061 = 0.9939$.

Hence, at the 10% level of significance, the null hypothesis is accepted based on the $t$-statistic of $t_0 = -2.523$ with $df = (300 - 6)$ and $p$-value $= 0.9939 > 0.10$. So, we can conclude the data does not support the hypothesis.

However, it is important to note that if $t_0 < 0$ for any right-sided hypothesis, it can be concluded directly that the null hypothesis is accepted, without taking into account its $p$-value, because $\widehat{C(k)} < 0$ belongs to the null hypothesis $H_0$: $C(k) \leq 0$ in (3.2).

## B.3  Testing a Left-Sided Hypothesis

A left-sided statistical hypothesis can be presented as follows:

$$H_0 : C(k) \geq 0 \text{ vs } H_1 : C(k) < 0, \text{ for each } k = 2, 3, \dots, 6 \tag{B.3}$$

and the conclusion of the testing of each hypothesis depends on the sign of the estimate of the parameter $\widehat{C(k)}$ which is the same as the sign of the observed value of the $t$-statistic. And for the $p$-value of the hypothesis, we have the following criteria:

1) If the observed value of the $t$-statistic is negative, or $t_0 < 0$, then the $p$-value $= Prob./2$.
2) If the estimate of the parameter $C(k) > 0$, which indicates $t_0 > 0$, then the $p$-value of the hypothesis should be computed as $p$-value $= 1 - (Prob./2)$.

**Example B.3**  *Testing the Hypothesis (3.3) for $k = 2$*
In this case, the proposed hypothesis is *Conditional for the level H = 2, the median of Y1 in G = 1 is smaller than in G = 2*, which is represented as $H_1$: $C(2) < 0$.
    Since the output presents the estimate of the parameter $\widehat{C(2)} = -0.500 < 0$, and $t_0 = -2.523 < 0$, the $p$-value of the hypothesis should be computed as $p$-value $= Prob./2 = 0.0122/2 = 0.0061$. Then, at the 1% level of significance, the null hypothesis is rejected based on the $t$-statistic of $t_0 = -2.523$ with $df = (300 - 6)$ and $p$-value $= 0.0061 < 0.01$. Hence, we can conclude that the data supports the hypothesis.

**Example B.4**  *Testing the Hypothesis (3.3) for $k = 6$*
In this case, the proposed hypothesis is *Conditional for the level G = 2, the median of Y1 in H = 3 is smaller than in H = 2*, which is represented as $H_1$: $C(6) < 0$.
    Since the output presents the estimate of the parameter $\widehat{C(6)} = 0.330 > 0$, and $t_0 = 1.620 > 0$, then the $p$-value of the hypothesis should be computed as $p$-value $= 1 - Prob./2 = 1 - 0.1062/2 = 1 - 0.0531 = 0.9469$. Then, at the 10% level of significance, the null hypothesis is accepted based on the $t$-statistic of $t_0 = 1.620$ with $df = (300 - 6)$ and $p$-value $= Prob. = 0.9469 > 0.10$. We can conclude that the data does not support the hypothesis.
    However, it is important to note that if $t_0 > 0$ for any left-sided hypothesis, we can conclude that the null hypothesis is accepted without taking into account its $p$-value. This is because $\widehat{C(k)} > 0$ belongs to the null hypothesis $H_0$: $C(k) \geq 0$ in (3.3).

# Appendix C

## Applications of Factor Analysis

In Chapter 10, we presented QRs of a baby latent variable (*BLV*) on alternative IVs, including a mother latent variable (*MLV*), based on the data in BBW.wf1. This appendix presents the specific factor analyses to generate the *BLV* and *MLV*.

## C.1  Generating the BLV

In this section, we will develop the *BLV*, using factor analysis, with the steps that follow:

1) With the data on-screen, we can obtain the dialog on the left side of Figure C.1 by selecting *Object/New Objects/Factor*. Then we can insert the three variables, *BBW*, *FUNDUS*, and *MUAC*.
2) Clicking the Estimation button produces the options on the right side of Figure C.1. If we select the *Principal factors* method, use 1 as the number of factors, and click *OK*, we obtain the output in Figure C.2.
3) We then can select *View/Eigenvalues* and click *OK* to see whether the single factor is acceptable, in the statistical sense, with an eigenvalue > 1. In this case, the output shows that a single latent variable *BLV* can be developed based on the three variables, *BBW*, *FUNDUS,* and *MUAC*.
4) After clicking *Proc/Name Factors,* we can insert the name *BLV* and click *OK*.
5) Finally, by selecting *Proc/Make Scores* and clicking *OK*, we make *BLV* directly available in the data file as a new additional variable.
6) With the output in Figure C.3 on-screen, clicking *View* obtains additional statistical results, such as the Eigenvalues Summary and Kaiser's Measure of Sampling Adequacy, as presented in Figures C.3 and C.4.

## C.2  Generating the MLV

By using the same method as presented for *BLV*, we can generate the *MLV* based on the two ordinal variables, *ED* (education level) and *SE* (socioeconomic level), and two dummy variables, *SMOKING* (*SMOKING* = 1 for the smoking mothers) and *NO_ANC* (*NO_ANC* = 1 for the mothers who don't have antenatal care). We obtain the outputs in Figures C.5 and C.6, which show a single *MLV* is an acceptable latent variable.

*Quantile Regression: Applications on Experimental and Cross Section Data Using EViews,* First Edition.
I Gusti Ngurah Agung.
© 2021 John Wiley & Sons Ltd. Published 2021 by John Wiley & Sons Ltd.

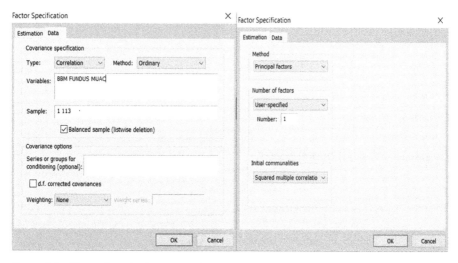

**Figure C.1** The options for factor specification.

Factor Method: Principal Factors
Date: 04/12/19  Time: 09:36
Covariance Analysis: Ordinary Correlation
Sample: 1 113
Included observations: 113
Number of factors: User-specified
Prior communalities: Squared multiple correlation

|  | Loadings | | |
|---|---|---|---|
|  | F1 | Communality | Uniqueness |
| BBW | 0.365700 | 0.133736 | 0.866264 |
| FUNDUS | 0.319634 | 0.102166 | 0.897834 |
| MUAC | 0.411562 | 0.169383 | 0.830617 |

| Factor | Variance | Cumulative | Difference | Proportion | Cumulative |
|---|---|---|---|---|---|
| F1 | 0.405285 | 0.405285 | --- | 1.000000 | 1.000000 |
| Total | 0.405285 | 0.405285 |  | 1.000000 |  |

|  | Model | Independence | Saturated |
|---|---|---|---|
| Discrepancy | 0.006949 | 0.094186 | 0.000000 |
| Parameters | 6 | 3 | 6 |
| Degrees-of-freedom | 0 | 3 | --- |

**Figure C.2** The output of the factor analysis in generating *BLV*.

Eigenvalues Summary
Eigenvalues of the Observed Matrix
Factor: Untitled
Date: 04/12/19  Time: 09:46

Eigenvalues: (Sum = 3, Average = 1)

| Number | Value | Difference | Proportion | Cumulative Value | Cumulative Proportion |
|---|---|---|---|---|---|
| 1 | 1.347882 | 0.463343 | 0.4493 | 1.347882 | 0.4493 |
| 2 | 0.884539 | 0.116960 | 0.2948 | 2.232421 | 0.7441 |
| 3 | 0.767579 | --- | 0.2559 | 3.000000 | 1.0000 |

**Figure C.3**  The eigenvalues summary of *BLV*.

Kaiser's Measure of Sampling Adequacy
Factor: Untitled
Date: 04/12/19  Time: 09:50

| | MSA |
|---|---|
| BBW | 0.564760 |
| FUNDUS | 0.592209 |
| MUAC | 0.548440 |
| Kaiser's MSA | 0.564041 |

Partial Correlation:

| | BBW | FUNDUS | MUAC |
|---|---|---|---|
| BBW | 1.000000 | | |
| FUNDUS | 0.083246 | 1.000000 | |
| MUAC | 0.202483 | 0.157700 | 1.000000 |

**Figure C.4**  The output of the Kaiser's measure.

Factor Method: Principal Factors
Date: 03/18/19 Time: 17:06
Covariance Analysis: Ordinary Correlation
Sample: 1 113
Included observations: 113
Number of factors: User-specified
Prior communalities: Squared multiple correlation

| | Loadings F1 | Communality | Uniqueness |
|---|---|---|---|
| ED | 0.530573 | 0.281508 | 0.718492 |
| SE | 0.491249 | 0.241326 | 0.758674 |
| SMOKING | 0.582860 | 0.339725 | 0.660275 |
| NO_ANC | 0.441722 | 0.195118 | 0.804882 |

| Factor | Variance | Cumulative | Difference | Proportion | Cumulative |
|---|---|---|---|---|---|
| F1 | 1.057677 | 1.057677 | --- | 1.000000 | 1.000000 |
| Total | 1.057677 | 1.057677 | | 1.000000 | |

| | Model | Independence | Saturated |
|---|---|---|---|
| Discrepancy | 0.010771 | 0.526708 | 0.000000 |
| Parameters | 8 | 4 | 10 |
| Degrees-of-freedom | 2 | 6 | --- |

**Figure C.5** The output of the factor analysis in generating *MLV*.

Eigenvalues Summary
Eigenvalues of the Observed Matrix
Factor: Untitled
Date: 03/18/19 Time: 17:05

Eigenvalues: (Sum = 4, Average = 1)

| Number | Value | Difference | Proportion | Cumulative Value | Cumulative Proportion |
|---|---|---|---|---|---|
| 1 | 1.880152 | 1.070877 | 0.4700 | 1.880152 | 0.4700 |
| 2 | 0.809275 | 0.105422 | 0.2023 | 2.689427 | 0.6724 |
| 3 | 0.703853 | 0.097133 | 0.1760 | 3.393280 | 0.8483 |
| 4 | 0.606720 | --- | 0.1517 | 4.000000 | 1.0000 |

**Figure C.6** The eigenvalues summary of *MLV*.

# References

Agung, I.G.N. (2008). Simple quantitative analysis but very important for decision making in business and management. Presented in the Third International Conference on Business and Management Research, Sanur Paradise, Bali, Indonesia (27–29 August, 2008).

Agung, I.G.N. (2009a). *Time Series Data Analysis Using EViews*. Singapore: Wiley.

Agung, I.G.N. (2009b). Simple quantitative analysis but very important for decision making in business and management. *The Ary Suta Center Series on Strategic Management* 3: 173–198.

Agung, I.G.N. (2009c). What should have a great leader done with statistics? *The Ary Suta Center Series on Strategic Management* 4: 37–47.

Agung, I.G.N. (2011a). *Cross Section and Experimental Data Analysis Using EViews*. Singapore: Wiley.

Agung, I.G.N. (2011b). *Manajemen Penulisan Skripsi, Tesis Dan Disertasi Statistika: Kiat-Kiat Untuk Mempersingkat Waktu Penulisan Karya Ilmiah Yang Bermutu*, 4e. Jakarta: PT RajaGrafindo Persada.

Agung, I.G.N. (2014). *Panel Data Analysis Using EViews*. UK: Wiley.

Agung, I.G.N. (2019). *Advanced Time Series Data Analysis: Forecasting Using EViews*. UK: Wiley.

Bliemel, F. (1973). Theil's forecast accuracy coeffcient: a clarification. *Journal of Marketing Research* X: 444–446.

Buchinsky, M. (1998). Recent advances in quantile regression models: a practical guideline for empirical research. *Journal of Human Resources* 33: 88–126.

Conover, W.J. (1980). *Practical Nonparametric Statistics*. New York: Wiley.

Daniel, H. (1954). Saddlepoint Approximations in Statistic. *Annals of Mathematical Statistics* 25: 631–650.

Davino, C., Furno, M., and Vistocco, D. (2014). *Quantile Regression: Theory and Application*. Wiley.

Gujarati, D.N. (2003). *Basic Econometric*. Boston: McGraw-Hill.

Hankel, J.E. and Reitsch, A.G. (1992). *Business Forecasting*. Boston: Allyn and Bacon.

Hao, L. and Naiman, D.Q. (2007). *Quantile Regression*. UK: Sage Publication Inc.

Hardle, W. (1999). *Applied Nonparametric Regression, Economic Society Monographs*. Cambridge University Press.

He, X. and Hu, F. (2002). Markov change marginal bootstrap. *Journal of American Statistical Association* 97 (459): 783–795.

Huitema, B.E. (1980). *The Analysis of Covariance and Alternatives*. New York: Wiley.

*Quantile Regression: Applications on Experimental and Cross Section Data Using EViews,* First Edition.
I Gusti Ngurah Agung.
© 2021 John Wiley & Sons Ltd. Published 2021 by John Wiley & Sons Ltd.

Kementa, J. (1980). *Element of Econometric*. New York: Macmillan Publishing Company.

Koenker, R. (2005). *Quantile Regression*. Cambridge University Press, Cambridge, New York, Melbourne, Cape Town, Singapore, Sao Pulo, Delhi, Dubai, Tokyo, Mexico City.

Koenker, R. and Hallock, K.F. (2001). Quantilee regression. *Journal of Economic Perspectives* 15 (4): 143–156.

Lepin, L.L. (1973). *Statistics for Modern Business Decisions*. Haarcourt Brace, Jovanovich, Inc.

Neter, J. and Wasserman, W. (1974). *Applied Statistical Models*. Homewood, Illinois: Richard D. Irwin, Inc.

Ramsey, J.B. (1969). Test for specification errors in classical linear least squares regression analysis. *Journal of the Royal Statistical Society Series* 8 (31): 350–371.

Theil, H. (1966). *Applied Econometric Forecast*. Amsterdam, North Holland: North-Holland Publishing Company.

Tsay, R.S. (2002). *Analysis of Financial Time Series*. Wiley.

Tukey, J.W. (1977). Modern techniques in data analysis. *Journal of Mathematical Finance* 7: 1.

Wilks, S.S. (1962). *Mathematical Statistics*. New York: Wiley.

Wilson, J.H. and Keating, B. (1994). *Business Forecasting*, 2e. Burr Ridge, Illinois: Richard D. Irwin, Inc.

Wooldridge, J.M. (2002). *Econometric Analysis of Cross Section and Panel Data*. Masschusetts: The MIT Press Cambridge.

# Index

## a

adjusted R-squared    66, 69, 74, 108, 112–116, 388–389
ANCOVA    34, 297, 341
ANOVA-QR    24–35, 65–66, 68, 70, 72, 74, 384–385, 387–390
appropriate    4
assumption    25, 207, 276
alternative    1, 3, 7

## b

baby latent variable (BLV)    465–467
BBW    381
BLV Forecast (BLVF)    385–387

## c

categorical variable    1, 6, 10
Chi-square    3, 9, 25–27, 30, 34, 211, 213–214, 293, 295–299, 313, 315, 320–322, 325–326, 331, 360, 347, 367, 375, 379, 382, 384, 386, 389, 409, 412, 415, 419–420, 424, 435, 437, 459
classification    1, 66, 69, 74
communality    466, 468
conditional for sample score    454–455
contrast    27, 33, 337, 342
correlation    381–382, 466, 468
criteria    461, 463

## d

default options    29–30, 32, 34, 292
derivative    453–455, 460
deviation    453–454, 460
dialog    1
dichotomous    3
difference-in-differences (DID)    28–29, 33–34, 71
directly    460, 461, 463, 465
distribution    1
Dropfirst    25–27, 31, 72, 269, 272–273, 291, 297–298, 334, 337, 385, 387
Droplast    31–32, 72–73, 299–300, 340

## e

eigenvalues    465, 467–468
endogenous variable    1
equality    1–6, 8–10
equality test    1–6, 8–10, 297, 299–300, 334, 337, 385, 390
equation specification (ES) 5–28, 31–34, 65, 70, 72–73, 107, 112, 114–116, 165–168, 171–172, 208–210, 214, 269, 271–272, 274, 291–292, 297–298, 334, 338–340, 382, 384, 387, 453–455, 461
estimation equation    25, 28, 33–34, 108, 112–113, 116, 269–271, 274, 298, 335, 338, 341–342, 388, 390
evaluation    66–68, 107–109, 112, 460

*Quantile Regression: Applications on Experimental and Cross Section Data Using EViews,* First Edition.
I Gusti Ngurah Agung.
© 2021 John Wiley & Sons Ltd. Published 2021 by John Wiley & Sons Ltd.

exogenous 1
expand(*) 25–27, 31–32, 65, 72–73, 267, 269,
    272–273, 291–294, 296–300, 334,
    337–340, 385

**f**

factor 25–26, 29, 33, 66, 68, 71, 73, 107, 165,
    170, 291–293, 295, 297–299, 338–340,
    465–468
forecast 65, 67–68, 107–109, 112, 334,
    336–337, 385–387, 460
forecast variable (FV) 110
FVPol 109, 112–113, 115–116

**g**

generated 1, 3–4, 7, 73, 387
group 1–2, 4, 6, 8, 10, 72–73

**h**

heterogeneous 170–171, 291–292, 294, 296,
    298
heterogeneous linear QR (HLQR) 292,
    295–300
HQR 291, 293–295, 297, 299

**i**

incomplete 4, 442–444, 446, 448, 458
independent identical distribution (IID) 25,
    453–455, 461
inequality 108, 112
interaction 165, 168, 171–174, 209–210, 212,
    214
interaction effect (IE) 66, 68–69, 71–74

**j**

joint effects 26–27, 33–34, 166–168, 170,
    172, 209, 211, 213–214, 216, 274, 299
jointly 30, 213

**k**

Kaiser's MSA 465, 467

**l**

latent variable 381–382, 384, 386, 388, 390
left-sided hypothesis 463, 465

likelihood ration (LR) 26, 166–168, 170,
    172–173, 211–212, 215–216, 298–299
linear programming 465
loadings 466, 468
logarithmic 115–116

**m**

mean regression (MR) 25, 453–455
median 1, 3–10
Mediansby(*) 109
minimization 455
mother latent variable (MLV) 381, 465,
    457–458

**n**

nonparametric (NP) 1, 3, 5, 7, 25, 295,
    390–391
nonparametric QR (NPQR) 270–276
normal 451–461

**o**

omitted variables test (OVT) 208–210
one-way ANOVA 384–385, 387–390
one-way ANOVA-QR 25–26, 28, 30, 32, 34,
    334
optimal solution 108, 112–113, 116, 166,
    269–271

**p**

parabolic 109
parameters 26, 28–29, 33–34, 293–297, 336,
    339–340, 342, 453–455, 458, 460–463,
    466, 468
parametric 382
polynomial 109, 112–115
population 1
predictors 1, 165–166, 168, 170, 172, 174,
    207–208, 210, 212, 214, 267–268, 283,
    304, 349, 362, 396, 402, 411, 420
principal factors 46–466, 468
pseudo R-squared 26, 108, 112–114, 116,
    167–168, 171, 272, 298, 335, 338,
    341–342, 388–389

## q

quantile process   107–109, 112–116, 385, 387, 399, 408, 413, 444, 455–460
quantile process estimate (QPE)   291, 292, 339, 387, 390
quantile slope equality test (QSET)   34, 291, 297, 299–300, 334, 336, 385, 456–458
quasi-likelihood ratio (QLR)   26, 211, 213, 273, 299, 334
Quasi-LR   26, 33–34, 108, 112–114, 116, 166–168, 170, 172–173, 211–212, 215–216, 269–270, 273–276, 298–299, 335, 385, 388, 390

## r

random   453–455
rank   3, 267–275, 292, 333, 381, 385
ranks (X1,a)   10
reduced model (RM)   170, 210–211, 276
redundant variables test (RVT)   26, 292, 298–299, 385
representations   28, 171–172
right-sided hypothesis   461, 463
Round   387–388

## s

semi-logarithmic   166, 171
semiparametric   25, 267
sum of absolute deviation (SAD)   454–455, 461

sum of squared deviation (SSD)   453–455, 461
Symmetric Quantile Test (SQT)   30, 293, 457, 459–460

## t

Theil inequality coefficient (TIC)   108, 112, 336, 339–340, 342
theoretical   171–172, 207, 210, 276
three-way interaction   26–27, 29, 207, 209–211, 213–214
translogarithmic   168–172, 174
two-sided hypothesis   461
two-way ANOVA   25–34, 334, 336, 338, 340–342
two-way interaction   165–166, 168, 170–174, 207–210, 21–214, 216

## u

unexpected   6, 10
unique optimal solution   108, 112–113, 116, 298, 335, 338, 341–342, 388–389
up-and-down relationship   165–166, 173, 207, 208, 210, 212
upper variable   112–113, 116, 165, 167–168, 172, 207, 210

## w

Wald test   4, 25–29, 33, 211, 213–214, 293, 295–300, 337, 340, 342
Whitney nonparametric test   5, 6
Wilcoxon nonparametric test   5, 6